Chemical Carcinogenesis and Mutagenesis II

Contributors

C.F. Arlett · D. Boettiger · R.D. Callander · D. Chalmers
C.S. Cooper · J.K. Cowell · S.W. Dean · M. Defais · J. DiGiovanni
A.M. Edwards · A. Hall · D.G. Harnden · C.C. Harris · A. Haugen
C.-H. Heldin · C.S.T. Hii · A.R. Lehmann · A.W. Murray
A.E. Pegg · M.J.O. Wakelam · B. Westermark

Editors

C.S. Cooper and P.L. Grover

Springer-Verlag Berlin Heidelberg New York
London Paris Tokyo Hong Kong

Colin S. Cooper, Ph. D.
Philip L. Grover, D. Sc.

The Institute of Cancer Research:
Royal Cancer Hospital
Chester Beatty Laboratories
Fulham Road
London SW3 6JB, Great Britain

With 16 Figures

ISBN-13: 978-3-642-74780-9 e-ISBN-13: 978-3-642-74778-6
DOI: 10.1007/978-3-642-74778-6

Library of Congress Cataloging-in-Publication Data (Revised for vol. 2). Chemical carcinogenesis and mutagenesis. (Handbook of experimental pharmacology; v. 94) Includes bibliographies and index. 1. Carcinogenesis. 2. Carcinogens. 3. Chemical mutagenesis. 4. Mutagens. I. Beland, F. A. (Frederick A.) II. Cooper, C. S. (Colin S.), 1954– . III. Grover, Philip L. [DNLM: 1. Carcinogens. 2. Mutagens. W1 HA51L v. 94/ QZ 202 C5174] QP905.H3 vol. 94 [RC268.5] 616.99'4071 89-11530
ISBN-13: 978-3-642-74780-9

Softcover reprint of the hardcover 1st edition 1990

2127/3130-543210 – Printed on acid-free paper

List of Contributors

C.F. Arlett, MRC Cell Mutation Unit, University of Sussex, Falmer, Brighton BN 1 9RR, Great Britain

D. Boettiger, Department of Microbiology G-2, School of Medicine, University of Pennsylvania, Philadelphia, PA 19104, USA

R.D. Callander, ICI, Central Toxicology Laboratory, Genetic Toxicology Group, Cell and Molecular Biology Section, Alderley Park, Macclesfield, Cheshire SK10 4TJ, Great Britain

D. Chalmers, Department of Microbiology G-2, School of Medicine, University of Pennsylvania, Philadelphia, PA 19104, USA

C.S. Cooper, Institute of Cancer Research, Royal Cancer Hospital, Chester Beatty Laboratories, 237, Fulham Road, London SW3 6JB, Great Britain

J.K. Cowell, I.C.R.F. Laboratory of Molecular Genetics, Department of Haematology and Oncology, 30, Guilford Street, London WC1 1EH, Great Britain

S.W. Dean, Microtest Research Unit, University Road, Heslington, York YO1 5DU, Great Britain

M. Defais, Laboratoire de Pharmacologie et de Toxicologie Fondamentales du Centre National de la Recherche Scientifique, 205, route de Narbonne, F-31077 Toulouse Cedex, France

J. DiGiovanni, Department of Carcinogenesis, The University of Texas, M.D. Anderson Cancer Center, Science Park – Research Division, P.O. Box 389, Smithville, TX 78957, USA

A.M. Edwards, Department of Medical Biochemistry, School of Medicine, Flinders University of South Australia, Bedford Park, South Australia 5042, Australia

A. Hall, The Institute of Cancer Research, Royal Cancer Hospital, Chester Beatty Laboratories, 237 Fulham Road, London SW3 6JB, Great Britain

D.G. Harnden, Paterson Institute for Cancer Research, Christie Hospital and Holt Radium Institute, Wilmslow Road, Manchester M29 9BX, Great Britain

C.C. Harris, Laboratory of Human Carcinogenesis, Division of Cancer Etiology, Building 37, Room 2C07, National Institutes of Health, National Cancer Institute, Bethesda, MD 20892, USA

A. Haugen, Department of Toxicology, National Institute of Occupational Health, P.O. Box 8149 Dep., N-0033 Oslo 1, Norway

C.-H. Heldin, Ludwig Institute for Cancer Research, Uppsala Branch, Biomedical Center, Box 595, S-751 23 Uppsala, Sweden

C.S.T. Hii, School of Biological Sciences, Flinders University of South Australia, Bedford Park, South Australia 5042, Australia

A.R. Lehmann, MRC Cell Mutation Unit, University of Sussex, Falmer, Brighton BN1 9RR, Great Britain

A.W. Murray, School of Biological Sciences, Flinders University of South Australia, Bedford Park, South Australia 5042, Australia

A.E. Pegg, Departments of Physiology and Pharmacology and Cancer Research Center, The Milton S. Hershey Medical Center, The Pennsylvania State University, P.O. Box 850, Hershey, PA 17033, USA

M.J.O. Wakelam, Molecular Pharmacology Group, Department of Biochemistry, University of Glasgow, Glasgow G12 8QQ, Great Britain

B. Westermark, Department of Pathology, University Hospital, S-751 85 Uppsala, Sweden

Dr. E. C. Miller

This volume is respectfully dedicated by its contributors to the memory of Dr. Elizabeth Cavert Miller, Professor of Oncology at the McArdle Laboratory for Cancer Research of the University of Wisconsin, who died on October 14, 1987, from the very disease to which she had devoted a lifetime of outstanding scientific research.

Foreword

I have been privileged to witness and participate in the great growth of knowledge on chemical carcinogenesis and mutagenesis since 1939 when I entered graduate school in biochemistry at the University of Wisconsin-Madison. I immediately started to work with the carcinogenic aminoazo dyes under the direction of Professor CARL BAUMANN. In 1942 I joined a fellow graduate student, ELIZABETH CAVERT, in marriage and we soon commenced a joyous partnership in research on chemical carcinogenesis at the McArdle Laboratory for Cancer Research in the University of Wisconsin Medical School in Madison. This collaboration lasted 45 years. I am very grateful that this volume is dedicated to the memory of Elizabeth. The important and varied topics that are reviewed here attest to the continued growth of the fields of chemical carcinogenesis and mutagenesis, including their recent and fruitful union with viral oncology. I feel very optimistic about the application of knowledge in these fields to the eventual solution of numerous problems, including the detection and estimation of the risks to humans of environmental chemical carcinogens and related factors.

JAMES A. MILLER
Van Rensselaer Potter
Professor Emeritus of Oncology
McArdle Laboratory for Cancer Research
University of Wisconsin Medical School
Madison, Wisconsin

Preface

In order to understand and hopefully to prevent the processes by which chemicals induce cancer in man, it will be necessary to achieve three goals. Firstly, the classes of chemicals that are responsible for human chemical carcinogenesis must be identified. Secondly, the detailed metabolism of these chemicals should be examined and their targets within susceptible cells defined. Thirdly, it is important to identify the cancer-specific changes that are induced in cells by chemical exposure and ultimately to discover how these changes lead to tumour induction. In this book we have attempted to bring together these three areas of cancer research which span several disciplines, ranging from epidemiology, through studies on metabolic activation, to cellular and molecular biology.

The International Agency for Cancer Research has listed many chemicals or mixtures of chemicals for which there is considered to be sufficient evidence of carcinogenicity in man. Despite this substantial tally, it remains a disturbing fact that the environmental, dietary or endogenous agents that are responsible for most of the major types of human cancer have still to be identified. In addition, even when carcinogenic substances, such as tobacco smoke, have been clearly implicated, there is still debate regarding the contributions of individual components of these complex mixtures to their overall biological effects. It is therefore not at all surprising that continued efforts will have to be made to identify potential carcinogens using a combination of approaches, including bacterial and mammalian cell mutation assays, epidemiological studies and the range of methods that can now be used to detect chemicals that have already become covalently attached to cellular macromolecules. From amongst this armoury the new ^{32}P-postlabelling procedure (Part I, Chapter 13) that enables extremely low levels of carcinogen-DNA adducts to be detected deserves particular mention as a technique that has the potential to provide completely fresh insights into the identity of chemicals that contribute to cancer development in man.

It is generally accepted that DNA is an important cellular target for chemical carcinogens. This belief arose, in part, because of the attractiveness of the somatic mutation model of cancer development that is reviewed in the introductory chapter and, in part, because of the correlations observed between the extents of covalent binding of chemicals to DNA and their carcinogenic potencies. As a consequence of the extensive interest in this area we now possess, for many important classes of chemical carcinogens, a detailed knowledge of their pathways of metabolic activation and of the mechanisms by which activated metabolites modify DNA. Both the specificity of interaction of the activated carcinogens with DNA and the cellular machinery responsible for repairing the lesions thus introduced have come under close scrutiny. Such studies have, in particular, demonstrated the central importance of DNA repair enzymes in determining the biological consequences of carcinogen exposure, as dramatically illustrated by the high incidence of some types of cancer in individuals with deficiencies in specific repair enzymes.

It is well established that the consequences of exposure to chemical carcinogens can be influenced by a variety of factors that may be collectively referred to as "modifiers of chemical carcinogenesis". These include the tumour promoters, which interact with specific cellular proteins, as well as a whole range of substances that can act as inhibitors of carcinogenesis. There are also occasions when chemicals may act synergistically with other classes of cancer-causing agents. Two examples are provided by (a) the 'cooperation' between fungal toxins and hepatitis B virus in the induction of hepatocellular carcinoma and (b) the proposed interaction between cigarette smoking and papilloma viruses in the development of cancer of the cervix. Despite the potential importance of such interactions in the induction of human cancer, surprisingly little is known about the molecular mechanisms involved in cooperation between viral and chemical agents. This is perhaps an area that should be targeted for particular attention in the future. Furthermore, an individual's genetic make-up, in addition to determining the status of the DNA repair enzymes as mentioned above, may have a key role in controlling that person's susceptibility to chemical exposure.

The new technologies of DNA transfection and molecular biology have resulted in significant advances (a) in the identification of cellular genes that are potential targets for chemical carcinogens and (b) in the characterization of the specific types of genetic alterations that may be involved in tumour induction. The potential targets now include several well-characterized protooncogenes, which may be activated by mutation to form oncogenes, as well as the less well characterized tumour-suppressor genes or anti-oncogenes that may be inactivated during tumour development. These molecular studies have lead to the unification of several areas of cancer research and cellular biology. Of particular note is the observation that the same genes are activated in a range of tumour types, indicating that there may exist common mechanisms of cancer development amongst histologically-diverse groups of tumours. Moreover the recognition that many of the protooncogenes encode proteins, such as growth factors and growth factor receptors, that are the normal components of cellular control pathways has caused a revolution in our perception of the ways in which alterations in cellular genes lead to transformation.

The newer molecular genetics studies also have implications for the more traditional areas of chemical carcinogenesis and mutagenesis. The confirmation that specific genetic alterations may be directly involved in tumour development justifies the extensive analysis of the interactions of carcinogens with DNA that has occurred over the past 20 years and provides a firm foundation for many of the bacterial and mammalian mutagenicity tests, which hitherto has rested precariously on the observed correlations between mutagenicity and carcinogenicity. It is equally important to appreciate the limitations of the molecular approaches used for the analysis of cancer induction. For example, these techniques, although extremely important, have so far provided few insights into the nature of the environmental and dietary components that contribute to the incidence of cancer in man, an observation that simply underscores the continued importance of the more traditional epidemiological and mutational studies. Indeed, it is our hope that the bringing together of chapters reviewing widely-separated aspects of chemical carcinogens in one book will serve to highlight both the virtues and the limitations of each area and will help to identify those lines of research that may, in the future, prove most fruitful.

Finally we wish to extend our gratitude to HELEN ANTON and AUDREY INGLEFIELD for their assistance with the organization and preparation of this book and to DORIS M. WALKER of Springer-Verlag for all her help and advice.

London

COLIN S. COOPER
PHILIP L. GROVER

Contents

Part I. Relationships Between Mutagenesis and Carcinogenesis

Part II. Modifiers of Chemical Carcinogenesis

CHAPTER 6

Tumour Promotion: Biology and Molecular Mechanisms

CHAPTER 12

The Role of Oncogene Activation in Chemical Carcinogenesis

CHAPTER 13

Growth Factors and Their Receptors

CHAPTER 14

Signal Transduction in Proliferating Normal and Transformed Cells

CHAPTER 15

Effect of Oncogenes on Cell Differentiation

Contents of Companion Volume 94, Part I

Part I. Relationships Between Mutagenesis and Carcinogenesis

CHAPTER 1

Use of Mutations in Bacteria as Indicators of Carcinogenic Potential

R. D. CALLANDER

A. Introduction

The use of bacterial mutation systems to detect chemical carcinogens began in the late 1940s, shortly after the demonstration that carcinogenic nitrogen and sulphur mustards induced gene mutations (and chromosome rearrangements) in the germ cells of *Drosophila melanogaster* (AUERBACH and ROBSON 1946). The rationale for such use stems from Boveri's theory of the mutational origin of cancer (see WOLF 1974), which in turn was refined by BAUER (1928) and gained increasing acceptance with successive comparative studies of mutagens and carcinogens in the early 1950s.

Use of bacteria in tests designed to detect chemically induced mutations (and by the above theory, potential carcinogens) is inherently attractive due to the numbers of cells that can be treated at one time. As cancer initiation is a relatively rare process at the cellular level (cf. PITOT et al. 1978), the assumption can be made that the mutational event involved also occurs at a very low spontaneous rate: therefore, large populations must be scored to observe significant numbers of such events. Treatment of 10^8 bacteria in a single petri dish allows the potential observation of a mutational event with a very low spontaneous rate of occurrence.

The mutational event(s) postulated to be cancer initiators are generally considered to be "forward" mutations, i.e. alteration of a normal gene to an abnormal one. The most widely used bacterial mutation assays (see below) have often been described as "reverse" mutation assays, where an already mutant locus is reverted to normality. This is a slight oversimplification, as forward mutation at one or more suppressor loci may also result in phenotypic reversion to prototrophy. Also, the actual mechanism of mutation is the same in either case: an alteration is induced in the sequence of bases in the DNA molecule. Thus, many of the arguments regarding the lack of relevance of using bacterial reverse mutation assays to detect potential carcinogens lose much of their impetus.

B. Historical Background

Early work in the field had already been pioneered by LURIA and DELBRUCK (1943): a modification of their growth fluctuation assay devised by RYAN (1955) in turn formed the basis of one of the main variants of the *Salmonella typhimurium* assay (the fluctuation test – see Sect. D.II.4). Thus, the genetic principles determined in the 1940s and 1950s (BURDETTE 1955) for the detection of rare mutational events in large populations of bacterial cells have constituted the basis for the subsequent development of a wide range of bacterial tests for mutagens (and therefore putative carcinogens). Appropriate strains of two species of enteric bacteria, *Escherichia coli* and *Salmonella typhimurium*, were selected, and techniques (such as that of RYAN 1955) were devised to treat the bacterial cells on individual petri dishes, e.g. in the spot test and treat-and-plate test (see Sects. D.II.2 and D.II.3). In the course of these developments, the testing of known carcinogens was, of course, paramount; the principle that any new assay should be validated against its ability to detect correctly known carcinogens still holds today. The direct-acting classes of chemicals such as the alkylating nitrogen and sulphur mustards were easily identified, but other types of carcinogens (e.g. aromatic amines, polycyclic hydrocarbons and dialkylnitrosamines) could not be efficiently detected. This apparent failure was eventually identified as being primarily due to the fact that the active mutagen/carcinogen was in fact a derivative of these chemicals, produced in the course of mammalian detoxification processes: a biochemical transformation not achieved by the metabolic processes indigenous to bacteria. Following on from this realisation, MALLING (1966) was able to activate dimethylnitrosamine to a mutagen for *Neurospora crassa* using a chemical oxidative mixture and subsequently used isolated mammalian organ fractions (containing microsomal enzymes) to obtain similar results with *S. typhimurium* (MALLING 1971). This work, combined with similar work by GARNER and co-workers (GARNER et al. 1972) and coupled to the design and isolation of various strains of *S. typhimurium* with specific deficiencies in their ability to synthesise the essential amino acid histidine, resulted in the development of the most widely used bacterial mutation assay: the *Salmonella* mutation test (AMES et al. 1975). This methods paper was the first to define a minimum acceptable battery of tester strains (see Table 1). The revision to the original published protocol (MARON and AMES 1983; see Sect. D.II.1) took into account advances in methodology and the development of new strains since 1975 and proposed an altered set of tester strains (Table 1).

A summary of the history of the Ames test is given in Table 1. The various stages in this development process are an indication of what might be expected in the development of a new, "alternative" test system, particularly the selection of a set of standard strains, the publication of a methods paper and then a "validation" paper, demonstrating the response of the assay to a number of defined mutagens.

Since these early developmental stages, a number of other bacterial assays have also been developed. Closely related to the basic *Salmonella* assay is a similar assay system using *Escherichia coli* strains (with a requirement for tryptophan rather than histidine: see BRIDGES et al. 1973; GREEN and MURIEL 1976). In some screening programmes, one or more of these *E. coli* WP2 strains is used in conjunction with the main battery of *S. typhimurium* TA strains (see Sect. C.I.1). Analogous assay systems using *Bacillus subtilis* strains were also developed over this same period of time, again primarily using reversion from amino-acid dependence (e.g. TANOOKA 1977).

Table 1. Steps in the development of the microsome-mediated *Salmonella* mutation assay (Ames test)

Stage	Development step	References
1	Selection of a standard set of tester strains	AMES (1971)
2	Construction of strains TA1535, TA1537, and TA1538 to form the basic tester set	AMES et al. (1973a)
3	Addition of a mammalian-derived activation system (S9-mix) to the assay protocol	AMES et al. (1973b)
4	Costruction of the more sensitive strains TA98 and TA100	MCCANN et al. (1975a)
5	Publication of the standard methods paper with a recommended minimum strain battery: TA1535, TA1537, TA1538, TA98, and TA100	AMES et al. (1975)
6	Publication of the first "validation" papers	MCCANN et al. (1975b); MCCANN and AMES (1976)
7	Publication of the revised methods paper, incorporating advances in methodology, etc. and a new recommended strain set: TA97, TA98, TA100, and TA102	MARON and AMES (1983)

As well as these "reverse" mutation assays, several forward mutation assays have also been developed using various bacterial species (e.g. *S. typhimurium*, RUIZ-VAZQUEZ et al. 1978; SKOPEK et al. 1978; *E. coli*, MOHN and ELLENBERGER 1977; *Klebsiella pneumoniae*, VOOGD et al. 1974). These are discussed briefly below (see Sect. D.III).

A number of other bacterial assays using alternative genetic endpoints have also been described in the literature. The endpoints include phage induction (MAMBER et al. 1984), SOS induction (QUILLARDET et al. 1982) and DNA-repair or "rec-assays" (KADA et al. 1972). The SOS chromotest has been proposed as a simpler, quicker replacement for the Ames *Salmonella* mutation assay (MARON and AMES 1983). Whilst various correlations between results in the two assays have been demonstrated (e.g. NAKAMURA et al. 1987), it must be stressed that induction of SOS activity is a non-specific response to DNA damage: assays based on such a response are, therefore, not necessarily detecting mutational events per se. It is conceivable that some of these assays may become useful adjuncts to a mutation assay for screening chemicals for genotoxic potential following suitable validation.

C. Use of Assays

I. General

The fundamental purpose behind testing a chemical in a bacterial mutation assay, of whatever type, falls into one of two broad areas. The test can either be used to assess the chemical's mutagenicity per se, as part of a toxicological screening battery, or else it can form part of a detailed mechanistic examination of the chemical's mutagenic potential. The protocols followed in the particular assay (even in the Ames test itself) can differ vastly depending on which of these two objectives is pursued. In the former case, the assay protocols employed are largely standardised (see below), with the bacterial assay being the primary screen of the intrinsic in vitro genotoxic activity of the chemical. Data from such screens may be qualified by the use of in vivo assay (see Sect. E.II).

1. Toxicology Screening

Use of an assay as part of the toxicological investigation of a compound can fulfil two purposes. The chemical can either be tested to assess its mutagenicity to bacteria as part of a general toxicological screen, or else the test can be performed to generate data destined to form part of a package of genetic (and other) toxicology data for submission to various governmental regulatory authorities. In the former case, specific assay protocols are not defined, although individual testing laboratories will normally have their own standard protocols. The data from such screening tests are normally used to define the genotoxic potential of the chemical, both mutagenic and carcinogenic, in conjunction with data from other short-term in vitro and in vivo screening tests. In the latter case, the choice of the particular test to be used and in many cases the specific protocol for that test are normally specified by the regulatory authority within its guidelines/test requirements (see BERRY and LITCHFIELD 1985, for a review of the various regulatory requirements). Without exception, the *Salmonella* assay (see Sect. D.II.1.a) forms part of the base-set of genotoxicity tests required, using the tester-strain battery first proposed in 1975: TA1535, TA1537, TA1538 (optional in most guidelines), TA98 and TA100. The inclusion in this test battery of one or more strains of *E. coli* may also be required by some authorities (e.g. Japan).

In many cases, both stages in testing will be required during the development of an industrial chemical, pharmaceutical product, pesticide, etc. The early use of a simple bacterial assay in such development allows the identification of chemicals that are genotoxic in vitro at a stage where it may be practical to choose an alternative candidate. Implicit in such decisions is the extrapolation of the known genotoxic results in the various short-term assays to the potential genotoxicity of the chemical in question to higher organisms. Such extrapolations may involve an assessment of the relationship between risk and exposure (see Sect. E.II.1).

2. Research Uses

Use of bacterial mutation assays for research purposes is, by definition, a less clear-cut topic. Each researcher will have his/her own particular reason for performing an assay, and, depending on the desired outcome, the standard protocol for that assay may well be modified to a greater or lesser extent. This can lead to potential problems: often the results of, for example, an Ames test are published, with brief details of the particular experimental conditions used. This result can then subsequently be re-reported by other authors as "compound X is Ames positive/negative", without also reporting that this result originated from a possibly highly modified variant of the basic test and was, perhaps, only conducted in one tester strain. This is particularly important in the case of negative results, for which the protocol used may have fallen short of that for a minimum acceptable assay, and therefore reporting the result as "Ames negative" is misleading.

Nevertheless, accumulation of such datasets is a necessary part of the establishment of a new assay, independent of whether it is destined for screening or research uses, and such pitfalls can arise in the early stages of the development of any test system.

The basic use of an assay in such a research-oriented role, other than simply to say whether or not the chemical in question is mutagenic in that assay, is normally to gain some insight into the mechanism of the mutational event. For example, the *Salmonella* assay can be carried out using a number of different *S. typhimurium* strains which have a number of different types of mutations at well-defined locations within the histidine operon. It is within this context that the newer strains, e.g. TA97 and TA102, have particular revelance. However, because of their restricted use such strains have not, as yet, been included in the tester-strain batteries specified by the various regulatory authorities. Table 2 gives a summary of the relevant genetic markers in the various strains. Depending on the pattern of results obtained from such a test, information about the type(s) of mutations being induced can be readily obtained. Similar use of a number of *E. coli* strains deficient in different DNA-repair enzymes can also be made (see VENITT et al. 1977; and also Sect. D.II.1.d.α). The genotypes of some of the more commonly used *E. coli* strains are also shown in Table 2. Although the results obtained in these types of analyses are too detailed to be adequately covered in this contribution, the use of the *Salmonella* assay for such purposes has been described elsewhere (HARTMAN et al. 1986).

Table 2. Genetic markers relevant to the *Salmonella* mutation assay and similar assay protocols

Amino acid operon mutations						Other markers		
Base-pair substitution		Frameshifts				LPS	Repair	R-Plasmid
Salmonella								
*his*G46	*his*G428	*his*C3076	*his*D3052	*his*D6610	*his*D6580			
TA1530		TA1532				*Δgal*	*ΔuvrB*	
TA1535[a]		TA1537[a]	TA1538[a]			*rfa*	*ΔuvrB*	
TA100[a,b]	TA104		TA98[a,b]	TA97[b]	TA96	*rfa*	*ΔuvrB*	pKM101
TA92	TA103		TA94			+	+	pKM101
	TA102[b,c]					*rfa*	+	pKM101
		*trp*D28						
		SO1007				+	+	pKM101
E. coli								
*trp*E								
WP2						+	+	
WP2*uvrA*						+	*ΔuvrA*	
WP2P*uvrA*						+	*ΔuvrA*	pKM101

[a] Tester strains used in the original recommended battery (AMES et al. 1975).
[b] Tester strains in the revised recommended minimum battery (MARON and AMES 1983).
[c] The *his*G428 mutation is carried on a multicopy plasmid (pAQ1) rather than on the bacterial DNA.

II. Limitations

The primary limitation in the use of bacterial mutation assays in the assessment of the carcinogenic potential of a chemical is inherent in the very nature of the assays themselves. Bacteria are single-cell organisms, with a single chromosome in which the DNA molecule is not closely bound to proteins, whereas cancer is an event that occurs in multi-cellular organisms. Thus, no test using bacteria can directly equate to cancer: as already discussed, extrapolations from mutagenesis to carcinogenesis must be made. This problem, which is common to all in vitro assay systems, has been addressed before (ICPEMC 1982), and repetition of this detailed discussion would be inappropriate. Two specific limitations have, however, been selected for discussion. These are the differences in metabolic potential between the whole animal and an in vitro test system, and the identification of chemicals that produce a carcinogenic response by mechanisms other than by direct interaction with DNA.

1. Metabolic Activation

The metabolism of chemical carcinogens has been previously addressed in detail (WRIGHT 1980) and is also discussed elsewhere in this volume. As has been discussed above (see Sect. B), bacteria lack many of the metabolic pathways that are found in higher organisms, and this can result in the inability of some chemicals to induce mutations in bacteria. Many of the developments of bacterial mutation assays have been a direct result of attempts to circumvent this limiting factor. A brief review of these developments is given below. The use of prokaryotic organisms is also complicated by prokaryotic-specific metabolic pathways. Some of the consequences of such metabolism are also discussed.

a) Use of Auxiliary Metabolising Systems

The realisation that many chemicals described as carcinogens are, in fact, only the precursors of the ultimate carcinogens and the identification of some of the metabolic pathways involved in the synthesis of these ultimate carcinogens led MALLING (1966) to develop an assay that could be used to demonstrate the mutagenic potential of dimethylnitrosamine towards *Neurospora*, bacteria and yeast. The success of this experiment highlighted the fact that differences in metabolic competence exist between mammalian cells and the organisms used in simple in vitro mutation assays.

Although a number of proposed metabolic activation systems were designed in the late 1960s/early 1970s (e.g. LITTERST et al. 1972; MALLING 1971), the system derived by GARNER et al. (1972) and refined by AMES et al. (S9-mix; 1973b, 1975) has become the basic activation system for all bacterial assays. This system uses a tissue fraction derived from homogenised rat liver tissue together with a cofactor mixture designed to create an active NADPH-generating system. The tissue homogenate is centrifuged (hence S9: Supernatant at 9000 g) to create a postmitochondrial fraction that is often misleadingly referred to as a microsomal preparation. Both the tissue and animal used to make S9 preparations can be varied as required. The work by LITTERST et al. (1972) showed that certain polychlorinated biphenyls (PCB) could induce particular enzymes in rat liver, and subsequently one particular PCB complex (Aroclor 1254) became a standard inducing agent for the animals to be used as the source of S9 fractions (AMES et al. 1975). Less hazardous enzyme-inducing agents, such as a mixture of phenobarbitone and β-naphthoflavone, have also been used (cf. GATEHOUSE 1987). Indeed, the use of one particular incuding agent is not mandatory, providing the efficacy of the particular agent chosen has been adequately demonstrated. Use of the S9-mix activation system as described by AMES et al. (1975) has recently been questioned, and the isocitrate/isocitrate dehydrogenase system as used in mammalian mutation assays (e.g. CLIVE et al. 1983) has been proposed as a more efficient NADPH-generating system (PAOLINI et al. 1987).

The concentration of the S9 fraction used in such exogenous activation systems is also an important factor in the efficacy of the system. Whilst the original *Salmonella* assay protocol (AMES et al. 1975) recommended use of between 4% and 10% S9 fraction in the mix (i.e. 20–50 µl S9 per plate using standard assay protocols – see Sect. D.II.1), the majority of practitioners routinely use 10% (50 µl/plate), as evidenced by the first international collaborative study (cf. BRIDGES et al. 1981 and individual reports in the same volume). The example of the hair dye component 2-(2′,4′-diaminophenoxy)ethanol being a mutagen to *Salmonella* only at high levels of S9 fraction in the S9-mix (VENITT et al. 1984) has been widely quoted as justification for such use of relatively high S9 levels as part of a screening assay. In contrast, for the routinely used control chemical 2-anthramine, increasing levels of the S9 fraction in the S9-mix reduced the apparent mutagenicity of the chemical (CALLANDER 1986). Thus, the optimum level of S9 fraction per plate for any given compound cannot be predicted (cf. AMES et al. 1975; VENITT et al. 1983) and if a single level is chosen for a routine assay, justification for the level chosen should be established before any assessments of mutagenicity are made.

b) Preparation of Auxiliary Metabolising Systems

The preparation of the tissue extracts (e.g. S9 fraction) should follow standard techniques, as outlined above, and reference to the methodology used should be quoted where appropriate. Minimisation of loss of efficacy through extended storage and incubation at elevated temperatures should be considered when choosing a suitable preparative method. Data have been presented (HUBBARD et al. 1985) that loss of some enzyme activities (e.g. ethylmorphine-*N*-demethylase) can occur in as little as 5 days at storage temperatures of −20 °C. Similarly, prolonged standing at +20 °C when thawing a frozen S9 sample can reduce the observed mutagenicity of a compound (HUBBARD et al. 1985). Although S9 fractions prepared by a standard technique from a consistent animal stock should give reproducible results for a known compound, monitoring of all enzyme pathways for every S9 preparation is not technically feasible: the possibility exists that the test chemical is activated by a metabolic pathway not checked by the normal control compounds. Repeat assays should be performed using different preparations where possible to provide a full examination of the reproducibility of the test response.

As discussed previously, the cofactor system used can affect the potential of the chosen activation system to produce mutagenic metabolites from the test chemical, as indeed can the choice of tissue extract itself. The work by PRIVAL et al. (1984) clearly confirms previous observations (e.g. ROBERTSON et al. 1982a) that, for benzidine at least, the rat is not the species of choice and furthermore that to assist the reduction of azo groups, additional cofactors are required in the S9-mix to optimise the activity of some benzidine-based azo-compounds. Thus, use of alternative exogenous activation systems derived from a different species of animal or containing additional cofactors may be relevant under certain circumstances. The use of intact hepatocytes rather than homogenised tissue (see HUBBARD et al. 1984; HASS et al. 1985) is a further alternative, although the data from the first international trial (cf. BRIDGES et al. 1981) suggest that this system may be more useful for research purposes.

c) Bacterial-Specific Metabolism

As mentioned above, the use of bacterial mutation tests as an indicator of mammalian carcinogenic potential is also complicated by the existence of highly active metabolic pathways particular to bacteria. The most widely studied examples of such bacterial-specific enzymes are a series of nitroreductases (cf. ROSENKRANZ and SPECK 1976; ROSENKRANZ et al. 1981). Whilst a series of both specific and non-specific nitroreductases has been identified (ROSENKRANZ et al. 1982) and it has been shown that these enzymes are responsible under normal assay conditions for the potent mutagenicity observed for a number of non-carcinogenic compounds that contain nitro groups including metronidazole (ROSENKRANZ and SPECK 1975) and nitrofurantoin (ROSENKRANZ and SPECK 1976), the bacterial specificity of such activation is not absolute. Assays used for the investigation of nitroreductase activity are conducted using strains of *Salmonella typhimurium* (derived from the standard Ames TA strains) which are deficient in the specific nitroreductases under investigation. Such strains are not mutated by nitro compounds when assays are conducted under aerobic conditions using the standard S9-mix. When incubated anaerobically, however, these nitroreductase-deficient strains do give positive results with a number of chemicals, an observation which supports the conclusion that the bacteria contain a second, non-specific, oxygen-labile nitroreductase (cf. ROSENKRANZ and SPECK 1976). When tested in the presence of an exogenous metabolism system (S9-mix), a mutagenic response with nitrofurantoin was also obtained under aerobic incubation, albeit much less potent than that observed with the parent nitroreductase-competent strains, indicating that such mammalian activation systems do contain some levels of the supposedly bacterial-specific enzymes.

The existence of specific and non-specific nitroreductase enzymes in bacteria complicates the conclusions drawn from the results of tests on a nitro-containing compound. Such results for a compound are often immediately ascribed to bacterial-specific nitroreductase activity; however, before such a judgement can be made, the patterns of activity with and without exogenous activation, any results from mammalian cell mutation assays and other genotoxicity tests, and data from metabolism experiments should all be considered. Even if all the other short-term tests prove negative, the positive bacterial assay result should not merely be dismissed as being due to nitroreductase activity. The possible involvement of the gut flora in the in vivo metabolism of a compound must also be considered (see below) before dismissing an apparent bacterial-specific response as being non-significant in terms of an overall hazard assessment.

d) Complex Metabolism

As well as the inherent differences between the metabolic processes in a bacterial cell and the mammalian system, the actual process of metabolic modification that a test chemical undergoes on exposure to the animals or humans can affect the ability of an in vitro assay to predict in the in vivo result. Although metabolism usually occurs in the liver, this is not by any means the only possibility. For example, hexachlorobutadiene (HCBD) undergoes significant metabolism in the kidney rather than in the liver (Nash et al. 1984). Also, metabolism of a compound during its passage through the mammalian system may well occur at more than one site. Such complex metabolism could include the gut bacteria (hence the warning about bacterial-specific metabolism above) and can be further complicated by enterohepatic recirculation. These sites have been implicated in the genotoxicity of dinitrotoluene, for example (Mirsalis et al. 1982). If such multi-cycle metabolism does occur in vivo, then it is unlikely that a simple in vitro bacterial assay will detect all the potential carcinogenic metabolites. Indeed, if such an assay gives a positive response, then it is more than likely that the mutagenicity is due to a separate mechanism, although this is not always the case. Similarly, use of an inappropriate tissue source for the auxiliary metabolism supplement to a bacterial assay is unlikely to be of any benefit in detecting mutagenic metabolites.

2. Non-Genotoxic Carcinogens

The inability to detect carcinogens which do not react with DNA is the most obvious deficiency in the use of bacterial mutation tests as predictors of carcinogenic potential. As the outcome of such tests depends on the interaction of the active chemical with DNA, then by definition the test will not be able to detect those chemicals whose mode of action is by means other than interaction with DNA. A recent review (Ashby and Tennant 1988) on the mutagenicity and carcinogenicity data for 222 chemicals, together with consideration of their chemical structures, has identified a number of "classes" of carcinogens which by a priori examination would not be expected to be detected by in vitro mutation assays. Inclusion of a significant proportion of such compounds in a range of chemicals being used to validate an assay will bias the observed percentage predictivity figures obtained from such a study.

Some of the alternative bacterial assays using endpoints other than reverse or forward mutations (see Sect. B for examples) may, after suitable development and validation against a defined endpoint, prove of benefit for the detection of such chemicals by short-term in vitro test systems. It must be emphasised, however, that the mechanisms of action of non-genotoxic carcinogens such as ethylthiourea and diethylhexylphthalate are multistage. Whilst such alternative assays may detect one or more of these steps, it is unlikely that any one such assay will be able to detect all the steps involved and therefore be suitable as a screening test. It is more likely that such assays will be of benefit in confirming details of a proposed mechanism.

Further discussion of this point is not relevant to the current topic, except to emphasise that it is important to identify the classes of compound that any proposed assay system will and will not detect before undue use of the assay as a general screening test occurs. A clear understanding of the genetic basis of the endpoint of such assays is also essential to allow rationalisation of any observed effects.

D. Methods

I. General Introduction

With the exception of host-mediated assays (see Sect. D.IV), all bacterial mutation assays involve the same fundamental methodology: the bacteria are cultured in vitro, and exposed at some point during this process to the chemical or treatment being investigated.

However, a number of distinct types of methodologies exist, and variations in specific methodologies can make fundamental differences to the scientific rationale behind the observed outcome of the individual mutation assays. For example, incubation of the bacteria in the presence of a chemical mutagen in a solid agar medium (with a limiting amount of a required growth factor present) will give rise to discrete colonies of bacterial cells, each arising from a single mutated cell. In contrast, similar incubation in a liquid medium will simply result in an increase in the number of mutated cells in the culture medium. These mutated cells may also have an altered growth rate in the medium with respect to the background of non-mutated cells, with the result that the ratio of mutated: non-mutated cells in the incubation medium may vary with time in a manner independent of the action of the test chemical. In addition, the chemical may itself have an inhibitory action on certain types of mutated cells, so that differential selection of cell types within the total population in the culture medium may render any firm conclusions regarding induced mutation impossible to achieve.

II. Outline Protocols for Reverse Mutation Assays

A recent review by GATEHOUSE (1987) outlines a number of the critical features in the detailed protocols for bacterial mutation assays in general. This review together with the existing United Kingdom Environmental Mutagenesis Society guidelines for bacterial mutation assays (VENITT et al. 1983; currently under revision) give full protocols for the performance of these assays. The following sections outline protocols for each of the main variants. As both of these documents and other reviews clearly state, the basic mutation assay is the *Salmonella his*$^-$ to *his*$^+$ mutation assay or Ames test. The performance of this assay in testing large numbers of compounds has been repeatedly and widely published (see references in GATEHOUSE 1987) and is the primary in vitro genotoxicity assay required by the various regulatory authorities as discussed above. The standard battery of *Salmonella* strains specified in most guidelines is still based on the original set of strains proposed by AMES et al. (1975) (TA1535, TA1537, TA98, and TA100), despite the publication some 5 years ago of the revised methods paper for the assay (MARON and AMES 1983) updating the proposed strain set (see Table 1). However, "other strains may be utilised when appropriate" (OECD 1983a). The rationale behind the protocols outlined below applies equally to the *E. coli* trp^- to trp^+ mutation assay (see Sect. D.II.1.d.α).

1. Plate-and-Treat Assays

These assays form the basic bacterial mutagen screening assay, as employed in most laboratories worldwide. As outlined above, the *Salmonella* mutation test (Ames test) is the most widely used example of this type. The assay can either be performed using a standard plate-incorporation technique or be extended by use of the pre-incubation modification of the plate assay. In either case, the rationale described below applies whether one or five tester strains are being used in a single experiment. In designing an experiment, thought should be given to the numbers of plates tested per dose level and the intervals between dose levels.

a) Plate-Incorporation Tests

The *Salmonella* mutation test (MARON and AMES 1983) employs the basic plate-incorporation technique in which about 10^8 histidine-requiring bacteria, an appropriate volume of a solution of the compound under examination and, when appropriate, an auxiliary metabolising system (e.g. S9-mix: see Sect. C.II.1) are added to an aliquot of molten soft agar containing a trace of histidine. This mixture is then poured evenly across the surface of a base agar plate containing glucose and a simple salts medium and is allowed to set. The plates are then incubated, inverted, in the dark at 37 °C for a period of 2–3 days. After this period of incubation, the plates are scored for the appearance of discrete bacterial colonies, each of which represents a mutated bacterium (see below). The basis of the assay is, therefore, to determine whether addition of graded doses of the test compound to a series of such plates induces a dose-related increase in the number of observed mutant colonies compared with that obtained on the plates treated only with the appropriate volume of the solvent.

The trace of histidine in the top agar layer allows the logarithmic growth of the histidine-requiring bacteria in the presence of the test compound and/or any of the metabolites of the compound that are generated by the S9-mix. This period of several generations of cell division is essential to allow the fixation of any promutagenic lesions that have occurred in the bacterial DNA, before exhaustion of the histidine supply halts the growth of the auxotrophic cells. Only those cells which have reverted to histidine-independence (i.e. to prototrophy) will continue to divide to form discrete, visible colonies randomly distributed across the test plate while the growth of the non-reverted cells forms a visible background lawn on the plate. Thinning or loss of this lawn is one indicator of compound-induced toxicity. Each visible colony is presumed to be the result of a single mutant bacterium, although the random distribution of such events on the test plate may result in two such colonies arising so closely together that they cannot be visually separated. Such possibilities must be considered if a detailed examination of mutant colonies is envisaged.

b) Pre-Incubation Tests

The standard plate-incorporation protocol is known to be suboptimal in the detection of a number of bacterial mutagens, including the aliphatic *N*-nitroso compounds (BARTSCH et al. 1976; YAHAGI et al. 1977) and 4-dimethylaminoazobenzene (butter yellow; see BRIDGES et al. 1981, and also individual reports in the same volume). Such limitations of the standard protocol have led to the addition of a pre-incubation step in which the bacteria, test-compound solution and S9-mix components are incubated at a temperature, usually between 30 °C and 37 °C, for a specific time, usually 20–60 min (YAHAGI et al. 1975; LEFEVRE and ASHBY 1981), before adding the soft agar and pouring as for the standard assay. Although it has been demonstrated that these "plate negative, pre-incubation positive" experiments are not always clear-cut (e.g. ROBERTSON et al. 1982 b; CALLANDER 1986), the need for a pre-incubation step for the expression of at least some mutagens, such as hexamethylmelamine is recognised (ASHBY et al. 1985). This variant of the standard screening test is perhaps best used after an initial plate-incorporation assay has given a negative or equivocal result.

It has further been demonstrated that a variant of this pre-incubation protocol, using untreated hamsters for the source of the S9 fraction, significantly enhances the mutagenic response observed for benzidine, and that the inclusion of additional azoreduction-supporting cofactors (i.e. flavin mononucleotide, exogenous glucose-6-phosphate dehydrogenase, NADH and an excess of glucose-6-phosphate) is required in certain cases to obtain a mutagenic response with some benzidine-based dyes, including Congo red (PRIVAL et al. 1984).

The routine inclusion of these steps into a standard "screening" testing protocol has yet to be fully justified, as the other tests in such a screening battery should detect any such compounds that the bacterial assay may miss. It is, however, relevant that the potential use of hamster S9 in place of, or in conjunction with, rat S9 has been examined in detail (ZEIGER et al. 1985). Analysis of an extensive *Salmonella* database covering 941 separate samples (799 discrete chemicals) looking at the optimum conditions for the detection of mutagenicity using four tester strains (TA1535, TA1537, TA98, and TA100) and three "activation systems" (no S9-mix and S9-mixes derived from Aroclor 1254-induced hamster and rat livers) with a 20-min pre-incubation period (ZEIGER and HAWORTH 1985), showed that 89% of the chemicals were correctly identified using only strains TA98 and TA100. In these studies the inclusion of hamster S9 was critical for the identification of a number of mutagens within the database.

c) Vapour- and Gaseous-Phase Testing

The basic Ames test protocol is designed for chemicals which can be added directly to the plate (i.e. either solids or liquids). Early attempts to assay volatile and gaseous compounds normally involved the isolation of the particulate component of, for example, complex mixtures or required condensation of the material on a filter (cf. KIER et al. 1974). Modifications of the basic protocol involving exposure of plates treated with the tester strain and S9-mix to various gas/air mixtures in a desiccator allowed the detection of vinyl chloride (RANNUG et al. 1974; BARTSCH et al. 1975) and monochlorodifluoromethane (LONGSTAFF and MCGREGOR 1978) as bacterial mutagens. Similar modifications demonstrated that most alkyl halides (even those with boiling points >175 °C) are more mutagenic when tested using the desiccator technique than in standard plate-incorporation assays (SIMMON 1981). Since then such techniques have been applied to the testing of a number of volatile and gaseous compounds, including bromochlorodifluoromethane (STYLES et al. 1985) and benzylchloride (FORSTER 1986), and a detailed protocol using dynamic flow exposure has recently been published (VICTORIN and STAHLBERG 1988). The warning expressed by SIMMON (1981) and re-iterated by FORSTER (1986) regarding the testing of liquids with apparently high boiling points should be borne in mind when assessing their bacterial mutagenicity. A review of this topic is currently being prepared (ASHBY 1989).

d) Other Alterations to the Standard Protocol

α) Alternative Strains. Although most bacterial mutagens will be detected by the standard tester strain set described above, in certain circumstances these strains are not necessarily the optimum choice. For example, TA1530, the precursor strain to TA1535 (see Table 2), has been demonstrated as being more sensitive to hydrazines than either TA1535 or TA100 (MALAVEILLE et al. 1983). Similarly, the cross-linking agent mitomycin C is toxic to the standard *ΔuvrB* strains, but it is detectable as a mutagen in repair-proficient strains such as TA92 or TA102 (LEVIN et al. 1982). Use of such non-standard strains may therefore be appropriate in particular situations, although the routine inclusion of all such strains in a screening assay would lead to practical problems through the sheer size of the resultant assay.

As stated above, the rationale behing the *Salmonella* mutation test applies equally to the trp^- to trp^+ assay using strains of *Escherichia coli*. The genotypes of the three strains currently listed as "acceptable" in those regulatory guidelines which include *E. coli* (e.g. OECD 1983b) are shown in Table 2. As previously stated, a number of alternative strains, with differing DNA-repair enzyme deficiencies, have also been described in the literature (see VENITT et al. 1977), and the use of such strains does not usually require any protocol modifications. At least one example of a chemical (caffeine) which is negative in *Salmonella*-based assays but is positive with *E. coli* has been described (CLARKE and WADE 1975). Such examples of species specificity, although rare, are support for the inclusion of at least one *E. coli* strain, in conjunction with the standard battery of *S. typhimurium* tester strains, as part of a full screen for bacterial mutagenicity.

β) Multi-Endpoint Strains. A new strain of *S. typhimurium*, SO1007, has recently been developed (BALBINDER and KERRY 1984) which carries the amber mutation *trp*D28 rather than a *his* mutation. The suggested assay procedure, which is basically similar to the Ames method (MARON and AMES 1983), selects for revertants on minimal agar plates supplemented with anthranilic acid. This procedure allows the recovery of two distinct types of revertants: anthranilate utilisers and true prototrophs. Although only a very limited demonstration of the performance of the assay has currently been published, the suggestions that such revertants can arise by both base substitution and frameshift mutations and that SO1007 could therefore be used as an "universal" tester strain in a screening assay merit further investigation. The genotype of this strain is listed in Table 2.

2. Spot Tests and Gradient-Plate Assays

Although the plate-incorporation assay forms the main protocol for the *Salmonella* mutation assay, the original precursor to the assay, the spot test (cf. AMES 1971), still offers the potential for a more rapid screening assay. Rather than testing a single compound over a range of doses, each requiring a number of plates, the spot test allows a number of compounds to be tested at one time by placing a small amount of neat material, either solid or liquid, on discrete areas of a single plate. A variant of the plate-incorporation assay has been developed in which a thin-layer chromatography (TLC) plate is incorporated into the base agar layer and then overlaid directly with the bacteria and S9-mix (BJORSETH et al. 1982). TLC separation of components of complex mixtures before this incorporation step effectively duplicates the spot test and potentially allows a qualitative examination of low levels of components present in such mixtures.

A variant of the plate-incorporation assay using a series of plates overlaid with a gradient of agar containing the test compound has also been developed (McMAHON et al. 1979). As with the spot test, this method allows more than one dose level to be tested per plate by forming a diffusion gradient across each test plate. More recently, a variant of this assay has been developed (HOUK et al. 1988), in which a continuous concentration gradient is created on a spiral plate, allowing testing of a 15-fold concentration range on a single plate.

Although both methods allow more rapid testing of compounds, the results obtained are strictly qualitative and the data produced cannot be analysed using conventional statistical techniques, primarily due to the inability to define the exact dose level at which a response is observed. Furthermore, these techniques are prone to generating either false negative or uninterpretable results when testing compounds with extremes of solubility and, therefore, diffusibility in water. Such assays do, however, form a potentially useful adjunct to the traditional plate-incorporation protocol by allowing pre-screening to identify any strong mutagens or to establish an approximate toxicity value, but they do not in themselves offer a protocol which would fulfil all the requirements for a "complete" mutation assay.

3. Treat-and-Plate Assays

Standard plate test protocols expose growing, dividing bacterial cells to the test compound. Whilst these protocols are highly sensitive to the mutagenic effects of a chemical, their design makes the quantitative determination of the toxic effects of the chemical difficult to achieve. Standard assays are also prone to generating false positive results when the test chemical is a compound which mimics a usable precursor of histidine (or the equivalent in species other than *Salmonella*). A graduated dose of such a compound will inevitably produce a dose-related increase in the number of colonies, simply by increasing the amount of growth that the lawn of auxotrophic cells will sustain, and thus increase the observed numbers of spontaneously arising revertant cells. For example, this phenomenon can be demonstrated by varying the amount of histidine in the top agar in the *Salmonella* assay (cf. GATEHOUSE 1987), while similar effects are observed when histidine itself is used as the test compound (CALLANDER, unpublished data presented at the XVIth European Environmental Mutagen Society Annual Meeting, Brussels, August 1986).

It is on occasion desirable to measure both mutagenic and toxic effects simultaneously, for example in the assessment of a known bactericidal agent suspected of having weak mutagenic activity. This can be achieved by using a treat-and-plate protocol, in which the bacteria are exposed to the test chemical in a non-nutrient medium for a short period of time and then washed free of the chemical before plating on selective medium for the assessment of mutagenicity.

At the same time, parallel cultures are plated on nutrient medium for an assessment of survival. This "viability" count allows correction for the growth-promoting effects of compounds such as histidine as well as correction for the toxic effects of a chemical when calculating a mutation frequency. It can be seen that this protocol has more in common with the protocols for mammalian gene mutation assays than with the standard bacterial mutation assay.

Due to this requirement for dual plating, such assays are more labour intensive than standard plate-and-treat protocols and are more suited to a detailed examination of the mechanisms of mutagenesis of model compounds than for use as a screening assay. They are also less sensitive than standard plate-incorporation assay due to the suboptimal treatment procedures and are prone to generating false positive results unless care is taken in the analysis of the data produced (cf. GREEN and MURIEL 1976).

4. Fluctuation Tests

As has previously been discussed the fluctuation test was originally derived by LURIA and DELBRUCK (1943) to determine whether bacterial variants were due to the appearance of random mutants rather than to adaptation to selective pressures. The modification to their technique proposed by RYAN (1955) was first used to measure induced mutation by CLARKE and WADE (1975).

The early fluctuation assay protocols were first adapted for mutagenicity screening by GREEN et al. (1976). As with the plate-incorporation assay, the auxotrophic bacteria are mixed with the test chemical, a metabolising system (e.g. S9-mix) if required and a trace of the appropriate amino acid. The main difference is that this mixture is prepared in a suitable liquid medium rather than on an agar plate. The mixture is then divided into a number of aliquots in test-tubes or other suitable containers, and incubated at 37 °C for several days. When the trace of amino acid in each tube is exhausted, only the revertant cells will continue to grow and the media in those tubes in which one or more revertants has arisen will become acid due to the metabolism of the glucose carbon source. Tubes containing revertants can therefore easily be identified by the addition of a suitable pH indicator. Subsequent modifications have included the use of disposable Microtitre plates (GATEHOUSE 1978) and the use of microsomal preparations (GATEHOUSE and DELOW 1979). These modifications increase the sensitivity of the assay by increasing the number of subcultures that can be examined at each dose level and by providing the metabolic activation step necessary for optimal activation of test compounds. A review of the protocols for the fluctuation assay has recently been published (HUBBARD et al. 1984). As referenced therein, the basic technique of the Microtitre assay is readily applicable to other test systems, including the induction of mutations in yeasts and L5178Y mouse lymphoma cells.

As with the plate-incorporation assays, the design of the fluctuation test allows the determination of a dose relationship. The normal statistical analysis compares the numbers of positive wells for test and control cultures. Although the number of positive wells does increase with dose, a better relationship is observed if the numbers of mutants are determined. Several methods can be used to approximate this value, the simplest of which is to use the proportion of wells that do not contain any mutants and the zero term of the Poisson distribution

$$p_0 = e^{-m},$$

where p_0 = the fraction of negative wells, and m = mean number of mutants per well. The complexity of this process of deriving actual numbers of induced mutants contrasts markedly with the simple process of counting numbers of colonies in a plate test.

Original reports (e.g. GREEN et al. 1976) stated that the fluctuation test was more sensitive than the Ames test, based on the detection of certain chemicals at lower concentrations than was possible with a standard plate incorporation assay. This quantitative difference in assay response does not, in itself, justify the inclusion of this additional and technically demanding assay in routine screening situations. Furthermore, these apparent differences in detection levels have, at least in part, been reduced by improvements over the past 10 years to the plate-incorporation protocol. More recently, it has been reported that the antineoplastic agent procarbazine is only detectable using the fluctuation test (GATEHOUSE and PAES 1983). However, recent debate over the in vitro genotoxicity of this chemical (see ASHBY 1986 and references therein) suggests that the reported negative Ames test results are due to the difficulties in creating a suitable metabolic activation system to detect procarbazine rather than to an inherent shortcoming of the assay protocol itself.

Nevertheless, the fluctuation test has at least one potential advantage over a standard Ames plate test. The normal S9-mix used to mimic mammalian metabolism (see Sect. D.II.1) is accepted as being suboptimal for some liver enzyme pathways, due to the disruptive preparation processes. Utilising a liquid medium-based assay such as the fluctuation test, intact hepatocytes can be more

Table 3. Summary of some bacterial forward mutation systems

Species and strain	Endpoint	References
S. typhimurium TA	*aza*^{s} to *aza*^{r}	SKOPEK et al. (1978)
S. typhimurium SV-3	*ara*^{s} to *ara*^{r}	RUIZ-VAZQUEZ et al. (1978)
E. coli K12	Various loci	MOHN and ELLENBERGER (1977)
B. subtilis	Sporulation loss	MACGREGOR and SACKS (1976)
Klebsiella pneumoniae *Citrobacter freundii*	*str*^{s} to *str*^{r}	VOOGD et al. (1974)

easily used as a metabolising system (HUBBARD et al. 1984). In situations in which such non-standard metabolism is required, preferential use of the fluctuation test may be justified.

III. Forward Mutation Assays

Forward mutation assays can detect mutagenic events that occur over a larger part of the genome (usually >1 gene) than those detected by reversion assays. Thus, such assays should theoretically be able to detect mutagenic agents which are not detected by standard single-locus reversion assays. Whilst a number of forward mutation systems have been developed (see Table 3 for a summary of some of the established assays, without exception the performance of such assays is technically more demanding than, for example, an Ames test. Further, the results from the few validation studies published to date do not show that such assays offer any advantage over properly conducted reversion assays for the routine screening of chemicals. However, limited use of some of the forward mutation assay protocols may be of relevance in particular investigations of molecular mechanisms of mutagenesis.

Some phage induction systems have also been referred to as forward mutation assays. As the endpoint of such systems is only an indirect method of measuring DNA damage, and several steps, mainly non-chromosomal, are required before active phage particles are produced, such a classification of this assay is misleading. As previously discussed, similar restrictions apply to assays using SOS-induction as an endpoint, even though forward mutation is one of the mechanisms by which SOS-induction may occur.

IV. Host-Mediated Assays

In contrast to the other bacterial mutagenicity assay protocols, the host-mediated assay (cf. GABRIDGE and LEGATOR 1969) exposes the tester strains in an in vivo rather than an in vitro environment. The indicator cells, whether a *Salmonella* strain or yeast cells, are injected into a host animal, usually a mouse, which is exposed to the test chemical by normal in vivo routes, for example by oral dosing. The route of introduction of the tester cells can either be by intravenous or intraperitoneal injection. After a set exposure period, the cells are recovered from the pleural cavity (CONNOR and LEGATOR 1984), the liver (see

BRUSICK 1980) or other organs, and then plated onto selective medium to determine the number of reverted cells. The combination of an in vitro tester organism with an in vivo exposure regime nominally overcomes the limitations regarding mimicking the biotransformation capacity of the living mammal. However, the practical problems in ensuring consistent recovery of cells per animal and the difficulty in assessing the exposure concentrations of test chemical to the tester cells and preventing bias from selective pressures limit the effective use of this assay as a screening test for bacterial mutations.

E. Use of the Results from Mutation Assays

I. Basic Meaning of Such Results

The results from a bacterial mutation assay reveal whether or not the compound being tested is mutagenic in a particular assay system. The use of multiple strains of bacteria in some assays (e.g. the *Salmonella* mutation assay) does not in itself necessarily add weight to the result since the compound is a bacterial mutagen whether one or all tester strains give a positive result. The use of strains with different genetic alterations merely serves to increase the chances of observing any induced mutations and also allows identification of the particular mutagenic event induced by the test compound (see Sect. C.I.2). Similarly, the use of exogenous metabolising systems such as the liver S9-mix in the Ames test simply reveals whether the compound requires metabolic conversion to unmask its mutagenic potential.

II. Use of the Results

Having determined that the chemical is a bacterial mutagen, the next step is an evaluation of the significance of this result. At this stage the apparent potency of the observed response becomes of some relevance (but see Sect. E.II.1 below). The assessment of mutagenicity in the presence and absence of exogenous metabolism is also important, whilst the possible classification of a chemical as a bacterial-specific mutagen (e.g. chlorodifluoromethane: LONGSTAFF et al. 1984) has already been discussed. If a chemical is mutagenic in the absence of auxiliary metabolism but this activity is abolished by the addition of such a metabolising system, then the potential for the detoxification of the chemical in vivo has been identified. Care must also be taken to ensure that the metabolic systems present in the in vitro system are a true reflection of the in vivo situation. The toxicological significance of a positive bacterial mutation assay result and the extrapolation of such results in terms of potential animal and human carcinogens are discussed below.

1. Extrapolation of the Results

Ostensibly the extrapolation of a positive bacterial mutation assay result in terms of the carcinogenic potential of the chemical should be a simple, quantitative calculation. The potency of the response in the bacterial assay, together with an assessment of the enhancement or nullification of this effect due to metabolic activation or detoxification, should be able to give a finite value for such carcinogenic potential. Such extrapolations have been best studied using radiation as the mutagen (see for example OFTEDAL 1979), but the same principle applies to any other definable mutagen.

Studies conducted in an attempt to validate the above theorem are most easily carried out in reverse, i.e. by testing the mutagenicity of a series of compounds of known carcinogenic potency. One such study (ASHBY et al. 1982) used a series of derivatives of the animal carcinogen 4-dimethylaminoazobenzene (DAB). The existing carcinogenicity data on each compound were used to derive a series of potency figures relative to DAB using the equation proposed by MILLER and MILLER (1953; see also ARCOS et al. 1968). Each compound was tested in the three main variants of the Ames test: the standard plate-incorporation assay, the pre-incubation assay and the fluctuation test. Each compound gave a positive result in at least two assays, despite the fact that four of them were apparently non-carcinogenic; thus, there was no qualitative correlation between the mutagenic and carcinogenic potencies of a series of related compounds.

A recent review on the genotoxicity to *Salmonella* of 222 chemicals tested for carcinogenicity by the U.S. National Toxicology Program has concluded that the Ames test, in conjunction with an assessment of structure-activity analysis, showed good qualitative correlation (~90%) with carcinogenicity (ASHBY and TENNANT 1988). It should be noted, however, that this same review showed that in the *Salmonella* assay positive results were obtained for 30% of the 83 non-carcinogens in the study, confirming previous observations that the *Salmonella* assay does not give a perfect correlation between mutagenicity and carcinogenicity (e.g. MCCANN and AMES 1976; TENNANT et al. 1987).

2. Toxicological Significance of the Results

The toxicological significance of a positive bacterial mutation result depends to some degree on the reason for testing the compound, as well as on the proposed use of the compound.

If the chemical is being tested as part of a submission of toxicology data to a regulatory authority, then a positive Ames test may be sufficient to have the use of the chemical restricted (cf. EEC 1983). The reasoning behind such decisions normally occurs on a case-by-case basis, depending on the results of the other toxicological tests that are also available. Evidence of bacterial mutagenicity in the absence of any other contradictory evidence should be taken as a warning of potential mutagenic carcinogenic effects. Such warnings may be modified appropriately by the existence of other short-term in vitro and in vivo genotoxicity test data.

Similarly, the potency of the response in terms of a minimum dose required to induce mutations, together with an assessment of the likely exposure to the chemical, will affect the significance of the result. A compound which induces a significant level of mutation at low doses and which is likely to come into direct contact with, for example, the skin will be considered to have a greater risk than a compound which just gives a positive result at the limits of detection of the assay and which is unlikely to have any direct contact with humans whatsoever.

F. Summary and Conclusions

A properly conducted, well-validated, bacterial mutagenicity assay can give a definite indication of the in vivo genotoxic potential of a test compound. The various modifications of the "basic" *Salmonella* mutation test (MARON and AMES 1983) that have been developed allow application of the assay to complex mixtures and to volatile compounds (e.g. VICTORIN and STAHLBERG 1988) and increase the potential use of the assay in hazard assessment.

The limitations of bacterial mutation assay systems in extrapolating to carcinogenic potential of are now well-defined and can therefore be used to rationalise the use of bacterial assays in screening for potential carcinogens. The future developments of bacterial mutation assays are likely to be less dramatic than those that have occurred over the past decade, but refinements to existing protocols are still being made.

References

Ames BN (1971) The detection of chemical mutagens with enteric bacteria. In: Hollaender A (ed) chemical mutagens: principles and methods for their detection, vol 1. Plenum, New York, pp 267–282

Ames BN, Lee FD, Durston WE (1973a) An improved bacterial test system for the detection and classification of mutagens and carcinogens. Proc Natl Acad Sci USA 70:782–786

Ames BN, Durston WE, Yamasaki E, Lee FD (1973b) Carcinogens are mutagens: a simple test system combining liver homogenates for activation and bacteria for detection. Proc Natl Acad Sci USA 70:2281–2285

Ames BN, McCann J, Yamasaki E (1975) Methods for detecting carcinogens and mutagens with the *Salmonella*/mammalian-microsome mutagenicity test. Mutat Res 31:347–364

Arcos JC, Argus MF, Wolf G (1968) Chemical induction of cancer, vol 1. Academic, New York

Ashby J (1986) Letter to the editor. Mutagenesis 1:309–317

Ashby J (1989) The evaluation of volatile chemicals for mutagenicity. Letter to the editor. Mutagenesis 4:160–162

Ashby J, Tennant RW (1988) Chemical structure, *Salmonella* mutagenicity, and extent of carcinogenicity as indicators of genotoxic carcinogenesis among 222 chemicals tested in rodents by the U.S. NCI/NTP. Mutat Res 204:17–115

Ashby J, Lefevre PA, Styles JA, Charlesworth J, Paton D (1982) Comparisons between carcinogenic potency and mutagenic potency to *Salmonella* in a series of derivatives of 4-dimethylaminoazobenzene (DAB). Mutat Res 93:67–81

Ashby J, Callander RD, Rose FL (1985) Weak mutagenicity to *Salmonella* of the formaldehyde-releasing anti-tumour agent hexamethylmelamine. Mutat Res 142:121–125

Auerbach C, Robson JM (1946) Production of mutations by allyl isothiocyanate. Nature 54:81–82

Balbinder E, Kerry D (1984) A new strain of *Salmonella typhimurium* reverted by mitomycin C and *N*-methyl-*N'*-nitro-*N*-nitrosoguanidine – a possible universal tester for mutagenic compounds. Mutat Res 130:315–320

Bartsch H, Malaveille C, Montesano R (1975) Human, rat and mouse liver-mediated mutagenicity of vinyl chloride in *S. typhimurium* strains. Int J Cancer 15:429–437

Bartsch H, Camus A-M, Malaveille C (1976) Comparative mutagenicity of *N*-nitrosamines in a semi-solid and in a liquid incubation system in the presence of rat or human tissue fractions. Mutat Res 37:149–162

Bauer KH (1928) Mutationstheorie der Geschwulstentstehung. Springer, Berlin Heidelberg New York

Berry DJ, Litchfield MH (1985) A review of the current regulatory requirements for mutagenicity testing. In: Ashby J, de Serres FJ, Draper M, Ishidate M Jr, Margolin BH, Matter BE, Shelby MD (eds) Evaluation of short-term tests for carcinogens. Report of the international programme on chemical safety's collaborative study on in vitro assays. Elsevier, Amsterdam, pp 727–740 (Progress in mutation research, vol 5)

Bjorseth A, Eidsa G, Gether J, Landmark L, Moller M (1982) Detection of mutagens in complex samples by the *Salmonella* assay applied directly on thin-layer chromatography plates. Science 215:87–89

Bridges BA, Mottershead RP, Green MHL, Gray WJH (1973) Mutagenicity of dichlorvos and methyl methanesulphonate for *Escherichia coli* WP2 and some derivatives deficient in DNA repair. Mutat Res 19:295–303

Bridges BA, Zeiger E, McGregor DB (1981) Summary report on the performance of bacterial mutation assays. In: de Serres FJ, Ashby J (eds) Evaluation of short-term tests for carcinogens. Report of the international collaborative program. Elsevier, New York, pp 49–67 (Progress in mutation research, vol 1)

Brusick D (1980) Principles of genetic toxicology. Plenum, New York, pp 232–235

Burdette WJ (1955) The significance of mutation in relation to the origin of tumours: a review. Cancer Res 15:201–226

Callander RD (1986) Observed convergence of the *Salmonella* plate and pre-incubation assays when employing varying levels of S9. Mutagenesis 1:439–443

Clarke CH, Wade MJ (1975) Evidence that caffeine, 8-methoxypsoralen and steroidal diamines are frameshift mutagens for *E. coli* K12. Mutat Res 28:123–125

Clive D, McCuen R, Spector JFS, Piper C, Mavournin KH (1983) Specific gene mutations in L5178Y cells in culture: a report of the US EPA Gene-Tox Program. Mutat Res 115:225–251

Connor TH, Legator MS (1984) The intraperitoneal host-mediated assay. In: Kilbey BJ, Legator M, Nichols W, Ramel C (eds) Handbook of mutagenicity test procedures, 2nd edn. Elsevier, Amsterdam, pp 643–654

EEC (1983) Commission directive 83/467/EEC: annex V to council directive 67/548/EEC on the classification, packaging and labelling of dangerous substances. Official Journal of the European Communities, no L 257, 16 September 1983, Brussels, pp 21–23

Forster R (1986) Mutagenicity of benzyl chloride to *Salmonella*. Mutagenesis 1:72–73

Gabridge MG, Legator MS (1969) A host-mediated microbial assay for the detection of mutagenic compounds. Proc Soc Exp Biol Med 130:831–834

Garner RC, Miller EC, Miller JA (1972) Liver microsomal metabolism of aflatoxin B_1 to a reactive derivative toxic to *Salmonella typhimurium* TA 1530. Cancer Res 32:2058–2066

Gatehouse D (1978) Detection of mutagenic derivatives of cyclophosphamide and a variety of other mutagens in a microtitre fluctuation test, without microsomal activation. Mutat Res 53:289–296

Gatehouse D (1987) Guidelines for testing of environmental agents: critical features of bacterial mutation assays. Mutagenesis 2:397–409

Gatehouse DG, Delow GF (1979) The development of a microtitre fluctuation test for the detection of indirect mutagens, and its use in the evaluation of mixed enzyme induction of the liver. Mutat Res 60:239–252

Gatehouse DG, Paes DJ (1983) A demonstration of the in vitro bacterial mutagenicity of procarbazine, using the microtitre fluctuation test and large concentrations of S9 fraction. Carcinogenesis 4:347–352

Green MHL, Muriel WJ (1976) Mutagen testing using trp^+ reversion in *Escherichia coli*. Mutat Res 38:2–32

Green MHL, Muriel WJ, Bridges BA (1976) Use of a simplified fluctuation test to detect low levels of mutagens. Mutat Res 38:33–42

Hartman PE, Ames BN, Roth JR, Barnes WM, Levin DE (1986) Target sequences for mutagenesis in *Salmonella* histidine-requiring mutants. Environ Mutagen 8:631–641

Hass BS, Heflich RH, Shaddock JG, Casciano DA (1985) Comparison of mutagenicities in a *Salmonella* reversion assay mediated by uninduced hepatocytes and hepatocytes from rats pretreated for 1 or 5 days with aroclor 1254. Environ Mutagen 7:391–403

Houk VS, Schalkowsky S, Claxton LD (1988) Development and validation of an automated approach to bacterial mutagenicity testing. Environ Molec Mutagen [Suppl 11] 11:49

Hubbard SA, Green MHL, Gatehouse D, Bridges JW (1984) The fluctuation test in bacteria. In: Kilbey BJ, Legator M, Nichols W, Ramel C (eds) Handbook of mutagenicity test procedures, 2nd edn. Elsevier, Amsterdam, pp 141–160

Hubbard SA, Brooks TM, Gonzalez LP, Bridges JW (1985) Preparation and characterisation of S9 fractions. In: Parry JM, Arlett CF (eds) Comparative genetic toxicology: the second UKEMS collaborative study. Macmillan, London, pp 413–438

ICPEMC (1982) International commission for protection against environmental mutagens and carcinogens: committee 2 final report. Mutagenesis testing as an approach to carcinogenesis. Mutat Res 99:73–91

Kada T, Tutikawa K, Sadaie Y (1972) In vitro and host-mediated "rec-Assay" procedures for screening chemical mutagens; and phloxine, a mutagenic red dye detected. Mutat Res 16:165–174

Kier LD, Yamasaki E, Ames BN (1974) Detection of mutagenic activity in cigarette smoke condensate. Proc Natl Acad Sci USA 71:4159–4163

Lefevre PA, Ashby J (1981) The effects of pre-incubation period and norharman on the mutagenic potency of 4-dimethylaminoazobenzene and 3′-methyl-4-dimethylaminoazobenzene. Carcinogenesis 2:927–931

Levin DE, Hollstein M, Christman MF, Schwiers EA, Ames BN (1982) A new *Salmonella* tester strain (TA102) with A·T base pairs at the site of mutation detects oxidative mutagens. Proc Natl Acad Sci USA 79:7445–7449

Litterst CL, Farber TM, Baker AM, Van Looen EJ (1972) Effect of polychlorinated biphenyls on hepatic microsomal enzymes in the rat. Toxicol Appl Pharmacol 23:112–122

Longstaff E, McGregor DB (1978) Mutagenicity of a halocarbon refrigerant monochlorodifluoromethane (R-22) in *Salmonella typhimurium*. Toxicol Lett 2:1–4

Longstaff E, Robinson M, Bradbrook C, Styles JA, Purchase IFH (1984) Genotoxicity and carcinogenicity of fluorocarbons: assessment by short-term in vitro tests and chronic exposure in rats. Toxicol Appl Pharmacol 72:15–31

Luria SE, Delbruck M (1943) Mutations of bacteria from virus sensitivity to virus resistance. Genetics 28:491–511

MacGregor JT, Sacks LE (1976) The sporulation system of *Bacillus subtilis* as the basis of a multi-gene mutagen screening test. Mutat Res 38:271–286

Malaveille C, Brun G, Bartsch H (1983) Studies on the efficiency of the *Salmonella*/rat hepatocyte assay for the detection of carcinogens as mutagens: activation of 1,2-dimethylhydrazine and procarbazine into bacterial mutagens. Carcinogenesis 4:449–455

Malling HV (1966) Mutagenicity of two potent carcinogens, dimethylnitrosamine and diethylnitrosamine, in *Neurospora crassa*. Mutat Res 3:537–540

Malling HV (1971) Dimethylnitrosamine: formation of mutagenic compounds by interaction with mouse liver microsomes. Mutat Res 13:425–429

Mamber SW, Bryson V, Katz SE (1984) Evaluation of the *Escherichia coli* K12 inductest for detection of potential chemical carcinogens. Mutat Res 130:141–151

Maron DM, Ames BN (1983) Revised methods for the *Salmonella* mutagenicity test. Mutat Res 113:173–215

McCann J, Ames BN (1976) Detection of carcinogens as mutagens in the *Salmonella*/microsome test: assay of 300 chemicals: discussion. Proc Natl Acad Sci USA 73:950–954

McCann J, Spingarn NE, Kobori J, Ames BN (1975a) Detection of carcinogens as mutagens: bacterial tester strains with R factor plasmids. Proc Natl Acad Sci USA 72:979–983

McCann J, Choi E, Yamasaki E, Ames BN (1975b) Detection of carcinogens as mutagens in the *Salmonella*/microsome test: assay of 300 chemicals. Proc Natl Acad Sci USA 72:5135–5139

McMahon RE, Cline JC, Thompson CZ (1979) Assay of 855 test chemicals in ten tester strains using a new modification of the Ames test for bacterial mutagens. Cancer Res 39:682–693

Miller JA, Miller EC (1953) The carcinogenic aminoazo dyes. Adv Cancer Res 1:339–396

Mirsalis JC, Hamm TE Jr, Sherrill JM, Butterworth BE (1982) The role of gut flora in the genotoxicity of dinitrotoluene. Nature 295:322–323

Mohn GR, Ellenberger J (1977) The use of *Escherichia coli* K12/343/113(λ) as a multi-purpose indicator strain in various mutagenicity testing procedures. In: Kilbey BJ, Legator M, Nichols W, Ramel C (eds) Handbook of mutagenicity test procedures. Elsevier, Amsterdam, pp 95–118

Nakamura S-I, Oda Y, Shimada T, Oki I, Sugimoto K (1987) SOS-inducing activity of chemical carcinogens and mutagens in *Salmonella typhimurium* TA1535/pSK1002: examination with 151 chemicals. Mutat Res 192:239–246

Nash JA, King LJ, Lock EA, Green T (1984) The metabolism and disposition of hexachloro-1,3-butadiene in the rat and its relevance to nephrotoxicity. Toxicol Appl Pharmacol 73:124–137

OECD (1983a) Guidelines for testing of chemicals: 471: Genetic toxicology: *Salmonella typhimurium*, reverse mutation assay. Organisation for Economic Co-operation and Development, Paris

OECD (1983b) Guidelines for testing of chemicals: 472: Genetic toxicology: *Escherichia coli*, reverse mutation assay. Organisation for Economic Co-operation and Development, Paris

Oftedal P (1979) Extrapolation of mutagenic test results to man. In: Paget GE (ed) Mutagenesis in sub-mammalian systems: status and significance. MTP Press, Lancaster, pp 189–194

Paolini M, Hrelia P, Corsi C, Bronzetti G, Biagi GL, Cantelli-Forti G (1987) NADPH as a rate-limiting factor for microsomal metabolism. An alternative and economic NADPH-generating system for microsomal monooxygenase in in vitro genotoxicity studies. Mutat Res 178:11–20

Pitot HC, Barsness L, Goldsworthy T, Kitagawa T (1978) Biochemical characterization of stages of hepatocarcinogenesis after a single dose of diethylnitrosamine. Nature 271:456–458

Prival MJ, Bell SJ, Mitchell VD, Peiperl MD, Vaughn VL (1984) Mutagenicity of benzidine and benzidine-congener dyes and selected monoazo dyes in a modified *Salmonella* assay. Mutat Res 136:33–47

Quillardet P, Huisman O, D'ari R, Hofnung M (1982) SOS chromotest, a direct assay of induction of an SOS function in *Escherichia coli* K-12 to measure genotoxicity. Proc Natl Acad Sci USA 79:5971–5975

Rannug U, Johansson A, Ramel C, Wachmeister CA (1974) The mutagenicity of vinyl chloride after metabolic activation. Ambio 3:194–197

Robertson J, Harris WJ, McGregor DB (1982a) Mutagenicity of azo-dyes in the *Salmonella*/activation test. Carcinogenesis 3:21–25

Robertson J, Harris WJ, McGregor DB (1982b) Factors affecting the response of *N,N*-dimethylaminoazobenzene in the Ames microbial mutation assay. Carcinogenesis 3:977–980

Rosenkranz HS, Speck WT (1975) Mutagenicity of metronidazole: activation by mammalian liver microsomes. Biochem Biophys Res Commun 66:520–525

Rosenkranz HS, Speck WT (1976) Activation of nitrofurantoin to a mutagen by rat liver nitroreductase. Biochem Pharmacol 25:1555–1556

Rosenkranz HS, McCoy EC, Mermelstein R, Speck WT (1981) A cautionary note on the use of nitroreductase-deficient strains of *Salmonella typhimurium* for the detection of nitroarenes as mutagens in complex mixtures including diesel exhausts. Mutat Res 91:103–105

Rosenkranz EJ, McCoy EC, Mermelstein R, Rosenkranz HS (1982) Evidence for the existence of distinct nitroreductases in *Salmonella typhimurium:* roles in mutagenesis. Carcinogenesis 3:121–123

Ruiz-Vazquez R, Pueyo C, Cera-olmedo E (1978) A mutagen assay detecting forward mutations in an arabinose-sensitive strain of *Salmonella typhimurium.* Mutat Res 54:121–129

Ryan FJ (1955) Spontaneous mutation in non-dividing bacteria. Genetics 40:726–738

Simmon VF (1981) Applications of the *Salmonella*/microsome assay. In: Stich HF, San RHC (eds) Short-term tests for chemical carcinogens. Springer, Berlin Heidelberg New York, pp 120–126

Skopek TR, Liber HL, Krolewski JJ, Thilly WG (1978) Quantitative forward mutation assay in *Salmonella typhimurium,* using 8-azaguanine resistance as a genetic marker. Proc Natl Acad Sci USA 75:410–414

Styles JA, Richardson CR, Callander RD, Cross MF, Bennett IP, Longstaff E (1985) Activity of bromochlorodifluoromethane (BCF) in three mutation tests. Mutat Res 142:187–192

Tanooka H (1977) Development and applications of *Bacillus subtilis* test systems for mutagens, involving DNA-repair deficiency and suppressible auxotrophic mutations. Mutat Res 42:19–32

Tennant RW, Margolin BH, Shelby MD, Zeiger E, Haseman JK, Spalding J, Caspary W, Resnick M, Stasiewicz S, Anderson B, Minor R (1987) Prediction of chemical carcinogenicity in rodents from in vitro genetic toxicity assays. Science 236:933–941

Venitt S, Bushell CT, Osborne M (1977) Mutagenicity of acrylonitrile (cyanoethylene) in *Escherichia coli.* Mutat Res 45:283–288

Venitt S, Forster R, Longstaff E (1983) Bacterial mutation assays. In: Report of the UKEMS sub-committee on guidelines for mutagenicity testing. Part 1: Basic test battery; minimal criteria; professional standards; interpretation; selection of supplementary assays. The United Kingdom Environmental Mutagen Society, Swansea, pp 5–40

Venitt S, Crofton-Sleigh C, Osborne MR (1984) The hair-dye reagent 2-(2′,4′-diaminophenoxy)ethanol is mutagenic to *Salmonella typhimurium.* Mutat Res 135:31–47

Victorin K, Stahlberg M (1988) A method for studying the mutagenicity of some gaseous compounds in *Salmonella typhimurium.* Environ Molec Mutagen 11:65–77

Voogd CE, van der Stel JJ, Jacobs J (1974) The mutagenetic action of nitroimidazoles. I. Metronidazole, nimorazole, dimetridazole and ronidazole. Mutat Res 26:483–490

Wright AS (1980) ICPEMC working paper 2/2: the role of metabolism in chemical mutagenesis and chemical carcinogenesis. Mutat Res 75:215–241

Wolf U (1974) Theodor Boveri and his book, *On the problem of the origin of malignant tumors.* In: German J (ed) Chromosomes and cancer. Wiley, New York, pp 3–20

Yahagi T, Degawa M, Seino Y, Matsushima T, Nagao M, Sugimura T, Hashimoto Y (1975) Mutagenicity of carcinogen azo dyes and their derivatives. Cancer Lett 1:91–96

Yahagi T, Nagao M, Seino Y, Matsushima T, Sugimura T, Okado M (1977) Mutagenicities of *N*-nitrosamines on *Salmonella.* Mutat Res 48:121–130

Zeiger E, Haworth S (1985) Tests with a preincubation modification of the *Salmonella*/microsome assay. In: Ashby J, de Serres FJ, Draper M, Ishidate M Jr, Margolin BH, Matter BE, Shelby MD (eds) Evaluation of short-term tests for carcinogens. Report of the international programme on chemical safety's collaborative study on in vitro assays. Elsevier, Amsterdam, pp 187–199 (Progress in mutation research, vol 5)

Zeiger E, Risko KJ, Margolin BH (1985) Strategies to reduce the cost of mutagenicity screening with the *Salmonella* assay. Environ Mutagen 7:901–911

CHAPTER 2

Mammalian Cell Mutations

C. F. Arlett

A. Introduction

The importance of in vitro assays of mutation in mammalian cells in the detection and study of carcinogens follows from a consideration of the role of mutation in carcinogenesis. An involvement of mutagenesis is implied from studies of the modulation of oncogenes where a G→T transversion in codon 12 of H-*ras* was seen in the activation of a human oncogene in bladder tumour carcinoma cells (Reddy et al. 1982). Point mutations at other positions within *ras* genes have also been detected. For example, point mutations at codon 13 have been observed in DNA from patients with acute myeloid leukaemia (Bos et al. 1985). A direct involvement of mutation is seen in studies in which mutations have been analysed in tumours from rodents treated with classical carcinogens. Here codon 12 of the H-*ras* gene was shown to be altered in mammary tumours induced by nitroso-methylurea in rats (Sukumar et al. 1983). In mouse liver, spontaneous tumours or tumours induced by treatment with furan and furfuranal were analysed (Reynolds et al. 1987). In the spontaneous tumours the activated H-*ras* mutations were always at codon 61. However, 40%–60% of the *ras* oncogenes detected in the furan- or furfural-induced tumours were at sites other than codon 61, suggesting that these novel mutations are a reflection of mutagen specificity. Two activated *raf* genes and four undefined oncogenes were detected in this study.

Cellular studies in support of a direct involvement of mutation follow from investigations such as those of Marshall et al. (1984) in which the H-*ras* proto-oncogene was linked to plasmid DNA and exposed in vitro to benzo[*a*]pyrene diol-epoxide, and the transforming *ras* was generated. It is of some interest to note that analysis of oncogene activation after direct induction of malignant transformation by X-rays in Syrian hamster and mouse embryo cells (C3H/10T/1/2) failed to detect any members of the *ras* family (Borek et al. 1987). This study also suggested that a set of novel oncogenes might well have been induced by the X-rays.

The role of chromosome changes in carcinogenesis can be seen at two levels: firstly, in examples of reciprocal translocation whereby deregulation of normal levels of transcriptional control leads to malignancy, as with the *myc* proto-oncogene in Burkitt's lymphoma (Croce 1986), or where an altered gene product may be involved, as in chronic myelogenous leukaemia (Fainstein et al. 1987); secondly, via non-reciprocal exchanges which may involve deletion or addition of chromosomal material. Deletions in chromosome 13q14.11 in retinoblastoma (STRONG et al. 1981) or 11p13 in Wilms' tumour (KOUFOS et al. 1985) are well-known. The involvement of both mutagenic and clastogenic events is emphasised in examples such as retinoblastoma (CAVENNE et al. 1983) and bilateral acoustic neurofibromatosis (ROULEAU et al. 1987), in which the tumour is thought to arise when, following a primary mutation, the normal allele is subsequently lost by deletion. In the case of multiple endocrine neoplasia type 2A, the primary defect maps to chromosome 10 (SIMPSON et al. 1987) and the "tumour-suppressing" sequences, which may be subject to deletion, to chromosome 1 (MATHEW et al. 1987).

The consistency with which the above changes can be detected in spontaneously occurring tumours, their induction following treatment with known carcinogens and their specificity provide a basis for the use of assays of genetic change in cultured mammalian cells in testing for carcinogens (MONTESANO et al. 1986; ROSENKRANZ 1988). The assays may, conveniently, be divided into two groups: those which test for chromosome damage (clastogenicity) and those which test for gene mutations. Currently, there is considerable debate (ASHBY 1986a, b; GATEHOUSE and TWEATS 1986; GARNER and KIRKLAND 1986; ENNEVER et al. 1987; ISHIDATE and HARNOIS 1987; TWEATS and GATEHOUSE 1988) regarding the efficiency, cost effectiveness and predictive capacity of both classes of assay as part of a larger debate concerning the relevance of all short-term tests (TENNANT et al. 1987; ROSENKRANZ 1988). Clastogenicity assays have been in use for many years (EVANS 1988), and a considerable body of data exists both publicly (ISHIDATE 1983) and privately. This large data base has generated considerable confidence amongst the students of clastogenicity that the use of these tests gives rise to sound data.

Given that we accept that both chromosomal and mutational events are involved in the ontogeny of malignancy, then it follows that a major disadvantage of clastogenicity assays is that they detect only chromosomal change. The substantial positive correlation between chromosome damage and lethality (DAVIES and EVANS 1966) means that the observations are made in cells that are probably destined to die and may thus be irrelevant to the production of a tumour. In addition, little is known of the mechanism of chromosome change. On the positive side, the assays encompass the whole genome, thus presenting a large target. Relatively few compounds exist which cause gene mutation without any accompanying clastogenic damage.

With regard to gene mutation assays, a much smaller data base is available (LAMBERT et al. 1986). The point of these clonal assays is that the changes are monitored in viable cells. Of particular interest are those assays which measure, or are capable of measuring, both chromosomal and gene mutations as end points. Here the examples are assays based upon the *tk* gene in L5178Y mouse lymphoma cells (HOZIER et al. 1985; MOORE et al. 1985b; COLE et al. 1986; CLIVE et al. 1988) and TK6 human lymphoblastoid cells (THILLY et al. 1980). The *hprt* gene is also capable of being exploited in this way (CLIVE et al. 1988). This dual testing capability may well prove to be the most significant single factor in defence of the use of cultured mammalian cells in testing for genotoxic and carcinogenic potential (ARLETT and COLE 1988).

B. Genetics and Principles of Experimental Design

I. Genetics

The basic assumption is that the in vitro response of cultured mammalian cells will reflect both the mutagenic susceptibility of the mammalian genome in vivo and the mutagenic capacity of the carcinogen under test.

The asexual nature of the cultured mammalian cell material under investigation precludes a formal Mendelian anaylsis of the yield of mutant colonies. This led to an early suggestion (HARRIS 1973; FOX and RADACIC 1978) that many of the variant cells may be the consequence of epigenetic phenomena rather than genetic events. To date the study of the nature of induced and spontaneous variants in endogenous genes has led to the belief that most of them have a genetic origin. However, the possibility that a proportion may have an epigenetic origin must not be overlooked (GEBARA et al. 1987; BEGG et al. 1986).

A number of systems are available which depend upon forward or reverse mutation. Most of these function by selecting rare variant cells which are able to survive in the presence of a toxic selective agent (COLE et al. 1983). If the gene product should act in a recessive mode, then changes in the gene can only be studied in the heterozygous or hemizygous states. The classical example here is the *hprt* gene on the X chromosome, which is hemizygous in the male and functionally hemizygous in the female as a consequence of lyonisation. The well-known system which depends upon selecting cells defective at the autosomal thymidine kinase (*tk*) locus in the mouse lymphoma L5178Y $TK^{+}/-$ line also depends upon the existence of a functional heterozygous cell derived from a homozygous parent (CLIVE et al. 1972)

Some systems utilise the semi-dominant action of the gene; here mutation in one of the two genes present in the functional diploid make it possible to detect the mutant phenotype. The classical example of this is ouabain resistance, which is due to changes in a gene coding for a membrane-bound Na^{+}/K^{+} ATPase (BAKER et al. 1974).

II. Experimental Design

The experiment requires the provision of a large bulk culture of cells to be divided into control and treated subpopulations.

It is recommended that true replicates are set up for control and each level of treatment. The size of the populations for study is influenced by the spontaneous mutant frequency. A convenient and often proposed rule is that each population should contain sufficient viable cells so that at least ten spontaneous mutants will survive treatment. The bulk population is maintained in such a way that a defined frequency of spontaneous mutants may be anticipated at the time of treatment. This spontaneous mutant frequency is a useful guide to stability in the system.

Stability may be maintained by recloning the cell line to establish a culture with a low spontaneous mutant frequency from which large stocks are frozen down in liquid nitrogen, and each experiment is then initiated from new stock from the freezer. Cells in regular subculture may also be diluted to a low density (less than 10 cells/ml) or subcultured into a medium containing various combinations of aminopterin or methotrexate and thymidine, cytosine, hypoxanthine and glycine (HAT, CHAT, THAG or THMG media) whose purpose is to purge the population of preexisting *hprt*$^-$ or *tk*$^-$ mutants. This purged population can then be expanded and frozen down for future study. Regular, weekly purges are not thought to represent good laboratory practice because of the potential disturbance of the nucleotide pools within treated cells.

Following mutagenic treatment, which may be in the presence or absence of a suitable activation system, the cells are washed and cultured in fresh medium. An assessment of the toxic effects of any treatment at this time is advised. The toxic effect may be estimated from the cloning efficiencies (ratio of colonies formed per cells plated) of control and treated samples. An alternative method is to undertake daily population counts during the period of growth following treatment (Clive et al. 1979).

The time after treatment at which the maximum frequency of newly induced mutants can be detected is dependent upon the cell line and selective system in use, and this "expression time" must be defined carefully. During this period the growth rate of cells is monitored and cells subcultured, if necessary, to prevent overgrowth and to maintain a dividing population. Mutants should be able to grow under these non-selective conditions with the same population-doubling times as non-mutants, otherwise there will be selection pressure against them. The expression time is influenced by phenotypic lag; the cells must have an opportunity to generate the mutant phenotype. In addition, the proteins that were synthesised from the gene before mutation may continue to be present in the cytoplasm. The mutant frequency is determined by cloning cells at a high density in medium containing a selective agent in which only mutant cells survive and, at a lower density, in non-selective medium to establish cloning efficiency. Colonies are counted and the mutant frequency derived from the ratio of the number of mutant colonies to the number of viable colony-forming cells plated. An appropriate form of statistical analysis has been prepared by the United Kingdom Environmental Mutagen Society.

The number of cells to be plated in selective agent is influenced by a number of factors. The prime requirement is to sample a sufficiently large population of cells to ensure detection of mutants. The actual number of cells which can be challenged by a selective agent in a single culture vessel differs between attached and suspension cultures. Metabolic co-operation, whereby toxic metabolites are transferred from one cell to another via gap junctions (van Zeeland et al. 1972), is an important factor with attached cells. Here, it is necessary to plate only such numbers of cells that individual cells are not in contact so that metabolic co-operation cannot occur. There is also a less well-defined "cell density" effect such that some selective agents cause an immediate cessation of cell division while, for others, a round of replication may be required for toxicity. An example of the former is 8-azaguanine and of the latter, 6-thioguanine. Toxic effects may also be modified by the metabolic products of dying cells. More cells can be plated in an individual dish with suspension cultures cloned in soft agar because of the third dimension and because metabolic co-operation is absent.

C. Cell Types

I. Assessment of Mutation in Rodent Cells

Primary, low-passage fibroblasts of rodent origin from either fetal or perinatal material have a reduced life span in culture (HAYFLICK and MOORHEAD 1961) and are not of general use in studies of mutagenesis. Cells suitable for such studies should show the following characteristics: (i) a stable karyotype at or near the diploid level; (ii) a reliable, high cloning efficiency (more than 50%); (iii) a stable, short generation time; and (iv) colony morphology which permits colony counting. In practice the established Chinese hamster lung V79, hamster ovary CHO, and mouse lymphoma L5178Y lines, which more or less fit these criteria, are in general use.

1. L5178Y Mouse Lymphoma Cells

These cells have probably been used most commonly because they are relatively easy to maintain in suspension culture. Two genetic variant forms of these cells have been used. The first is the so-called wild-type L5178Y ($TK^{+}/_{+}$) cells, and examples of the use of these cells are given by COLE and ARLETT (1976) and COLE et al. (1976). Experiments using colony formation in soft agar containing selective agents such as 6-thioguanine, ouabain, excess thymidine (COLE and ARLETT 1976) and cytosine arabinoside (ROGERS et al. 1980) are possible. These cells may also be utilised in assays based upon modifications of the Lauria-Delbruck fluctuation test (COLE et al. 1976, 1983). The second and most frequently used variant is the L5178Y $TK^{+}/_{-}$ line cell which has been selected to be hemizygous at the thymidine kinase locus (CLIVE et al. 1979). Mutants resistant to triflurothymidine (TFT) or bromodeoxyuridine (BUdr) may be selected to provide cells defective in thymidine kinase. Colonies are selected in soft agar and fall into two categories, large and small (MOORE et al. 1985a; HOZIER et al. 1985). The former appears to represent a class of variants which have point mutations or small deletions within the gene; the latter consists of variants with larger deletions including the whole gene or chromosome aberrations involving the *tk* gene (MOORE et al. 1985). This latter class of variants represents the chromosome-breakage class of mutants.

The L5178Y $TK^{+}/_{-}$ TFT resistance system has been used most frequently as the mammalian cell in vitro test in the battery of short-term tests designed to detect carcinogens and mutagens (CLIVE et al. 1979). The large data base gives considerable validity to the test, but there are critics who suggest that it is supersensitive, giving positive scores with many non-carcinogens such as urea, glucose and dimethyl sulphoxide (WAGENHEIM and BOLESFOLDI 1988). Spurious positives can arise because of dubious decision-making. A positive response in the assay has often been based upon an arbitary assignment of "foldness". That is, a two- or fourfold increase at some dose level over the control level has been reported to represent a positive result. Newly devised statistical packages, which also demand an improved experimental design, should obviate the need for such arbitrary decisions and will probably reduce the frequency of spurious positive results. A second cause of such results may well reside in any treatment mode being positive at concentrations at which a considerable physiological imbalance is achieved, for example, by disturbing pH or osmolarity (Brusick 1986). Detailed knowledge of the cell biology of the system is the only way these problems can be overcome (Cole et al. 1986).

The L5178Y $TK^{+}/-$ cell system can provide the basis of an integrated assay based upon the assessment of the frequency of large and small TFT-resistant colonies, ouabain resistance and micronucleus induction combined within a 48-h expression time. There is also the opportunity to assess the frequency of "classical" chromosome aberrations and 6-thioguanine resistance (at a 7-day expression time) in the same treated and control populations. Thus, clastogenicity is assessed by micronucleus production and, possibly, by metaphase chromosome aberrations and from the proportion of small TFT-resistant colonies. Large colony TFT-resistant mutants and, in addition, ouabain- and 6-thioguanine-resistant mutants can be used to assess the extent of gene or point mutagenic change (Arlett and Cole 1988).

2. Chinese Hamster V79 and CHO Cells

These cells are more commonly employed in fundamental studies than as a basis of short-term tests, although they do have a role in testing programmes. The test system is usually based upon resistance to 6-thioguanine or 8-azaguanine. As an assay system they are often regarded as insensitive, but in certain circumstances this might be regarded as a virtue. Possible explanations for why this *hprt* system might be less sensitive than the *tk*-based L5178Y assay include:

(i) The spontaneous mutant frequency to 6-thioguanine resistance is generally around an order of magnitude lower than TFT resistance. Thus, many more cells must be treated, maintained in culture and tested to obtain a similar level of statistical precision.

(ii) The long expression time required for the selection of 6-thioguanine-resistant mutants (6–9 days) may mean that a class of slow growing mutants (which would be detected in the *tk*-based assay with its short, 2–3-day expression time) is lost. This could, in theory, be overcome by the use of 8-azaguanine as a selective agent (Arlett 1977), although this agent cannot be used for mouse cells.

(iii) The *hprt* gene may be less mutable than the *tk* gene, or the mutations may not be detected by current methodology.

(iv) Some classes of chromosome events such as very large deletions, loss of whole chromosome, recombination and gene conversion may be detected in systems based on the autosomal *tk* gene but not on the X-linked *hprt* gene.

The relative contribution of these possibilities cannot be resolved at present. However, the expanded protocols developed by McGregor et al. (1987) indicate that the insensitivity in the V79 and CHO cells is intrinsic rather than a consequence of the limitations of the size of experiment determined by cells growing attached to surfaces. The insensitivity may arise as a consequence of the location of the gene of interest close to a vital gene or genes on the X chromosome. Target size does not seem an appropriate explanation since the *hprt* gene is ~40 kb long in comparison with the *tk* gene at ~10 kb; these estimates of size do not, of course, take into account the relative size of the coding sequences.

An explanation of differences in sensitivity, which can be invoked on those occasions when knowledge of the relevant metabolism exists, is that a cell may be defective in a particular activation system.

At the fundamental level the Chinese hamster cells have been of considerable value. The analysis of mutants has, until recently, been hindered by the absence of molecular probes for specific genes. Probes for *aprt* (LOWY et al. 1980), *hprt* (JOLLY et al. 1983; KONECKI et al. 1982), *tk* (BRADSHAW and DEININGER 1984) and dihydrofolate reductase (*dhfr*) (URLAUB et al. 1985) are now available.

Earlier observations were relatively crude. They included evidence which suggested that sparsely ionising radiation does not induce base-pair changes or frame-shift mutations because an increase in the frequency of ouabain resistance (in an essential gene) is not observed following treatment of the cells with gamma or X-rays (ARLETT et al. 1975; THACKER et al. 1978).

With the *hprt* locus, which has been used most extensively, cloned *hprt* cDNA probes (BRENNAND et al. 1982; JOLLY et al. 1983; KONECKI et al. 1982) have made it possible to undertake an analysis of the structural changes in individual mutants. FUSCOE et al. (1983) showed that ten spontaneous and one of nine ultraviolet light-induced mutants exhibited major deletions affecting the locus. No changes in the restriction fragments were observed in the remainder, suggesting that they carried very small deletions or point mutations. Chromosome analysis showed correlated deletions of the functional gene and rearrangements involving the X chromosome. This study was enlarged to include molecular analysis of mutants induced by X-rays in the V79 cells (FUSCOE et al. 1986). Here, the frequency of *hprt* gene deletions was higher; four of seven amongst induced mutations and one of twelve spontaneous mutants had deletions. VRIELING et al. (1985) estimated that 70%–80% of X-ray-induced mutations were caused by large deletions. THACKER (1986) also showed that the largest class of gamma ray- or alpha particle-induced V79 *hprt*-deficient cells carried large deletions of the gene. BREIMER et al. (1986), in addition to confirming the deletion-like nature of ionising radiation-induced *hprt*-deficient mutants, found that, in two of twelve mutants examined, putative insertional mutants exist since the locus appears to have enlarged. In summary, approximately half the ionising radiation-induced mutants had lost all the coding sequences detectable in the assays available in this system (BREIMER 1988). Amongst ionising radiation-induced *hprt*-minus mutants in Chinese hamster cells, the frequency of correlated, cytologically detectable changes in the X-chromosome appears to be reduced (FUSCOE et al. 1986; BROWN and THACKER 1984) in comparison with the situation reported for human diploid fibroblasts (COX and MASSON 1978).

While most observations have been performed with ionising radiation, in addition to the results from UV treatment (FUSCOE et al. 1983), KING and BROOKES (1984) analysed benzo[*a*]pyrene diol-epoxide-induced mutants, and their results indicate the induction of base-substitution mutants.

The *aprt* locus coding for a non-essential enzyme in CHO cells provides an especially useful system for mutant analysis since the locus is small (3.9 kb) and contains a sufficient number of sites for endonuclease recognition so that mapping is possible down to 20–30 bp (MEUTH and ARRAND 1982; NALBANTOGLU et al. 1983). Again ionising radiation has proved to be the most popular agent for the induction of mutants for analysis, but, unlike the situation with *hprt*, they usually have (BREIMER et al. 1986) restriction patterns similar to spontaneously occurring mutants (NALBANTOGLU et al. 1987), indicating that they represent base-pair changes or deletions and insertions less than the 25-bp sizing which is possible. Essentially similar results were obtained by GROSOVSKY et al. (1986) with gamma irradiation in this system. GLICKMAN et al. (1987) and DROBETSKY et al. (1987) showed that the majority (26 of 34) of UV-induced mutants were single base-substitution mutants, mostly G:C→A:T transitions. The majority (27 of 30) of spontaneous mutants also involves point mutations. Analysis at this particular locus would seem to hold out the greatest potential for the study of specific gene changes induced by chemical agents (MEUTH et al. 1987). In addition to the CHO cells utilised in the studies of MEUTH and GLICKMAN reported above, strains of mouse lymphoma L5178Y $TK^{+}/-$ cells which are also hemizygous at the *aprt* locus are now available (PAERATAKUL and TAYLOR 1986), opening up the potential for combining two different autosomal assay systems within one cell line.

CHO cells have also been used to assess genetic changes at the dihydrofolate reductase (*dhfr*) locus (GRAF and CHASIN 1982; URLAUB et al. 1983). Resolution with this system is limited to 100 bp, and the characteristics of spontaneous, UV- and gamma-ray-induced mutants have been assessed (MITCHELL et al. 1986; URLAUB et al. 1985). Both UV and gamma irradiation appear to induce substantial deletions, unlike the *aprt* system.

A system based on CHO cells but which measures mutation on human chromosome 11 in a hamster-human hybrid (A_L-J1) has also been reported (WALDREN et al. 1986; PUCK and WALDREN 1987; WALDREN and PUCK 1987).

3. Mouse Lymphocytes

The ability to culture splenocytes and to measure the frequency of 6-thioguanine-resistant mutants in these cells has opened up the possibility of assaying mutant induction in the animal. Dose-dependent induction kinetics have been reported following whole body exposure to X-rays (DEMPSEY and MORLEY 1986) and exposure to ethylnitrosourea (JONES et al. 1987a). Analysis of seventeen, spontaneously occurring, 6-thioguanine-resistant clones showed them to be deficient in HPRT activity, and they all showed surface markers characteristic of the major class of T lymphocytes (JONES et al. 1987b). Twelve of these seventeen had alterations in the *hprt* gene; of these, two represented deletions, seven were partial deletions, and three represented unidentified alterations. More detailed and specific analysis of both spontaneous and induced mutants must await the application of the polymerase chain reaction (PCR) to these cells (SAIKI et al. 1985).

II. Human Cells

Cells of human origin are available in culture in a number of forms. Primary cultures are represented by fibroblasts and keratinocytes from skin or T lymphocytes from blood. Tumour-derived cells such as HeLa (GEY et al. 1952) or from testicular or bladder tumours (HOGAN et al. 1977; MASTERS et al. 1986) are available, but while still primary in origin, possible changes in response to DNA-damaging agents suggests that they should not be categorised with primary cultures from normal tissue.

A third category of cell includes virally transformed material. Here primary cells have been transformed by, in the case of fibroblasts, SV40 (HUSCHTSCHA and HOLLIDAY 1983) or its genetically sterilised derivates such as the ori$^-$ virus (MURNANE et al. 1985). In addition, there is now a panel of transformed fibroblasts which have been established following transfection with the plasmid pSV3 that contains the selectable markers *gpt* (the selection agent is mycophenolic acid) or *neo* (the selection agent is G418) (MAYNE et al. 1986). Cells from the B-cell lineage are available as Epstein-Barr virus-transformed lymphoblastoids (POVEY et al. 1973).

The system of greatest general use with human cells is based upon the *hprt* locus. Its relevance is made immediately apparent because of the existence of the Lesch-Nyhan syndrome (KELLY and WYNGAARDEN 1983), in which individuals are defective in the HPRT enzyme. Studies of these patients show that there is considerable heterogeneity when analysis is performed at the molecular level (YANG et al. 1984; WILSON et al. 1983, 1986; GIBBS and CASKEY 1987). Positive features of the system include the ease with which mutants may be selected and characterised as having a defective protein. The gene is localised on the X chromosome. However, there may be four pseudogenes in the human genome (PATEL et al. 1986). The gene is large, ~40 kb, the cDNA probe is 1.3 kb, and the limit of resolution is around 500 bp. Thus, there are some considerable difficulties in analysis with this system. GIBBS and CASKEY (1987) describe two systems, the ribonuclease A cleavage technique and PCR, which potentially allow the detection of most point mutations. PCR is of particular relevance where mutants have been isolated in a culture with more than one X chromosome, avoiding possible problems with activation of a lyonised chromosome (NADON et al. 1986).

1. Untransformed Fibroblasts

With untransformed human fibroblasts, a considerable body of data exists (LAMBERT et al. 1986). These studies are confined to establishing dose-response curves in cells from normal, xeroderma pigmentosum, hereditary cutaneous malignant melanoma or dysplastic naevus syndrome patients. The list of agents employed is limited but includes heat, azaserine, ionising radiation (of various qualities), 254-nm UV light, long-wavelength UV light plus psoralens, light from sun lamps, ethyl-*N*-nitrosourea, 4-nitroquinoline 1-oxide, benzo[*a*]pyrene and its diol-epoxide, *N*-methyl-*N'*-nitro-*N*-nitrosoguanidine, *N*-acetyloxy-2-acetylaminofluorene, ICR-191 and aflatoxin B. Most of the results have been generated using the *hprt* system, although resistance to diphtheria toxin has also been employed with success. The results show (i) that human fibroblasts are mutated at the *hprt* locus by known carcinogens and (ii) that cells from the cancer-prone xeroderma pigmentosum (XP) patients are hypermutable when treated with 254-nm UV and agents which may be anticipated to be "UV-like" with respect to the lesions in DNA which they induce. Indeed, the hypermutability of cells from excision or daughter-strand repair-defective XP patients provides the main argument for the concept that cancer in such individuals is a direct consequence of the defects in repair of damage inflicted by sunlight (KRAEMER 1980; KRAEMER et al. 1987).

This interpretation will probably prove to be too simple. Recent observations with cells from patients suffering from the autosomal recessive disease trichothiodystrophy (TTD) (STEFANINI et al. 1986) have shown that they are (i) hypersensitive to and hypermutable by 254-nm UV, (ii) have a substantial reduction in excision repair and (iii) fall into complementation group D of XP. Despite having the cellular characteristics of XP patients, the clinical features associated with this disease are quite distinct; they may or may not be photosensitive (STEFANINI et al. 1987), and there is no evidence of an elevated frequency of tumours in these patients. A possible explanation which accommodates this discrepancy follows from the suggestion that increased mutability is not in itself a sufficient condition and that there is a possible further defect in XP patients perhaps related to immune surveillance (BRIDGES 1981; MORISON et al. 1985). It is of some interest to note that cells from patients with the UV-sensitive condition Cockayne's syndrome (CS) are also hypermutable (LEHMANN 1987), but the patients show no evidence of increased tumour incidence.

The ionising radiation-hypersensitive syndrome, ataxia-telangiectasia (A-T) (BRIDGES and HARNDEN 1982), in which individuals are unambiguously cancer prone (WALDMANN et al. 1983; MORRELL et al. 1986), has also provided fibroblasts for mutation studies. In general, although not inevitably, A-T cells are found to be short-lived and clone less well than their normal counterparts (HOAR 1975; SHILOH et al. 1982b). Thus, mutation experiments on A-T fibroblasts are made difficult by these characteristics. It is particularly difficult to achieve statistically acceptable data even using large populations of cells. The observations of ARLETT and HARCOURT (1982) suggest that the cells are refractory to mutation induction by gamma radiation. The results of SIMONS (1982) indicate that while they may be of normal mutability, they are certainly not hypermutable. If we accept that the cells are hypomutable, it is then possible to suppose that they are defective in an error-prone repair process which, in normal cells, permits survival at the price of mutation.

Although A-T cells are hypersensitive to the lethal effects of ionising radiation, it does not seem reasonable to suppose that radiation is important with respect to the increased incidence of cancer since the individuals are not exposed to significant radiation doses. Rather, it may well be that the greater frequency of spontaneous, and increased, chromosome breakage (BRIDGES and HARNDEN 1982), which is also characteristic of the disease, allows expression of cancer genes.

A-T fibroblasts have been shown to be hypersensitive to the lethal action of other DNA-damaging agents such as bleomycin and neocarsinostatin (SHILOH et al. 1982a), which suggests that the increased incidence of cancer may represent a response to as yet unidentified environmental agents.

The study of mutability and repair in this syndrome is of continued interest because of the reports of increased cancer incidence in the heterozygotes (SWIFT et al. 1987; PIPPARD et al. 1988). A study of gamma ray-induced mutagenesis in primary fibroblasts from A-T heterozygotes showed them to be indistinguishable from normals (ARLETT and HARCOURT 1983).

Fibroblasts from a single hereditary dysplastic naevus syndrome patient have been claimed to be hypermutable following treatment with broad spectrum (290–400 nm) UV light from an artificial sun lamp (HOWELL et al. 1984). Cells from a second patient were not shown to be hypermutable. The hypersensitivity of the fibroblasts of such patients to the lethal action of 254-nm UV and the chemical carcinogen 4-nitroquinoline 1-oxide requires confirmation (SMITH et al. 1982, 1983). PERERA et al. (1986) found hypermutability to 254-nm UV in a set of three lymphoblastoid lines developed from hereditary dysplastic naevus patients. However, these same cultures were not hypersensitive to its lethal effects using both a growth assay and a microtitre well assay. The absence of any recognisable defect in DNA repair following exposure to UV light suggests that the hereditary DNS cells may have an alternative, error-prone pathway. The relationship of this hypothetical defect to the increased incidence of melanoma has been discussed by PERERA et al. (1986). In particular the detection of a mutated *ras* oncogene in these same melanoma cell lines may provide the genetic basis (ALBINO et al. 1984; PADUA et al. 1984) of the necessary instabilities.

2. Transformed Fibroblasts

These cells are particularly easy to work with, being convenient to maintain in culture and having, usually, a reproducibly high cloning efficiency. Since the cells are immortalised, they have the potential for being exploited in studies of the molecular nature of any mutational changes. A limited data base is available for mutation studies in SV40-tranformed cells (ARLETT and COLE 1986). In particular, the mutability of such cells may not be identical with that observed in the primary, untransformed parental culture. Thus, in the case of A-T cells, the transformed fibroblasts appear to be mutable by ionising radiation in contrast to the primary cultures. Modulation of the behaviour of cells following transformation has been reported for a number of endpoints.

Enhanced resistance to the lethal effects of ionising radiation has been noted in transformed cells from both normal individuals and A-T patients (GREEN et al. 1987). No changes in response to UV-C have been detected (MAYNE et al. 1986), but there is a high frequency of conversion ($\sim 30\%$) to the so-called mer minus phenotype (DAY et al. 1980), whereby transformed cells lose their methyl transferase activity and become hypersensitive to the lethal action of alkylating agents. Thus, the original suggestion that XP cells are defective in the repair of alkylation damage (TEO et al. 1983) is shown, ultimately, to be a consequence of the cultures under test being transformed by SV40 virus. The point to be made here is that transformed cells may well behave differently to untransformed cells, and until this aspect of their behaviour is quantified, we must accept mutation data from such cells with some caution.

3. Lymphoblastoid Cell Lines

Human lymphoblastoid cell lines developed following infection of B cells with Epstein-Barr virus or from individuals with mononucleosis (POVEY et al. 1973) have been used at two levels in the investigation of mutagenesis. The culture TK6 developed by THILLY et al. (1980) has been proposed as a human alternative to the mouse lymphoma $TK^{+}/-$ cells for use in mutagenicity assays. The data base is less substantial than for L5178Y cells, but the proposal has the attraction of relevance to the human condition. These cells, like the mouse lymphoma, have the potential advantage that multiple endpoints including TFT, ouabain and thioguanine resistance can be assessed in a single experiment, together with metaphase analysis or the detection of micronuclei (YANDELL et al. 1986; LIBER et al. 1986).

Lymphoblastoid cell lines established from the cancer-prone syndromes XP, A-T, and CS are available for study (HENDERSON and RIBECKY 1980). With A-T, the lymphoblastoid cells are mutable (TATSUMI and TAKEBE 1984), albeit less so than normal lymphoblastoids, in contrast to the observations on untransformed A-T fibroblasts (ARLETT and HARCOURT 1982).

As is the case with the SV40-transformed fibroblasts, lymphoblastoids frequently have a mer^{-} phenotype (ARITA et al. 1988).

While lymphoblastoid cell lines have the potential for the study of ouabain and thioguanine resistance, they may also have the potential for the generation of variants of the major histocompatibility complex antigens. In a study using gamma rays KAVATHAS et al. (1980) were able to exploit the linkage between HLA haplotypes and the glyoxalase alleles on chromosome 6 to demonstrate the induction of deletions, mitotic recombinations, inversions and translocations by this agent.

4. T Lymphocytes

T lymphocytes may be utilised in two ways. Firstly, mutants may be assessed in lymphocytes taken directly from the individual. Here a number of laboratories (ALBERTINI 1985; MESSING and BRADLEY 1985; MORLEY et al. 1985; HENDERSON et al. 1986) have been able to measure mutant frequency to 6-thioguanine resistance. The major features of this system are: (i) that there is an age-related response which may, in part, reflect the progressive inactivation of the X chromosome; (ii) smoking has a profound influence on mutant frequency giving a 30%–40% enhancement over age-matched controls; and (iii) blood from DNA repair defective, cancer-prone individuals with XP or A-T show elevated mutant frequencies. The system might be used for population monitoring since individuals in populations exposed to hazardous working environments or who adopt a hazardous life style may be compared with control populations. It should be pointed out, however, that the effect of smoking is so substantial that it is likely to obliterate anything but the most dramatic effects unless compensated for by a rigorous experimental design.

Since these mutants have the ability to grow in culture, it has been possible to examine the molecular nature of the defect which gives rise to the 6-thioguanine-resistant phenotype (ALBERTINI 1985; TURNER et al. 1985; NICKLAS et al. 1987).

Mutants have been characterised as being deletions or point mutations by Southern blotting (BRADLEY et al. 1987), and the PCR technique offers an increased potential for analysis with this material. By exploiting the existence of probes for the T-cell receptor markers, alpha, beta and gamma chains, together with other rearrangements, it is possible to determine the time at which a particular mutant was induced during haematopoesis and whether clonal expansion has occurred (ALBERTINI et al. 1987). It is possible to use T lymphocytes in a study of in vitro mutagenesis using the *hprt* system (MORLEY et al. 1985; SANDERSON et al. 1984; SKULIMOWSKI et al. 1986). JANATIPOUR et al. (1988) have shown that it is also possible, using the HLA-A locus, to detect mutants in both circulating lymphocytes and induced mutants following in vitro exposure to X-rays or mitomycin C. Clearly these methodologies are only relevant to individuals who are heterozygous at the HLA-A locus.

5. Erythrocytes

Two other systems are available for population monitoring in humans. Both utilise erythrocytes and therefore should be regarded as methods for detecting variants but not mutants, since these cells obviously cannot be propagated. The glycophorin A locus (MN alleles) has been utilised in the study of "mutant" frequency amongst atomic bomb survivors (LANGLOIS et al. 1987).

The system depends upon the co-dominant expression of alleles on chromosome 4 and the glycophorin loss assay utilises antibodies for the different allelic forms from heterozygous donors. It is not of universal applicability, therefore. On the other hand, it is possible, using monoclonal antibodies, to identify the rare circulating erythrocytes carrying variant haemoglobin A (VERWOERD et al. 1987) in all individuals. Since the frequency of such variant erythrocytes carrying amino-acid-substituted haemoglobin is very low ($>/10^7$) it will be necessary to apply modern techniques of computerised image analysis to scoring (STAMATOYANNOPOULOS et al. 1984).

D. Use of Recombinant Shuttle Vectors

In recent years attempts have been made to overcome the formidable technical problems encountered in any programme devoted to the analysis of mutation in endogenous genes in mammalian cells. It is clear that the PCR techniques are likely to have a dramatic effect on efforts in this field. However, systems have been developed in which mutations in "foreign" genes in a mammalian host cell can be transferred ("shuttled") into a prokaryotic host for analysis (DUBRIDGE and CALOS 1988).

I. Integrated Shuttle Vectors

Here the system depends upon the ability of the vector sequence to integrate into the genome of the target cells. The resulting transformants must be analysed for the number of integrated sequences (copy number), their arrangement and stability. We have found this analysis to be particularly time consuming since the integrated vectors are rarely maintained in a stable fashion (GEBARA et al. 1987). In a comparative study of reversion in the endogenous *hprt* gene and the integrated bacterial *gpt* gene in a set of pSV_2 gpt transformants of CHO $hprt^-$ cell line, TINDALL et al. (1986) showed that, while the *hprt* gene gave consistent reversion frequencies, the results obtained with the *gpt* gene were very variable.

Two schemes have been developed to overcome the problem of stability. In the first, GLAZER et al. (1986) added a second selectable marker (*neo*) to the vector, which was used to maintain stability but not as a target gene. In the second, a retroviral packaging system was used to introduce the vector into the mammalian cells (ASHMAN et al. 1986). This system also included a second gene (*neo*) for continuous selection of plasmid sequences. Sequence data for mutation have been generated using these improved vectors (ASHMAN et al. 1986; GLAZER et al. 1986).

In order to provide sequence data it is necessary to recover the mutated vectors from the mammalian cells into *E. coli*. The technique is based upon fusion with monkey COS cells which are permissive for SV40. Unfortunately, this process has itself proved to be mutagenic (ASHMAN and DAVIDSON 1984). A modified recovery technique (GLAZER et al. 1986) that involves packaging the vector from the mammalian genome into bacteriophage particles yielded a very low frequency of spontaneous mutant phage. Analysis of UV-induced mutants in mouse L cells revealed that they were at pyrimidine-cytosine sequences and largely C→T transitions.

II. Extrachromosomal Shuttle Vectors

Most such vectors are based upon SV40 and are designed so that the vector replicates extrachromosomally in the mammalian cell followed by isolation of low molecular weight DNA and direct transformation of *E. coli* with this DNA. SV40-based plasmids will replicate very rapidly over a period of 2–3 days to reach a copy number of 10^4–10^5 per cell. Such replication is lethal, and these systems may only be regarded as transient.

Elevated spontaneous mutation frequencies (ASHMAN and DAVIDSON 1985; CALOS et al. 1983; MACGREGOR et al. 1987; RAZZAQUE et al. 1983, 1984; SARKAR et al. 1984) appear to be characteristic of the systems as initially designed and vitiate their use for the study of mutation spectra. It is supposed that during passage through the cytoplasm to the nucleus of the target, all the plasmids are exposed to host cell nuclease activity. A new generation of SV40-based plasmids with larger amounts of redundant or non-essential sequences in addition to the target gene have been developed. Such plasmids have sufficiently low spontaneous frequencies to allow the investigation of induced mutations.

Sequence data on spontaneous and induced mutants in a variety of cell types are available. Amongst spontaneous mutants HAUSER et al. (1987) revealed a high proportion of mutants with multiple base changes from vectors recovered from monkey cells. With UV light (HAUSER et al. 1986) multiple base changes were also observed. The plasmid pZ189 has been used for the study of UV-induced mutation in monkey cells (SEIDMAN et al. 1985) and human cells from normal individuals and patients suffering from XP (BREDBERG et al. 1986;

SEETHARAM et al. 1987). When plasmids damaged by UV light were propagated in XP cells from complementation group A or D, more mutants were observed than in similar experiments with normal cells. Fewer multiple mutation sites and a decrease in the number of sites were also noted, but two unique mutational hotspots (one for group D and one for group A) were identified. The proportion of G:C to A:T transitions was increased in XP cell lines from both complementation groups compared with normal cells. Interestingly, the frequency of multiple base substitutions is significantly reduced in the XP cells, suggesting that the multiple base substitutions are characteristic of the normal excision repair pathway (SEETHARAM et al. 1987). When the incision step of excision repair, which is believed to be the defect in XP A and D cells, is minimised by the introduction of a single-stranded nick in the plasmid, a large increase in the frequency of multiple base mutations could be seen (SEIDMAN et al. 1985).

Plasmids based upon those viruses which occur naturally in an extrachromosomal state have also been utilised as shuttle vectors. Limited studies with plasmids based on bovine papillomavirus have been reported (ASHMAN and DAVIDSON 1985; MACGREGOR and BURKE 1987). The greatest difficulty here is the limited host range. ASHMAN and DAVIDSON (1985) encountered a high spontaneous mutation frequency, and MACGREGOR and BURKE (1987) recorded a high frequency of rearranged plasmid sequences. Shuttle vectors based upon Epstein-Barr virus (EBV) (YATES et al. 1984) hold out much greater promise since they replicate extrachromosomally in a variety of cell types from different species (YATES et al. 1984). In addition these plasmids can be maintained in a stable copy number, but in contrast to the SV40-based plasmids, the EBV-based plasmids are present in a copy number of 1–40.

Using the EBV-based plasmid pHet with the herpes simplex virus thymidine kinase gene as a target, DRINKWATER and KLINEDINST (1986) were able to demonstrate a low spontaneous frequency and a dose-dependent induction of mutants with ethylnitrosourea. The kinetics of plasmic mutation induction were paralleled by induction of mutation in the host genome *hprt* gene. DUBRIDGE et al. (1987), using different EBV-based plasmids with the bacterial *lac*I gene as target, again showed a low spontaneous mutation frequency (less than 10^{-5}). Following exposure to *N*-nitroso-*N*-methylurea, more than 95% of both independent nonsense and missense mutations were shown to be G:C to A:T transitions.

E. Conclusions

The importance of the study of mutation in our understanding of carcinogenesis is seen when activated oncogenes arise from specific base changes either in DNA isolated from tumours or as a consequence of the exposure of proto-oncogenes to known carcinogens. The detection of mutagens and carcinogens using cultured mammalian cells has given rise to a number of assays based on a limited number of forward mutation systems and normally limited to a few cell types of nonhuman origin. Nevertheless, a substantial data base using the mouse L5178Y $TK^{+}/-$ system is available. While this assay system can be used, in combination with a general genotoxicity profile, to reveal the potential of a particular compound to mutate mammalian cells, it can tell us little about mechanisms. We need to know a great deal more about the changes which are actually involved in human carcinogenesis. Attempts to define mutation spectra using the endogenous *aprt* gene are labour intensive and are severely limited in their application.

The exploitation of molecular biological analysis of mutation in either integrated or extrachromosomally maintained shuttle vectors, with which genetic analysis is performed in bacteria, has given rise to a number of promising avenues of research. It is especially gratifying to note that, in those limited number of cases examined, the mutant spectra are not grossly dissimilar to those described for endogenous mammalian genes. It would seem to me, however, that the polymerase chain reaction procedure is a tool of unique power and application and is likely to revitalise our interest in mutational changes in endogenous genes.

Acknowledgements. I am indebted to Drs. B. A. Bridges and J. Cole for many helpful comments.

References

Albertini RJ (1985) Somatic gene mutations in vivo as indicated by the 6-thioguanine-resistant T-lymphocytes in human blood. Mutation Res 150:411–422

Albertini RJ, Nicklas JA, Sullivan LM, Hunter TC, O'Neill JP (1987) *hgprt* Mutation in vivo in human T lymphocytes: quantitative considerations. In: Moore MM, De Marini DM, De Serres FJ, Imdall KR (eds) Mammalian cell, mutagenesis. Cold Spring Harbor, pp 139–148 (Banbury Report 28)

Albino AP, Le Strange RL, Oliff AI, Furth ME, Old LJ (1984) Transforming *ras* genes from human melanoma: a manifestation of tumour heterogeneity? Nature 308:69–72

Arita I, Tatsumi I, Tachibana A, Toyoda M, Takebe H (1988) Instability of Mex$^-$ phenotype in human lymphoblastoid cell lines. Mutation Res 208:167–172

Arlett CF (1977) Mutagenicity testing with V79 Chinese hamster cells. In: Kilby BJ, Legator M, Nichols W, Ramel C (eds) Handbook of mutagenicity test procedures. Elsevier, Amsterdam, pp 175–192

Arlett CF, Cole J (1986) Mutation studies in cells established from human cancer prone syndromes. In: Ramel C, Lambert B, Magnusson J (eds) Genetic toxicology of environmental chemicals. Part A: principles and mechanisms of action. Liss, New York, pp 237–244

Arlett CF, Cole J (1988) The role of mammalian cell mutation assays in mutagenicity and carcinogenicity testing. Mutagenesis 6:455–458

Arlett CF, Harcourt SA (1982) Variation in response to mutagens amongst normal and repair-defective human cells. In: Lawrence CW (ed) Induced mutagenesis molecular mechanisms and their implications for environmental protection. Plenum, New York, pp 249–266

Arlett CF, Harcourt SA (1983) The mutagen sensitivity response of cells from individuals heterozygous for DNA repair deficiency genes. In: Castellani A (ed) The use of human cells for the evaluation of risk from physical and chemical agents. Plenum, New York, pp 155–167

Arlett CF, Turnbull D, Harcourt SA, Lehmann AR, Colella CM (1975) A comparison of the 8-azaguanine and ouabain resistance systems for the selection of induced mutant Chinese hamster cells. Mutation Res 33:261–278

Ashby J (1986 a) Letter to the editor. Mutagenesis 1:309–317

Ashby J (1986 b) The prospects for a simplified and internationally harmonized approach to the detection of possible human carcinogens and mutagens. Mutagenesis 1:3–16

Ashman CR, Davidson RL (1984) High spontaneous mutation frequency in shuttle vector sequences recovered from mammalian cellular DNA. Mol Cell Biol 4:2266–2272
Ashman CR, Davidson RL (1985) High spontaneous mutation frequency of BPV shuttle vector. Somatic Cell Mol Genet 11:499–504
Ashman CR, Jagadeeswaran P, Davidson RL (1986) Efficient recovery and sequencing of mutant genes from mammalian chromosomal DNA. Proc Natl Acad Sci USA 83:3356–3368
Baker RM, Brunette DM, Mankovitz R, Thompson LH, Whitmore GF, Siminovitch L, Till TE (1974) Ouabain-resistant mutants of mouse and hamster cells in culture. Cell 1:9–21
Begg AH, Axelman J, Migeon BR (1986) Reactivation of X-linked genes in human fibroblasts transformed by origin-defective SV40. Somatic Cell Mol Genet 12:585–594
Borek C, Ong A, Mason H (1987) Distinctive transforming genes in X-ray-transformed mammalian cells. Proc Natl Acad Sci USA 84:794–798
Bos JL, Toksoz D, Marshall CJ, Verlaan-de Vries M, Veeneman GH, Van Der Eb AJ, Van Boom JH, Janssen WG, Steenvoorden CM (1985) Amino-acid substitutions at codon 13 of the N-*ras* oncogene in human acute myeloid leukaemia. Nature 315:726–730
Bradley WEC, Gareau JLP, Seifert AM, Messing K (1987) Molecular characterisation of 15 rearrangements among 90 in vivo somatic mutants shows that deletions predominate. Mol Cell Biol 7:956–960
Bradshaw HDJ, Deininger P (1984) Human thymidine kinase gene: molecular cloning and nucleotide sequence of a cDNA expressible in mammalian cells. Mol Cell Biol 4:2316–2320
Bredberg A, Kraemer KH, Seidman MM (1986) Restricted ultraviolet mutational spectrum in a shuttle vector propagated in xeroderma pigmentosum cells. Proc Natl Acad Sci USA 83:8273–8277
Breimer LH (1988) Ionising radiation-induced mutagenesis. Br J Cancer 57:6–18
Breimer LH, Nalbantoglu J, Meuth M (1986) Structure and sequence of mutations induced by ionizing radiation at selectable loci in Chinese hamster ovary cells. J Mol Biol 192:669–674
Brennand J, Chinault AC, Konecki DS, Melton DW, Caskey CT (1982) Cloned cDNA sequences of the hypoxanthine/guanine phosphoribosyltransferase gene from a mouse neuroblastoma cell line found to have amplified genomic sequences. Proc Natl Acad Sci USA 79:1950–1954
Bridges BA (1975) Genetic effects of UV on *Escherichia coli* – a model for prokaryotes. In: Nygaard OF, Adler HI, Sinclair WK (eds) Radiation research, biomedical, chemical, and physical perspectives. Academic, New York, pp 626–631
Bridges BA (1976) Mutation induction. In: Macdonald KD (ed) Second intern symp genetics of indust microorgs. Academic, London
Bridges BA (1981) How important are somatic mutations and immune control in skin cancer? Reflections on xeroderma pigmentosum. Carcinogenesis 2:471–472
Bridges BA, Harnden DG (1982) Ataxia-telangiectasia – a cellular and molecular link between cancer, neuropathology and immune deficiency. Wiley, London
Brown R, Thacker J (1984) The nature of mutants induced by ionising radiation in cultured cells. I. Isolation and initial characterisation of spontaneous, ionising radiation induced, and EMS induced mutants resistant to 6-TG. Mutation Res 129:269–281
Brusick D (1986) Genotoxic effects in cultured mammalian cells produced by low pH treatment conditions and increased ion concentrations. Environ Mutagen 8:879–886
Calos MP, Lebkowski JS, Botchan MR (1983) High mutation frequency in DNA transfected into mammalian cells. Proc Natl Acad Sci USA 80:3015–3019
Cavenee W, Dryja TP, Phillips RA, Benedict WF, Godbout R, Gallie BL, Murphree AL, Strong LC, White RL (1983) Expression of recessive alleles by chromosomal mechanisms in retinoblastoma. Nature 305:779–784
Clive D, Flamm WG, Machesko MR, Berhnheim NJ (1972) A mutational assay system using the thymidine kinase locus in mouse lymphoma cells. Mutation Res 16:77–87
Clive D, Johnson JFS, Spector AG, Batson AG, Brown MM (1979) Validation and characterization of the L5178Y/TK Mouse lymphoma mutagen assay system. Mutat Res 59:61–108

Clive D, Turner N, Krehl R (1988) Procarbazine is a potent mutagen at the heterozygous thymidine kinase ($tk^+/_-$) locus of mouse lymphoma assay. Mutagenesis 3:83–87
Cole J, Arlett CF (1976) Ethyl methanesulphonate mutagenesis with L5178Y mouse lymphoma cells: a comparison of ouabain, thioguanine and excess thymidine resistance. Mutat Res 34:507–526
Cole J, Arlett CF, Green MHL (1976) The fluctuation test as a more sensitive system for determining induced mutation in L5178Y mouse lymphoma cells. Mutat Res 41:377–386
Cole J, Arlett CF, Green MHL, Lowe J, Muriel W (1983) A comparison of the agar cloning and microtitration techniques for assaying cell survival and mutation frequency in L5178Y mouse lymphoma cells. Mutat Res 111:371–386
Cole J, Muriel WJ, Bridges BA (1986) The mutagenicity of sodium fluoride to L5178Y [wild-type and TK+/− (3.7.2C)] mouse lymphoma cells. Mutagenesis 1:157–167
Cox R, Masson WK (1978) Do radiation-induced thioguanine-resistant mutants of cultured mammalian cells arise by HGPRT gene mutation or X-chromosome rearrangement? Nature 276:629–630
Croce CM (1986) Chromosome translocation and human cancer. Cancer Res 46:6019–6023
Davies DR, Evans HJ (1966) The role of genetic damage in radiation induced cell lethality. Adv Radiat Biol 2:243–353
Day RS, Ziolkowsky CHJ, Scudiero DA, Meyer SH, Lubinieck AS, Giradi AJ, Galloway SM, Bynum GD (1980) Defective repair of alkylated DNA by human tumour and SV40-transformed human cell strains. Nature 228:724–727
Dempsey JL, Morley AA (1986) Measurement of in vivo mutant frequency in lymphocytes in the mouse. Environ Mutagen 8:385–391
Drinkwater N, Klinedinst DK (1986) Chemically induced mutagenesis in a shuttle vector with a low background mutant frequency. Proc Natl Acad Sci USA 83:3402–3406
Drobetsky EA, Grosovsky AJ, Glickman BW (1987) The specificity of UV-induced mutations at an endogenous locus in mammalian cells. Proc Natl Acad Sci USA 84:9103–9107
DuBridge R, Tang P, Hsia HC, Leong P-M, Miller JH, Calos MP (1987) Analysis of mutation in human cells by using an Epstein-Barr virus shuttle system. Mol Cell Biol 7:379–387
DuBridge RB, Calos MP (1988) Recombinant shuttle vectors for the study of mutation in mammalian cells. Mutagenesis 3:1–10
Ennever FK, Noonan TJ, Rosenkranz HS (1987) The predictivity of animal bioassays and short-term genotoxicity tests for carcinogenicity and non-carcinogenicity to humans. Mutagenesis 2:73–78
Evans HJ (1988) Mutation cytogenetics: past, present and future. Mutat Res 204:355–363
Fainstein E, Marcelle C, Rosner A, Canaani E, Gale RP, Dreazen O, Smith SD, Croce CM (1987) A new fused transcript in Philadelphia chromosome positive acute lymphocytic leukaemia. Nature 330:386–391
Fox M, Radacic M (1978) Adaptational origin of some purine-analogue resistant phenotypes in cultured mammalian cells. Mutat Res 49:275–296
Fuscoe JC, Fenwick JRG, Ledbetter DH, Caskey CT (1983) Deletion and amplification of the HGPRT locus in Chinese hamster cells. Mol Cell Biol 3:1086–1096
Fuscoe JC, Ockey CH, Fox M (1986) Molecular analysis of X-ray-induced mutants at the HPRT locus in V79 Chinese hamster cells. Int J Radiat Biol 49:1011–1020
Garner RC, Kirkland DJ (1986) Reply. Mutagenesis 1:233–235
Gatehouse DG, Tweats DJ (1986) Letter to the editor. Mutagenesis 1:307–308
Gebara MM, Drevon C, Harcourt SA, Steingrimsdottir H, James MR, Burke JF, Arlett CF, Lehmann AR (1987) Inactivation of a transfected gene in human fibroblasts can occur by deletion, amplification, phenotypic switching, or methylation. Mol Cell Biol 7:1459–1464
Gey GO, Coffman WD, Kubicek WT (1952) Tissue culture studies of the proliferative capacity of cervical carcinoma and normal epithelium. Cancer Res 12:264–265
Gibbs RA, Caskey CT (1987) Molecular analysis of human *hprt* mutations. In: Moore MM, De Marini DM, De Serres FJ, Tindall KR (eds) Mammalian cell mutagenesis. Banbury Report 28, Cold Spring Harbor, pp 237–247

Glazer PM, Sarkar SN, Summers WC (1986) Detection and analysis of UV-induced mutations in mammalian cell DNA using a lambda phage shuttle vector. Proc Nat Acad Sci USA 83:1041–1044

Glickman BW, Drobetsky EA, Grososvsky AJ (1987) A study of the specificity of spontaneous and UV-induced mutation at the endogenous *aprt* gene of Chinese hamster overy cells. In: Moore MM, De Marini DM, De Serres FJ, Tindall KR (eds) Mammalian cell mutagenesis. Banbury Report 28, Cold Spring Harbor, pp 167–182

Green MHL, Lowe JE, Arlett CF, Harcourt SA, Burke JF, James MR, Lehmann AR, Povey SM (1987) A gamma-ray resistant derivative of an ataxia-telangiectasia cell line obtained following DNA-mediated gene transfer. J Cell Sci [Suppl] 6:127–137

Grosovksy AJ, Drobetsky EA, De Jong PJ, Glickman BW (1986) Southern analysis of genomic alterations in gamma ray induced $APRT^-$ hamster cell mutants. Genetics 113:405–415

Harris M (1973) Anomalous patterns of mutation in cultured mammalian cells. Genetics 73:181–185

Hauser J, Seidman MM, Sidur K, Dixon K (1986) Sequence specificity of point mutations induced during passage of a UV-irradiated shuttle vector plasmid in monkey cells. Mol Cell Biol 6:277–285

Hauser J, Levine AS, Dixon K (1987) Unique pattern of point mutations arising after gene transfer into mammalian cells. EMBO J 6:63–67

Hayflick L, Moorhead PS (1961) The serial cultivation of human diploid cell strains. Exp Cell Res 25:585–621

Henderson EE, Ribecky R (1980) DNA repair in lymphoblastoid cell lines established from human genetic disorders. Chem Biol Interact 33:63–82

Henderson L, Cole H, Cole J, James SE, Green MHL (1986) Detection of somatic amutations in man: evaluation of the microtiter cloning assay for T-lymphocytes. Mutagenesis 1:195–200

Hoar DI (1975) Phenotypic manifestations of ataxia-telangiectasia. Lancet ii:1048

Hogan B, Fellous M, Avner P, Jacob F (1977) Isolation of a human teratoma cell line which expresses F9 antigen. Nature 270:515–518

Howell JN, Greene MH, Corner RC, Maher VM, McCormick JJ (1984) Fibroblasts from patients with hereditary cutaneous malignant melanoma are abnormally sensitive to the mutagenic effect of simulated sunlight and 4-nitroquinoline-1-oxide. Proc Natl Acad Sci USA 81:1179–1183

Hozier J, Sawyer J, Clive D, Morre M (1985) Chromosome 11 aberrations in small colony L5178Y −/− mutants early in their clonal history. Ann NY Acad Sci 107:423–425

Huschtscha LI, Holliday R (1983) Limited and unlimited growth of SV40-transformed cells from human diploid MRC-5 fibroblasts. J Cell Sci 63:77–99

Ishidate M Jr, Harnois MC (1987) The clastogenicity of chemicals in cultured mammalian cells. Letter to the editor. Mutagenesis 2:240–243

Ishidate MJ (1983) The data book of chromosomal aberration tests in vitro. Realize, Tokyo

Janatipour M, Trainor KJ, Kutlaca R, Bennett G, Hay J, Turner DR, Morley AA (1988) Mutations in human lymphocytes studied by an HLA selection system. Mutat Res 198:221–226

Jolly DJ, Okayama H, Berg P, Esty AC, Filpula D, Bohlen P, Johnson GG, Shively JE, Hunkapillar T, Friedmann T (1983) Isolation and characterization of a full-length expressible cDNA for human hypoxanthine phosphoribosyltransferase. Proc Natl Acad Sci USA 80:477–481

Jones IM, Burkhart-Schultz K, Crippen TL (1987a) Cloned mouse lymphocytes permit analysis of somatic mutations that occur in vivo. Somatic Cell Mol Genet 13:325–333

Jones IM, Burkhart-Schultz K, Strout CL, Crippen TL (1987b) Factors that affect the frequency of thioguanine-resistant lymphocytes in mice following exposure to ethylnitrosourea. Environ Mutagen 9:317–329

Kavathas P, Bach FH, DeMars R (1980) Gamma ray-induced loss of expression of HLA and glyoxalase 1 alleles in lymphoblastoid cells. Proc Natl Acad Sci USA 77:4251–4255

Kelly WN, Wyngaarden JB (1983) Clinical syndromes associated with hypoxanthine-guanine phosphoribosyltransferase deficiency. In: Stanbury JB, Wyngaarden JB, Fredrickson DS, Goldstein JL, Brown MS (eds) The metabolic basis of inherited disease, 5th edn. McGraw-Hill, New York, pp 1115–1143

King HWS, Brookes P (1984) On the nature of mutations induced by the diolepoxide of benzo(a)pyrene in mammalian cells. Carcinogenesis 5:965–970

Konecki DS, Brennand J, Fuscoe JC, Caskey CT, Chinault AC (1982) Hypoxanthine-guanine phosphoribosyltransferase genes of mouse and Chinese hamster: construction and sequence analysis of cDNA recombinants. Nucleic Acids Res 10:6753–6775

Koufos A, Hansen MF, Copeland N, Jenkins NA, Lampkin BC, Cavenee WK (1985) Loss of heterozygosity in three embryonal tumours suggests a common pathogenetic mechanism. Nature 316:330–334

Kraemer KH (1980) Xeroderma pigmentosum. In: Demis DJ, Dobson RL, McGuire J (eds) Clinical dermatology, vol 4. Harper and Row, Hagerstown, pp 1–33

Kraemer KH, Lee MM, Scotto J (1987) Xeroderma pigmentosum. Cutaneous, ocular and neurologic abnormalities in 830 published cases. Arch Dermatol 123:241–250

Lambert B, Chu EHY, De Carli L, Ehling UH, Evans HJ, Hayashi M, Thilly WG, Vainio H (1986) Assays for genetic changes in mammalian cells. IARC Sci Publ 83:167–243

Langlois RG, Bigbee WL, Kyoizumi S, Nakamura N, Bean MA, Akiyama M, Jensen RH (1987) Evidence for increased somatic cell mutations at the glycophorin A locus in atomic bomb survivors. Science 236:445–448

Lehmann AR (1987) Cockayne's syndrome and trichothiodystrophy: defective repair without cancer. Cancer Rev 7:82–103

Liber HL, Leong PK, Terry VH, Little JB (1986) X-rays mutate human lymphoblast cells at genetic loci that should respond only to point mutagens. Mutat Res 163:91–97

Lowy I, Pellicer A, Jackson JF, Sim G-K, Silverstein S, Axel R (1980) Isolation of transforming DNA: cloning the hamster *aprt* gene. Cell 22:817–823

MacGregor GR, Burke JF (1987) Stability of a bacterial gene in a bovine papillomavirus-based shuttle vector maintained extrachromosomally in mammalian cells. J Gen Virol 68:247–252

MacGregor GR, James MR, Arlett CF, Burke JF (1987) Analysis of mutations occurring during replication of a SV40 shuttle vector in mammalian cells. Mutat Res 183: 273–278

Marshall CJ, Vousden KM, Phillips DM (1984) Activation of c-Ha-*ras*-1 proto-oncogene by in vitro modification with a chemical carcinogen, benzo(a)pyrene diol-epoxide. Nature 310:586–589

Masters JRW, Hepburn PJ, Walker L, Highman W, Trejdosiewicz LK, Povey S, Parkar M, Hill BT, Riddle PR, Franks LM (1986) Tissue culture model of transitional cell carcinoma: characterisation of twenty-two human urothelial cell lines. Cancer 46: 3630–3636

Mathew CGP, Smith BA, Thorpe K, Wong K, Royle NJ, Jeffreys AJ, Ponder BAJ (1987) Deletion of genes on chromosome 1 in endocrine neoplasia. Nature 328:524–526

Mayne LV, Priestley A, James MR, Burke JF (1986) Efficient immortalization and morphological transformation of human fibroblasts by transfection with SV40 DNA linked to a dominant marker. Exp Cell Res 162:530–538

McGregor D, Riach C, Cattanach P, Caspary W (1987) Development and investigation of an expanded CHO/*hgprt* locus assay. Mutagenesis 2:303

Messing K, Bradley WEC (1985) In vivo mutant frequency rises among breast cancer patients after exposure to high doses of gamma-irradiation. Mutat Res 152:107–112

Meuth M, Arrand JE (1982) Alterations of gene structures in EMS-induced mutants of mammalian cells. Mol Cell Biol 2:1459–1462

Meuth M, Nalbantoglu J, Phear G, Miles C (1987) Molecular basis of genome rearrangements at the hamster *aprt* locus. In: Moore MM, De Marini DM, De Serres FJ, Tindall KR (eds) Mammalian cell mutagenesis. Banbury Report 28, Cold Spring Harbor, pp 183–191

Mitchell PJ, Urlaub G, Chasin L (1986) Spontaneous splicing mutations at the dihydrofolate reductase locus in Chinese hamster ovary cells. Mol Cell Biol 6:1926–1935

Montesano R, Bartsch H, Vainio H, Wilbourn J, Yamasaki H (1986) Long-term and short-term assays for carcinogens: a critical appraisal. IARC Sci Publ 83. IARC, Lyon

Moore MM, Clive D, Howard BE, Batson AG, Turner NT (1985a) In situ analysis of trifluorothymidine-resistant (TFT^r) mutants of L5178Y/$TK^{+/-}$ mouse lymphoma cells. Mutat Res 151:147–159

Moore MM, Clive D, Hozier JC, Howard BE, Batson AG, Turner NT, Sawyer J (1985b) Analysis of trifluorothymidine-resistant (TFT^r) mutants of L5178Y/$TK^+/-$ mouse lymphoma cells. Mutat Res 151:161–174

Moore MM, De Marini DM, De Serres FJ, Tindall KR (1987) Mammalian cell mutagenesis. Banbury Report 28:1–385

Morison WL, Bucana C, Hashem N, Kripke ML, Cleaver JE, German JL (1985) Impaired immune function in patients with xeroderma pigmentosum. Cancer Res 45:3929–3931

Morley AA, Trainor KJ, Dempsey JL, Seshadri RS (1985) Methods for study of mutations and mutagenesis in human lymphocytes. Mutat Res 147:363–367

Morrell D, Cromartie E, Swift M (1986) Mortality and cancer incidence in 263 patients with ataxia-telangiectasia. JNCI 77:89–92

Murnane JP, Fuller LF, Painter RB (1985) Establishment and characterization of a permanent pSV ori^--transformed ataxia-telangiectasia cell line. Exp Cell Res 158:119–126

Nadon N, Sekhon G, Brown LJ, Korn N, Petersen JW, Strandtmann J, Chang C, DeMars R (1986) Derepression of *HPRT* locus on inactive X chromosome of human lymphoblastoid cell line. Somatic Cell Mol Genet 12:541–554

Nalbantoglu J, Goncalves O, Meuth M (1983) Structure of mutant alleles at the *aprt* locus of CHO cells. J Mol Biol 167:575–594

Nalbantoglu J, Phear G, Meuth M (1987) DNA sequence analysis of spontaneous mutations at the *aprt* locus of hamster cells. Mol Cell Biol 7:1445–1449

Nicklas JA, O'Neill JP, Albertini RJ (1986) Use of T-cell receptor gene probes to quantify the in vivo hprt mutations in human T-lymphocytes. Mutat Res 173:67–72

Padua RA, Barrass N, Currie GA (1984) A novel transforming gene in a human malignant melanoma cell line. Nature 311:671–673

Paeratakul U, Taylor MW (1986) Isolation and characterisation of mutants at the APRT locus in the L-5178Y TK^+/TK^- mouse lymphoma cell line. Mutat Res 160:61–69

Patel PI, Framson PE, Caskey CT, Chinault AC (1986) Fine structure mapping of the human hypoxanthine phosphoribosyltransferase gene. Mol Cell Biol 6:393–403

Perera MIR, Um KI, Grenne MH, Waters HL, Bredberg A, Kraemer KH (1986) Hereditary dysplastic nevus syndrome: lymphoid cell ultraviolet hypermutability in association with increased melanoma susceptibility. Cancer Res 4:1005–1009

Pippard EC, Hall AJ, Barker DJP, Bridges BA (1988) Cancer in homozygotes and heterozygotes of ataxia-telangiectasia and xeroderma pigmentosum in Britain. Cancer Res 48:2929–2933

Povey S, Gardiner SE, Watson B, Mowbray S, Harris H, Arthur E, Steel CM, Blenkinsop C, Evans HJ (1973) Genetic studies on human lymphoblastoid lines: isozyme analysis on cell lines from forty-one different individuals and on mutants produced following exposure to a chemical mutagen. Ann Hum Genet 36:247–266

Puck TT, Waldren CA (1987) Mutation in mammalian cells: theory and implications. Somatic Cell Mol Genet 13:405–409

Razzaque A, Mizusawa H, Seidman M (1983) Rearrangement and mutagenesis of a shuttle vector plasmic after passage in mammalian cells. Proc Natl Acad Sci USA 80:3010–3014

Razzaque A, Chakrabarti S, Joffee S, Seidman M (1984) Mutagenesis of a shuttle vector plasmid in mammalian cells. Mol Cell Biol 4:435–441

Reddy EP, Reynolds RK, Santos E, Barbacid M (1982) A point mutation is responsible for the acquisition of transforming properties by the T24 human bladder carcinoma oncogene. Nature 300:149–152

Reynolds SH, Stowers SJ, Patterson RM, Maronpot RR, Aaronson SA, Anderson MW (1987) Activated oncogenes in B6C3F1 mouse liver tumours: implications for risk assessment. Science 237:1309–1316

Rogers AM, Hill R, Lehmann AR, Arlett CF, Burns VW (1980) The induction and characterization of mouse lymphoma L5178Y cell lines resistant to 1-β-D-arabinofuranosyl-cytosine. Mutat Res 69:139–148

Rosenkranz HS (1988) Strategies for the deployment of batteries of short-term tests. Mutat Res 205:1–426

Rouleau GA, Wertelecki W, Haines JL, Hobbs WJ, Trofatter JA, Seizinger BR, Martuza RL, Superneau DW, Conneally PM, Gusella JF (1987) Genetic linkage of bilateral acoustic neurofibromatosis to a DNA marker on chromosome 22. Nature 329:246–248

Saiki RK, Scharf S, Faloona F, Mullis KB, Horn GT, Erlich HA, Arnheim N (1985) Enzymatic amplification of B-globin genomic sequences and restriction site analysis for diagnosis of sickle cell anemia. Science 230:1350–1354

Sanderson BJS, Dempsey JL, Morley AA (1984) Mutations in human lymphocytes: effect of X- and UV-irradiation. Mutat Res 140:223–227

Sarkar S, DasGupta UB, Summers WC (1984) Error-prone mutagenesis detected in mammalian cells by a shuttle vector containing the *supF* gene of *Escherichia coli.* Mol Cell Biol 4:2227–2230

Seetharam S, Protic-Sabljic M, Seidman MM, Kraemer KH (1987) Abnormal ultraviolet mutagenic spectrum in plasmid DNA replicated in cultured fibroblasts in a patient with the skin cancer-prone disease, xeroderma pigmentosum. J Clin Invest 80:1613–1617

Seidman MM, Dixon K, Razzaque A, Zagursky RJ, Berman ML (1985) A shuttle vector plasmic for studying carcinogen-induced point mutations in mammalian cells. Gene 38:233–237

Shiloh Y, Tabor E, Becker Y (1982a) The response of ataxia-telangiectasia homozygous and heterozygous skin fibroblasts to neocarzinostatin. Carcinogenesis 3:815–820

Shiloh Y, Tabor E, Becker Y (1982b) Colony-forming ability of ataxia-telangiectasia skin fibroblasts is an indicator of their early senescence and increased demand for growth factors. Exp Cell Res 140:191–199

Simons JWIM (1982) Studies on survival and mutation in ataxia-telangiectasia cells after X-irradiation under oxic and anoxic conditions. In: Bridges BA, Harnden DG (eds) Ataxia-telangiectasia – cellular and molecular link between cancer, neuropathology and immune deficiency. Wiley, Chichester, pp 165–167

Simpson NE, Kidd KK, Goodfellow PJ, McDermic H, Myers S, Kidd JR, Jackson CE, Duncan AMV, Farrer LA, Brasch K, Castiglione C, Genel M, Gertner J, Greenberg CR, Gusella JF, Holden JJA, White BN (1987) Assignment of multiple endocrine neoplasia type 2A to chromosome 10 by linkage. Nature 328:528–530

Skulimowski AW, Turner DR, Morley AA, Sanderson BJS, Haliandros M (1986) Molecular basis of X-ray-induced mutation at the HPRT locus in human lymphocytes. Mutat Res 162:105–112

Smith PJ, Greene MH, Devlin DA, McKeen EA, Paterson M (1982) Abnormal sensitivity to UV-radiation in cultured skin fibroblasts from patients with hereditary cutaneous malignant melanoma and dysplastic nevus syndrome. Int J Cancer 30:39–45

Smith PJ, Greene MH, Adams D, Paterson M (1983) Abnormal responses to the carcinogen 4-nitroquinoline 1-oxide of cultured fibroblasts from patients with dysplastic nevus syndrome and hereditary cutaneous malignant melanoma. Carcinogenesis 4:911–916

Stamatoyannopoulos GP, Nute E, Lindsley D, Farquhar M, Brice M, Nakamoto N, Papayannopoulu T (1984) Somatic-cell mutation monitoring system based on human hemoglobin mutants. In: Ansari AA, de Serres FJ (eds) Single-cell mutation monitoring systems, methodologies and applications. Plenum, New York, pp 1–29 (Topics in chemical mutagenesis 2)

Stefanini M, Lagomarsini P, Arlett CF, Marinoni S, Borrone C, Crovato F, Trevisan G, Cordone G, Nuzzo F (1986) Xeroderma pigmentosum (complementation group D) mutation is present in patients affected by trichothiodystrophy with photosensitivity. Hum Genet 74:107–112

Stefanini M, Lagomarsini P, Giorgi R, Nuzzo F (1987) Complementation studies in cells from patients affected by trichothiodystrophy with normal or enhanced photosensitivity. Mutat Res 191:117–119

Strong LC, Riccardi VM, Ferrell RE, Sparkes RS (1981) Familial retinoblastoma and chromosome 13 deletion transmitted via an insertional translocation. Science 213:1501–1503
Sugden B, Marsh K, Yates J (1985) A vector that replicates as a plasmid and can be efficiently selected in B-lymphoblasts transformed by Epstein-Barr virus. Mol Cell Biol 5:410–413
Sukumar S, Notario V, Martin-Zanca D, Barbacid M (1983) Induction of mammary carcinomas in rats by nitroso-methylurea involves malignant activation of H-*ras*-1 locus by single point mutations. Nature 306:658–661
Swift M, Reitnauer PJ, Morrel D, Chase CL (1987) Breast and other cancers in families with ataxia-telangiectasia. N Engl J Med 316:1289–1294
Tatsumi K, Takebe H (1984) Gamma-irradiation induces mutation in ataxia-telangiectasia lymphoblastoid cells. Gann 75:1040–1043
Tennant RW, Margolin BH, Shelby MD, Zeiger E, Haseman JK, Spalding J, Caspary W, Resnick M, Stasiewicz S, Anderson B, Minor R (1987) Prediction of chemical carcinogenicity in rodents from in vitro genetic toxicity assays. Science 236:933–941
Teo IA, Lehmann AR, Muller R, Rajwesky MF (1983) Similar rate of O^6-ethylguanine elimination from DNA in normal human fibroblast and xeroderma pigmentosum cell strains not transformed by SV40. Carcinogenesis 4:1075–1077
Thacker J (1986) The nature of mutants induced by ionising radiation in cultured hamster cells. III. Molecular characterisation of HPRT-deficient mutants induced by gamma rays or alpha particles showing that the majority have deletions of all or part of the *hprt* gene. Mutat Res 160:267–275
Thacker J, Stephens MA, Stretch A (1978) Mutation to ouabain resistance in Chinese hamster cells: induction by ethyl methanesulphonate and lack of induction by ionising radiation. Mutat Res 51:255–270
Thilly WG, DeLuca JG, Furth EE, Hoppe H IV, Kaden DA, Krolewski JJ, Liber HL, Skopek TR, Slapikoff SA, Tizard RJ, Penman BW (1980) Gene-locus mutation assays in diploid human lymphoblast lines. In: de Serres FJ, Hollaender A (eds) Chemical mutagens, vol 6. Plenum, New York, pp 331–364
Tindall KR, Stankowski LF, Machanoff R, Hsie AW (1986) Analyses of mutation in pSV2gpt-transformed cells. Mutat Res 160:121–131
Todd PA, Glickman BW (1982) Mutational specificity of uv light in *E. coli:* indications for a role for DNA secondary structure. Proc Natl Acad Sci USA 79:4123–4127
Turner DR, Morley AA, Haliandros AA, Kutlaca R, Sanderson BJ (1985) In vivo somatic mutations in human lymphocytes frequently result from major gene alterations. Nature 315:343–345
Tweats DJ, Gatehouse DG (1988) Further debate of testing strategies. Mutagenesis 3:95–102
Urlaub G, Kas E, Carothers AM, Chasin LA (1983) Deletions of the diploid dihydrofolate reductase locus from cultured mammalian cells. Cell 33:405
Urlaub G, Carothers AM, Chasin LA (1985) Efficient cloning of single-copy genes using specialised cosmid vectors: Isolation of mutant dehydrofolate reductase genes. Proc Natl Acad Sci USA 82:1189–1193
Van Zeeland AA, Van Diggelen MCE, Simons JWIM (1972) The role of metabolic cooperation in selection of hypoxanthine-guanine-phosphoribosyl-transferase (HGPRT)$^-$ deficient mutants from diploid mammalian cell strains. Mutat Res 14:355–363
Verwoerd NP, Bernini LF, Bonnet J, Tanke HJ, Natarajan AT, Tates AD, Sobels FH, Ploem JS (1987) Somatic cell mutations in humans detected by image analysis of immunofluorescently stained erythrocytes. In: Burger G, Ploem JS, Goerttler K (eds) Clinical cytometry and histometry. Academic, London, pp 465–469
Vrieling H, Simons JWIM, Arwert F, Natarajan AT, Van Zeeland AA (1985) Mutations induced by X-rays at the *hprt* locus in Chinese hamster cells are mostly large deletions. Mutat Res 144:281–286
Wagenheim J, Bolesfoldi G (1988) Mouse lymphoma L5178Y thymidine kinase locus assay of 50 compounds. Mutagenesis 3:193–205
Waldmann TA, Misiti J, Nelson DL, Kraemer KH (1983) Ataxia-telangiectasia: a multisystem hereditary disease with immunodeficiency, impaired organ maturation, X-ray hypersensitivity, and a high incidence of neoplasia. Ann Intern Med 99:367–379

Waldren C, Correll L, Sognier MA, Puck TT (1986) Measurement of low levels of X-ray mutagenesis in relation to human disease. Proc Natl Acad Sci USA 83:4839–4843

Waldren CA, Puck TT (1987) Steps toward experimental measurement of total mutations relevant to human disease. Somatic Cell Mol Genet 13:411–414

Wilson JM, Young AB, Kelley WN (1983) Hypoxanthine-guanine phosphoribosyltransferase deficiency. The molecular basis of the clinical syndrome. N Engl J Med 309:900–910

Wilson JM, Stout JT, Palella TD, Davidson BL, Kelley WN, Caskey CT (1986) A molecular survey of hypoxanthine-guanine phosphoribosyltransferase deficiency in man. J Clin Invest 77:188–195

Yandell DW, Dryja TP, Little JB (1986) Somatic mutation at a heterozygous autosomal locus in human cells occur more frequently by allele loss than by intragenic structural alterations. Somatic Cell Mol Genet 12:255–263

Yang TP, Patel PI, Chinault AC, Stout JT, Jackson LG, Hildebrand BM, Caskey CT (1984) Molecular evidence for new mutations at the *hprt* locus in Lesch-Nyhan patients. Nature 310:412–414

Yates J, Warren N, Reisman D, Sugden B (1984) A *cis*-acting element from the Epstein-Barr viral genome that permits stable replication of recombinant plasmids in latently infected cells. Proc Natl Acad Sci USA 81:3806–3810

CHAPTER 3

Mechanisms of Repair in Mammalian Cells

M. DEFAIS

A. Introduction

The environment represents a constant threat to the genetic material of all living organisms. Radiation and many chemical agents interact with DNA and cause a great variety of lesions. In addition, DNA undergoes spontaneous damage such as base loss, chemical alteration of bases and replication errors. All these lesions affect the coding integrity of the DNA and may trigger chromosome changes such as mutations, recombination, rearrangements, chromosome aberrations or gene amplification. The cell's capacity to repair or tolerate these lesions is essential for genomic stability. Various enzymatic mechanisms are involved in the assurance of the integrity of DNA (HANAWALT et al. 1979; FRIEDBERG 1985). A number of hereditary diseases predisposing to malignancy in humans involve deficiencies in processing DNA lesions (see Chap. 4). Moreover, it is becoming quite clear that unrepaired lesions which lead to mutations play a role in the activation of proto-oncogenes (see Chap. 12).

Before starting to review some of the different mechanisms that mammalian cells have developed in order to repair DNA lesions, the complexity of the mammalian genome has to be considered. DNA is associated with both histones and non-histone proteins to form chromatin. The association of DNA with histones creates repeating units, the nucleosomes, that are separated by linker DNA sequences of variable length which are sensitive to nucleases (KORNBERG 1977). The nucleosomes are themselves organised by higher levels of folding which allows packaging of 6×10^9 base pairs of DNA, representing a linear 180 cm, into 46 chromosomes in a human cell, the total length of which is 200 μm (STRYER 1981). This complex structure influences the distribution of certain lesions; for example, in damage produced by particular chemicals such as the psoralens, adducts occur selectively in linker regions, while lesions caused by UV seem to be uniformly distributed in both linker DNA and in the nucleosomes (for a review see HANAWALT et al. 1979). In addition, the accessibility of the lesions to the repair enzymes can also be influenced by nucleosome structure and thus could affect an essential early step in repair. Pyrimidine dimers in mammalian chromatin do not appear to be equally accessible to repair enzymes (WILKINS and HART 1974). This accessibility can be increased by disruption of the DNA-histone association by the use of solutions of high ionic strength (VAN ZEELAND et al. 1981).

In addition to the factors introduced by the complexity of genomic organisation, many distinct mechanisms are known to be implicated in the repair of various lesions in mammalian cells.

Even though most of these mechanisms are far from being understood in detail, a great improvement in our knowledge of them has come from developments in molecular biology. I will not deal in this review with the repair effected on modified or unusual bases in DNA by DNA glycosylases or by AP endonucleases, but both types of enzymes have been described in mammalian cells (for reviews see LINDAHL 1982; FRIEDBERG 1985; SANCAR and SANCAR 1988). I will first discuss repair mechanisms that are able to restore the genetic information to its original state, such as excision repair and repair of alkylated bases. Secondly, the mechanisms of tolerance by which cells can survive the continued presence of lesions in their DNA will be considered. I will also then review the evidence for mismatch repair in mammalian cells.

B. Excision Repair

Excision repair acts on bulky lesions, the most representative of which are pyrimidine dimers that result from UV irradiation. Excision repair can be divided into three main stages: (a) incision, (b) excision and repair synthesis and (c) liga-

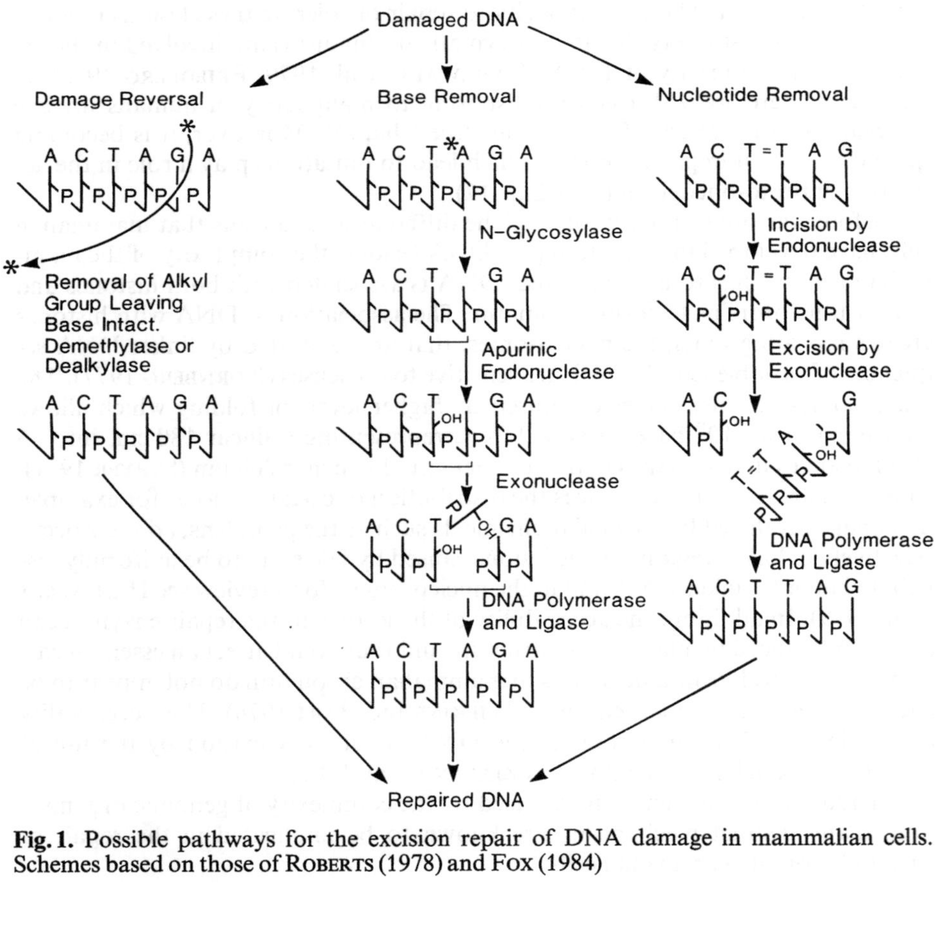

Fig. 1. Possible pathways for the excision repair of DNA damage in mammalian cells. Schemes based on those of ROBERTS (1978) and FOX (1984)

tion, that will be described separately (Fig. 1). The heterogeneity of repair, due either to the chromatin structure or to preferential repair in active genes, will also be considered.

I. Incision

Most of the information on this first step of excision repair in mammalian cells has come from studies on cells from xeroderma pigmentosum (XP) patients, who are UV sensitive and deficient in the excision of pyrimidine dimers and of various types of chemical damage (SETLOW 1978). When fusion between pairs of cells from different patients was carried out in various combinations, certain combinations appeared normal in their repair capacity. Cell fusion has revealed at least nine complementation groups, which suggests that at least nine genes could be involved in excision repair of bulky lesions (FISHER et al. 1985). There is evidence that the defect in most XP cells involves the incision step. Normal human cells rapidly accumulate strand breaks in DNA after UV irradiation, as detected by the alkaline elution technique (FORNACE et al. 1976). These strand breaks disappear with time, indicating that repair has taken place. Cells from XP complementation groups A–D do not accumulate breaks. The incision step can be restored by means of specific enzymes such as bacteriophage T4 endonuclease V, which recognises only pyrimidine dimers. If the enzyme is introduced into permeabilised XP cells, normal levels of repair synthesis can then take place (TANAKA et al. 1975; SMITH and HANAWALT 1978).

The repair deficiencies of xeroderma pigmentosum cells will be analysed in more depth in the following chapter, but the data reviewed above already indicate that several genes seem to be necessary for the incision step of the excision repair of bulky lesions to take place in mammalian cells. The incision step could be performed either by a system analogous to the *E. coli* uvrABC incision process (SANCAR and SANCAR 1988), or, alternatively, some of the genes involved could be necessary in order to destabilise chromatin and make the lesions accessible for incision by other mechanisms.

Excision repair mutants have been isolated from Chinese hamster ovary (CHO) cells (THOMPSON et al. 1981; WOOD and BURKI 1982). Five complementation groups have been characterised and each group has been analysed for its capacity to incise DNA during post-UV irradiation incubation. Cells of all of the complementation groups produce very low levels of incision compared with the wild type cells (THOMPSON and CARRANO 1983). The relation between the defects in human cells and in rodent cells will be examined later.

II. Excision and Repair Synthesis

Excision of bulky lesions in mammalian cells is closely coupled with resynthesis. The details of the mechanisms involved are completely unknown, but the results are detected as unscheduled DNA synthesis or repair synthesis. This DNA synthesis does not occur in S phase cells and was first observed by autoradiography (RASMUNSSEN and PAINTER 1964; CLEAVER and THOMAS 1981). Excision repair can also be detected by measuring the loss of sites sensitive to enzymes that are specific for pyrimidine dimers (PATERSON et al. 1973; WILLIAMS and CLEAVER 1979) or by the loss of antibody binding sites (CORNELIS 1978). Repair synthesis gives rise to newly synthesised sequences of about 30–100 nucleotides (HANAWALT et al. 1979).

Recently, extracts from human lymphoid cell lines have been shown to promote repair synthesis of short sequences of nucleotides on UV-irradiated or psoralen-damaged plasmid DNA (WOOD et al. 1988). This repair synthesis was absent in XP cell extracts. However, the repair activity could be reconstituted by mixing extracts of XPA and XPC cells. This is one of the first examples of in vitro excision repair in mammalian cells.

III. Enzymes Possibly Involved in Excision Repair

To my knowledge no enzyme has yet been described as being unequivocally involved in excision repair in mammalian cells. Until recently, the known mammalian polymerases seemed to be devoid of associated exonuclease activity, so that excision repair involving these enzymes would have to rely on independent exonucleases. DNase IV, which has been purified from rabbit tissues, shows a preference for double-stranded DNA and is able to excise dimers from incised DNA (LINDAHL et al. 1969; LINDAHL 1971). Other exonucleases which also have $5' \rightarrow 3'$ thymine dimer excision activity have been purified from human KB cells (COOK and FRIEDBERG 1978), two of which could be heterogeneous forms of the same enzyme. Some $5' \rightarrow 3'$ exonucleases from human placenta have been also described and purified (DONINGER and GROSSMAN 1976; PEDRINI and GROSSMAN 1983). They resemble the activities extracted from human KB cells; they degrade single-stranded DNA and are able to excise thymine dimers from incised duplex DNA.

Mammalian DNA polymerases α and β both seem to be involved in repair synthesis. The α polymerase is known to be the replicase in mammalian cells, while the β polymerase is generally considered as a repair enzyme. However, the relative roles in repair of these two polymerases are still controversial. Work has been reported in which blocking the activity of the α polymerase either by the use of differentiated cells (HÜBSCHER et al. 1978) or by the use of specific inhibitors like aphidicolin (PEDRALI-NOY and SPADARI 1979) did not prevent repair synthesis in UV-irradiated DNA, suggesting that the β polymerase was responsible for this activity. Other reports described an inhibition of both semiconservative replication and repair synthesis by aphidicolin, implying that the α polymerase is involved in both replication and repair (HANOAKA et al. 1979). In addition a mutant of CHO cells that is resistant to aphidicolin, and which may therefore contain an altered α polymerase, is UV sensitive (LIU et al. 1983).

It is thus likely that both polymerases are involved in the repair of DNA lesions caused by irradiation or by chemicals and that they are able to replace each other, as is the case in *E. coli* (for a review see FRIEDBERG 1985). In addition, a newly discovered eukaryotic DNA polymerase, DNA polymerase δ, which has an associated $3' \rightarrow 5'$ exonuclease activity, may also be involved in excision repair (NISHIDA et al. 1988; DRESLER and FRATTINI 1986).

IV. Ligation

Polynucleotide ligase has been purified from several mammalian tissues (LINDAHL and EDELMAN 1968; SÖDERHÄLL and LINDAHL 1973). Its activity depends on the presence of double-stranded DNA with adjacent 3′-OH and 5′-P termini, ATP and Mg^{2+}. Two forms of DNA ligase have been purified, from human cells (SPADARI et al. 1971; PEDRALI-NOY et al. 1973) and from calf thymus (SÖDERHÄLL and LINDAHL 1975).

It has been shown that ligase activity increases in exponentially growing cultures as compared with confluent cultures and that UV irradiation results in an increased ligase activity in confluent cells (MEZZINA and NOCENTINI 1978).

V. Heterogeneity of Damage and Repair

As mentioned earlier, the chromatin structure influences the interaction of damaging agents with DNA and the accessibility of the lesions that are produced to repair enzymes (for a review see BOHR et al. 1987).

Chromatin is sensitive to digestion with specific nucleases such as micrococcal nuclease or staphylococcal nuclease, which preferentially degrade linker DNA as core DNA is more resistant to digestion. This sensitivity has been used to determine the distribution of repair synthesis in mammalian cells exposed to DNA damage. By comparing the sensitivity to nucleases of the DNA labelled during repair synthesis with that of the bulk DNA, it is possible to determine where lesions and repair take place. For example, treatment of mammalian cells with UV or with chemicals that produce DNA damage randomly in chromatin results in repair synthesis stretches which are initially more sensitive to digestion than the bulk DNA. This sensitivity disappears with time (SMERDON and LIEBERMAN 1978). These results were interpreted as showing that preferential repair in the linker region was followed gradually by repair in nucleosomes leading to a randomisation of the repair patches with respect to core DNA. It has also been proposed that nuclease sensitivity represents the loss of nucleosome structure due to the repair (OLESON et al. 1979). The fact that the nuclease sensitivity of chromatin did not always occur at linker DNA was shown by ZOLAN et al. (1982a). Randomly distributed pyrimidine dimers and linker-specific angelicin adducts were found to be repaired with the same kinetics. In addition, the sensitivity to nuclease of the repair patches was independent of the initial distribution of the lesions in the DNA. Moreover, the repair patches that resulted from angelicin adducts were randomly distributed in the chromatin, implying that nucleosomes did not necessarily return to their original sites after DNA repair.

Heterogeneity in repair has also been studied in repetitive sequences of DNA. No difference in repair synthesis was found in rodent cells between highly repetitive satellite DNA and bulk DNA after UV irradiation (LIEBERMAN and POIRIER 1974). However, heterochromatic α DNA sequences of cultured African green monkey cells show a deficiency in the repair of chemical adducts (ZOLAN et al. 1982b). Repair synthesis represented 30%–60% of that of bulk DNA, depending on the chemical. This repair is influenced both by the status of the cells with respect to growth and by the cell cycle stage (LEADON and HANAWALT 1986). It is possible that the highly condensed chromatin structure of α DNA in confluent cells is inaccessible to the repair enzymes that act on bulky lesions (SMITH 1987).

VI. Preferential DNA Repair in Active Genes

The idea that repair could occur in DNA sequences that were particularly important to cell survival came essentially from two observations: (1) low levels of repair of bulky lesions were reported in heterochromatic, non-transcribed DNA (ZOLAN et al. 1982b, 1984; LEADON et al. 1983) and (2) rodent cells in culture seem deficient in repair since they remove only 20% of UV-irradiation damage in 24 h and still survive (BOHR et al. 1985, 1986). This last estimation was based on repair measurement over the entire genome, which would not detect preferential repair in small sequences. A new methodology was used to determine repair in defined genomic sequences and was first applied to the dihydrofolate reductase (DHFR) gene in CHO cells (BOHR et al. 1985). Cellular DNA was digested with an appropriate restriction enzyme and then treated with T4 endonuclease V to detect the frequency of pyrimidine dimers (GANESAN et al. 1981). The fraction of molecules without dimers was determined by Southern hybridization with specific probes of the analysed gene. Parental DNA was separated from bromodeoxyuridine-labelled, replicated DNA by a cesium chloride gradient. This step was necessary since replicated DNA, which is expected to be free of dimers, would case an overestimation of the molecules without lesions.

It was shown that two-thirds of the dimers were removed from a 14.1-kb restriction fragment of the amplified DHFR gene within 26 h after UV irradiation. Little repair occurred in the upstream fragments, and 15% of the lesions were removed from the overall genome.

Repair also occurred more rapidly in the human DHFR gene than in non-transcribed α DNA sequences or in bulk DNA (MELLON et al. 1986). Even more important was the discovery that selective removal of the lesion occurred in the transcribed DNA strand in rodent or human cells (MELLON et al. 1987). Repair was also shown to be efficient in a transcriptionally active proto-oncogene, the mouse *c-abl* gene, and inefficient in the inactive *c-mos* gene (MADHANI et al. 1986).

These results imply that repair is more essential to facilitate transcription than to allow cells to replicate their DNA. Another implication of these results could be an accumulation of premutational sites in the non-transcribed strand which would then be transmitted to the progeny.

VII. Human Genes for Repair

Excision repair in humans probably requires at least nine genes for the early incision step since there are at least nine complementation groups of XP patients. In CHO cells five complementation groups have been described (WOOD and BURKI 1982; THOMPSON et al. 1981; THOMPSON and CARRANO 1983). CHO repair mutants were used to characterise the first human excision repair gene (WESTERVELD et al. 1984). This *ERCC-1* gene has been cloned on the basis of its capacity to correct the repair defect of the CHO mutant 43-3B of complementation group 1 (previously called group 2, BOHR et al. 1988), which is sensitive to UV light and mitomycin C. The *ERCC-1* gene has a size of 12 kb and is located on chromosome 19 (VAN DUIN et al. 1986). Parts of this gene were also cloned by RUBIN et al. (1985). The *ERCC-1* gene codes for a 297-amino acid protein translated from a 1.1-kb transcript. Only this protein confers UV and mitomycin C resistance to the CHO mutant. The *ERCC-1* gene was compared with other repair genes and showed homology to the yeast Rad10 repair protein and in its longer C-terminal part; it has similarities with part of the *E. coli uvrA* and *uvrC* genes products (VAN DUIN et al. 1988). Mouse *ERCC-1* gene has also been isolated. Sequence analyses show the same homology with Rad10 and a comparable C-terminal extension as in the human gene. This C-terminal part of the *ERCC-1* gene seems to be very important for its repair function since the capacity of the protein to restore UV resistance disappears if this part is deleted. This region shows a significant homology with the C-terminus of the uvrC protein of *E. coli*, in which a mutation also abolishes the repair function (DOOLITTLE et al. 1986). This suggests that both proteins may share a similar function. The homology in repair proteins of prokaryotes and lower and higher eukaryotes suggests that DNA repair enzymes may have been well conserved during evolution.

Recently, the *ERCC-1* gene was reported to restore only preferential repair of the DHFR active gene in repair-deficient mutant CHO cells (BOHR et al. 1988). Further investigations are needed to confirm these results. Purification of the *ERCC-1* protein will be necessary to precisely determine its function.

The *ERCC-2* human gene correcting excision repair defects in CHO UV5 mutants of complementation group 2 (previously group 1) has been cloned and characterised (WEBER et al. 1988). This gene increases the UV resistance of the CHO mutant and allows an elevated rate of strand incision to occur after UV irradiation. It is also carried by chromosome 19. Another human gene, *ERCC-5*, has also been isolated and involved in excision repair (STRNISTE et al., UCLA meeting 1988). The *ERCC-1* gene product is not able to correct XP defects. A possible correction is under study for the two other genes.

VIII. Repair of DNA Strand Breaks

Ionising radiation introduces various lesions into DNA including single- and double-strand breaks that occur either directly or as a result of enzymatic activity on damaged bases.

Six X-ray-sensitive mutants of CHO cells have been isolated (JEGGO and KEMP 1983). These mutants are sensitive to ionising radiation and to the radiomimetic drug bleomycin. They show a markedly decreased ability to repair double-strand breaks produced by γ irradiation. They belong to the same complementation group and are recessive to the wild type (JEGGO 1985).

These mutants have allowed two human genes involved in the repair of DNA breaks to be cloned: the *XRCC-1* gene, which is located on chromosome 19 (THOMPSON, UCLA meeting 1988), and the *XRCC-2* gene, which complements the deficiency of *xrs6* mutants by its ability to repair double-strand breaks (STRNISTE, UCLA meeting 1988).

C. Repair of Alkylated Bases

Alkylating agents, which represent an important group of environmental chemicals, are toxic and mutagenic to cells in general and are powerful carcinogens in mammalian cells. O^6-methylguanine is the main mutagenic lesion which appears in DNA after treatment with alkylating agents, while 3-methyladenine is the main toxic lesion (for reviews, see LINDAHL et al. 1988; FRIEDBERG 1985).

Numerous studies have demonstrated that mammalian cells contain enzymatic activities able to remove these lesions. It has long been known that various tissues remove O^6-alkylguanine from DNA with different efficiencies and that this repair capacity is inversely correlated with the production of tumours (GOTH and RAJEWSKY 1974; PEGG 1978; PEGG and BALOG 1979). The cellular enzymatic repair activity rapidly removes O^6-methylguanine and O^6-ethylguanine by transfer of the alkyl group to a cysteine residue of the acceptor protein, forming a S-methyl cysteine during the reaction (TEO and KARRAN 1982; METHA et al. 1981; MEDCALF and LAWLEY 1981). The enzyme does not act on *N*-methylpurines. This methyltransferase exists in a variety of mammalian tissues, with liver cells possessing the highest activity and human cells having more than rodent cells (PEGG et al. 1982; WALDSTEIN et al. 1982; FOOTE et al. 1983; HARRIS et al. 1983). The "enzyme" has a molecular weight of 24000 and is consumed during the reaction.

Thus it appears that mammalian methyltransferase is similar to *E. coli* methyltransferase. However, the *E. coli* Ada protein of molecular weight 39000 has two acceptor cysteines in two different domains of the protein. One in the C-terminal domain of the protein receives a methyl group from O^6-methylguanine and the other in the N-terminal domain receives a methyl group from methylphosphotriesters. In contrast, the mammalian methyltransferase is unable to act on methylphosphotriesters (KATAOKA et al. 1986; YAROSH et al. 1984). The double methylation of the *E. coli* Ada protein transforms it into an activator of transcription of its own gene and other genes of the adaptive response (for review, see LINDAHL et al. 1988).

Several studies designed to demonstrate the inducibility of mammalian methyl transferase have been attempted. CHO cells and SV40-transformed human skin fibroblasts were reported to become resistant to the induction of sister chromatid exchanges (considered as an indicator of persistent lesions) after exposure to very low doses of *N*-methyl-*N'*-nitro-*N*-nitrosoguanidine (SAMSON and SCHWARTZ 1980). The cells were also better able to survive a second treatment by higher doses of the same drug. An increased survival of Chinese hamster V79 cells pretreated with methylnitrosourea has also been described (DURRANT et al. 1981). However, in neither study were mutations decreased by the adaptive treatment. A cross reactivity in the cytotoxicities resulting from different pretreatments in rat hepatoma cells was ascribed to an adaptive response but again with no effect on mutations (LAVAL and LAVAL 1984).

Rats chronically fed with diethylnitrosamine show a loss of O^6-ethylguanine from the DNA of liver cells after a challenge with a larger dose (PEGG and BALOG 1979). The pretreatment provoked an increase of the enzymatic activity in the cells that is responsible for the repair of O^6-ethylguanine (MONTESANO et al.

1980). However, this activity seems also to be induced by other, non-specific treatments (PEGG and PERRY 1981).

Several groups have looked for evidence of inducibility of the mammalian methyl transferase in human fibroblasts without success (KARRAN et al. 1982; YAROSH et al. 1984; FROSINA et al. 1984). Thus, even if the function of the mammalian methyltransferase appears to be very similar to that of *E. coli*, its regulation is very likely to be different. Proof of its inducibility does not appear to be convincing. In addition, mammalian cells have a high constitutive level of the enzyme compared with that which is present in *E. coli*.

There is an apparent similarity between the C-terminal domain of *E. coli* Ada protein and the mammalian methyl transferase. When the former is expressed in mutant cell lines deficient in alkylation repair (Mer^- phenotype), the cells recover their ability to repair O^6-methylguanine (KATAOKA et al. 1986; SAMSON et al. 1986; BRENNAND and MARGISON 1986). These results will be discussed in the following chapters.

Like *E. coli*, mammalian cells also contain another enzymatic activity, a 3-methyladenine DNA glycosylase which removes N3 and N7 alkylpurines (GALLAGHER and BRENT 1982, 1984). In *E. coli* this activity also repairs lesions such as *O*-alkyl pyrimidines, (O^2- and O^2-alkylcytosine). The same enzyme in mammalian cells is unable to repair these lesions, which appear to be repaired by an as yet unknown mechanism (BRENT et al. 1988).

D. Mechanisms of Tolerance

There are indications that pyrimidine dimers represent a block to DNA replication (for reviews, see HALL and MOUNT 1981; HANAWALT et al. 1979). Experiments involving photoreactivation of pyrimidine dimers showed a post-irradiation recovery of DNA synthesis in marsupials (KRISHNAN and PAINTER 1973). However, the complexity of mammalian DNA organisation renders most experiments on replication after UV irradiation difficult to analyse or interpret. The mammalian genome normally replicates as small replicons of poorly defined size and number (EDENBERG and HUBERMAN 1975; HAND 1978). These replicons eventually meet, thus allowing joining up of the newly synthesised strand. Thus, the observation of discontinuous DNA synthesis is difficult to interpret and does not necessarily indicate gaps in front of lesions (FRIEDBERG 1985). Nevertheless, attempts have been made to understand the effect of DNA damage on replication initiation and elongation (Fig. 2).

I. DNA Synthesis on Damaged Templates

One indication of the blockage of DNA replication initiation came from the work of PARK and CLEAVER (1979a, b). In irradiated human cells the total amount of DNA synthesised per unit time and the size of the newly synthesised DNA initially decrease compared with these of the non-irradiated cells and then recover. But even when DNA elongation had completely recovered, the total

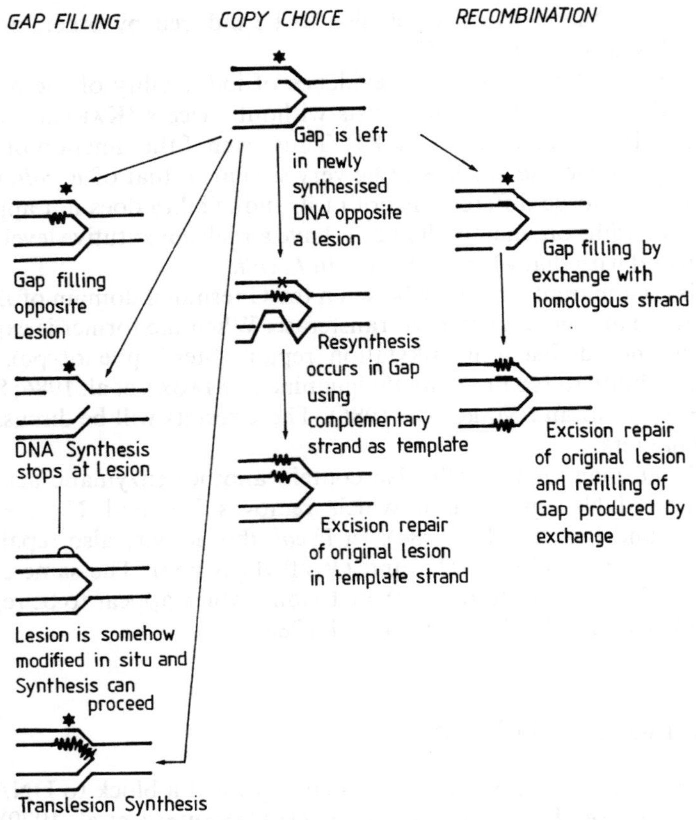

Fig. 2. Possible mechanisms of tolerance in mammalian cells and their viruses. Schemes based on those of Roberts (1978) and Fox (1984)

amount of DNA synthesis per unit time is still lower than in the control cells, probably indicating that some replicons fail completely to replicate. The same type of results have been obtained by employing chemicals instead of UV (Painter 1977).

A decrease in the elongation rate of DNA caused by the presence of lesions could also explain the overall diminution of DNA synthesis. It is generally accepted that, in the first hour post-irradiation, the decrease in DNA replication depends on the average number of lesions per replicon. With a few lesions per replicon, initiation is predominantly affected, while chain elongation is blocked with higher numbers of lesions per replicon (Cleaver et al. 1983).

A response resembling daughter-strand gap repair in *E. coli* has been described in mammalian cells (Lehmann 1972). The incubation of mouse lymphoma cells in radioactive medium after UV irradiation produced DNA of smaller molecular weight than control DNA in alkaline sucrose gradients. The size of the DNA increased with further incubation. In these experiments the DNA was fragmented before centrifugation to avoid artifacts due to the high molecular weight of mammalian DNA. Nevertheless, interpretation is difficult because of the presence of multiple replicons which generate, by themselves, discontinuities in the newly synthesised strand. However, on occasions when the size of the newly synthesised DNA corresponds to the interdimer distance, it is conceivable that the discontinuities really are a consequence of the presence of dimers. Experiments in which cells were post-incubated in BrdU and submitted to BU photolysis to recreate the small DNA fragments suggest that the gaps that result from DNA damage may be filled by de novo synthesis (LEHMANN 1972). However, other work suggests that lesions in the DNA do not result in the formation of gaps (PARK and CLEAVER 1979 b).

The filling of gaps by DNA strand exchanges, i.e. by a recombinational event, appears to be infrequent (FUJIWARA and TATSUMI 1977; WATERS and REGAN 1976). However, the role of recombination in repair is well documented in animal viruses. The phenomenon known as multiplicity reactivation, in which several UV-irradiated viruses survive better than a single virus through recombination of their genomes, has been described in human cells (DAS GUPTA and SUMMERS 1978; SELSKY et al. 1979).

Another possible mechanism for filling gaps, which would explain the difficulties encountered in detecting them is the presence of trans-lesion synthesis or, in other words, of the replicative bypass of the damage. This has been demonstrated in monkey cells infected by a temperature-sensitive SV40 mutant which allows synchronised viral DNA replication (SARASIN and HANAWALT 1980). When the infected cells were UV irradiated, the newly synthesised viral DNA fragments had a size approximately equivalent to the inter-dimer distance in the template strands. If the cells were further incubated, viral molecules became longer as if replication, after some delay, had started again in a continuous mode. No dimers could be detected in the nascent SV40 DNA strands.

Thus, it appears that the mechanisms by which mammalian cells tolerate damage in their DNA are not, as yet, at all clear and need further study.

II. Inducible DNA Processing

A complex inducible response, the SOS response, has been described in *E. coli* (RADMAN 1974; WALKER 1984). As part of this response, an error-prone, DNA-processing mechanism helps bacteria to survive when their DNA replication is transiently blocked or slowed down. Viral probes for repair have been used as one way of studying this system (DEFAIS et al. 1983).

Weigle reactivation is the increased survival of UV-irradiated phage resulting from the induction of the SOS response in their hosts (WEIGLE 1953; DEFAIS et al. 1971; RADMAN 1975). This increased survival is accompanied by an increased mutagenesis of the viral progeny.

The phenomenon of Weigle reactivation has been reported in studies carried out in mammalian cells using a variety of UV-irradiated viruses and after different inducing treatments have been applied to the host cells. This phenomenon has been called enhanced virus reactivation and was first described with herpes virus infected, UV-irradiated monkey cells (BOCKSTAHLER and LYTLE 1970). It has been confirmed in human cells infected with the same virus and treated with carcinogens (LYTLE et al. 1978). Other viruses such as SV40 are subject to enhanced reactivation when their hosts are pretreated with either radiation or carcinogens (SARASIN and HANAWALT 1978; for review, see DEFAIS et al. 1983). Treatment with drugs inhibiting normal DNA replication such as hydroxyurea or cycloheximide also promotes SV40 reactivation (SARASIN and HANAWALT 1978). It has been proposed that the inhibition of normal DNA replication is a direct or indirect signal for the induction of a replicative complex that is able to replicate UV-irradiated DNA (SARASIN and HANAWALT 1980). The expression of virus reactivation is independent of excision repair in human cells since it has also been observed in XP cells (LYTLE et al. 1976).

Weigle reactivation in bacteria is highly mutagenic (DEFAIS et al. 1976). Since enhanced virus reactivation resembles this phenomenon, it is relevant to look for a mutagenic response that accompanies enhanced virus reactivation.

Reactivation of herpes virus in UV-irradiated monkey cells was shown to be accompanied by an increased mutation frequency in the viral progeny (DAS GUPTA and SUMMERS 1978). Increased reversion of a UV-irradiated, temperature-sensitive mutant of SV40 has been observed in UV-irradiated, monkey kidney cells (SARASIN and BENOIT 1980). Mutagenesis increased with the UV flux that was applied to the virus. An increase in the mutation frequency of a SV40 shuttle vector in mitomycin C-treated monkey cells has also been reported (ROILIDES et al. 1988).

The multiplicity of infection seems to play a major, but unclear, role in the induction of mutations. In the case of herpes virus, enhanced mutagenesis may be correlated with a high multiplicity of infection (LYTLE et al. 1980). However, enhanced mutagenesis has also been shown in SV40 after infection of mitomycin C-treated host cells with a low multiplicity of infection (SARASIN and BENOIT 1986).

Inducible host function has also been investigated in parvoviruses (CORNELIS et al. 1982). These single-stranded DNA viruses cannot remove DNA lesions by excision or repair them by recombination. They therefore have to replicate damaged DNA to survive after irradiation. Enhanced mutagenesis has been observed at low multiplicities of infection in non-irradiated and in irradiated viruses replicating in UV-treated hosts, therefore implying that the cellular mutator activity is not targeted on the viral genome.

Thus, it seems that error-prone replication occurs in mammalian cells following treatment by agents producing damage or breaks in DNA. The detailed mechanisms are unknown, and much more work is needed in order to understand whether mutagenesis and increased survival are linked and are really inducible.

E. Mismatch Repair

DNA containing mismatched, or unpaired, bases occurs in vivo as a result of several genetic processes including (a) recombination between homologous but non-identical DNA sequences, (b) replication errors and (c) deamination of 5-methylcytosine to thymine. Correction of mismatched bases may have a double role in living cells: in the conservation of genetic information by repair of replicative errors before transmission to the progeny and in the diversification of genetic information. Historically, it was first suggested that clustered recombinational events, known as high negative interference in bacteriophage (AMATI and MESELSON 1965) and non-reciprocal recombination, or gene conversion, in fungi (HOLLIDAY 1964), could result from a mismatch repair. Mismatch repair is well documented in prokaryotes, and the mechanisms are clearly understood (for reviews, see RADMAN and WAGNER 1986; CLAVERYS and LACKS 1986; MODRICH 1987).

Several studies appear to indicate the existence of mismatch repair in mammalian cells. 5-Methylcytosine is the most abundant modified base in the DNA of mammalian cells (DOERFLER 1983; RIGGS and JONES 1983). Its presence is generally correlated with the suppression of transcription of certain genes. However, it represents a general mutagenic hazard to the cells because of its spontaneous deamination to thymine, leading to a mismatch base pair G-T. This suggests the need for a mismatch repair system similar to the "very short patch mismatch repair" existing in *E. coli* (LIEB 1983; LIEB et al. 1986; JONES et al. 1987a, b).

HARE and TAYLOR (1985) transfected monkey cells with hemimethylated heteroduplex DNA containing either a G-T or an A-C mismatch. The results suggested that methylation may direct repair to occur in the unmethylated strand, as it does in *E. coli*. They also showed that the presence of single-strand nicks was able to direct repair, implying that the correction was of the newly synthesised strand. However, they observed that in the absence of either signal for discriminating between strands, repair still occurred. In addition, there is no real proof that a repair is taking place in these experiments. Alternative processes such as recombination or strand loss could also be involved.

When heteroduplexes containing about 8% of mismatched bases were introduced into monkey cells, a wide variety of molecules carrying a patchwork of parental sequences were observed, indicating that a cellular processing of these heteroduplexes had occurred (ABASTADO et al. 1984). However, this processing could either be repair or recombination. In mammalian cells, genomic DNA contains highly repetitive sequences such as those found in the multigene families. It is often suggested that these sequence can undergo recombination and gene conversion. It has thus been proposed that mismatch repair would be responsible for the diversification of genes in multigene families such as the histocompatibility genes (KOURILSKY 1983) and the variable region of the immunoglobulin genes (RADMAN 1983).

Evidence for some specificity in the repair of mismatched bases in mammalian cells appeared recently (BROWN and JIRICNY 1987, 1988). Specific G-T mispairs were introduced in synthetic duplexes ligated into SV40 DNA between two restriction sites. The sequences were such that the correction of the mismatch would lead to the creation of two different new restriction sites following the corrected base. The analysis of the DNA collected from SV40 plaques showed 90% of corrections in favour of guanine and only 8% in favor of thymine, suggesting the existence of a specific mismatch repair protecting mammalian cell DNA from the loss of 5-methylcytosine. This study has been extended to all base/base mispairs in order to characterise potential repair specificity and efficiency. All mismatches appear to be corrected but with different efficiencies, and the correction seems to be influenced by the sequences flanking the mismatch (BROWN and JIRICNY 1988).

Homologous recombination may also be important in the repair of DNA damage. For example, repair of double-strand breaks has been shown to result in homologous recombination between exogenous DNAs (BRENNER et al. 1986). Co-transferred plasmids carrying mutations in the herpes simplex virus gene *tk* encoding for thymidine kinase, one of which contained a double-strand gap, were repaired by homologous recombination to give a functional *tk* gene in mouse cells. This recombination was shown to be a non-reciprocal exchange.

It is apparent from these studies that much more information is needed in order to confirm the existence of the mechanisms of mismatch repair in mammalian cells.

F. Concluding Remarks

Several systems for repairing or tolerating DNA lesions are obviously present in mammalian cells even though their detailed mechanisms have not yet been elucidated, mainly due to the complexity of mammalian DNA organisation.

Following the processing that it undergoes DNA damage can give rise to mutations or more complex rearrangements of DNA such as translocations, amplifications or transpositions. These genomic modifications provoked by the presence or the processing of damage may play an important role in carcinogenesis. Deficiencies in DNA damage processing have been correlated with several human hereditary diseases that result in a predisposition to cancer.

In addition, genomic responses in cells to different stresses may also be important in modulating the effects of radiotherapy and chemotherapy. Thus a better understanding of DNA damage processing could improve the results of cancer therapy.

Finally, even though no correlation between DNA damage and aging has yet been established, it is conceivable that some kinds of lesions may be involved in senescence (ADELMAN et al. 1988).

References

Abastado JP, Cami B, Huynh Dinh T, Igolen J, Kourilsky P (1984) Processing of complex heteroduplexes in *Escherichia coli* and *cos-1* monkey cells. Proc Natl Acad Sci USA 81:5792–5796

Adelman R, Saul RL, Ames BN (1988) Oxidative damage to DNA: relation to species metabolic rates and life span. Proc Natl Acad Sci USA 85:2706–2708

Amati P, Meselson M (1965) Localized negative interference in bacteriophage lambda. Genetics 51:369–379

Bockstahler LE, Lytle CD (1970) UV light enhanced reactivation of a mammalian virus. Biochem Biophys Res Comm 41:184–189

Bohr VA, Smith CA, Okumoto DS, Hanawalt PC (1985) DNA repair in an active gene: removal of pyrimidine dimers from the DHFR gene of CHO cells is much more efficient than in the genome overall. Cell 40:359–369

Bohr VA, Okumoto DS, Hanawalt PC (1986) Survival of UV-irradiated mammalian cells correlates with efficient DNA repair in an essential gene. Proc Natl Acad Sci USA 83:3830–3833

Bohr VA, Phillips DH, Hanawalt PC (1987) Heterogeneous DNA damage and repair in the mammalian genome. Cancer Res 47:6426–6436

Bohr VA, Chu EHY, van Duin M, Hanawalt PC, Okumoto DS (1988) Human repair gene restores normal pattern of preferential DNA repair in repair defective CHO cells. Nucleic Acid Res 16:7397–7403

Brennand J, Margison GP (1986) Reduction of the toxicity and mutagenicity of alkylating agents in mammalian cells harboring the *Escherichia coli* alkyltransferase gene. Proc Natl Acad Sci USA 83:6292–6296

Brenner DA, Smigocki AC, Camerini-Otero RD (1986) Double-strand gap repair results in homologous recombination in mouse L cells. Proc Natl Acad Sci USA 83:1762–1766

Brent TP, Dolan ME, Fraenkel-Conrat H, Hall J, Karran P, Laval F, Margison GP, Montesano R, Pegg AG, Potter PM, Singer B, Swenberg JA, Yarosh DB (1988) Repair of *O*-alkylpyrimidines in mammalian cells: a present consensus. Proc Natl Acad Sci USA 85:1759–1762

Brown TC, Jiricny J (1987) A specific mismatch repair event protects mammalian cells from loss of 5-methylcytosine. Cell 50:945–950

Brown TC, Jiricny J (1988) Different base/base mispairs are corrected with different efficiencies and specificities in monkey kidney cells. Cell 54:705–711

Claverys JP, Lacks SA (1986) Heteroduplex deoxyribonucleic acid base mismatch repair in bacteria. Microbiol Rev 50:133–165

Cleaver JE, Thomas GH (1981) Measurement of unscheduled synthesis by autoradiography. In: Friedberg EC, Hanawalt PC (eds) DNA repair: a laboratory manual of research procedures, vol I, part B. Dekker. New York, pp 277–287

Cleaver JE, Kaufman WK, Kapp LN, Park SD (1983) Replicon size and excision repair as factors in the inhibition and recovery of DNA synthesis from ultraviolet damage. Biochem Biophys Acta 739:207–215

Cook KH, Friedberg EC (1978) Multiple thymine dimer excising nuclease activities in extracts of human KB cells. Biochemistry 17:850–857

Cooper CS (1989) The role of oncogene activation in chemical carcinogenesis. In: Cooper CS, Grover PL (eds) Carcinogenesis and mutagenesis. II. Springer, Berlin Heidelberg New York (Handbook of experimental pharmacology, vol 94/II)

Cornelis JJ (1978) Characterization of excision repair of pyrimidine dimers in eukaryotic cells as assayed with antidimer sera. Nucleic Acids Res 5:4273–4281

Cornelis JJ, Su ZZ, Rommelaere J (1982) Direct and indirect effects of ultraviolet light on the mutagenesis of parvovirus H-1 in human cells. EMBO J 1:693–699

Das Gupta UB, Summers WC (1978) Ultraviolet reactivation of herpes simplex virus is mutagenic and inducible in mammalian cells. Proc Natl Acad Sci USA 75:2378–2381

Defais M, Fauquet P, Radman M, Errera M (1971) Ultraviolet reactivation and ultraviolet mutagenesis of λ in different genetic systems. Virology 43:495–503

Defais M, Caillet-Fauquet P, Fox MS, Radman M (1976) Induction kinetics of mutagenic DNA repair activity in *E. coli* following ultraviolet irradiation. Mol Gen Genet 148:125–130

Defais M, Hanawalt PC, Sarasin AR (1983) Viral probes for DNA repair. In: Lett JT (ed) Advances in radiation biology, vol 10. Academic, New York, pp 1–37

Doerfler W (1983) DNA methylation and gene activity. Ann Rev Biochem 52:93–124

Doninger J, Grossman L (1976) Human correxonucleases. Purification and properties of a DNA repair exonuclease from human placental nuclei. J Biol Chem 251:4579–4587

Doolittle RF, Johnson MS, Husain I, Van Houten B, Thomas DC, Sancar A (1986) Domainal evolution of a procaryotic DNA repair protein and its relationship to active-transport proteins. Nature 323:451–453

Dresler SL, Frattini MG (1986) DNA replication and UV-induced DNA repair synthesis in human fibroblasts are much less sensitive than DNA polymerase α to inhibition by butylphenyl-deoxyguanosine triphosphate. Nucleic Acid Res 14:7093–7102

Durrant LG, Margison GP, Boyle JM (1981) Pretreatment of Chinese hamster V79 cells with MNU increases survival without affecting DNA repair or mutagenicity. Carcinogenesis 2:55–60

Edenberg H, Huberman J (1975) Eukaryotic chromosome replication. Ann Rev Genet 9:245–284

Fisher E, Keijzer W, Thielman HW, Popanda O, Bohnert E, Edler EG, Jung EG, Bootsma D (1985) A ninth complementation group in *Xeroderma pigmentosum* XP-1. Mutat Res 145:217–225

Foote RS, Pal BC, Mitra S (1983) Quantitations of O6-methylguanine DNA methyltransferase in HeLa cells. Mutat Res 119:221–228

Fornace AJ Jr, Kohn KW, Kann HE Jr (1976) DNA single-strand breaks during repair of UV damage in human fibroblasts and abnormalities in *Xeroderma pigmentosum*. Proc Natl Acad Sci USA 73:39–43

Fox M (1984) Drug resistance and DNA repair. In: Fox BW, Fox M (eds) Antitumour drug resistance. Springer, Berlin Heidelberg New York, pp 335–369 (Handbook of experimental pharmacology, vol 72)

Friedberg EC (1985) DNA repair. Freeman, New York

Frosina G, Bonatti S, Abbondandolo A (1984) Negative evidence for an adaptive response to lethal and mutagenic effects of alkylating agents in V79 Chinese hamster cells. Mutat Res 129:243–250

Fujiwara Y, Tatsumi M (1977) Low levels of DNA exchanges in normal human and *Xeroderma pigmentosum* cells after UV irradiation. Mutat Res 43:279–290

Gallagher PE, Brent TP (1982) Partial purification and characterization of 3-methyladenine DNA glycosylase from human placenta. Biochemistry 21:6404–6409

Gallagher PE, Brent TP (1984) Further purification and characterization of human 3-methyladenine DNA glycosylase. Evidence for specificity. Biochim Biophys Acta 782:394–401

Ganesan AK, Smith CA, Van Zeeland AA (1981) Measurement of the pyrimidine dimer content of DNA in permeabilized bacterial or mammalian cells with endonuclease V of bacteriophage T4. In: Friedberg EC, Hanawalt PC (eds) DNA repair – a laboratory manual of research procedures, vol I, part A. Dekker, New York, pp 89–97

Goth R, Rajewsky M (1974) Persistence of O^6-ethylguanine in rat brain DNA: correlation with nervous system-specific carcinogenesis by ethylnitrosourea. Proc Natl Acad Sci USA 71:639–643

Hall JD, Mount DM (1981) Mechanisms of DNA replication and mutagenesis in ultraviolet-irradiated bacteria and mammalian cells. Prog Nucleic Acid Res Mol Biol 25:53–126

Hanawalt PC, Cooper PK, Ganesan AK, Smith CA (1979) DNA repair in bacteria and mammalian cells. Ann Rev Biochem 48:783–836

Hand R (1978) Eukaryotic DNA: organization of the genome for replication. Cell 15:317–325

Hanoaka F, Kato H, Ikegami S, Ohashi M, Yamada MA (1979) Aphidicolin does inhibit repair replication in HeLa cells. Biochem Biophys Res Comm 87:575–587

Hare JT, Taylor H (1985) On role of DNA methylation in vertebrate cells in strand discrimination in mismatch repair. Proc Natl Acad Sci USA 82:7350–7354

Harris A, Karran P, Lindahl T (1983) O^6-methyl-guanine-DNA methyltransferase of human lymphoid cells. Cancer Res 43:3247–3252

Holliday R (1964) A mechanism for gene conversion in fungi. Genet Res 5:282–304

Hübscher U, Kuenzle CC, Limacher W, Schoner R, Spadari S (1978) Functions of DNA polymerases α, β, and γ in neurons during development. CSH Symp Quant Biol 43:625–629

Jeggo P (1985) Genetic analysis of X-ray sensitive mutants of the CHO cell line. Mutat Res 146:265–270

Jeggo P, Kemp LM (1983) X-ray sensitive mutants of Chinese hamster ovary cell line isolation and cross sensitivity to other DNA-damaging agents. Mutat Res 112: 313–327

Jones M, Wagner R, Radman M (1987a) Mismatch repair of deaminated 5-methylcytosine. J Mol Biol 194:155–159

Jones M, Wagner R, Radman M (1987b) Mismatch repair and recombination in *E. coli*. Cell 50:621–626

Karran P, Arlett CF, Broughton BC (1982) An adaptive response to the cytotoxic effects of *N*-methyl-*N*-nitrosourea is apparently absent in normal human fibroblasts. Biochimie 64:717–721

Kataoka H, Hall J, Karran P (1986) Complementation of sensitivity to alkylating agents in *Escherichia coli* and Chinese hamster ovary cells by expression of a cloned bacterial DNA repair gene. EMBO J 5:3195–3200

Kornberg RD (1977) Structure of chromatin. Ann Rev Biochem 46:931–954

Kourilsky P (1983) Genetic exchanges between partially homologous nucleotide sequences: possible implications for multigene families. Biochimie 65:85–93

Krishnan D, Painter RB (1973) Photoreactivation and repair replication in rat kangaroo cells. Mutat Res 17:213–222

Laval F, Laval J (1984) Adaptive response in mammalian cells: cross reactivity of different pretreatments on cytotoxicity as compared to mutagenicity. Proc Natl Acad Sci USA 81:1062–1066

Leadon SA, Hanawalt PC (1986) Cell-cycle dependent repair of damage in α and bulk DNA of monkey cells. Mutat Res 166:71–76

Leadon SA, Zolan ME, Hanawalt PC (1983) Restricted repair of aflatoxin B1 induced damage in α DNA of monkey cells. Nucleic Acids Res 11:5675–5689

Lehmann AR (1989) Cancer-prone human disorders with defects in DNA repair. In: Cooper CS, Grover PL (eds) Carcinogenesis and mutagenesis. II. Springer, Berlin Heidelberg New York (Handbook of experimental pharmacology, vol 94/II)

Lehmann AR (1972) Post-replication repair of DNA in UV-irradiated mammalian cells. J Mol Biol 66:319–337

Lieb M (1983) Specific mismatch correction in bacteriophage lambda crosses by very short patch repair. Mol Gen Genet 191:118–125

Lieb M, Allen E, Reed D (1986) Very short patch mismatch repair in phage lambda: repair sites and strength of repair tracts. Genetics 114:1041–1060

Lieberman MW, Poirier MC (1974) Intragenomal distribution of DNA repair synthesis: repair in satellite and mainband DNA in cultured mouse cells. Proc Natl Acad Sci USA 71:2461–2475

Lindahl T (1971) Excision of pyrimidine dimers from ultraviolet-irradiated DNA by exonucleases from mammalian cells. Eur J Biochem 18:407–414

Lindahl T (1982) DNA repair enzymes. Ann Rev Biochem 51:61–87

Lindahl T, Edelman GM (1968) Polynucleotide ligase from myeloid and lymphoid tissues. Proc Natl Acad Sci USA 61:680–687

Lindahl T, Gally JA, Edelman GM (1969) Deoxyribonuclease IV: a new exonuclease from mammalian tissues. Proc Natl Acad Sci USA 62:597–603

Lindahl T, Sedgwick B, Sekiguchi M, Nakabeppu Y (1988) Regulation and expression of the adaptive response to alkylating agents. Ann Rev Biochem 57:133–157

Liu PK, Chang CC, Trosko JE, Dube DK, Martin GM, Loeb LA (1983) Mammalian mutator mutant with an aphidicolin resistant DNA polymerase α. Proc Natl Acad Sci USA 80:797–801

Lytle CD, Day RS III, Hellman KB, Bockstahler LE (1976) Infection of UV-irradiated normal human and XP fibroblasts by herpes simplex virus: studies on capacity and weigle reactivation. Mutat Res 36:257–264

Lytle CD, Coppey J, Taylor WD (1978) Enhanced survival of ultraviolet-irradiated herpes simplex virus in carcinogen-pretreated cells. Nature 272:60–62

Lytle CD, Goddard JG, Lin C (1980) Repair and mutagenesis of herpes simplex virus in UV-irradiated monkey cells. Mutat Res 70:139–149

Madhani HD, Bohr VA, Hanawalt PC (1986) Differential DNA repair in transcriptionally active and inactive proto-oncogenes: c-*abl* and c-*mos*. Cell 45:417–423

Medcalf ASC, Lawley PD (1981) Time course of O^6-methylguanine removal from DNA of N.methyl.N.nitroso urea-treated human fibroblasts. Nature 289:796–798

Mellon IM, Bohr VA, Smith CA, Hanawalt PC (1986) Preferential DNA repair of an active gene in human cells. Proc Natl Acad Sci USA 83:8878–8882

Mellon IM, Spivak GS, Hanawalt PC (1987) Selective removal of transcription blocking DNA damage from the transcribed strand of the mammalian DHFR gene. Cell 51:241–249

Metha JR, Ludlum DB, Renard A, Verly W (1981) Repair of O^6-ethylguanine in DNA by a chromatin fraction from rat liver: transfer of the ethyl group to an acceptor protein. Proc Natl Acad Sci USA 78:6766–6770

Mezzina M, Nocentini S (1978) DNA ligase activity in UV-irradiated monkey kidney cells. Nucleic Acids Res 5:4317–4328

Modrich P (1987) DNA mismatch correction. Ann Rev Biochem 56:435–466

Montesano R, Bresil H, Planche-Martel G, Margison GP, Pegg AE (1980) Effect of chronic treatment of rats with dimethylnitrosamine on the removal of O^6-methylguanine from DNA. Cancer Res 40:452–458

Nishida C, Reinhard P, Linn S (1988) DNA repair synthesis in human fibroblasts requires DNA polymerase δ. J Biol Chem 263:501–510

Oleson FB, Mitchell BL, Dipple A, Lieberman MV (1979) Distribution of DNA damage in chromatin and its relation to repair in human cells treated with 7-bromoethylbenz(a)anthracene. Nucleic Acids Res 7:1343–1361

Painter RB (1977) Rapid test to detect agents that damage human DNA. Nature 265:650–651

Park SD, Cleaver JE (1979 a) Recovery of DNA synthesis after ultraviolet irradiation of *Xeroderma pigmentosum* cells depends on excision repair and is blocked by caffeine. Nucleic Acids Res 6:1151–1159

Park SD, Cleaver JE (1979 b) Postreplication repair: questions of its definition and possible alterations in XP cell strains. Proc Natl Acad Sci USA 76:3927–3931

Paterson MC, Lohman PHM, Sluyter ML (1973) Use of a UV endonuclease from *Micrococcus luteus* to monitor the progress of DNA repair in UV-irradiated human cells. Mutat Res 19:245–256

Pedrali-Noy GCF, Spadari S (1979) Effect of aphidicolin on viral and human DNA polymerases. Biochem Biophys Res Comm 88:1194–1202

Pedrali-Noy GCF, Spadari S, Ciarrochi G, Pedrini AM, Falaschi A (1973) Two forms of DNA ligase of human cells. Eur J Biochem 39:343–351

Pedrini AM, Grossman L (1983) Purification and characterization of DNase VIII, a $5' \rightarrow 3'$ directed exonuclease from human placental nuclei. J Biol Chem 258:1536–1543

Pegg AE (1978) Enzymatic removal of O^6-methylguanine from DNA by mammalian cell extracts. Biochem Biophys Res Comm 84:166–173

Pegg AE, Balog B (1979) Formation and subsequent excision of O^6-ethylguanine from DNA of rat liver following administration of diethylnitrosamine. Cancer Res 39:5003–5009

Pegg AE, Perry W (1981) Stimulation of transfer of methyl groups from O^6-methylguanine in DNA to protein by rat liver extracts in response to hepatoxins. Carcinogenesis 2:1195–1200

Pegg AE, Roberfroid M, Von Barth C, Foote RS, Mitra S, Bresil H, Likhachev A, Montesano R (1982) Removal of O^6-methylguanine from DNA by human liver fractions. Proc Natl Acad Sci USA 79:5162–5165

Radman M (1974) Phenomenology of an inducible mutagenic DNA repair pathway in *Escherichia coli:* SOS repair hypothesis. In: Prakash L, Sherman F, Miller M, Lawrence C, Tabor HW (eds) Molecular and environmental aspects of mutagenesis. Thomas, Springfield, pp 128–142

Radman M (1975) SOS repair hypothesis: phenomenology of an inducible DNA repair which is accompanied by mutagenesis. In: Hanawalt PC, Setlow RB (eds) Molecular mechanisms for repair of DNA, part A. Plenum, New York, pp 355–367

Radman M (1983) Diversification and conservation of genes by mismatch repair: a case for immunoglobulin genes. In: Friedberg EC, Bridges BA (eds) Cellular responses to DNA damage. Liss, New York, pp 287–298

Radman M, Wagner R (1986) Mismatch repair in *Escherichia coli.* Ann Rev Genet 20:523–538

Rasmunssen RE, Painter RB (1964) Evidence for repair of ultraviolet damaged deoxyribonucleic acid in cultured mammalian cells. Nature 203:1360–1362

Riggs AD, Jones PA (1983) 5-methylcytosine, gene regulation and cancer. Adv Cancer Res 40:1–30

Roberts JJ (1978) The repair of DNA modified by cytotoxic, mutagenic and carcinogenic chemicals. Adv Radiat Biol 7:211–436

Roilides E, Munson PJ, Levine AS, Dixon K (1988) Use of a simian virus 40-based shuttle vector to analyze enhanced mutagenesis in mitomycin C-treated monkey cells. Mol Cell Biol 8:3943–3946

Rubin JS, Prideaux VR, Huntington FW, Dulhanty AM, Whitmore GF, Bernstein A (1985) Molecular cloning and chromosomal localization of DNA sequences associated with a human DNA repair gene. Mol Cell Biol 5:398–405

Samson L, Schwartz JL (1980) Evidence for an adaptive DNA repair pathway in CHO and human skin fibroblast cell lines. Nature 287:861–863

Samson L, Defler B, Waldstein EA (1986) Suppression of human DNA alkylation-repair defects by *Escherichia coli* DNA-repair genes. Proc Natl Acad Sci USA 83:5607–5610

Sancar A, Sancar GB (1988) DNA repair enzymes. Ann Rev Biochem 57:29–67

Sarasin A, Benoit A (1980) Induction of an error-prone mode of DNA repair in UV-irradiated monkey kidney cells. Mutat Res 70:71–81

Sarasin A, Benoit A (1986) Enhanced mutagenesis of UV-irradiated simian virus 40 occurs in mitomycin C-treated host cells only at a low multiplicity of infection. Mol Cell Biol 6:1102–1107

Sarasin AR, Hanawalt PC (1978) Carcinogens enhance survival of UV irradiated simian virus 40 in treated monkey cells: induction of a recovery pathway? Proc Natl Acad Sci USA 75:346–350

Sarasin AR, Hanawalt PC (1980) Replication of ultraviolet-irradiated simian virus 40 in monkey kidney cells. J Mol Biol 138:299–319

Selsky CA, Henson P, Weichselbaum RR, Little JB (1979) Defective reactivation of ultraviolet light-irradiated herpes virus by a Bloom's syndrome fibroblast line. Cancer Res 39:3392–3396

Setlow RB (1978) Repair deficient human disorders and cancer. Nature 27:713–717

Smerdon MJ, Lieberman MW (1978) Nucleosome rearrangements in human chromatin during UV-induced DNA repair synthesis. Proc Natl Acad Sci USA 75:4238–4241

Smith CA (1987) DNA repair in specific sequences in mammalian cells. J Cell Sci [Suppl] 6:225–241

Smith CA, Hanawalt PC (1978) Phage T4 endonuclease V stimulates DNA repair replication in isolated nuclei from ultraviolet irradiated human cells, including *Xeroderma pigmentosum* fibroblasts. Proc Natl Acad Sci USA 75:2598–2602

Söderhäll S, Lindahl T (1973) Two DNA ligase activities from calf thymus. Biochem Biophys Res Comm 53:910–916

Söderhäll S, Lindahl T (1975) Mammalian DNA ligases. Serological evidence for two separate enzymes. J Biol Chem 250:8438–8444

Sparadi S, Ciarrochi G, Falaschi A (1971) Purification and properties of a polynucleotide ligase from human cell cultures. Eur J Biochem 22:75–78

Stryer L (1981) Biochemistry, 2nd edn. Freeman, San Francisco

Tanaka K, Sekiguchi M, Okada Y (1975) Restoration of ultraviolet-induced unscheduled DNA synthesis of *Xeroderma pigmentosum* cells by the concomitant treatment with bacteriophage T4 endonuclease V and HVJ (Sendai virus). Proc Natl Acad Sci USA 72:4071–4085

Teo IA, Karran P (1982) Excision of O^6-methylguanine from DNA by human lymphocytes to remove O^6-methylguanine from DNA. Carcinogenesis 3:923–928

Thompson LH, Carrano AV (1983) Analysis of mammalian cell mutagenesis and DNA repair using in vitro selected CHO cell mutants. In: Friedberg EC, Bridges BA (eds) Cellular responses to DNA damage. Liss, New York, pp 125–143

Thompson LH, Busch DB, Brookman KW, Mooney CL, Glaser DA (1981) Genetic diversity of UV-sensitive DNA repair mutants of Chinese hamster ovary cells. Proc Natl Acad Sci USA 78:3734–3737

van Duin M, de Wit J, Odijk H, Westerveld A, Yasui A, Koken MHM, Hoeijmakers JHJ, Bootsma D (1986) Molecular characterization of the human excision repair gene *ERCC-1*: cDNA cloning and amino acid homology with the yeast DNA repair gene *RAD10*. Cell 44:913–923 van Duin M, van den Tol J, Warmerdam P, Odijk H, Meijer D, Westerveld A, Bootsma D, Hoeijmakers JHJ (1988) Evolution and mutagenesis of the mammalian excision repair gene *ERCC-1*. Nucleic Acids Res 16:5305–5322

Van Zeeland AA, Smith CA, Hanawalt PC (1981) Sensitive determination of pyrimidine dimers in DNA of UV-irradiated mammalian cells. Introduction of T4 endonuclease V into frozen and thawed cells. Mutat Res 82:173–189

Walker GC (1984) Mutagenesis and inducible responses to deoxyribonucleic acid damage in *Escherichia coli*. Microbiol Rev 48:60–93

Waldstein EA, Cao EH, Bender MA, Setlow RB (1982) Abilities of extracts of human lymphocytes to remove O^6-methylguanine from DNA. Mutat Res 95:405–416

Waters R, Regan JD (1976) Recombination of UV induced pyrimidine dimers in human fibroblasts. Biochem Biophys Res Comm 72:803–807

Weber CA, Salazar EP, Stewart SA, Thompson LH (1988) Molecular cloning and biological characterization of a human gene, *ERCC2*, that corrects the nucleotide excision repair defect in CHO UV5 cells. Mol Cell Biol 8:1137–1146

Weigle JJ (1953) Induction of mutations in a bacterial virus. Proc Natl Acad Sci USA 39:628–636

Westerveld A, Hoeijmakers JHJ, van Duin M, de Wit J, Odijk H, Pastink A, Wood RD, Bootsma D (1984) Molecular cloning of a human DNA repair gene. Nature 310:425–428

Wilkins RJ, Hart RW (1974) Preferential DNA repair in human cells. Nature 247:35–36

Williams JL, Cleaver JE (1979) Removal of T4 endonuclease V-sensitive sites from SV40 DNA after exposure to ultraviolet light. Biochim Biophys Acta 562:429–437

Wood RD, Burki HJ (1982) Repair capability and the cellular age response for killing and mutation induction after UV. Mutat Res 95:505–514

Wood RD, Robins P, Lindahl T (1988) Complementation of the *Xeroderma pigmentosum* DNA repair defect in cell free extracts. Cell 53:97–106

Yarosh DB, Rice M, Day RS III, Foote RS, Mitra S (1984) O^6-methylguanine-DNA methyltransferase in human cells. Mutat Res 131:27–36

Zolan ME, Smith CA, Calvin NM, Hanawalt PC (1982a) Rearrangement of mammalian chromatin structure following excision repair. Nature 299:462–464

Zolan ME, Cortopassi GA, Smith CA, Hanawalt PC (1982b) Deficient repair of chemical adducts in α DNA of monkey cells. Cell 28:613–619

Zolan ME, Smith CA, Hanawalt PC (1984) Formation and repair of furocoumarin adducts in α DNA and bulk DNA of monkey cells. Biochemistry 23:63–69

CHAPTER 4

Cancer-Prone Human Disorders with Defects in DNA Repair

A. R. LEHMANN and S. W. DEAN

A. Introduction

Previous chapters have provided ample evidence that the initiation step of the carcinogenic process involves the interaction of an activated carcinogen with cellular DNA. Whether the alteration in the DNA produced by this interaction will eventually result in neoplastic transformation of the damaged cell is dependent on a variety of factors. If the DNA damage is rapidly, efficiently and accurately repaired or processed, the normal status of the cell will be regained. If not, the cell may die or may be mutated, and, in rare instances, it may progress further along the carcinogenic pathway. The crucial role of DNA repair in the avoidance of carcinogenesis is demonstrated by the identification of several autosomal recessive human genetic disorders (DNA-repair syndromes) in which a deficiency in the repair or processing of DNA damage is associated with an enhanced frequency of cancer, as well as with a variety of other clinical abnormalities. The prototype disease in this category is xeroderma pigmentosum (XP) in which a clearly identifiable defect in the repair of UV damage in DNA is associated with a high frequency of UV-induced mutations in cultured cells and with a greatly increased incidence of skin cancer in affected patients. A variety of other disorders have subsequently been shown to involve cellular hypersensitivity to the lethal or clastogenic effects of carcinogenic, DNA-damaging agents. This cellular hypersensitivity is thought to result from defects in the ability of the cells to repair or process damage produced by the appropriate agent. Thus ataxia-telangiectasia (A-T) cells are sensitive to ionising radiation, Fanconi's anaemia (FA) cells to cross-linking agents and Cockayne's syndrome (CS) and trichothiodystrophy (TTD) cells to UV. The relationship between the molecular defects and the clinical symptoms of these diseases is much less apparent than with XP. CS and TTD are not, in fact, cancer-prone disorders, despite having many cellular features in common with XP. As discussed below, TTD cells can have a molecular defect indistinguishable from that in XP despite having very different clinical features. Thus, it may transpire that the link between defective DNA repair and carcinogenesis may be considerably more complex than had originally been suggested from studies on XP.

The DNA repair syndromes are complex multisystem disorders, which implies that DNA repair and processing enzymes may be important not only for the prevention of carcinogenesis but also for many aspects of differentiation and development. In this chapter we review the molecular and cellular properties of cultured cells from patients with these disorders, and we attempt to provide some possible explanations for the relationships of the cellular abnormalities to the clinical symptoms. For earlier reviews, see ARLETT and LEHMANN (1978), SETLOW (1978), FRIEDBERG et al. (1979), and HANAWALT and SARASIN (1986).

B. Xeroderma Pigmentosum

I. Clinical Symptoms

The clinical symptoms of XP have been reviewed in detail by KRAEMER and co-workers (KRAEMER 1980; KRAEMER et al. 1987). Briefly, the earliest symptoms are acute sun sensitivity and the appearance of pigmented macules (freckles) in sun-exposed areas of the skin.

This is followed by telangiectases, atrophy and dryness of the skin and subsequently benign growths, skin carcinomas and melanomas. The median age of onset of the first tumour is 8 years. The eyes are affected in a similar way to the skin. In some, but not all, patients there are neurological abnormalities and mental handicap. The progressive mental deterioration in these patients is attributed to loss of neurons. Onset of neurological abnormalities may occur at varying ages. Death results at an early age from the malignancies.

II. Cellular Sensitivities

Cultured fibroblasts or lymphocytes from XP patients are hypersensitive to the toxic effects of UV light. The degree of hypersensitivity is dependent on the genetic complementation group of the patient's cells (ANDREWS et al. 1978), varying from only slight to some 15-fold hypersensitivity. The hypersensitivity to UV light is associated with similar sensitivity to a variety of chemical carcinogens which have in common the ability to produce bulky adducts in cellular DNA (see Part 1, Chaps. 9 and 10). Examples include polycyclic aromatic hydrocarbon derivatives (MAHER et al. 1977), aromatic amines (HEFLICH et al. 1980) and 4-nitroquinoline-1-oxide derivatives. The hypersensitivity of XP cells to these agents is paralleled by a comparable inability to reactivate infecting viruses damaged by the same class of agents (a process known as host-cell reactivation, HCR). Viruses that have been used in such studies include SV40 (ABRAHAMS and VAN DER EB 1976), adenoviruses (DAY 1974), herpes simplex (COPPEY and MENEZES 1981; SELSKY and GREER 1978) and Epstein-Barr viruses (HENDERSON 1978). XP cells show a reduced ability to carry out HCR of all these viruses after UV irradiation. In extensions of this type of study, cells have been transfected with either viral or plasmid (PROTIC-SABLJIC and KRAEMER 1986) DNA exposed to UV light or carcinogen treatment. Again the ability of XP cells to reactivate the functions expressed by the transfecting DNA was much reduced when compared with normal cells.

III. DNA Repair

1. Excision Repair

The molecular defect in the majority of XP cells is a reduced ability to remove, from cellular DNA, damaged sites produced by UV light or by chemical car-

Table 1. Some examples of DNA-damaging agents which give rise to alterations whose repair is defective in xeroderma pigmentosum

Agent	Test	Complementation group[a]	References
UV (pyrimidine dimers)	Several	A–I	Numerous
UV (6–4 photoproducts)	Immunochemical	A, C, D	MITCHELL et al. (1985)
UV (DNA-protein cross-links)	Alkaline elution	A	FORNACE and KOHN (1976)
Aflatoxin B1	Repair replication	A	SARASIN et al. (1977)
Anthramycin	Adducts, UDS	A	HURLEY et al. (1979)
Aromatic amides	Adducts, survival, repair replication	A	HEFLICH et al. (1980)
Benzo[*a*]pyrene diol-epoxides	Adducts, UDS	A	DAY et al. (1978)
8-Methoxypsoralen/UV	Alkaline elution	A	BREDBERG et al. (1982)
	Adducts	A, D	VUKSANOVIC and CLEAVER (1987)
4-Nitroquinoline-1-oxide	UDS	A–G	ZELLE and BOOTSMA (1980)
		A–D	TANAKA et al. (1980)
Polycyclic hydrocarbons	Survival, mutation	A, C, V	MAHER et al. (1977); AUST et al. (1980)

[a] This column shows the complementation groups in which the defective response has been demonstrated. V, variants; UDS, unscheduled DNA synthesis.

cinogens. Many studies have demonstrated a reduced ability to carry out repair synthesis following damage with the particular agent under study, but in some cases an inability to remove a specific lesion has been demonstrated directly. Some examples of the alterations which are inefficiently repaired in XP cells are given in Table 1. The common feature of this type of DNA damage is the formation of bulky distortions in cellular DNA. Damage produced by ionising radiation or simple alkylating agents which produce much smaller distortions in the DNA is repaired normally in XP cells.

The well-established defect in the repair of UV damage in DNA has provided a method for carrying out genetic complementation analysis. To date ten distinct complementation groups have been identified, of which nine (groups A–I) are defective in excision repair of UV damage. Members of each complementation group have characteristics which in general distinguish them from those of other groups. These are summarised in Table 2. Members of group A have the most pronounced defect in all aspects of the response to UV light, including the degree of hypersensitivity to its toxic effects and the magnitude of the defect in excision repair. In a recent study KANTOR and HULL (1984) used non-dividing cells to follow excision repair over a period of many days. With this system it has been shown that even group A cells are capable of excising a significant proportion of pyrimidine dimers over a long period of time. The principal deficiency is a complete absence of a relatively rapid excision mechanism that is responsible for removing the majority of the damage from normal cells within the first day after irradiation.

Cells in group C do have residual excision repair, at a level of 10%–30% of that in normal cells at early times after irradiation. Recent work has shown that this repair occurs in clusters in the DNA rather than being randomly distributed

Table 2. The complementation groups of xeroderma pigmentosum

Complementation group	Features	UV sensitivity[a]	UDS (% of normal)
A	Most sensitive in every respect Most have severe neurological abnormalities. Complementation is rapid	5	0.5
B	The single member also has CS	4	5–10
C	Repair only of a specific fraction of the DNA associated with the nuclear matrix Patients rarely have neurological abnormalities	3	10–30
D	More sensitive than expected from level of repair Defect in modification of damage Defect in Ap endonuclease Patients usually have neurological abnormalities	4	15–40
E	Mild form. 50% of normal repair. Mild symptoms	2	50
F	Repair slow but prolonged	3	10–20
G	Severe. Little repair	4	0–10
H	Single member also has CS	4	30
I	Single member[b]	4	15
Variant		1–2	100

[a] Colony-forming ability expressed on a crude scale of 1 (normal) to 5 (most sensitive). Exceptions exist within the groups.
[b] It is now known that this cell strain actually belongs in group C. Group I no longer exists! (BOOTSMA et al. 1989).

throughout the genome. Evidence suggests that DNA is attached to the nuclear matrix at points along its length and that repair in XP-C cells is confined to these attachment points (MULLENDERS et al. 1986).

Cells in group D also have substantial excision repair, repair synthesis levels being 15%–40% of that in normal cells, but they are nevertheless considerably more sensitive to the lethal effects of UV than are group C cells. This has remained a mystery for many years. Recently, PATERSON and co-workers (1987) have shown that in XP-D and XP-A cells some 15%–20% of the pyrimidine dimers which remain in cellular DNA are converted into an altered form in which the phosphodiester bond between the two pyrimidines of the cyclobutane dimer is broken. They suggested that this structure was an aborted intermediate of the normal excision repair process (WEINFELD et al. 1986).

Cells in group F showed unusual kinetics of excision repair (ZELLE et al. 1980; FUJIWARA et al. 1985). Although the rate of repair synthesis shortly after treatment is rather low (10%–20% of normal), it is maintained for longer periods, so that quite a substantial portion of the damage is removed by 1 day after UV exposure.

The single members of groups B and H both have the clinical and cellular features of CS in addition to those of XP. This will be discussed below.

Several groups have provided convincing evidence that the defect in excision repair in XP cells lies in or prior to the initial incision step of the repair process

Table 3. Enzyme activities reported to be deficient in xeroderma pigmentosum

Enzyme activity	Complementation groups showing defect[a]	References
Dimer excision in chromatin	A	MORTELMANS et al. (1976)
	A, C, G (not D)	KANO and FUJIWARA (1983)
Endonuclease for OsO_4 damage	A, C	HELLAND et al. (1984)
Single-strand binding protein	A (not D)	KUHNLEIN et al. (1983)
Anthramycin endonuclease	A	LEE et al. (1982)
Altered nucleases	A	OKORODUDU et al. (1982)
AP endonuclease	D	KUHNLEIN et al. (1978)
Catalase	C, D, H	VUILLAUME et al. (1986)
Photoreactivating enzyme	A–E	SUTHERLAND et al. (1975)
Alkylation damage-endonuclease activity	A	WITTE and THIELMANN (1979)

[a] This column shows the complementation groups in which the defect has been observed. "not D" implies that the activity has been investigated in group D, but no defect was found. AP, apurinic endonuclease.

(see "correction of the defect" below). Attempts to pinpoint the precise nature of the enzymological defect in XP cells have so far, however, met with limited success. Some enzyme abnormalities have been detected, but these have not been definitively linked to the cellular defects. These are summarised in Table 3. An important observation made by MORTELMANS et al. (1976) and confirmed more recently by KANO and FUJIWARA (1983) was that extracts of XP cells from some complementation groups were able to excise pyrimidine dimers from naked DNA but not from the DNA in chromatin. This suggested that the defect in these XP cells might lie in the ability of repair enzymes to recognize damage in chromatin or to gain access to the damaged DNA rather than in the actual removal process. Unfortunately, this work has not been pursued further. There have also been reports of altered endonucleases or binding proteins in some XP cells. As yet these remain as isolated reports awaiting confirmation. Recently WOOD et al. (1988) have developed a human cell free system which can carry out excision-repair of UV damaged plasmids. This system is defective in XP cells and therefore provides a means to characterize the XP defects at the enzymological level. A intriguing finding is the deficiency in catalase activity in several XP lines (VUILLAUME et al. 1986). It might be expected that catalase deficiency would result in hypersensitivity to oxidative damage, but there is no evidence to suggest that this is indeed the case with XP cells.

2. Correction of the Defect

Alternative approaches to identifying the defects in XP cells have involved attempts to correct the defect with either protein or DNA from other sources. Studies by TANAKA and co-workers (1975) and DE JONGE et al. (1985) showed that introduction of the T4 or *M. luteus* UV-endonuclease preparations (either by cell permeabilisation or by micro-injection) were able to correct the defect in all the XP complementation groups. These enzymes are specifically able to nick DNA at pyrimidine dimer sites by first carrying out a glycosylase reaction, destroying the glycosylic bond of the 5′ pyrimidine of the dimer and then effecting an endonucleolytic nick in the DNA between the two pyrimidines. The net result is the production of an "unhooked" pyrimidine dimer with a nick 5′ to the attached pyrimidine. Thus the gene products of all nine complementation groups appear to be required to reach this stage during excision repair in intact human cells. The human repair enzyme complex is thus a much more sophisticated machine than the T4 or *M. luteus* enzymes. In contrast to these results, micro-injection of XP cells with a mixture of the uvrA,B,C and D proteins from *Escherichia coli* was not able to correct the defect in complementation groups A or C (ZWETSLOOT et al. 1986) even though these proteins are capable of excising pyrimidine dimers from DNA both in vitro and in *E. coli* cells.

Using micro-injection of crude extracts from HeLa cells, VERMEULEN et al. (1986) were able to correct the defect in all XP complementation groups. This correcting activity was shown to be a high molecular weight protein. Similar observations were reported for correction of XP-A cells by YAMAIZUMI et al. (1986) – the "Group A correcting factors" were proteins of 160 K and 190 K.

Many laboratories have attempted to correct the defect in XP-A cells by transfection with DNA from normal cells. An early report claimed successful correction (TAKANO et al. 1982), but subsequent efforts have been unsuccessful (GANESAN et al. 1983; ROYER-POKORA and HASELTINE 1984; SCHULTZ et al. 1985; LEHMANN 1985). Very recently, using transfection techniques TANAKA et al. (1989) have succeeded in cloning the mouse gene which corrects the XP-A defect. On the other hand micro-injection of XP cells with poly(A)$^+$ RNA from normal cells was able to correct the defect in XP-A and XP-G cells (LEGERSKI et al. 1984). The correcting RNA was found in the 11*S* and 13*S* fractions, respectively. Partial restoration of normal UV sensitivity has also been obtained by microcell fusion with mouse/human hybrids containing single human chromosomes. With this technique it may prove possible to identify the human chromosomes which are responsible for the correction of the XP defect (SCHULTZ et al. 1987).

IV. XP Variants

The tenth group of XP cells are the so-called XP variants, which have near normal sensitivity to the lethal effects of UV irradiation and normal excision repair. These cells have a defect in an alternative repair process, variously known as daughter-strand repair, post-replication repair or recovery, or bypass repair. This is an ill-characterised process whereby, during replication, cells are able to tolerate damage remaining in their DNA. XP variants have a defect, which is manifested as a reduced molecular weight of newly synthesised DNA in UV-irradiated cells and a delay in the production of intact high molecular weight DNA strands following UV irradiation. The exact nature of the defect is not clear, but only one complementation group has been identified so far (JASPERS et al. 1981). Excision-defective XP cells also have a defect in this process, but it is less marked than in XP variants.

V. Mutagenesis

Following UV irradiation or treatment with carcinogens, all XP cells tested, both of the excision-defective and variant type, have greatly elevated frequencies of induced mutations (e.g. MAHER et al. 1977; GLOVER et al. 1979; ARLETT et al. 1980). The extent of the hypermutability is at least as great, and sometimes considerably greater (ARLETT and HARCOURT 1982; GROSOVSKY and LITTLE 1983; TATSUMI et al. 1987), than the degree of hypersensitivity. This hypermutability is found at several different loci, and it is paralleled by an increased sensitivity to UV-induced transformation to anchorage independence (MAHER et al. 1982; MCCORMICK et al. 1986).

VI. UV-Inducible Functions

Exposure of cells to UV light results in the induction of a number of activities including the synthesis of plasminogen activator (MISKIN and BEN-ISHAI 1981) as well as a series of other new proteins (SCHORPP et al. 1984). In XP cells the induced level of plasminogen activator was much higher than in normal cells. If irradiated cells are infected with UV-irradiated virus, the survival of the virus is higher than with infection of non-irradiated cells (enhanced reactivation), and more mutations are induced in the resulting viruses (enhanced mutagenesis). In XP cells these functions are induced by much lower doses of UV than are required for corresponding induction in normal cells, presumably reflecting the persistence, in XP cells, of unexcised damage. In XP variants the actual amount of enhanced reactivation of herpes simplex virus was some tenfold higher than in normal or excision-deficient XP cells according to COPPEY and MENEZES (1981), but this observation was not reproduced by ABRAHAMS et al. (1984). The induction responses of XP cells are shown in Table 4.

Table 4. Inducible activities showing altered responses in xeroderma pigmentosum or Cockayne's syndrome (CS)

Inducible activity	Complementation groups	References
Plasminogen activator	A, C, D, variant	MISKIN and BEN-ISHAI (1981)
Enhanced reactivation of herpes simplex virus	A, C, D and CS	COPPEY and MENEZES (1981)
	A, C, D	ABRAHAMS et al. (1984)
Proteins	CS	SCHORPP et al. (1984)
Recombination in herpes simplex virus	A	DASGUPTA and SUMMERS (1980)

VII. Relationship to Clinical Symptoms

The hypermutability of XP cells provides strong evidence for the hypothesis that defective DNA repair in the skin of XP cells results in an increased frequency of mutations in skin cells, which in turn can account for the elevated incidence of skin cancers associated with XP. This is supported by recent studies showing that activation of *ras* oncogenes can result from a specific base change which has been shown to be a direct consequence of interaction of the DNA with carcinogens. This basic scheme is undoubtedly correct, but the actual relationships between repair, mutagenesis and carcinogenesis are much more complicated, as illustrated by the other human repair-deficient syndromes discussed below. Furthermore, other factors undoubtedly contribute to the increased frequency of skin cancer in XP. It was originally suggested by BRIDGES (1981) and subsequently shown by MORISON et al. (1985) that UV-induced immune suppression in the skin cells was a contributory factor to the production of skin cancers and that XP individuals were much more sensitive to UV-induced immune suppression in the skin than were normal controls. NORRIS et al. (1989), on the other hand, found that the adaptive immune response was normal in 5 XP patients, but these patients had a specific defect in Natural Killer cell activity.

Apart from the diverse skin symptoms which can all be attributed, directly or indirectly, to sunlight-induced damage of cellular DNA, the other principal symptom of a minority of XP patients is progressive neurological degeneration caused by a generalised loss of neurons. Robbins and co-workers have suggested that this neuronal death results from the accumulation of DNA damage in neuronal DNA subsequent to exposure to endogenous carcinogens (e.g. see ROBBINS 1978). ROBBINS was able to correlate the sensitivity of XP cells to UV (and by implication to carcinogens) with the age of onset of neurological abnormalities. Thus, XP-A cells were most sensitive, and in general there were associated neurological abnormalities before the age of 7. XP-D cells are not quite so sensitive as XP-A cells, and neurological abnormalities appear somewhat later. Patients from other complementation groups do not usually develop neurological abnormalities, and their cells are less sensitive to the lethal effects of UV and chemical carcinogens.

C. Cockayne's Syndrome

I. Clinical Symptoms

The clinical features of CS have been reviewed by GUZZETTA (1972) and include dwarfism with severe physical and mental retardation. Wasting (cachexia) results from loss of cutaneous fat, and together with microcephaly, beaked nose and sunken eyes this gives a characteristic senile appearance. There is, in addition, progressive retinal and neurological degeneration, skeletal abnormalities and, as in XP, extreme sun sensitivity, although the skin lesions are different from those in XP, and there are no reports of skin cancer being associated with CS.

II. Cellular and Molecular Studies

As with XP, the solar sensitivity is reflected as cellular hypersensitivity of cultured fibroblasts and lymphoblastoid cells to the lethal effects of UV irradiation and chemical carcinogens (e.g. WADE and CHU 1979) and as a deficiency in the ability of CS cells to reactivate irradiated viruses. Unlike XP, this hypersensitivity cannot be attributed to gross defects in either excision repair or daughter-strand repair, as these appear to be normal by conventional assays. Nevertheless, following exposure to UV light, the rates both of replicative DNA synthesis and of RNA synthesis, which are initially depressed in all cells, recover in normal cells but remain depressed in CS cells (and also in most XP cell lines) (LEHMANN et al. 1979; MAYNE and LEHMANN 1982). The failure of RNA synthesis to recover following UV irradiation represents a biochemical abnormality in CS which is manifest very soon after irradiation. It provides indirect evidence for a defect in CS of some kind of DNA-repair or DNA-processing event. It is now known that UV damage in actively transcribing regions of DNA is repaired preferentially, and it has been shown that CS cells are deficient in this preferential repair process, while remaining proficient in the slower process which removes UV damage from the bulk of the DNA (MAYNE et al., 1988).

The defect in RNA synthesis has provided a means for carrying out complementation tests on CS cells, and three complementation groups have been identified (LEHMANN 1982a). Two patients with XP, who are the single members of complementation groups B (ROBBINS et al. 1974) and H (MOSHELL et al. 1983), also have the clinical features of CS. As discussed previously (LEHMANN 1987), this implies that the two pathways which give rise to the normal phenotype with respect to XP and CS may contain at least two common steps. A mutation in either of the genes controlling these two steps will result in symptoms of both disorders.

The limited amount of mutagenesis data that exists for CS suggests that the fibroblasts are hypermutable by UV light (ARLETT and HARCOURT 1982).

III. Relationship to Clinical Symptoms

The neat and coherent relationship between DNA damage, cellular mutagenesis and UV carcinogenesis in XP is confounded by the lack of such relationships in CS. The cellular features of the two disorders are very similar, at least in fibroblasts and lymphoblastoid cells, yet the molecular defect is manifestly different. Furthermore, whereas the cellular features of XP give rise to readily explainable skin lesions including multiple skin cancers, such lesions are absent in CS in which sun sensitivity is but one of an array of different dysfunctions. It is difficult even to speculate about the relationship between a putative DNA-repair deficiency and clinical symptoms such as cachectic dwarfism and skeletal abnormalities.

D. Trichothiodystrophy

The recent discovery of a DNA-repair defect in another disorder, trichothiodystrophy (TTD), has added further complexity to the relationship between DNA repair and carcinogenesis. Patients with this disorder have sulphur-deficient, brittle hair. The condition is accompanied by physical and mental handicaps of varying severity, scaling of the skin (ichthyosis), and about half the patients show severe photosensitivity. The symptoms are very heterogeneous, and patients with ichthyosis from birth appear not to be photosensitive, whereas those in whom the ichthyosis develops later are photosensitive.

Cells from photosensitive patients are, like XP cells, deficient in excision-repair of UV damage, and recently STEFANINI and co-workers (1986) have shown that this deficiency is not complemented by cells from XP complementation group D. The simplest interpretation of these data is that different mutations in the same gene (XP-D) can give rise to the symptoms either of XP or of the quite different disorder TTD. This is an intriguing paradox, since, although there is some overlap of symptoms, many features of XP are not found in TTD, in particular the skin lesions and the multiple skin cancers, and conversely many symptoms of TTD are not associated with XP, e.g. the ichthyosis and the brittle hair. Furthermore, the group of patients with TTD who are not sun-sensitive but who otherwise have similar symptoms to the photosensitive TTD patients show no cellular abnormalities in their response to UV light (LEHMANN 1987). A two-mutation hypothesis to resolve this paradox has been proposed by LEHMANN and NORRIS (1989).

E. Ataxia-Telangiectasia

I. Clinical Symptoms

A-T is a multisystem disorder characterised by progressive cerebellar ataxia, oculocutaneous telangiectases, elevated levels of alpha-fetoprotein, deficiencies in both cellular and humoral immune systems (WALDMANN et al. 1983a) and a greatly increased incidence of cancer (reviewed by SEDGWICK and BODER 1972; WALDMANN et al. 1983b; BODER 1985). Some 80% of the tumours are found in the reticuloendothelial system, the other 20% being carcinomas in various organs. Death results from progressive neurological degeneration, from pulmonary infections or from the tumours. The sensitivity of A-T patients to ionising radiation was revealed when three patients showed an extreme reaction to conventional doses of radiotherapy. Many studies over the past 12 years have attempted to pinpoint the molecular defect in A-T cells, but as yet this goal has not been achieved. Much of the clinical, cellular and molecular work on A-T has been reported in two books (BRIDGES and HARNDEN 1982; GATTI and SWIFT 1985) and is summarised in a recent review (MCKINNON 1987).

II. Cellular Sensitivities

The sensitivity of these patients to ionising radiation is reflected at the cellular level as hypersensitivity of cultured A-T fibroblasts, lymphoblastoid cells and T lymphocytes to the lethal effects of various types of ionising radiation. Cross-sensitivity to chemical agents is much more restricted than in XP, hypersensitivity being limited to those agents such as bleomycin and neocarzinostatin which damage DNA via a free radical mechanism (SHILOH et al. 1983). The sensitivity to UV and most carcinogens is close to normal. The increased sensitivity of A-T cells to ionising radiation can be amplified in experiments which measure recovery from potentially lethal damage (PLD). In conventional survival assays, cells are plated at low density immediately after irradiation, and their colony-forming ability is measured. In PLD experiments non-dividing cells are irradiated and then maintained in a quiescent state for varying periods of time prior to plating. During this time, PLD is repaired, and the survival rate of the cells increases as a consequence. This increase in survival rate is seen in normal cells but not in A-T cells. This implies that A-T cells are deficient in the ability to repair PLD.

If the cellular sensitivity of A-T cells is attributable to defective DNA repair, it may also be manifest as a decrease in the ability to carry out HCR of irradiated viruses. Normal HCR was, however, found with A-T cells infected with irradiated adenovirus (RAINBOW 1978), herpes simplex (HENDERSON and BASILIO 1983) and parvovirus H1 (HILGERS et al. 1987). These results imply that, if there is a defect in DNA repair in A-T cells, it is specifically related to repair of damaged chromatin rather than to simple DNA structures.

III. DNA Repair

Ionising radiation produces single- and double-strand breaks in cellular DNA as well as a heterogeneous mixture of types of base damage. Repair of many of these types of damage has been compared in normal and A-T cells. In most cases no defect has been detected in the A-T cells (reviewed by HUANG and SHERIDAN 1981; LEHMANN 1982c). For example, in most studies the rates and extents of rejoining of single- and double-strand breaks were similar in normal and A-T cells. In cases in which indications of a defect have been obtained (e.g. COQUERELLE et al. 1987), this has not been reproduced in other laboratories. Thus for example, an early finding of PATERSON et al. (1976) suggested that a class of A-T cell lines was deficient in the removal of certain types of base damage, which could be detected by an enzymatic assay. These observations have not, however, been reproduced in more recent work from another laboratory, in which lower radiation doses were used (FORNACE et al. 1986).

The activities of a number of DNA repair enzymes have also been measured, and again reports of observed differences have proved controversial and difficult to reproduce (reviewed in HUANG and SHERIDAN 1981). A particularly interesting finding is the report by INOUE et al. (1981) that A-T cell extracts are deficient in an activity which "cleans up" radiation-induced strand breaks by converting them into priming sites for DNA polymerase I. This activity was present in normal cell extracts but was much reduced in several A-T extracts. Again, however, these findings have not yet been satisfactorily reproduced in other laboratories.

Recently, a different approach has been used to study the defect in A-T cells using plasmids digested with restriction enzymes (COX et al. 1986). Plasmid pSV2gpt was digested with enzymes which cut the plasmid either in, or close to, the *gpt* gene. These plasmids were then transfected into normal or A-T cell lines and the activity of the *gpt* gene measured. A functional *gpt* gene will only be produced if the double-strand breaks created in the plasmid by the restriction enzymes are accurately repaired in the cells. When an A-T cell line was compared with several normal cell lines, the A-T cell line appeared to be deficient in the ability to restore the activity of the *gpt* gene. This was interpreted as a deficiency in the accuracy of rejoining of double-strand breaks. These exciting results have so far been obtained with only one A-T cell line and require further confirmation. Similar results were not obtained in analogous experiments using a different plasmid (GREEN and LOWE 1987).

IV. DNA Synthesis and the Cell Cycle

Ionising radiation and also bleomycin cause an inhibition of the initiation of DNA synthesis and a delay in the G2 phase of the cell cycle. In A-T cells the inhibition of DNA synthesis is reduced relative to normal cells (i.e. the rate of synthesis remains close to the level in undamaged cells). This aberrant DNA synthesis has been observed in some thirty A-T cell lines, and it has provided a method for detecting genetic complementation following cell fusion. Four complementation groups (AB, C-E) have been identified (JASPERS et al., 1988). The finding in every case of A-T of aberrant DNA synthesis associated with gamma-ray hypersensitivity suggests a causal relationship between these two phenomena. This hypothesis is also supported by the finding of an abnormal response of DNA synthesis in A-T cells to bleomycin (CRAMER and PAINTER 1981) and neocarzinostatin (SHILOH and BECKER 1982), agents to which A-T cells show specific hypersensitivity, whereas the response to other carcinogenic agents is normal for both cell killing (ARLETT et al. 1982) and DNA synthesis (JASPERS et al. 1982). Several lines of evidence do not, however, support the contention that reduced inhibition of DNA synthesis is the basic defect in A-T cells. These include (1) the deficiency in the repair of PLD (see above) which is observed in non-dividing cells that are not synthesising DNA, (2) increased chromosomal breakage in cells which are in the G1 phase of the cell cycle (CORNFORTH and BEDFORD 1985), and (3) the isolation of a radioresistant derivative of an A-T cell line which has regained normal sensitivity to the lethal effects of ionising radiation but has retained most of the aberrant DNA synthesis characteristic of A-T cells (LEHMANN et al. 1986).

Recently, MOHAMED and LAVIN (1986) developed a permeabilised cell system which retained the aberrant DNA synthesis response of A-T cells. They found that introduction of A-T cell extracts into permeabilised normal cells prevented the radiation-induced inhibition of DNA synthesis. This finding opens up a route for purifying this "A-T factor". Some caution must, however, be exercised in the interpretation of these results, since they imply that A-T should be a dominant disorder whereas, from both clinical and cellular genetic studies, it is known to be recessive.

The picture that emerges from these studies is that, following irradiation of normal cells, damage in DNA triggers some type of signal which causes DNA synthesis and to be slowed down. This control process allows the cells time to recover from some of the deleterious effects of the radiation damage. In A-T cells, this control process does not function. The failure in A-T cells to transmit a signal subsequent to DNA damage may be related to the inability of A-T cells to carry out the process of "enhanced reactivation". If UV- or gamma- (or X)-irradiated viruses are used to infect cells which themselves have been previously irradiated, the survival of the irradiated virus is higher when infecting damaged cells than undamaged cells. This phenomenon, termed enhanced reactivation, is deficient in A-T cells when the infecting virus is adenovirus 2 (JEEVES and RAINBOW 1986) or single-stranded parvovirus H1 (HILGERS et al. 1987). This may be a further example of failure of a damage-dependent signal to be transmitted in A-T cells.

V. Cytogenetics

A-T is one of three "chromosome-breakage syndromes" (HECHT and MCCAW 1977), the others being FA and Bloom's syndrome. Lymphocytes and cultured fibroblasts contain elevated levels of chromosome aberrations, and, in addition, lymphocytes but not fibroblasts contain characteristic rearrangements which nearly always involve two of four specific breakpoints located on chromosomes 7 or 14 at the locations 7p13, 7q34, 14q11–12, 14q32 (TAYLOR 1982). These breakpoints are in the genes for the T-cell receptor gamma-, beta- and alpha-chains and close to the gene for the immunoglobulin IgH chain, respectively. Recent experiments have shown that, in several tumours from A-T patients with clones of cells containing rearrangements including chromosome 14, the breakpoint at 14q11–12 is actually in the T-cell receptor alpha-chain gene (e.g. see KENNAUGH et al. 1986). We may interpret these observations as implying that during the normal rearrangements of the T-cell receptor genes which occur during development of the immune response, errors are made at relatively high frequency in A-T cells, resulting in aberrant chromosome rearrangements.

A-T blood and medium from cultured A-T fibroblasts, but not from A-T lymphoblastoid cells, contain a "clastogenic factor" which induces chromosome aberrations in normal cells cultured in its presence. The clastogenic factor from A-T fibroblasts is a small peptide (molecular weight 500–1000) (SHAHAM and BECKER 1981).

The induction of chromosome aberrations in A-T lymphocytes (BENDER et al. 1985) or fibroblasts by ionising radiation or bleomycin (SHAHAM et al. 1983) is greatly enhanced when compared with normal cells. Furthermore, whereas irradiation of normal cells in the G0 or G1 phase results only in chromosome-type aberrations, similar treatment of A-T cells results in both chromosome and chromatid-type damage (e.g. TAYLOR et al. 1976). This observation is interpreted as a deficiency in A-T cells in the ability to effect the complete and accurate repair of some types of single-strand damage.

The hypersensitivity of A-T cells to the clastogenic action of ionising radiation has provided another assay for genetic complementation, and four groups have been found among seven Australian A-T patients (CHEN et al. 1984). As yet, the relationship between these complementation groups and those of the European and American patients has not yet been established.

VI. Mutagenesis

The measurement of mutations in A-T fibroblasts has proven to be extremely difficult, for reasons which are essentially technical. ARLETT and HARCOURT (1982) found that A-T primary fibroblasts were not mutable by ionising radiation. Mutations could, however, be induced in SV40-transformed A-T fibroblasts and EBV-transformed lymphoblastoid lines (TATSUMI and TAKEBE 1984; ARLETT, personal communication). The induced mutation frequency was lower than that in corresponding cell types from normal donors. Thus, in contrast to the enhanced UV-mutability of XP cells, A-T cells are not hypermutable by gamma-irradiation, and they may, in fact, be hypomutable, implying that they may have a deficiency in an error-prone repair process. It is now possible to measure the frequency of mutants present in the circulating lymphocytes of people. The mutant frequency in several A-T patients is much higher than in normal individuals (COLE et al. 1989).

VII. Growth Properties

1. Growth Rate and Lifespan

There have been several reports of the abnormal growth and structural properties of A-T cells. These include observations that A-T fibroblasts grew more slowly than normal cells (ELMORE and SWIFT 1976), that lymphocytes from A-T patients were difficult to stimulate with mitogens or to transform with Epstein-Barr virus (HENDERSON and BASILIO 1983) and that A-T lymphoblastoid cell lines had a longer cell cycle time than normal lines (COHEN and SIMPSON 1980). This last property could be attributed to a twofold increase in the length of the S phase of the cell cycle. THOMPSON and HOLLIDAY (1983) have suggested that the lifespan in culture of A-T cells is shorter than that of normal cells. Cultured A-T fibroblasts have been used in this unit for over 10 years by C. F. ARLETT and S. A. HARCOURT, and although the growth properties have not been systematically investigated, it is their impression (personal communication) that some A-T cells do demonstrate slow growth while others have growth features and senescence which are indistinguishable from normal cells. Similar findings were reported by JASPERS (1985).

A careful study by SHILOH et al. (1982) showed that, under optimum culture conditions, the doubling times of most A-T fibroblast cultures was similar to that of normal cells, and the plating efficiency of the cells was also similar. The A-T cells did however senesce somewhat earlier than the normal cultures. These workers also found that A-T cells were much more dependent than normal cells on growth-promoting factors in the serum used to supplement the growth medium.

2. Cellular Proteins

Altered properties of cellular proteins in A-T have been described by MCKINNON and BURGOYNE (1984, 1985a, b). At high cell density in several (but not all) A-T lymphoblastoid lines, the levels of 43 KD actin were reduced and replaced by a 37 KD protein which was probably a degradation product. This was not observed in fibroblasts. The same authors suggested that the fibroblasts had an unusual cell morphology with altered microfilament stress fibre patterns (McKinnon and Burgoyne 1985a). Related observations on the structure of the cytoskeleton and A-T fibroblasts were reported by Becker (1986). Murnane and Painter (1983) also reported some altered protein synthesis patterns in some A-T fibroblasts, with the overproduction of several secreted proteins including fibronectin. In contrast Lavin and Seymour (1984) found a deficiency in fibronectin production in A-T lymphoblastoid lines.

The conclusion from these diverse studies is that A-T cells may have altered cell surface morphology, but the reported anomalies do not appear to be found consistently in all cell lines. The relationship of these abnormalities to the aberrant response to radiation remains to be determined.

A recent report suggests that A-T cells may have an abnormality in sulphur metabolism. After treatment of cells with diethyl maleate to deplete intracellular glutathione, normal levels of glutathione were rapidly restored in normal cells, whereas recovery was much slower in A-T cells (Meredith and Dodson 1987). Unfortunately, we have not been able to confirm these findings (Dean et al. 1988).

VIII. Relationship to Clinical Symptoms

An attempt to relate the molecular and cellular defects in A-T, which are in many cases controversial, to the very diverse clinical symptoms of the disease presents a formidable task. Many authors have speculated on these relationships (e.g. Cox et al. 1986; Bridges and Harnden 1982; Gatti and Swift 1985; Becker 1986); there is as yet, however, a paucity of hard data. We shall briefly review the various hypotheses.

At the cellular level the evidence for hypersensitivity of A-T cells to the lethal and clastogenic effects of ionising radiation and bleomycin is incontrovertible, and this hypersensitivity is associated with aberrant control of DNA synthesis. At the molecular level there is some evidence for some type of aberration in joining of strand breaks, but this remains to be proven, and for a failure in some kind of signal transmission to control DNA synthesis. In undamaged cells there is evidence for altered growth properties which can be attributed to altered cellular morphology. The precise nature of these anomalies is as yet unclear. A-T cells contain, in addition, a high frequency of chromosome aberrations, and T lymphocytes contain chromosome rearrangements with break-points close to the genes for the T-cell receptor proteins.

The basic defect in A-T cells could reside in an enzyme, possibly of a regulatory nature, which is involved in the processing of DNA breaks, such that in A-T some breaks may be incorrectly rejoined and the signal to slow down DNA synthesis not produced. The defective enzyme responsible for this aberrant rejoining may, in lymphocytes, be involved, as its normal function, in effecting the breakage and rejoining events in rearrangements of immunoglobulin and T-cell receptor genes, which occur during development of the immune response. Inaccurate rejoining in T cells could, as suggested above, result in the characteristic chromosome rearrangement seen in A-T cells, and this may in turn be responsible for some of the immune deficiencies and for the development of lymphoid malignancies in A-T.

The relationship between defective DNA repair and neurological abnormalities has been discussed above in relation to XP. Similar models could presumably be invoked for A-T. As discussed above, effective functional repair of neuronal DNA may be required for maintenance of the integrity of the central nervous system. Thus, neuronal death following exposure to endogenous radiomimetic agents may occur much more rapidly in A-T individuals. What is unexplained, however, is the quite different specificities of the neurological abnormalities in XP and A-T. The aetiology of the telangiectases in A-T is perhaps the most obscure feature. Recently, however, BECKER (1986) has reported that A-T cell lines secreted a "telangiectasia factor" which stimulated bovine aorta cells to develop into large colonies. This may be related to the altered cellular growth and morphology discussed above. The relationship of this factor to the other cellular features of A-T is unknown. The increased frequency of lymphoid tumours in A-T patients may be partly attributable to the immunodeficiency associated with the disease. Some 20% of tumours found in A-T patients are, however, carcinomas, and similar solid tumours are never found in patients with other immunodeficiency diseases. It is thus necessary to seek some other contributory factor. The chromosomal instability associated with A-T may be an important factor in disposing the patients towards cancer. It is well-known that tumour cells are usually associated with chromosomal alterations, although the cause-effect relationship has not yet been established.

F. Fanconi's Anaemia

I. Clinical Symptoms

The wide spectrum of clinical abnormalities in FA has been comprehensively reviewed (BEARD 1976; GLANZ and FRASER 1982; AUERBACH 1984). The general characteristics include a progressive pancytopenia, microcephaly, abnormal skin pigmentation, short stature, genital hypoplasia and skeletal defects, particularly of the upper limbs, although the last occurs in only about 40% of cases.

Patients are also prone to early development of malignant diseases such as hepatoma and squamous cell carcinoma at various sites. Marrow failure tends to occur between 5 and 10 years of age, nearly all patients developing aplastic anaemia with 15%–20% also developing leukaemia. Death is generally early and due to haemorrhage or infection.

II. Cellular Sensitivities

As with many cellular parameters, it is difficult to summarise accurately the cell killing effects of cytotoxic agents in FA cells since a variety of cell lines and treatment protocols have been used. FA fibroblasts and lymphoblasts are, however, very consistently hypersensitive to the toxic effects of bifunctional alkylating agents. Diagnostic of the FA phenotype, hypersensitivity to mitomycin-C (MMC) can be anything from twofold (WEKSBERG et al. 1979; NAGASAWA and LITTLE 1983) to greater than 300-fold (FUJIWARA et al. 1984) depending on which

FA lines are tested and which normal cell lines are used for comparison. Generally, one would expect to see between 5- and 30-fold hypersensitivity to MMC (ISHIDA and BUCHWALD 1982; POLL et al. 1984; FUJIWARA et al. 1977). FA cells are also approximately 10-fold hypersensitive to nitrogen mustard (NM) (SASAKI 1975; ISHIDA and BUCHWALD 1982; DEAN and FOX 1983) and a little more sensitive, about 20-fold, to diepoxybutane (ISHIDA and BUCHWALD 1982; FUJIWARA et al. 1984).

Hypersensitivity has also been demonstrated to psoralen plus UV (WEKSBERG et al. 1979; WUNDER and FLEISCHER-REISCHMANN 1983; POLL et al. 1984; FUJIWARA et al. 1984; VIJAYALAXMI et al. 1985), cisplatinum (ISHIDA and BUCHWALD 1982; FUJIWARA et al. 1984; PLOOY et al. 1985), cyclophosphamide and bifunctional nitrosoureas (ISHIDA and BUCHWALD 1982). No such hypersensitivity has been seen with other drugs such as monofunctional alkylating agents (WEKSBERG et al. 1979; HENDERSON and RIBECKY 1980; ISHIDA and BUCHWALD 1982).

The cytotoxic effects of physical agents are less consistent. In both lymphoblastoid and fibroblast cultures, sensitivity to X-ray or gamma irradiation has been found to lie in the normal range (SASAKI 1975; FUJIWARA et al. 1977; FORNACE et al. 1979; WEISCHELBAUM et al. 1980; HENDERSON and RIBECKY 1980), though ARLETT and HARCOURT (1980) reported a 1.5 to 2.3-fold hypersensitivity of one cell strain to gamma irradiation. The majority of studies using UV irradiation suggest normal sensitivity (FUJIWARA et al. 1977; LEHMANN et al. 1977; HENDERSON and RIBECKY 1980; SMITH and PATERSON 1981).

To summarise, FA cells are markedly hypersensitive to cross-linking agents and show normal, or near normal, sensitivity to monofunctional drugs and to X-ray or UV radiation.

III. Cytogenetics

The spontaneous frequency of chromosome aberrations in lymphocytes from FA patients is up to 15-fold higher than that in normal cells (AUERBACH et al. 1981; BERGER et al. 1980; COHEN et al. 1982). Treatment with cross-linking agents causes marked chromosome damage in FA cells at concentrations of the drug which induce far fewer aberrations in normal cells (COHEN et al. 1982; SASAKI and TONOMURA 1973; BERGER et al. 1980; MIURA et al. 1983). Indeed, MMC- and particularly diepoxybutane-induced clastogenic stress are used as indicators of the FA phenotype in amniotic material for prenatal diagnosis of the disease (VOSS et al. 1981; AUERBACH et al. 1981; CERVENKA et al. 1981).

Interestingly, gamma- or X-ray-induced chromosome aberrations appear to be more frequent in FA than in normal cells (HIGURASHI and CONEN 1971; BIGELOW et al. 1979; SASAKI and TONOMURA 1973) in contrast to the apparently normal sensitivity to the lethal effects of X-rays. Increased chromosome aberrations have also been reported for UV, decarbamoyl mitomycin C, psoralen plus UV, actinomycin D (SASAKI and TONOMURA 1973) and cisplatinum (KWEE et al. 1983; POLL et al. 1982). A normal, or only very slight, sensitivity to induced chromosome damage is observed after monofunctional alkylating agents (LATT et al. 1975; HEDDLE et al. 1978; SASAKI and TONOMURA 1973). There is a good correlation therefore between the induction of chromosome aberrations and cell killing by many of the cytotoxic agents tested. However, the increased clastogenicity of agents such as X-rays is inconsistent with this relationship.

IV. DNA Synthesis and the Cell Cycle

DNA synthesis in FA cells is slower than in normal cells (DUTRILLAUX et al. 1982; SASAKI 1975) yet exhibits a similar rate of bidirectional, replication-fork migration as determined using fibre autoradiography (HAND 1977). Reports indicated increased inhibition of DNA synthesis in FA cells exposed to MMC compared with normal cells (KAISER et al. 1982; SASAKI 1975) and a delay in entry of FA cells into mitosis (NOVOTNA et al. 1979; KAISER et al. 1982; LATT et al. 1975, 1982; MIURA et al. 1983; CERVENKA et al. 1981). CLAASSEN et al. (1986) demonstrated, however, that the MMC-induced inhibition of DNA synthesis in FA is due to some cells not synthesising DNA, whereas a fraction do so at the normal rate. Another report, using cytofluorometric analysis, demonstrated that two FA fibroblast cell lines showed no inhibition of DNA synthesis or cell cycle progression following treatment with NM (DEAN and FOX 1983). Inhibition in normal, but not in FA, cells was observed at both equitoxic and equal NM concentrations and at levels of NM allowing up to 80% survival.

One interesting observation pertains to recovery of DNA synthesis as measured by [^{3}H]thymidine incorporation after inhibition by psoralen plus UV treatment (MOUSTACCHI and DIATLOFF-ZITO 1985). Two classes of FA cell lines were identified. In the first group the cells showed normal kinetics of inhibition and recovery of DNA synthesis, whereas in the second group there was little or no post-treatment recovery (MOUSTACCHI et al. 1987). There is no correlation between recovery of DNA synthesis and cytogenetic or clinical responses.

Such varied observations are open to more than one interpretation. The reports of increased inhibition of DNA synthesis and mitotic delay in FA after rather cytotoxic treatment with MMC may simply reflect an increased proportion of dead cells and their accumulation prior to mitosis. This would be less likely in the work with NM since the concentrations used were far less cytotoxic. Indeed, the presence of two classes of response to psoralen plus UV may explain the different results with other drugs (see below).

V. DNA Repair

In view of the DNA cross-linking characteristics of the drugs to which FA cells are sensitive, considerable attention has been given to the induction and consequent repair of such lesions. The removal of MMC-induced cross-links by FA cells has been shown by alkaline elution, hydroxyapatite chromatography and alkaline sucrose gradient sedimentation to follow normal kinetics (POLL et al. 1984; SOGNIER and HITTELMAN 1983; FORNACE et al. 1979). In contrast Fujiwara and his co-workers report deficient removal of cross-links in several FA lines (FUJIWARA 1982; FUJIWARA and TATSUMI 1975; FUJIWARA et al. 1977). DNA cross-linking and other aspects of DNA repair, such as unscheduled DNA synthesis, after treatment with other cross-linking agents have also revealed normal responses in FA cells (SASAKI and TONOMURA 1973; KAISER et al. 1982;

POLL et al. 1984; GRUENERT et al. 1985; SOGNIER and HITTELMAN 1983; FORNACE et al. 1979; PLOOY et al. 1985), although again there are some contrasting reports suggesting that drug-induced cross-links are more persistent in FA (PLOOY et al. 1985; GRUENERT and CLEAVER 1985; FUJIWARA et al. 1977).

A significant amount of work has been carried out using UV or X-rays as the cytotoxic agents, again with most reports failing to detect any defect in repair in FA (AHMED and SETLOW 1978; COQUERELLE and WEIBEZAHN 1981; LEHMANN et al. 1977; FUJIWARA et al. 1977; SANYAL 1977). The significance of some early reports showing repair defects in FA in response to X-ray- or UV-induced damage is not clear since FA cells are not characteristically sensitive to these agents. In summary, there are many contradictory reports suggesting that FA cells are or are not repair-deficient, whether using cross-linking drugs, to which FA cells are hypersensitive, or agents to which they exhibit normal cellular sensitivity.

The reader may be forgiven for finding this situation more than a little confusing. This may result in part from the large number of different cell lines in culture and from the use of material directly from patients.

Evidence for genetic heterogeneity in FA comes from complementation analysis (ZAKRZEWSKI and SPERLING 1980; DUCKWORTH-RYSIECKI et al. 1985) and from recent observations of MOUSTACCHI et al. (1987), who found two categories of response to 8-methoxypsoralen and near UV. One group of FA cells (group A) showed little or no post-treatment recovery of DNA synthesis whereas the second (group B) behaved in the normal manner, exhibiting rapid recovery. Both groups showed a reduced capacity for incision of psoralen-induced cross-links, as measured by alkaline elution, group A cells being least proficient. Furthermore, both failure to recover DNA synthesis and reduced incision of cross-links correlated well with cell survival (PAPADOPOULO et al. 1987). These observations, using single representatives from each group, demand that a comprehensive complementation analysis of commonly used FA lines be undertaken. It is also necessary to carry out an extensive, coordinated study, using several cell lines in each laboratory, each looking at a different biochemical endpoint and using more than the usual one or two cytotoxic agents. We may then be in a position to explain the diversity in biochemical observations in terms of genetic differences between the (two) classes of FA.

VI. Other Cellular Aspects

Cultured cells from FA patients have characteristic growth parameters. They have a variable and often very low plating efficiency, occasionally as low as 1% or 2% (FORNACE et al. 1979; ARLETT and HARCOURT 1980; WEICHSELBAUM et al. 1980). Cell doubling time is also increased, sometimes exceeding twice the normal range (WEKSBERG et al. 1979; DEAN and FOX 1983); all the cell cycle phases are prolonged. FA fibroblast cell lines also tend to suffer early senescence in culture (AUERBACH et al. 1980). A SV40-transformed FA cell line (GM6914), does, however, show 15%–30% plating efficiency (comparable with normal SV40-transformed cell lines), normal growth rate and indefinite lifespan (DUCKWORTH-RYSIECKI et al. 1986).

Due to the potential of oxygen species to interfere with chromosomal integrity, JOENJE and co-workers assessed the effect of O_2 tension on spontaneous and drug-induced chromosome aberrations. There was evidence of increased spontaneous damage with increased O_2 tension in FA cells (JOENJE and OOSTRA 1983), and O_2 potentiated the damaging effect of MMC in FA cells (JOENJE et al. 1987). Superoxide dismutase, involved in O_2 radical scavenging, was reduced in erythrocytes from FA patients (JOENJE et al. 1979), which led other workers to examine the effects of superoxide dismutase and other potential scavengers on drug-induced damage. Incubation with superoxide dismutase did indeed reduce MMC-induced chromosome damage (NORDENSON 1977; RAJ and HEDDLE 1980), although a similar effect was seen in normal cells (RAJ and HEDDLE 1980). Of most interest is the observation that cell survival after MMC could be enhanced by superoxide dismutase in two FA lines but not in normal cells (NORDENSEN 1977).

VII. Mutagenesis

Very little data exist on the mutability of FA cells. Spontaneous 6-thioguanine resistance was reportedly greater than double the normal rate in five of seven FA patients tested (VIJAYALAXMI et al. 1985). TAKEBE et al. (1987) have recently found a lymphoblastoid FA cell line to be hypermutable after treatment with concentrations of diepoxybutane which were only slightly cytotoxic, whereas the spontaneous mutation frequency was within the normal range. Mutation to diphtheria toxin resistance in one fibroblast cell line, both spontaneous and induced by MNNG, was comparable with the normal rate according to GUPTA and GOLDSTEIN (1980).

G. Bloom's Syndrome

Bloom's syndrome (BS) is a cancer-prone disorder which, like A-T and FA, is associated with a high frequency of chromosome aberrations in circulating lymphocytes. It has therefore long been considered a possible candidate for having a defect in DNA repair. Cellular studies have not, however, provided any convincing evidence for hypersensitivity to DNA-damaging agents or for a defect in DNA repair. Space does not allow for a detailed review of the clinical and cellular characteristics of BS in this chapter.

Very recently though, it was shown by WILLIS and LINDAHL (1987), WILLIS et al. (1987) and CHAN et al. (1987) that the activity of DNA ligase I (one of the two DNA ligases present in mammalian cells) is reduced in most BS cell lines, and that in two other BS cell lines although the activity of DNA ligase I is normal, the enzyme has anomalous chromatographic properties. WILLIS et al. (1987) showed in addition that the residual DNA ligase I activity was temperature-sensitive in BS lines, suggesting that a mutation in the structural gene of DNA ligase I is the primary defect in BS cells. This represents the first instance in which a definitive defect in a DNA-repair enzyme has been identified in a DNA-repair disorder.

H. Heterozygotes

The genetic disorders described above occur with frequencies in the general population of the order of 1 in 10^5 for each disorder. Heterozygotes for these disorders will therefore occur at a frequency of about 1% (see SWIFT et al. 1986 for detailed discussion). As this is a significant proportion of the population, it becomes a question of extreme importance as to whether these heterozygotes show any of the symptoms of the homozygotes. This question has been addressed for XP, FA and in detail for A-T by studies on families with affected homozygotes. There is little evidence for any symptoms in the unaffected members of XP families. An early report of increased cancer incidence in FA families was not substantiated in later work. Extensive studies of SWIFT and co-workers (see SWIFT et al. 1987) have shown, however, that A-T heterozygotes are cancer prone. This was most marked for breast cancer, the relative risk being 6.8 in A-T heterozygotes. This implies that about 8% of patients with breast cancer in the American white population are likely to be heterozygous for A-T. It is possible that A-T heterozygotes may also have increased risk of other types of cancer, but the scarcity of the diseases renders the epidemiological data too uncertain for definitive conclusions to be drawn. The situation could be clarified if a cellular test were devised to identify A-T heterozygotes in the general population. Several laboratories are currently engaged in double-blind studies aimed at identifying A-T heterozygotes (NAGASAWA et al. 1987; M. C. Paterson, C. F. Arlett, J. Nove, personal communications). At the time of writing, all these studies were able to identify A-T heterozygotes as a group as being more radiosensitive than normal individuals, using several different endpoints. Unfortunately, however, a substantial minority of individuals within the group did not conform to the expected pattern, i.e. some A-T heterozygotes showed a response in the normal range, and some normals showed a response in the sensitive range. Cloning and mapping of the A-T genes would greatly facilitate the identification of A-T heterozygotes by linkage studies.

I. Future Prospects: Cloning the Genes

The cellular and biochemical studies described in this chapter have provided a substantial body of information concerning the cellular defects associated with these disorders, and they have demonstrated the complexity of DNA repair processes in human cells. With XP we know that the deficiency lies in an early stage of excision repair of bulky lesions, but the involvement of at least nine gene products corresponding to the nine complementation groups remains mysterious, although one can speculate on a number of schemes (e.g. see LEHMANN 1982b). In no instance has the gene product which is defective in XP cells been identified. With the other disorders cellular sensitivies and biochemical abnormalities have been identified, but the nature of the primary defects remains elusive. As an alternative approach, several laboratories have initiated attempts to clone the genes which are defective in these disorders. This task has proved extraordinarily difficult. The initial step has involved attempts to correct the cellular sensitivies by transfection with DNA from normal cells. The results of these efforts with XP cells have been described in Sect. B2 see p. 76, and similar problems have been encountered with A-T (GREEN et al. 1987) and FA cells (DIATLOFF-ZITO et al. 1986; BUCHWALD et al. 1987). Human cell lines appear to be rather poor recipients for the integration of large quantities of DNA, which is a prerequisite for the success of this approach. Undoubtedly, however, improvements in techniques and the adoption of alternative strategies will eventually lead to the successful cloning of the genes, which in turn will permit the analysis of the mutations associated with the disorders, rapid complementation tests, identification of heterozygotes and characterisation of the gene products. This will aid our understanding of the mechanisms of DNA repair, the molecular basis of the diseases and the relationship of DNA repair to differentiation, development and carcinogenesis.

References

Abrahams PJ, Van Der Eb AJ (1976) Host-cell reactivation of ultraviolet-irradiated SV40 DNA in five complementation groups of xeroderma pigmentosum. Mutat Res 35:13–22

Abrahams PJ, Huitema BA, Van Der Eb AJ (1984) Enhanced reactivation and enhanced mutagenesis of herpes simplex virus in normal human and xeroderma pigmentosum cells. Mol Cell Biol 4:2341–2346

Ahmed FE, Setlow RB (1978) Excision repair in ataxia-telangiectasia, Fanconi's anemia, Cockayne syndrome, and Bloom's syndrome after treatment with ultraviolet radiation and *N*-acetoxy-2-acetylaminofluorene. Biochim Biophys Acta 521:805–817

Andrews AD, Barrett SF, Robbins JH (1978) Xeroderma pigmentosum neurological abnormalities correlate with colony-forming ability after ultraviolet radiation. Proc Natl Acad Sci USA 75:1984–1988

Arlett CF, Harcourt SA (1980) Survey of radio-sensitivity in a variety of human cell strains. Cancer Res 40:926–932

Arlett CF, Harcourt SA (1982) Variation in response to mutagens amongst normal and repair-defective human cells. In: Lawrence CW (ed) Induced mutagenesis. Plenum, New York, pp 249–266

Arlett CF, Lehmann AR (1978) Human disorders showing increased sensitivity to the induction of genetic damage. Ann Rev Genet 12:95–115

Arlett CF, Harcourt SA, Lehmann AR, Stevens S, Ferguson-Smith MA, Morley WN (1980) Studies on a new case of xeroderma pigmentosum (XP3BR) from complementation group G with cellular sensitivity to ionizing radiation. Carcinogenesis 1:745–751

Arlett CF, Harcourt SA, Teo IA, Broughton BC (1982) The response of ataxia-telangiectasia fibroblasts to the lethal effects of an array of DNA-damaging agents. In: Bridges BA, Harnden DG (eds) Ataxia-telangiectasia – a cellular and molecular link between cancer, neuropathology, and immune deficiency. Wiley, Chichester, pp 169–176

Auerbach AD (1984) Diagnosis of diseases of DNA synthesis and repair that affect the skin using cultured amniotic fluid cells. Semin Dermatol 3:172–184

Auerbach AD, Wolman SR, Chaganti RSK (1980) A spontaneous clone of Fanconi anemia fibroblasts with chromosome abnormalities and increased growth potential. Cytogenet Cell Genet 28:265–270

Auerbach AD, Adler B, Chaganti RSK (1981) Prenatal and postnatal diagnosis and carrier detection of Fanconi anemia by a cytogenetic method. Pediatrics 67:128–135

Aust AR, Falahee KJ, Maher VM, McCormick JJ (1980) Human-cell mediated benzo(a)pyrene cytotoxicity and mutagenicity in human diploid fibroblasts. Cancer Res 40:4070–4075

Beard MEJ (1976) Fanconi anaemia. In: Porter R, Fitzsimmons DW (eds) Congenital disorders of erythropoiesis. Elsevier, Amsterdam, pp 103–114 (Ciba Foundation Symposium 37)
Becker Y (1986) Cancer in ataxia-telangiectasia patients: analysis of factors leading to radiation-induced and spontaneous tumors. Anticancer Res 6:1021–1032
Bender MA, Rary JM, Kale RP (1985) G_2 chromosomal radiosensitivity in ataxia telangiectasia lymphocytes. Mutat Res 152:39–47
Berger R, Bernheim A, Le Coniat M, Vecchione D, Schaison G (1980) Nitrogen mustard-induced chromosome breakage: a tool for Fanconi's anemia diagnosis. Cancer Genet Cytogenet 2:269–274
Bigelow SB, Rary JM, Bender MA (1979) G_2 chromosomal radiosensitivity in Fanconi's anemia. Mutat Res 63:189–199
Boder E (1985) Ataxia-telangiectasia: an overview. In: Gatti RA, Swift M (eds) Ataxia-telangiectasia: genetics, neuropathology, and immunology of a degenerative disease of childhood. Liss, New York, pp 1–63
Bootsma D, Keijzer W, Jung EG, Bohnert E (1989) Xeroderma pigmentosum complementation group XP-I withdrawn. Mutat Res 218:149–151
Bredberg A, Lambert B, Söderhäll S (1982) Induction and repair of psoralen cross-links in DNA of normal human and xeroderma pigmentosum fibroblasts. Mutat Res 93:221–234
Bridges BA (1981) How important are somatic mutations and immune control in skin cancer? Reflections on xeroderma pigmentosum. Carcinogenesis 2:471–472
Bridges BA, Harnden DG (1982) Ataxia-telangiectasia – a cellular and molecular link between cancer, neuropathology and immune deficiency. Wiley, London
Buchwald M, Ng J, Clarke C, Duckworth-Rysiecki G (1987) Studies of gene transfer and reversion to mitomycin C resistance in Fanconi anemia cells. Mutat Res 184:153–159
Cervenka J, Arthur D, Yasis C (1981) Mitomycin C test for diagnostic differentiation of idiopathic aplastic anemia and Fanconi anemia. Pediatrics 67:119–127
Chan JYH, Becker FF, German J, Ray JH (1987) Altered DNA ligase I activity in Bloom's syndrome cells. Nature 325:357–359
Chen P, Imray FP, Kidson C (1984) Gene dosage and complementation analysis of ataxia telangiectasia lymphoblastoid cell lines assayed by induced chromosome aberrations. Mutat Res 129:165–172
Claassen E, Kortbeek H, Arwert F (1986) Effects of mitomycin C on the rate of DNA synthesis in normal and Fanconi anaemia cells. Mutat Res 165:15–19
Cohen MM, Simpson SJ (1980) Growth kinetics of ataxia telangiectasia lymphoblastoid cells: evidence for a prolonged S period. Cytogenet Cell Genet 28:24–33
Cohen MM, Simpson SJ, Honig GR, Maurer HS, Nicklas JW, Martin AO (1982) The identification of Fanconi anemia genotypes by clastogenic stress. Am J Hum Genet 34:794–810
Cole J, Arlett CF, Green MHL, James SE, Henderson L, Cole H, Sala-Trepat M, Benzi R, Price ML, Bridges BA (1989) Measurement of mutant frequency to 6-thioguanine resistance in circulating T-lymphocytes for human population monitoring. In: Jolles G (ed) New trends in genetic risk assessment. Academic, London, pp 175–203
Coppey J, Menezes S (1981) Enhanced reactivation of ultraviolet-damaged herpes virus in ultraviolet pretreated skin fibroblasts of cancer prone donors. Carcinogenesis 2:787–793
Coquerelle TM, Weibezahn KF (1981) Rejoining of DNA double-strand breaks in human fibroblasts and its impairment in one ataxia telangiectasia and two Fanconi strains. J Supramolec Struct Cell Biochem 17:369–376
Coquerelle TM, Weibezahn KF, Lücke-Huhle C (1987) Rejoining of double strand breaks in normal human and ataxia-telangiectasia fibroblasts after exposure to ^{60}Co gamma-rays, ^{241}Am alpha-particles or bleomycin. Int J Radiat Biol 51:209–218
Cornforth MN, Bedford JS (1985) On the nature of a defect in cells from individuals with ataxia-telangiectasia. Science 227:1589–1591
Cox R, Debenham PG, Masson WK, Webb MBT (1986) Ataxia-telangiectasia – a human mutation giving high frequency misrepair of DNA double-stranded scissions. Mol Biol Med 3:229–244

Cramer P, Painter RB (1981) Bleomycin-resistant DNA synthesis in ataxia telangiectasia cells. Nature 291:671–672
Dasgupta UB, Summers WC (1980) Genetic recombination of herpes simplex virus, the role of the host cell and UV-irradiation of the virus. Molec Gen Genet 178:617–623
Day RS (1974) Studies on repair of adenovirus 2 by human fibroblasts using normal, xeroderma pigmentosum, and xeroderma pigmentosum heterozygous strains. Cancer Res 34:1965–1970
Day RS, Scudiero D, Dimattina M (1978) Excision repair by human fibroblasts of DNA damaged by r-7, t-8-dihydroxy-t-9,10-oxy-7,8,9,10-tetrahydrobenzo(a)pyrene. Mutat Res 50:383–394
Dean SW, Fox M (1983) Investigation of the cell cycle response of normal and Fanconi's anaemia fibroblasts to nitrogen mustard using flow cytometry. J Cell Sci 64:265–279
Dean SW, Sykes HR, Cole J, Jaspers NGJ, Linssen P, Verkerk A (1988) Correspondence re: Michael J. Meredith and Marion L. Dodson. Impaired glutathione biosynthesis in cultured ataxia-telangiectasia cells. Cancer Res 48:5374–5375
De Jonge AJR, Vermeulen W, Keijzer W, Hoeijmakers JHJ, Bootsma D (1985) Microinjection of *Micrococcus luteus* UV-endonuclease restores UV-induced unscheduled DNA synthesis in cells of 9 xeroderma pigmentosum complementation groups. Mutat Res 150:99–105
Diatloff-Zito C, Papadopoulo D, Averbeck D, Moustacchi E (1986) Abnormal response to DNA crosslinking agents of Fanconi anaemia fibroblasts can be corrected by transfection with normal human DNA. Proc Natl Acad Sci USA 83:7034–7038
Duckworth-Rysiecki G, Cornish K, Clarke CA, Buchwald M (1985) Identification of two complementation groups in Fanconi anemia. Somatic Cell Molec Genet 11:35–41
Duckworth-Rysiecki G, Toji L, Clarke JNC, Buchwald M (1986) Characterization of a simian virus 40-transformed Fanconi anaemia fibroblast cell line. Mutat Res 166:207–214
Dutrillaux B, Aurias A, Dutrillaux A-M, Buriot D, Prieur M (1982) The cell cycle of lymphocytes in Fanconi anemia. Hum Genet 62:327–332
Elmore E, Swift M (1976) Growth of cultured cells from patients with ataxia-telangiectasia. J Cell Physiol 89:429–432
Fornace AJ, Kohn KW (1976) DNA-protein cross-linking by ultraviolet radiation in normal human and xeroderma pigmentosum fibroblasts. Biochim Biophys Acta 435:95–103
Fornace AJ, Little JB, Weichselbaum RR (1979) DNA repair in a Fanconi's anemia fibroblast cell strain. Biochim Biophys Acta 561:99–109
Fornace AJ, Kinsella TJ, Dobson PP, Mitchell JB (1986) Repair of ionizing radiation DNA base damage in ataxia-telangiectasia cells. Cancer Res 46:1703–1706
Friedberg EC, Ehmann UK, Williams JI (1979) Human diseases associated with defective DNA repair. Adv Radiat Biol 8:85–174
Fujiwara Y (1982) Defective repair of mitomycin C crosslinks in Fanconi's anemia and loss in confluent normal human and xeroderma pigmentosum cells. Biochim Biophys Acta 699:217–225
Fujiwara Y, Tasumi M (1975) Repair of mitomycin C damage to DNA in mammalian cells and its impairment in Fanconi's anemia cells. Biochem Biophys Res Comm 66:592–599
Fujiwara Y, Tatsumi M, Sasaki M (1977) Cross-link repair in human cells and its possible defect in Fanconi's anemia cells. J Mol Biol 113:635–649
Fujiwara Y, Kano Y, Yamamoto Y (1984) DNA interstrand crosslinking, repair, and SCE mechanism in human cells in special reference to Fanconi anemia. Basic Life Sci 29:787–800
Fujiwara Y, Uehara Y, Ichihashi M, Nishioka K (1985) Xeroderma pigmentosum complementation group F: more assignments and repair characteristics. Photochem Photobiol 41:629–634
Ganesan A, Spivak G, Hanawalt P (1983) Expression of DNA repair genes in mammalian cells. In: Nagley P, Linnane AW, Peacock WJ, Pateman JA (eds) Manipulation and expression of genes in eukaryotes. Academic, Sydney, pp 45–54

Gatti RA, Swift M (1985) Ataxia-telangiectasia. Genetics, neuropathology, and immunology of a degenerative disease of childhood. Liss, New York
Glanz A, Fraser FC (1982) Spectrum of anomalies in Fanconi anaemia. J Med Genet 19:412–416
Glover TW, Chang C-C, Trosko JF, Li SS-I (1979) Ultraviolet light induction of diphtheria toxin-resistant mutants in normal and xeroderma pigmentosum human fibroblasts. Proc Natl Acad Sci USA 76:3982–3986
Green MHL, Lowe JE (1987) Failure to detect a DNA repair-related defect in the transfection of ataxia-telangiectasia cells by enzymatically restricted plasmid. Int J Radiat Biol 52:437–446
Green MHL, Lowe JE, Arlett CF, Harcourt SA, Burke JF, James MR, Lehmann AR, Povey SM (1987) A gamma-ray resistant derivative of an ataxia-telangiectasia cell line obtained following DNA-mediated gene transfer. J Cell Sci Suppl 6:127–137
Grosovksy AJ, Little JB (1983) Mutagenesis and lethality following S phase irradiation of xeroderma pigmentosum and normal human diploid fibroblasts with ultraviolet light. Carcinogenesis 4:1389–1393
Gruenert DC, Cleaver JE (1985) Repair of psoralen-induced cross-links and monoadducts in normal and repair-deficient human fibroblasts. Cancer Res 45:5399–5404
Gruenert DC, Kapp LN, Cleaver JE (1985) Inhibition of DNA synthesis by psoralen-induced lesions in xeroderma pigmentosum and Fanconi's anemia fibroblasts. Photochem Photobiol 41:543–550
Gupta RS, Goldstein S (1980) Diphtheria toxin resistance in human fibroblast cell strains from normal and cancer-prone individuals. Mutat Res 73:331–338
Guzzetta F (1972) Cockayne-Neill-Dingwall syndrome. In: Vinken PJ, Bruyn GW (eds) Handbook of clinical neurology, vol 13. Elsevier, Amsterdam, pp 431–440
Hanawalt PC, Sarasin A (1986) Cancer-prone hereditary diseases with DNA processing abnormalities. Trends Genet 2:124–129
Hand R (1977) Human DNA replication: fiber autoradiographic analysis of diploid cells from normal adults and from Fanconi's anemia and ataxia telangiectasia. Hum Genet 37:55–64
Hecht F, McCaw BK (1977) Chromosome instability syndromes. In: Mulvihill JJ, Miller RW, Fraumeni JF (eds) Genetics of human cancer. Raven, New York, pp 105–123
Heddle JA, Lue CB, Saunders EF, Benz RD (1978) Sensitivity to five mutagens in Fanconi's anaemia as measured by the micronucleus method. Cancer Res 38:2983–2988
Heflich RH, Hazard RM, Lommel L, Scribner JD, Maher VM, McCormick JJ (1980) A comparison of the DNA binding, cytotoxicity and repair synthesis induced in human fibroblasts by reactive derivatives of aromatic amide carcinogens. Chem Biol Interact 29:43–56
Helland D, Kleppe R, Lillehaug JR, Kleppe K (1984) Xeroderma pigmentosum: in vitro complementation of DNA repair endonuclease. Carcinogenesis 5:833–836
Henderson EE (1978) Host cell reactivation of Epstein-Barr virus in normal and repair-defective leukocytes. Cancer Res 38:3256–3263
Henderson EE, Basilio M (1983) Transformation and repair replication in lymphocytes from ataxia telangiectasia. Proc Soc Expl Biol Med 172:524–534
Henderson EE, Ribecky R (1980) DNA repair in lymphoblastoid cell lines established from human genetic disorders. Chem Biol Interact 33:63–82
Higurashi M, Conen PE (1971) In vitro chromosomal radiosensitivity in Fanconi's anemia. Blood 38:336–342
Hilgers G, Chen YQ, Cornelis JJ, Rommelaere J (1987) Deficient expression of enhanced reactivation of parvovirus H-1 in ataxia telangiectasia cells irradiated with X-rays or u.v. light. Carcinogenesis 2:315–319
Huang PC, Sheridan RB III (1981) Genetic and biochemical studies with ataxia telangiectasia. Hum Genet 59:1–9
Hurley LH, Chandler C, Garner TF, Petrusek R, Zimmer SG (1979) DNA binding, induction of unscheduled DNA synthesis, and excision of anthramycin from DNA in normal and repair-deficient human fibroblasts. J Biol Chem 254:605–608

Inoue T, Yokoiyama A, Kada T (1981) DNA repair enzyme deficiency and in vitro complementation of the enzyme activity in cell-free extracts from ataxia-telangiectasia fibroblasts. Biochim Biophys Acta 655:49–53

Ishida R, Buchwald M (1982) Susceptibility of Fanconi's anaemia lymphoblasts to DNA cross-linking and alkylating agents. Cancer Res 42:4000–4006

Jaspers NGJ (1985) DNA synthesis in ataxia telangiectasia. PhD Thesis, Erasmus University, Rotterdam, Netherlands

Jaspers NGJ, Gatti RA, Baan C, Linssen PCML, Bootsma D (1988) Genetic complementation analysis of ataxia-telangiectasia and Nijmegen breakage syndrome: a survey of 50 patients. Cytogenet Cell Genet 49:259–263

Jaspers NGJ, Jansen-van De Kuilen G, Bootsma D (1981) Complementation analysis of xeroderma pigmentosum variants. Exp Cell Res 136:81–90

Jaspers NGJ, De Wit J, Regulski MR, Bootsma D (1982) Abnormal regulation of DNA replication and increased lethality in ataxia telangiectasia cells exposed to carcinogenic agents. Cancer Res 42:335–341

Jeeves WP, Rainbow AJ (1986) An aberration in gamma-ray-enhanced reactivation of irradiated adenovirus in ataxia telangiectasia fibroblasts. Carcinogenesis 7:381–387

Joenje H, Oostra AB (1983) Effect of oxygen tension on chromosomal aberrations in Fanconi anemia. Hum Genet 65:99–101

Joenje H, Frants RR, Arwert F, De Bruin GJM, Kostense PJ, Van De Kamp JJP, De Koning J, Eriksson AW (1979) Erythrocyte superoxide dismutase deficiency in Fanconi's anaemia established by two independent methods of assay. Scand J Clin Lab Invest 39:759–764

Joenje H, Nieuwint AWM, Taylor AMR, Harnden DG (1987) Oxygen toxicity and chromosomal breakage in ataxia telangiectasia. Carcinogenesis 8:341–344

Kaiser TN, Lojewski A, Dougherty C, Juergens L, Sahar E, Latt SA (1982) Flow cytometric characterization of the response of Fanconi's anemia cells to mitomycin C treatment. Cytometry 2:291–297

Kano Y, Fujiwara Y (1983) Defective thymine dimer excision from xeroderma pigmentosum chromatin and its characteristic catalysis by cell-free extracts. Carcinogenesis 4:1419–1424

Kantor GJ, Hull DR (1984) The rate of removal of pyrimidine dimers in quiescent cultures of normal human and xeroderma pigmentosum cells. Mutat Res 132:21–31

Kennaugh AA, Butterworth SV, Hollis R, Baer R, Rabbitts TH, Taylor AMR (1986) The chromosome breakpoint at 14q32 in an ataxia telangiectasia t(14;14) T cell clone is different from the 14q32 breakpoint in Burkitts and an inv(14) T cell lymphoma. Hum Genet 73:254–259

Kraemer KH (1980) Xeroderma pigmentosum. In: Demis DJ, Dobson RL, McGuire J (eds) Clinical dermatology, vol 4. Harper and Row, Hagerstown, pp 1–33

Kraemer KH, Lee MM, Scotto J (1987) Xeroderma pigmentosum. Cutaneous, ocular and neurologic abnormalities in 830 published cases. Arch Dermatol 123:241–250

Kuhnlein U, Lee B, Penhoet EE, Linn S (1978) Xeroderma pigmentosum fibroblasts of the D group lack an apurinic DNA endonuclease species with a low apparent K_m. Nucleic Acids Res 5:951–960

Kuhnlein U, Tsang SS, Lokken O, Tong S, Twa D (1983) Cell lines from xeroderma pigmentosum complementation group A lack a single-stranded-DNA-binding activity. Biosci Rep 3:667–674

Kwee ML, Poll EHA, Van de Kamp JJP, De Koning H, Erikkson AW, Joenje H (1983) Unusual response to bifunctional alkylating agents in a case of Fanconi anaemia. Hum Genet 64:384–387

Latt SA, Stetten G, Juergens LA, Buchanan GR, Gerald PS (1975) Induction by alkylating agents of sister chromatid exchanges and chromatid breaks in Fanconi's anemia. Proc Natl Acad Sci USA 72:4066–4070

Latt SA, Kaiser TN, Lojewski A, Dougherty C, Juergens L, Brefach S, Sahar E, Gustashaw K, Schreck RR, Powers M, Lalande M (1982) Cytogenetic and flow cytometric studies of cells from patients with Fanconi's anemia. Cytogenet Cell Genet 33:133–138

Lavin MF, Seymour GJ (1984) Reduced levels of fibronectin in ataxia-telangiectasia lymphoblastoid cells. Int J Cancer 33:359–363

Lee DE, Okorodudu AO, Lambert WC, Lambert MW (1982) Defective DNA endonuclease activity on anthramycin treated DNA in xeroderma pigmentosum and mouse melanoma cells. Biochem Biophys Res Comm 107:395–402

Legerski RJ, Brown DB, Peterson CA, Robberson DL (1984) Transient complementation of xeroderma pigmentosum cells by microinjection of poly(A)$^+$ RNA. Proc Natl Acad Sci USA 81:5676–5679

Lehmann AR (1982a) Three complementation groups in Cockayne syndrome. Mutat Res 106:347–356

Lehmann AR (1982b) Xeroderma pigmentosum, Cockayne syndrome and ataxia-telangiectasia: disorders relating DNA repair to carcinogenesis. In: Bodmer WF (ed) Cancer surveys 1. Oxford University Press, Oxford, pp 93–118

Lehmann AR (1982c) The cellular and molecular responses of ataxia-telangiectasia cells to DNA damage. In: Bridges BA, Harnden DG (eds) Ataxia-telangiectasia – a cellular and molecular link between cancer, neuropathology and immune deficiency. Wiley, Chichester, pp 83–101

Lehmann AR (1985) Use of recombinant DNA techniques in cloning DNA repair genes and in the study of mutagenesis in mammalian cells. Mutat Res 150:61–67

Lehmann AR (1987) Cockayne's syndrome and trichothiodystrophy: defective repair without cancer. Cancer Rev 7:82–103

Lehmann AR, Norris PG (1989) DNA repair and cancer: speculations based on studies with xeroderma pigmentosum, Cockayne's syndrome and trichothiodystrophy. Carcinogenesis 10:1353–1356

Lehmann AR, Kirk-Bell S, Arlett CF, Harcourt SA, De Weerd-Kastelein EA, Keijzer W, Hall-Smith P (1977) Repair of ultraviolet light damage in a variety of human fibroblast cell strains. Cancer Res 37:904–910

Lehmann AR, Kirk-Bell S, Mayne L (1979) Abnormal kinetics of DNA synthesis in ultraviolet light-irradiated cells from patients with Cockayne's syndrome. Cancer Res 39:4237–4241

Lehmann AR, Arlett CF, Burke JF, Green MHL, James MR, Lowe JE (1986) A derivative of an ataxia-telangiectasia (A-T) cell line with normal radiosensitivity but A-T like inhibition of DNA synthesis. Int J Radiat Biol 49:639–643

Maher VM, McCormick JJ, Grover P, Sims P (1977) Effect of DNA on the cytotoxicity and mutagenicity of polycyclic hydrocarbon derivatives in normal and xeroderma pigmentosum human fibroblasts. Mutat Res 43:117–138

Maher VM, Rowan LA, Silinskas KC, Kateley SA, McCormick JJ (1982) Frequency of UV-induced neoplastic transformation of diploid human fibroblasts is higher in xeroderma pigmentosum cells than in normal cells. Proc Natl Acad Sci USA 79:2613–2617

Mayne LV, Lehmann AR (1982) Failure of RNA synthesis to recover after UV-irradiation: an early defect in cells from individuals with Cockayne's syndrome and xeroderma pigmentosum. Cancer Res 42:1473–1478

Mayne LV, Mullenders LHF, Van Zeeland AA (1988) Cockayne's syndrome: a UV sensitive disorder with a defect in the repair of transcribing DNAA but normal overall excision repair. In: Friedberg, E, Hanawalt, P (eds) Mechanisms and consequences of DNA damage processing. Liss, New York, pp 349–353

McCormick J, Kateley-Kohler S, Watanabe M, Maher V (1986) Abnormal sensitivity of human fibroblasts from xeroderma pigmentosum variants to transformation to anchorage independence by ultraviolet radiation. Cancer Res 56:489–492

McKinnon PJ (1987) Ataxia-telangiectasia: an inherited disorder of ionizing-radiation sensitivity in man. Hum Genet 75:197–208

McKinnon PJ, Burgoyne LA (1984) Regulation of the cellular actin levels in response to changes in the cell density in ataxia-telangiectasia lymphoblastoid cells. Biochem Biophys Res Comm 119:561–566

McKinnon PJ, Burgoyne LA (1985a) Altered cellular morphology and microfilament array in ataxia-telangiectasia fibroblasts. Eur J Cell Biol 39:161–166

McKinnon PJ, Burgoyne LA (1985b) Evidence for the existence of an actin-derived protein in ataxia-telangiectasia lymphoblastoid cell lines. Exp Cell Res 158:413–422

Meredith MJ, Dodson ML (1987) Impaired glutathione biosynthesis in cultured human ataxia-telangiectasia cells. Cancer Res 47:4576–4581

Miskin R, Ben-Ishai R (1981) Induction of plasminogen activator by ultraviolet light in normal and xeroderma pigmentosum fibroblasts. Proc Natl Acad Sci USA 78:6236–6240

Mitchell DL, Haipek CA, Clarkson JM (1985) (6–4) Photoproducts are removed from the DNA of UV-irradiated mammalian cells more efficiently than cyclobutane pyrimidine dimers. Mutat Res 143:109–112

Miura K, Morimoto K, Koizumi A (1983) Proliferative kinetics and mitomycin C-induced chromosome damage in Fanconi's anemia lymphocytes. Hum Genet 63:19–23

Mohamed R, Lavin MF (1986) Ataxia-telangiectasia cell extracts confer radioresistant DNA synthesis on control cells. Exp Cell Res 163:337–348

Morison WL, Bucana C, Hashem N, Kripke ML, Cleaver JE, German JL (1985) Impaired immune function in patients with xeroderma pigmentosum. Cancer Res 45:3929–3931

Mortelmans K, Friedberg EC, Slor H, Thomas G, Cleaver JE (1976) Defective dimer excision by cell-free extracts of xeroderma pigmentosum cells. Proc Natl Acad Sci USA 8:2757–2761

Moshell AM, Ganges MB, Lutzner MA, Coon HG, Barrett SF, Dupuy JM, Robbins JH (1983) A new patient with both xeroderma pigmentosum and Cockayne syndrome establishes the new xeroderma pigmentosum complementation group H. In: Friedberg EC, Bridges BA (eds) Cellular responses to DNA damage. Liss, New York, pp 209–213

Moustacchi E, Diatloff-Zito C (1985) DNA semi-conservative synthesis in normal and Fanconi anemia fibroblasts following treatment with 8-methoxypsoralen and near ultraviolet light or with X-rays. Hum Genet 70:236–242

Moustacchi E, Papadopoulo D, Diatloff-Zito C, Buchwald M (1987) Two complementation groups of Fanconi's anemia differ in their phenotypic response to a DNA-crosslinking treatment. Hum Genet 75:45–47

Mullenders LHF, Van Kesteren AC, Bussmann CJM, Van Zeeland AA, Natarajan AT (1986) Distribution of u.v.-induced repair events in higher-order chromatin loops in human and hamster fibroblasts. Carcinogenesis 7:995–1002

Murnane JP, Painter RB (1983) Altered protein synthesis in ataxia-telangiectasia fibroblasts. Biochemistry 22:1217–1222

Nagasawa H, Little JB (1983) Suppression of cytotoxic effect of mitomycin-C by superoxide dismutase in Fanconi's anemia and dyskeratosis congenita fibroblasts. Carcinogenesis 4:795–798

Nagasawa H, Kraemer KH, Shiloh Y, Little JB (1987) Detection of ataxia-telangiectasia heterozygous cell lines by postirradiation cumulative labelling index: measurements with coded samples. Cancer Res 47:398–402

Nordenson I (1977) Effect of superoxide dismutase and catalase on spontaneously occurring chromosome breaks in patients with Fanconi's anemia. Hereditas 86:147–150

Norris PG, Limb GA, Hamblin AS, Hawk JLM (1988) Impairment of natural-killer-cell activity in xeroderma pigmentosum. N Engl J Med 319:1668–1669

Novotna B, Goetz P, Surkova NI (1979) Effects of alkylating agents on lymphocytes from controls and from patients with Fanconi's anemia. Hum Genet 48:41–50

Okorodudu AO, Lambert WC, Lambert MW (1982) Nuclear deoxyribonuclease activities in normal and xeroderma pigmentosum lymphoblastoid cells. Biochem Biophys Res Comm 198:576–584

Papadopoulo D, Averbeck D, Moustacchi E (1987) The fate of 8-methoxypsoralen-photoinduced DNA interstrand crosslinks in Fanconi's anemia cells of defined genetic complementation groups. Mutat Res 184:271–280

Paterson MC, Smith BP, Lohman PHM, Anderson AK, Fishman L (1976) Defective excision repair of gamma-ray-damaged DNA in human (ataxia telangiectasia) fibroblasts. Nature 260:444–447

Paterson MC, Middlestadt MV, MacFarlane SJ, Gentner NE, Weinfeld M (1987) Molecular evidence for cleavage of intradimer phosphodiester linkage as a novel step in excision repair of cyclobutyl pyrimidine photodimers in cultured human cells. J Cell Sci Suppl 6:161–176

Plooy ACM, Van Dijk M, Berends F, Lohman PHM (1985) Formation and repair of DNA interstrand cross-links in relation to cytotoxicity and unscheduled DNA synthesis induced in control and mutant human cells treated with *cis*-diamminedichloroplatinum(II). Cancer Res 45:4178–4184

Poll EHA, Arwert F, Joenje H, Eriksson AW (1982) Cytogenetic toxicity of antitumor platinum compounds in Fanconi's anemia. Hum Genet 61:228–230

Poll EHA, Arwert F, Kortbeek HT, Eriksson AW (1984) Fanconi anaemia cells are not uniformily deficient in unhooking of DNA interstrand crosslinks, induced by mitomycin C or 8-methoxypsoralen plus UVA. Hum Genet 68:228–234

Protic-Sabljic M, Kraemer KH (1986) Host cell reactivation by human cells of DNA expression vectors damaged by ultraviolet radiation or by acid-heat treatment. Carcinogenesis 10:1765–1770

Rainbow AJ (1978) Production of viral structural antigens by irradiated adenovirus as an assay for DNA repair in human fibroblasts. In: Hanawalt PC, Friedberg EC (eds) DNA repair mechanisms. Academic, New York, pp 541–545

Raj AS, Heddle JA (1980) The effect of superoxide dismutase, catalase and L-cysteine on spontaneous and on mitomycin C induced chromosomal breakage in Fanconi's anemia and normal fibroblasts as measured by the micronucleus method. Mutat Res 78:59–66

Robbins JH (1978) Significance of repair of human DNA: evidence from studies of xeroderma pigmentosum. JNCI 61:645–656

Robbins JH, Kraemer KH, Lutzner MA, Festoff BW, Coon HG (1974) Xeroderma pigmentosum, an inherited disease with sun sensitivity, multiple cutaneous neoplasms, and abnormal DNA repair. Ann Intern Med 80:221–248

Royer-Pokora B, Haseltine WA (1984) Isolation of UV-resistant revertants from a xeroderma pigmentosum complementation group A cell line. Nature 311:390–392

Sanyal AB (1977) Study of ultraviolet repair and spontaneous DNA strand breaks in Fanconi's anemia cells. Indian J Exp Biol 15:976–979

Sarasin AR, Smith CA, Hanawalt PC (1977) Repair of DNA in human cells after treatment with activated aflatoxin B1. Cancer Res 37:1786–1793

Sasaki MS (1975) Is Fanconi's anaemia defective in a process essential to the repair of DNA cross links? Nature 256:501–503

Sasaki MS, Tonomura A (1973) A high susceptibility of Fanconi's anemia to chromosome breakage by DNA cross-linking agents. Cancer Res 33:1829–1836

Schorpp M, Mallick U, Rahmsdorf HJ, Herrlich P (1984) UV-induced extracellular factor from human fibroblasts communicates the UV response to non-irradiated cells. Cell 37:861–868

Schultz RA, Barbis DP, Friedberg EC (1985) Studies on gene transfer and reversion to UV resistance in xeroderma pigmentosum cells. Somatic Cell Molec Genet 11:617–624

Schultz RA, Saxon PJ, Glover TW, Friedberg EC (1987) Microcell-mediated transfer of a single human chromosome complements xeroderma pigmentosum group A fibroblasts. Proc Natl Acad Sci USA 84:4176–4179

Sedgwick RP, Boder E (1972) Ataxia-telangiectasia. In: Vinken PJ, Bruyn GW (eds) Handbook of clinical neurology. Elsevier, Amsterdam, pp 267–339

Selsky CA, Greer S (1978) Host-cell reactivation of UV-irradiated and chemically-treated herpes simplex virus-1 by xeroderma pigmentosum, XP heterozygotes and normal skin fibroblasts. Mutat Res 50:395–405

Setlow RB (1978) Repair deficient human disorders and cancer. Nature 271:713–717

Shaham M, Becker Y (1981) The ataxia telangiectasia clastogenic factor is a low molecular weight peptide. Hum Genet 58:422–424

Shaham M, Becker Y, Lerer I, Voss R (1983) Increased level of bleomycin-induced chromosome breakage in ataxia telangiectasia skin fibroblasts. Cancer Res 43:4244–4247

Shiloh Y, Becker Y (1982) Reduced inhibition of replicon initiation and chain elongation by neocarzinostatin in skin fibroblasts from patients with ataxia telangiectasia. Biochim Biophys Acta 721:485–488

Shiloh Y, Tabor E, Becker Y (1982) Colony-forming ability of ataxia-telangiectasia skin fibroblasts is an indicator of their early senescence and increased demand for growth factors. Exp Cell Res 140:191–199

Shiloh Y, Tabor E, Becker Y (1983) Abnormal response of ataxia-telangiectasia cells to agents that break the deoxyribose moiety of DNA via a targeted free radical mechanism. Carcinogenesis 4:1317–1322

Smith PJ, Paterson MC (1981) Abnormal responses to mid-ultraviolet light of cultured fibroblasts from patients with disorders featuring sunlight sensitivity. Cancer Res 41:511–518

Sognier MA, Hittelman WN (1983) Loss of reparability of DNA interstrand crosslinks in Fanconi's anemia cells with culture age. Mutat Res 108:383–393

Stefanini M, Lagomarsini P, Arlett CF, Marinoni S, Borrone C, Crovato F, Trevisan G, Cordone G, Nuzzo F (1986) Xeroderma pigmentosum (complementation group D) mutation is present in patients affected by trichothiodystrophy with photosensitivity. Hum Genet 74:107–112

Sutherland BM, Rice M, Wagner EK (1975) Xeroderma pigmentosum cells contain low levels of photoreactivating enzyme. Proc Natl Acad Sci USA 72:103–107

Swift M, Morrell D, Cromartie E, Chamberlin AR, Skolnick MH, Bishop DT (1986) The incidence and gene frequency of ataxia-telangiectasia in the United States. Am J Hum Genet 39:573–583

Swift M, Reitnauer PJ, Morrel D, Chase CL (1987) Breast and other cancers in families with ataxia-telangiectasia. N Engl J M 316:1289–1294

Takano T, Noda M, Tamura T-A (1982) Transfection of cells from a xeroderma pigmentosum patient with normal human DNA confers UV resistance. Nature 296: 269–270

Takebe H, Tatsumi K, Tachibana A, Nishigori C (1987) High sensitivity to radiation and chemicals in relation to cancer and mutation. In: Fielden, EM, Fowler, JF, Hendry, JH, Scott D (eds) Radiation research, vol 2. Taylor and Francis, London, pp 443–448

Tanaka K, Sekiguchi M, Okada Y (1975) Restoration of ultraviolet-induced unscheduled DNA synthesis of xeroderma pigmentosum cells by the concomitant treatment with bacteriophage T4 endonuclease V and HVJ (Sendai virus). Proc Natl Acad Sci USA 72:4071–4075

Tanaka K, Takebe H, Okada Y (1980) Unscheduled DNA synthesis induced by 4-nitroquinoline-1-oxide in xeroderma pigmentosum cells and their complementing heterodikaryons. Somatic Cell Genet 6:739–749

Tanaka K, Satokata I, Ogita Z, Uchida T, Okada Y (1989) Molecular cloning of a mouse DNA repair gene that complements the defect of group-A xeroderma pigmentosum. Proc Natl Acad Sci USA 86:5512–5516

Tatsumi K, Takebe H (1984) Gamma-irradiation induces mutation in ataxia-telangiectasia lymphoblastoid cells. Gann 75:1040–1043

Tatsumi K, Toyoda M, Hashimoto T, Furuyama J-I, Kurihara T, Inoue M, Takebe H (1987) Differential hypersensitivity of xeroderma pigmentosum lymphoblastoid cell lines to ultraviolet light mutagenesis. Carcinogenesis 8:53–57

Taylor AMR (1982) Cytogenetics of ataxia-telangiectasia. In: Bridges BA, Harnden DG (eds) Ataxia-telangiectasia – a cellular and molecular link between cancer, neuropathology and immune deficiency. Wiley, Chichester, pp 53–81

Taylor AMR, Metcalfe JA, Oxford JM, Harnden DG (1976) Is chromatid-type damage in ataxia telangiectasia after irradiation at G_0 a consequence of defective repair? Nature 260:441–443

Thompson KVA, Holliday R (1983) Genetic effects on the longevity of cultured human fibroblasts. II. DNA repair deficient syndromes. Gerontology 29:83–88

Vermeulen W, Osseweijer P, De Jonge AJR, Hoeijmakers JHJ (1986) Transient correction of excision repair defects in fibroblasts of 9 xeroderma pigmentosum complementation groups by microinjection of crude human cell extracts. Mutat Res 165:199–206

Vijayalaxmi, Wunder E, Schroeder TM (1985) Spontaneous 6-thioguanine-resistant lymphocytes in Fanconi anemia patients and their heterozygous parents. Hum Genet 70:264–270

Voss R, Kohn G, Shaham M, Benzur Z, Arnon J, Ornoy A, Yaffe H, Golbus M, Auerbach AD (1981) Prenatal diagnosis of Fanconi anemia. Clin Genet 20:185–190

Vuillaume M, Calvayrac R, Best-Belpomme M, Tarroux P, Hubert M, Decroix Y, Sarasin A (1986) Deficiency in catalase activity of xeroderma pigmentosum cell and simian virus 40-transformed human cell extracts. Cancer Res 46:538–544

Vuksanovic L, Cleaver JE (1987) Unique cross-link and monoadduct repair characteristics of a xeroderma pigmentosum revertant cell line. Mutat Res 184:255–263

Wade MH, Chu EHY (1979) Effects of DNA damaging agents on cultured fibroblasts derived from patients with Cockayne syndrome. Mutat Res 59:49–60

Waldmann TA, Broder S, Goldman CK, Frost K, Korsmeyer SJ, Medici MA (1983a) Disorders of B cells and helper T cells in the pathogenesis of the immunoglobulin deficiency of patients with ataxia telangiectasia. J Clin Invest 71:282–295

Waldmann TA, Misiti J, Nelson DL, Kraemer KH (1983b) Ataxia-telangiectasia: a multisystem hereditary disease with immunodeficiency, impaired organ maturation, X-ray hypersensitivity, and a high incidence of neoplasia. Ann Int Med 99:367–379

Weichselbaum RR, Nove J, Little JB (1980) X-ray sensitivity of fifty-three human diploid fibroblast cell strains from patients with characterized genetic disorders. Cancer Res 40:920–925

Weinfeld M, Gentner NE, Johnson LD, Paterson MC (1986) Photoreversal-dependent release of thymidine and thymidine monophosphate from pyrimidine dimer-containing DNA excision fragments isolated from ultraviolet-damaged human fibroblasts. Biochemistry 25:2656–2664

Weksberg R, Buchwald M, Sargent P, Thompson MW, Siminovitch L (1979) Specific cellular defects in patients with Fanconi anemia. J Cell Physiol 101:311–324

Willis AE, Lindahl T (1987) DNA ligase I deficiency in Bloom's syndrome. Nature 325:355–357

Willis AE, Weksberg R, Tomlinson S, Lindahl T (1987) Structural alterations of DNA ligase I in Bloom's syndrome. Proc Natl Acad Sci USA 84:8016–8020

Witte I, Thielmann HW (1979) Extracts of xeroderma pigmentosum group A fibroblasts introduce less nicks into methyl methanesulfonate-treated DNA than extracts of normal fibroblasts. Cancer Lett 6:129–136

Wood RD, Robins P, Lindahl T (1988) Complementation of the xeroderma pigmentosum DNA repair defect in cell-free extracts. Cell 53:97–106

Wunder E, Fleischer-Reischmann B (1983) Response of lymphocytes from Fanconi's anemia patients and their heterozygous relatives to 8-methoxy-psoralene in a cloning survival test system. Hum Genet 64:167–172

Yamaizumi M, Sugano T, Asahina H, Okada Y, Uchida T (1986) Microinjection of partially purified protein factor restores DNA damage specifically in group A of xeroderma pigmentosum cells. Proc Natl Acad Sci USA 83:1476–1479

Zakrzewski S, Sperling K (1980) Genetic heterogeneity of Fanconi's anemia demonstrated by somatic cell hybrids. Hum Genet 56:81–84

Zelle B, Bootsma D (1980) Repair of DNA damage after exposure to 4-nitroquinoline-1-oxide in heterokaryons derived from xeroderma pigmentosum cells. Mutat Res 70:373–381

Zelle B, Berends F, Lohman PHM (1980) Repair of damage by ultraviolet radiation in xeroderma pigmentosum cell strains of complementation groups E and F. Mutat Res 73:157–169

Zwetsloot JCM, Barbeiro AP, Vermeulen W, Arthur HM, Hoeijmakers JHJ, Backendorf C (1986) Microinjection of *Escherichia coli* UvrA, B, C and D proteins into fibroblasts of xeroderma pigmentosum complementation groups A and C does not result in restoration of UV-induced unscheduled DNA synthesis. Mutat Res 166:89–98

CHAPTER 5

DNA Repair and Carcinogenesis by Alkylating Agents

A. E. PEGG

A. Introduction

There is now general acceptance that most, but not all, chemical carcinogens initiate a change to neoplastic growth by virtue of their interaction with DNA. This DNA damage then leads to the alteration of the activity of certain critical genes by inducing either mutations or an increased frequency of gene rearrangement which, in turn, results in abnormal regulation of gene expression. Although multiple mutations may be required to produce neoplastic growth, and the rate and frequency of progression towards neoplasia are greatly influenced by the extent of exposure to tumor promoters, there is good evidence that the formation and persistence of DNA damage correlates with the potency of a number of chemical carcinogens. These studies suggest that DNA repair processes, if sufficiently active and rapid, can provide a means to reduce the likelihood that exposure to a carcinogen might initiate a tumor. Numerous laboratory investigations, some of which are discussed in this chapter, are consistent with this hypothesis. There is also abundant evidence from epidemiological studies of the increased incidence of cancer in individuals suffering from heritable diseases involving defects in DNA repair. Such findings are consistent with the concept that such repair processes play a major role in minimizing the risk associated with environmental exposure to carcinogens.

The experimental work described in this chapter will be confined to studies of the formation and repair of DNA adducts produced by carcinogens that generate simple alkylating agents. This limitation is necessary in order to reduce a vast and rapidly expanding field to a manageable topic and to provide a more focussed presentation. For a number of reasons, which include the potential environmental significance of some of the carcinogens, the comparative simplicity of the chemistry involved in their synthesis, biochemical activation, and the DNA adducts formed, there is a relatively large body of experimental work on these alkylating carcinogens and on the repair of alkylated DNA. Although studies in a considerable number of laboratories are described, this chapter is by no means intended to be a comprehensive review of the many references in the field, and more extensive documentation can be found in a number of reviews (BARTSCH and MONTESANO 1984; BARTSCH et al. 1987; LAWLEY 1976; MAGEE et al. 1976; PEGG 1977a, 1983, 1984; PREUSSMANN and STEWART 1984; SAFFHILL et al. 1985; SINGER 1984, 1986; SINGER and KUSMIEREK 1982; SINGER and GRUNBERGER 1983; SWENBERG et al. 1985).

B. Alkylating Carcinogens

I. Classes of Compounds

A brief list of carcinogens representative of the wide range of carcinogenic chemicals that lead to the production of alkylated bases in DNA is given below. Only the compounds which contain methyl groups and act as methylating agents are shown, but, in many cases, the higher homologues have also been studied and have provided informative results. These compounds can be divided into three classes: direct acting agents such as dimethyl sulfate, methyl methanesulfonate, and *N*-methyl-*N*-nitrosourea which react without the need for activation; those that require a low molecular weight catalyst (e.g., a thiol group) for effective decomposition to form an alkylating species such as *N*-methyl-*N*′-nitro-*N*-nitrosoguanidine; and those that require metabolic activation such as dimethylnitrosamine, 1,2-dimethylhydrazine, and 4-(*N*-methyl-*N*-nitrosamino)-1-(3-pyridyl)-1-butanone (NNK). A more comprehensive list of the myriad *N*-nitroso compounds which are carcinogenic and may be converted to alkylating species are given in reviews by LAWLEY (1976), MAGEE et al. (1976), PEGG (1983), PREUSSMANN and STEWART (1984), LIJINSKY (1987), BARTSCH et al. (1987), and HECHT and HOFFMANN (1988). The reviews also discuss in detail the metabolic activation systems required for their conversion to alkylating agents. These are now becoming much better understood at the enzyme level and in some cases have been fully characterized. For example, it is now clear that dimethylnitrosamine is converted to a methylating agent by the action of cytochrome P-450j, also known as P-450ac (THOMAS et al. 1987; HONG et al. 1987). The distribution of these activating systems is obviously a key factor in the organ specificity of carcinogens requiring such activation and determines the initial extent of alkylation, but subsequent DNA repair may greatly modify the levels of adducts.

II. Spectrum of DNA Adducts

At least 13 sites for alkylation of DNA are known to occur when alkylating agents react with DNA under physiological conditions (LAWLEY 1976; PEGG 1977a, 1983, 1984; PREUSSMANN and STEWART 1984; SAFFHILL et al. 1985; SINGER 1984, 1986; SINGER and KUSMIEREK 1982; SINGER and GRUNBERGER 1983). These include the alkylphosphate triesters which occur in both the Rp and the Sp diastereoisomers and may involve any of the 16 different possible dinucleotide pairs. These are usually considered as a single adduct, and little is known about their physiological significance and repair in mammalian cells. Alkylation of the ring nitrogen atoms of the bases occurs at most of the potential sites, and 1-methyladenine, 3-methyladenine, 7-methyladenine, 3-methylguanine, 7-methylguanine, 3-methylcytosine, and 3-methylthymine have all been detected. In addition, there is alkylation on the exo-oxygen atoms of the bases, forming O^6-methylguanine, O^2-methylthymine, O^4-methylthymine, and O^2-methylcytosine. Different alkylating agents form different proportions of these products. Comparisons of the relative carcinogenic potency of these agents, the miscoding potential of the various adducts, and other evidence that are all discussed in a number of comprehensive reviews (LAWLEY 1976; PEGG 1977a, 1983, 1984;

PREUSSMANN and STEWART 1984; SAFFHILL et al. 1985; SINGER 1984, 1986; SINGER and KUSMIEREK 1982; SINGER and GRUNBERGER 1983; BASU and ESSIGMANN 1988; TOPAL 1988) have focussed attention on the oxygen adducts as the most probable causes of mutations which initiate mutagenesis. However, the other adducts certainly cannot be ruled out completely. They may contribute at a lower frequency or may enhance the effects of the oxygen adducts by stimulating regenerative or patch repair DNA synthesis which could lead to the incorporation of the incorrect base when a site containing an oxygen adduct is reached.

It should also be noted that, in addition to the direct alkylation of DNA, it is theoretically possible for alkylated bases to be introduced into DNA by means of the alkylation of the nucleotide pool and the subsequent use of the alkylated deoxynucleotide triphosphate by DNA polymerase (TOPAL and BAKER 1982; TOPAL 1988). Such a pathway for O^6-methylguanine would probably lead to its incorporation opposite thymine rather than cytosine, and accurate repair of the O^6-methylguanine would then result in a mutation (TOPAL et al 1986), whereas such repair of the DNA alkylated directly in which the methylated guanine is opposite cytosine would have a protective effect. However, the bulk of the evidence suggests that this indirect route for the formation of alkylated bases in DNA occurs rarely, if at all (PEGG 1983; SNOW and MITRA 1987; BASU and ESSIGMANN 1988).

III. Structure of Adducted DNA and Mutagenesis

There is convincing data from experiments with isolated enzymes and in viral, bacterial, and mammalian cells that the presence of O^6-methylguanine in DNA leads to the induction of mutations which are $G \cdot C \rightarrow A \cdot T$ transitions (SINGER and KUSMIEREK 1982; SINGER and GRUNBERGER 1983; PEGG 1984; PREUSSMANN and STEWART 1984; SAFFHILL et al. 1985; SINGER 1984, 1986; BASU and ESSIGMANN 1988). This is in agreement with the original model proposed by LOVELESS (1969) that the alkylation at the O^6-position of guanine would fix the enol tautomer of the base and facilitate base pairing with thymine. Similarly, the presence of O^4-alkylthymine leads to the misincorporation of guanine and the production of $A \cdot T \rightarrow G \cdot C$ transitions (SINGER and KUSMIEREK 1982; SINGER and GRUNBERGER 1983; SAFFHILL et al. 1985; SINGER 1984, 1986; BASU and ESSIGMANN 1988). Direct evidence supporting the concept has been obtained in recent elegant studies in which O^6-methylguanine or O^4-alkylthymine have been inserted into defined sites in viral of plasmid DNA. This DNA was then reintroduced into a suitable host cell and the incidence of mutations at this site determined (LOECHLER et al. 1984; CHAMBERS et al. 1985; BHANOT and RAY 1986; TOPAL et al. 1986; PRESTON et al. 1986, 1987; BASU and ESSIGMANN 1988).

However, although it is clear that the mutations observed are in agreement with the enol tautomer hypothesis (LOVELESS 1969), more recent studies in which the structure of self-complementary oligodeoxynucleotides containing these adducts opposite various bases has been investigated by high resolution, two-dimensional NMR techniques (PATEL et al. 1985, 1986a–c; KALNIK et al. 1988a, b) do not support this model. Instead, from the analysis of a set of four self-complementary dodecadeoxynucleotides, in each one of which O^6-methylguanine is placed opposite a different base, it appears that the hydrogen bonding between O^6-methylguanine and thymine is much weaker than was originally thought and that this structure is, in fact, no more favored than the O^6meG · A pair which does not introduce a major distortion in the helix (PATEL et al. 1986a–c). A more relevant parameter may be that, with the O^6meG · T pair, there was little or no distortion of the phosphodiester backbone (PATEL et al. 1985). This could be an important factor in favoring the incorporation of thymine opposite O^6-methylguanine by DNA polymerases.

Similarly, NMR studies of the structure of dodecadeoxynucleotides containing G · O^4meT and A · O^4meT (KALNIK et al. 1988a, b) show clearly that miscoding due to O^4-methylthymine is not explained by the presence of a base pair between O^4meT and G stabilized by two hydrogen bonds (SINGER and KUSMIEREK 1982; SINGER 1986). In fact, the A · O^4meT lesion shows a structure in which wobble pairing allows the methyl group in the syn position relative to the N3 of the base to be accommodated within the double helix, whereas the wobble pairing for G · O^4meT does not occur, and a structure stabilized by only one, short, intramolecular hydrogen bond is indicated (KALNIK et al. 1988a, b). The explanation for the observed mutations is, therefore, elusive, but it was suggested that the greater lipophilicity of O^4-methylthymine may be important in holding the G · O^4meT together within the active site of the DNA polymerase (KALNIK et al. 1988b).

Further physicochemical studies of the mechanism of mutagenesis are clearly needed, but the concept that accurate repair of the O^6-methylguanine or O^4-alkylthymine lesions prior to DNA replication will prevent the fixation of damage caused by alkylating agents reacting at these sites is supported by the experimental evidence, irrespective of the mechanism of miscoding. Direct demonstration that this is the case in *Escherichia coli* was provided by data showing that the number of mutations obtained from the constructs with O^6-methylguanine or O^4-alkylthymine was greater in strains lacking the ability to carry out this repair (LOECHLER et al. 1984; CHAMBERS et al. 1985; BHANOT and RAY 1986; PRESTON et al. 1986, 1987; BASU and ESSIGMANN 1988).

C. Factors Influencing Potency and Specificity of Alkylating Carcinogens

It is clear that a number of factors can have a major influence on the carcinogenic potential of these agents. These include: the ability of the compound to be taken up or to penetrate into cells; the rate of metabolic activation and of detoxification reactions; the presence of compounds catalyzing or retarding the conversion to an alkylating species; the presence of alternative target molecules for alkylation which might reduce the interaction with DNA; the types of DNA adducts which are formed; their distribution within the genome; the rate of DNA synthesis; and the extent and accuracy of DNA repair. It is extremely unlikely that only one of these factors needs to be taken into account in order to understand the relative extent of the initiation of tumors in different organs or between different species. Furthermore, many of the factors listed above can be broken down into subclasses, and our ability to examine many of these processes does not always extend to a level of resolution which may be needed for a full understanding. It is, therefore, not at all surprising that many apparent exceptions to the hypothesis that different repair capacity may contribute to the chances of initiating a tumor have been reported. The capacity for DNA repair is only one factor which must be taken into account, and even within this area there are levels at which analysis cannot yet be carried out. Some examples of the detailed analysis which may prove necessary are discussed below in Sect. F and G.

D. Repair of Alkylated DNA in Mammalian Cells

I. DNA Repair In Vivo

Studies of the persistence of alkylated lesions in DNA with time in cultured cells or in intact animals have indicated that all of the alkylation products are lost from DNA at rates which exceed that of cell death or of the spontaneous breakdown of the DNA (PEGG 1977a; SAFFHILL et al. 1985; SINGER and KUSMIEREK 1982; SINGER 1984, 1986; PEGG 1983). This loss must therefore indicate that some enzymatic process is available to remove all of the products, but these reactions have been characterized for only a few of the adducts. It may be misleading to equate disappearance of an adduct from DNA in vivo with DNA repair when the mechanism underlying this loss is unknown. The DNA may not have been restored to its original state merely because the adduct itself cannot be detected. There is, therefore, an urgent need for a more complete understanding of the biochemical nature of the processes which underlie the loss of alkylation products from DNA in mammalian cells.

II. Properties of DNA Repair Proteins

Many detailed reviews of DNA repair processes are available including those by LINDAHL (1982), STRAUSS (1985), and SANCAR and SANCAR (1988). At present, only three mammalian proteins have been fully characterized which remove alkylation products from DNA. These are 3-alkyladenine DNA glycosylase, formamidopyrimidine-DNA glycosylase, and O^6-alkylguanine-DNA-alkyltransferase (AGT). The 3-alkyladenine DNA glycosylase from mammalian cells has been purified extensively. It is not specific for 3-methyladenine but also clearly acts on 7-methylguanine and 3-methylguanine, and probably on 7-methyladenine as well (MARGISON and PEGG 1981; GALLAGHER and BRENT 1984; MALE et al. 1985, 1987). It releases the free base from the DNA leaving behind an apurinic site. This reaction also occurs spontaneously, so the rate of loss of these products from DNA is the sum of the enzymatic and nonenzymatic reactions. The removal of 3-methyladenine seems to be enhanced to the greatest extent by the enzyme. The resulting apurinic sites are then subject to repair by the action of an AP endonuclease that excises a region of DNA containing the site, which is then filled in by the action of DNA polymerase and ligase to form a repair patch.

Exposure of 7-methylguanine in DNA to an alkaline pH leads to the opening of the imidazole ring forming the two rotameric forms of 2,6-diamino-4-

hydroxy5-(*N*-methyl-formamido)pyrimidine. This product is removed by the action of formamidopyrimidine-DNA glycosylase, and this enzyme has been detected and assayed in crude extracts of rat and hamster liver (MARGISON and PEGG 1981). The formamidopyrimidine-DNA glycosylase from *E. coli* has been purified to homogeneity, cloned, and expressed (BOITEAUX et al. 1987), but there are no reports of its purification from mammalian cells. The extent to which 7-methylguanine in DNA is actually converted to the ring-opened form is unclear since a relatively high pH is needed for this reaction, and reports of the presence of the ring-opened adduct have been limited. One group has described significant amounts of this product in mammalian DNA after treatment with alkylating agents (BERANEK et al. 1983; KADLUBAR et al. 1984), but the possibility of artifactual generation of this adduct during isolation and analysis of the DNA has not been entirely ruled out. The formamidopyrimidine-DNA glycosylase, which is widely distributed, may be more important for the repair of DNA in which the purine rings have been opened by γ-irradiation, but its possible role in response to alkylation damage should not be overlooked since the 2,6-diamino-4-hydroxy 5-(*N*-methylformamido)pyrimidine is clearly non-coding (BOITEAUX and LAVAL 1983) and is persistent in DNA (BERANEK et al. 1983), once formed.

The O^6-alkylguanine adduct is not repaired by the glycosylase/AP endonuclease pathway but instead is restored in a single step by the action of the AGT protein (LINDAHL 1982; PEGG 1983; YAROSH 1985; PEGG and DOLAN 1987). This protein catalyzes the transfer of the alkyl group to a cysteine acceptor site within its polypeptide sequence. The alkyl acceptor site is not regenerated, and the protein therefore acts stoichiometrically rather than catalytically, and it is not really an enzyme. The number of O^6-alkylguanine lesions that can be repaired rapidly by the cell is thus limited to the number of molecules of the AGT protein (PEGG and HUI 1978; PEGG 1978, 1983; PEGG et al. 1983, 1984). When the supply of the protein is exhausted, further repair depends on the de novo synthesis of the AGT. This mechanism at first sight seems highly inefficient, since a whole protein molecule is used up for every alkyl group removed from DNA, but, it should be noted that the AGT is a small protein (MW about 22000) and that no additional enzymes nor the synthesis of DNA in repair patches are required for the correction of the O^6-alkylguanine.

No other pathways for repair of alkylation lesions in DNA in mammalian cells have been characterized at the biochemical level. A far more complete understanding of the repair of alkylation damage in *E. coli* is available (see LINDAHL 1982 and SANCAR and SANCAR 1988). It is clear that, in *E. coli*, nucleotide excision repair can also remove O^6-alkylguanine and O^4-alkylthymine from DNA (CHAMBERS et al. 1985; SAMSON et al. 1988). Such excision repair may become increasingly important as the size of the alkyl group increases. It may also be the case that excision repair contributes to the loss of alkylation adducts from DNA in mammalian cells, and a relatively inefficient and slow removal of most of the products by this mechanism might account for the loss of all of the alkylation products described above.

However, it should be emphasized that there are significant differences between bacteria and mammalian cells in the handling of alkylation damage. This is particularly obvious in the case of the alkylated pyrimidines. In *E. coli*, the alkA gene product (3-alkyladenine DNA glycosylase II), which is the equivalent of the mammalian 3-alkyladenine DNA glycosylase, is also able to act on O^2-methylpyrimidines (McCarthy et al. 1984), but attempts to demonstrate a mammalian glycosylase activity for these products have failed (Brent et al. 1988). Similarly, the *E. coli* ada gene product is an inducible alkyltransferase which repairs O^6-methylguanine, but this protein can also act on O^4-methylthymine using the same cysteine acceptor site and on the Sp configuration of methylphosphate triesters using an different acceptor site (McCarthy et al. 1983; Weinfeld et al. 1985). The mammalian AGT does not act on either O^4-methylthymine or alkylphosphate triesters (Yarosh 1985; Brent et al. 1988; Dolan et al. 1988a).

The mammalian DNA repair proteins have been mainly studied with respect to their abilities to remove methyl adducts, but it is clear that they do work on DNA containing longer alkyl groups, although the rate of removal does decrease as the size of the alkyl group increases (Pegg 1983; Pegg et al. 1984; Singer 1984, 1986; Morimoto et al. 1985; Pegg and Dolan 1987). This has been studied in detail for the AGT protein. Ethyl, *n*-propyl, 2-chloroethyl, and even, possibly, *n*-butyl groups are removed sufficiently rapidly by the protein for it to be probable that this is the major route of their removal in cells which contain AGT. However, branched chain adducts such as *iso*-propyl or *iso*-butyl are only removed very slowly by AGT, and these products, and possibly also 2-hydroxyethyl and *n*-butyl adducts, may be handled predominantly by other pathways which are likely to be those for general excision repair (Chambers et al. 1985; Saffhill et al. 1985).

III. Distribution of Mammalian DNA Repair Activities

There have been very few studies on the distribution of the 3-methyladenine DNA glycosylase in mammalian cells, and no descriptions of the distribution of the formamidopyrimidine-DNA glycosylase have been reported. It appears that both enzymes are ubiquitous, having been found in all species and cell types examined. Craddock and Henderson (1982) reported that there were only small differences in the content of 3-methyladenine DNA glycosylase between organs of the rabbit, but recently Washington et al. (1988) have found much greater variations between tissues in the mouse, with stomach having the most activity and brain, ovary, and liver, the least. Further examination of this question seems warranted. There were also significant differences in activity between two mouse stocks (Washington et al. 1988), and rabbits and cats had more activity than rodents (Craddock and Henderson 1982). The release of 7-methylguanine from DNA was faster when hamster liver was used as a source of extract than when rat liver was used, suggesting that these species may differ in their content of 3-methyladenine DNA glycosylase (Margison and Pegg 1981), but comparisons of the purified glycosylase enzymes from these tissues have not been made.

The distribution of AGT activity shows a very marked cell, organ, and species specificity (reviewed by Pegg 1983; Pegg and Dolan 1987). The content of AGT in human tissues is, in general, an order of magnitude greater than that present in the equivalent rodent cells (Pegg et al. 1982, 1983; Wiestler et al. 1984; Graf-

STROM et al. 1984). There are also major differences in the amount of AGT between different organs. In all species examined, the AGT level is highest in the liver and lowest in the nervous system (PEGG 1983; PEGG et al. 1985; GRAFSTROM et al. 1984; CRADDOCK and HENDERSON 1984; JUN et al. 1986). The amounts of AGT activity measured in various organs correlate well with the persistence of O^6-alkylguanine in these organs, confirming the key role which this protein plays in the repair of the O^6-adduct (PEGG 1983, 1984, 1985). As might be expected from these results, there are also marked differences in AGT between different cell types making up an organ (see Sect. F). The content of AGT in rat hepatocytes is much higher than that present in the nonparenchymal cells (SWENBERG et al. 1982), and in the lung the activity in macrophages is considerably greater than that in Clara cells (DEILHAUG et al. 1985; BELINSKY et al. 1988).

E. Correlation of Adduct Persistence with Tumor Initiation

I. Single-Dose Experiments

Goth and Rajewsky's pioneering observations on the persistence of alkylated bases in the DNA of rats treated with *N*-ethyl-*N*-nitrosourea suggested that the rate of repair of certain lesions might contribute towards the explanation for the striking organ specificity of tumor induction by this agent. Single doses of *N*-ethyl-*N*-nitrosourea given to pregnant rats lead to virtually a 100% incidence of tumors in the nervous system in the offspring even though the initial extent of alkylation of DNA in the brain is no greater than in non-target tissues such as liver. GOTH and RAJEWSKY (1974) observed that the hepatic content of O^6-ethylguanine is rapidly reduced whereas the content of this adduct declines very slowly in the brain. Other ethylation products including 7-ethylguanine, 3-ethyladenine, O^4-ethylthymine, and ethylphosphate triesters showed a similar rate of disappearance in all organs examined. These observations focussed attention on the possibility that the presence of O^6-ethylguanine in DNA at the time of DNA replication was a cause of the initiation of neoplastic growth and that the repair of this adduct provided a protective mechanism against this initiation. Work by KLEIHUES and MARGISON (1974) indicated that a similar effect was seen after administration of *N*-methyl-*N*-nitrosourea, and it was found that after weekly treatments with *N*-methyl-*N*-nitrosourea, O^6-methylguanine accumulated in rat brain DNA but not in liver and to an intermediate extent in kidney (MARGISON and KLEIHUES 1975). This persistence correlates with the sensitivity of these tissues to tumor induction.

These findings were confirmed and extended in a number of laboratories (reviewed by PEGG 1983, 1984), and there is now general agreement that this mechanism contributes towards the carcinogenicity and organ and species specificity of carcinogenesis by alkylating agents. However, paradoxically and as discussed in Sect. H, the activated oncogene characterized in a nervous system tumor induced by *N*-ethyl-*N*-nitrosourea is the *neu* oncogene which contains an $A \cdot T \rightarrow T \cdot A$ transversion, a mutation which does not involve alkylation of guanine. Furthermore, subsequent experiments have shown that accumulation of

O^6-alkylguanine occurs in the brains of gerbils, mice, and hamsters treated with these nitrosourea derivatives and that these species have only a very low susceptibility to brain tumor initiation (KLEIHUES and RAJEWSKY 1984). It is, therefore, clear that other factors contribute substantially to the species differences in sensitivity to brain tumor induction by alkylating agents.

Treatment of rats with single doses of dimethylnitrosamine leads to the induction of kidney tumors (MAGEE et al. 1976; PREUSSMANN and STEWART 1984). The initial level of DNA alkylation is much greater in the liver than in the kidney because the liver contains a much higher content of cytochrome P450j which is responsible for the metabolism of dimethylnitrosamine to the alkylating species (SWANN and MAGEE 1986). This discrepancy was resolved by studies of the persistence of O^6-methylguanine in these tissues. These revealed that the repair of this product occurred much more completely in the liver (NICOLL et al. 1975). The O^6-methylguanine adduct content of the kidney showed a small but quite rapid fall and then remained at a high level for several days. This saturation of repair in the kidney was later explained by the exhaustion of the AGT content in this organ (PEGG 1978, 1983). In both tissues, the contents of 7-methylguanine and 3-methyladenine in DNA did not show a difference in either rate or extent of removal. These results suggest that it is unrepaired O^6-methylguanine that is responsible for the initiation of tumors by dimethylnitrosamine and that the liver is protected from the carcinogenic insult by having a higher content of the AGT protein.

When dimethylnitrosamine is administered to rats in continuous feeding experiments, the resulting tumors do occur in the liver (MAGEE et al. 1976; PETO et al. 1984; PREUSSMANN and STEWART 1984). Although at first sight this appears to be inconsistent with the greater sensitivity of the kidney described above, it can readily be explained by taking into account the pharmacokinetics of the treatment. The capacity of the liver to metabolize dimethylnitrosamine is such that small daily doses of the carcinogen given in the drinking water are completely metabolized in the liver (DIAZ GOMEZ et al. 1977; PEGG and PERRY 1981). This first pass clearance prevents the interaction with other organs such as the kidney which contain lower content of AGT and are less able to repair O^6-methylguanine.

The ability of the liver to repair O^6-methylguanine also shows saturation kinetics when the existing AGT is exhausted. When doses of alkylating agents are used which produce more O^6-methylguanine than the number of molecules of AGT, this product accumulates in the liver DNA (PEGG 1977b, 1978, 1983; PEGG and HUI 1978; PEGG et al. 1984). The incidence of liver tumors after such doses is much higher than that found when lower doses are used (PETO et al. 1984). However, even the lower doses, which do not saturate the repair by AGT, do lead to an increase in liver tumor induction. This can be explained by several factors. Firstly, adducts other than O^6-methylguanine may contribute to the initiation of tumors. Even though this contribution is small when O^6-methylguanine is not repaired, it may be relatively more important under conditions in which most of the O^6-methylguanine is repaired by AGT. Secondly, the repair of O^6-methylguanine by AGT does not occur instantaneously, and whenever this product is formed in DNA, there is a probability that DNA replication will occur before repair. There is an increased incidence of liver tumors induced by alkylating agents in animals in which the liver is regenerating after partial hepatectomy or in new-born animals in which the rate of cell division in the liver is greater than in adults (CRADDOCK 1971; MAGEE et al. 1976; PEGG 1983; PREUSSMANN and STEWART 1984). Finally, it is possible that some sites of production of O^6-methylguanine are repaired much more slowly than others and that the small, remaining fraction of this product, even in the presence of AGT, may be sufficient to induce a low incidence of tumors.

Syrian golden hamsters are more sensitive to liver tumor induction by dimethylnitrosamine than rats (MAGEE et al. 1976; MONTESANO 1981), but the extent of DNA alkylation in the liver was similar in both species (STUMPF et al. 1979). The rate of repair of 3-methyladenine and 7-methylguanine was somewhat faster in hamster liver, which may be related to a higher content of the glycosylase (MARGISON and PEGG 1981), but studies of the persistence of O^6-methylguanine in these species indicated that the repair of this base in hamster liver was saturated at considerably lower levels than in the rat (STUMPF et al. 1979; MONTESANO 1981). This difference may explain the greater susceptibility of hamsters to liver tumor initiation and is due to a lower content of AGT in this species and to the induction of AGT in the rat after exposure to dimethylnitrosamine (see Sect. E.III).

The bulk of the experimental evidence does support the concept that, under some circumstances, the presence of unrepaired O^6-alkylguanine in tissue DNA may lead to the initiation of tumors. However, although this may be necessary for initiation to occur, it is clearly not by itself sufficient. In *Xenopus*, there is little repair of the O^6-methylguanine that is formed in liver or kidney DNA after administration of *N*-methyl-*N*-nitrosourea, but no tumors were observed in these organs (HODGSON et al. 1980).

II. Chronic Dose Experiments

Although experiments in which single doses or a few daily doses of the carcinogen have much more convenient experimental protocols for studying the formation of DNA adducts and their repair, most carcinogenesis experiments and potential environmental exposures involve long-term exposure to lower doses. Detailed studies by Swenberg and colleagues have been carried out on the content of alkylated bases in the liver DNA of rodents treated with dimethylnitrosamine, diethylnitrosamine, and 1,2-dimethylhydrazine under these conditions (BEDELL et al. 1982; LINDAMOOD et al. 1982, 1984; LEWIS and SWENBERG 1983; SWENBERG et al. 1982, 1984, 1985; RICHARDSON et al. 1985; DYROFF et al. 1986; BOUCHERON et al. 1987).

Exposure of rats to 30 ppm of 1,2-dimethylhydrazine in the diet leads to the accumulation of O^6-methylguanine in the nonparenchymal cells, which are the major source of liver tumors induced by this carcinogen (93% incidence), but there was much less accumulation of this adduct in the hepatocytes (BEDELL et al. 1982; LEWIS and SWENBERG 1983). The nonparenchymal cells also have a higher rate of cell division. Crude calculations of the theoretical probability of tumor initiation, made by taking the product of the cell replication rate and the content of

O^6-methylguanine divided by the amount of DNA in the respective cell population, indicated that the probability was 17-fold higher for angiosarcomas as compared with hepatocellular carcinomas (SWENBERG et al. 1985). However, hepatocellular carcinomas are induced in 40% of the animals. This discrepancy could relate to other unknown factors rendering the hepatocytes more sensitive to tumor initiation or may indicate the contribution of other DNA adducts. These might include O^4-methylthymine. Although this is formed initially at levels of less than 1% of the O^6-methylguanine formed, it was present in hepatocytes at levels similar to O^6-methylguanine after 28 days' exposure (RICHARDSON et al. 1985). This reflects the much slower removal of O^4-methylthymine, which was lost from the DNA with a half-life of about 20 h after exposure to a 20 mg/kg dose of 1,2-dimethylhydrazine (RICHARDSON et al. 1985).

Studies in which mice were exposed chronically to dimethylnitrosamine also indicated that O^6-methylguanine was maintained at moderate levels in the hepatocyte DNA (about 5 pmol/mg) but progressively accumulated in the non-parenchymal cells, reaching 50 pmol/mg after 1 month's exposure (LINDAMOOD et al. 1982, 1984). The relative initiation index, calculated from these data as described above, indicated a 2.5 times greater probability of initiation in the non-parenchymal cells, which is in reasonable agreement with the tumor incidence (SWENBERG et al. 1985).

Exposure of rats to diethylnitrosamine at 40 ppm for up to 11 weeks showed that O^4-ethylthymine accumulated in the hepatocytes to levels that were at least 50 times greater than the steady state level of O^6-ethylguanine (SWENBERG et al. 1984; DYROFF et al. 1986). This accumulation of O^4-ethylthymine is consistent with studies showing that this adduct is lost very slowly from hepatic DNA with a half-life of about 11 days (RICHARDSON et al. 1985). Therefore, the combination of the rapid repair of O^6-ethylguanine by the AGT protein, the induction of AGT after exposure to diethylnitrosamine (see Sect. E.III), and the lack of efficient repair of O^4-ethylthymine results in a change in the ratio of these adducts from the initial 4:1 (O^6-ethylguanine: O^4-ethylthymine) to 1:50 after 4 weeks' exposure. These results suggest that O^4-ethylthymine may be the critical unrepaired lesion leading to the initiation of tumors in rat hepatocytes after exposure to diethylnitrosamine (SWENBERG et al. 1984, 1985). Further studies with various diethylnitrosamine concentrations (BOUCHERON et al. 1987) in which the alkylation, persistence of alkylated bases, and cell proliferation were measured in various lobes of the liver (RICHARDSON et al. 1986) support these conclusions but indicate that the differences in diethylnitrosamine activation and in the induction of DNA replication may also be important factors in the dose response (SWENBERG et al. 1987). It seems clear that the rapid removal of O^6-ethylguanine by the AGT protein protects the liver from the potential initiation of tumors by this adduct, and this renders O^4-ethylthymine, which is formed in smaller amounts, the primary persistent lesion that leads to tumors.

Investigations of the accumulation and persistence of methylated DNA adducts in the respiratory tract of rats treated with NNK have been carried out by Belinsky and colleagues (BELINSKY et al. 1986, 1987a, b, 1988; DEVEREUX et al. 1988). After a single treatment with NNK, the content of alkylated DNA was greatest in the nasal mucosa (a major target for NNK carcinogenicity), followed by the liver and the lung. However, after 12 days of treatment, the content of O^6-methylguanine in the liver had declined significantly, due to induction of the AGT repair mechanism (see Sect. E.III). In contrast, the content in nasal mucosa declined by only about twofold, probably due to toxicity to this tissue which led to a decrease in carcinogen activation (BELINSKY et al. 1986). The high incidence of tumors derived from the olfactory mucosa can be explained by the extensive formation of O^6-methylguanine in this tissue and the increase in cell proliferation resulting from the alkylation damage (BELINSKY et al. 1987a).

However, in the lung, which is also a major target for NNK carcinogenicity, the content of O^6-methylguanine increased over the 12 days of treatment, and there was also some accumulation of O^4-methylthymine to a level about 5% of that of O^6-methylguanine. When treatment was stopped, there was a rapid loss of O^4-methylthymine from the lung DNA, but the O^6-methylguanine persisted (BELINSKY et al. 1986). More detailed studies of the alkylation of the lung DNA by NNK revealed that the content of O^6-methylguanine was not evenly distributed in the different cell types but was much greater in the Clara cells and that alkylation of the lung was also nonlinear with dose (BELINSKY et al. 1987a, b). At low doses of NNK, the O^6-methylguanine to dose ratio was much greater than at high doses. This effect was also most marked in the Clara cells, suggesting that there is a high-affinity pathway for activation of NNK leading to alkylation of these cells. This contrasts strikingly with the metabolism of dimethylnitrosamine which does not show such specificity (DEVEREUX et al. 1988). In addition to showing a high level of alkylation at low doses of NNK, Clara cells were less active than macrophages, alveolar type II cells, and alveolar small cells in carrying out the removal of O^6-methylguanine from their DNA when treatment with NNK ceased (BELINSKY et al. 1988). This correlates with the observation that the AGT content of the Clara cells is lower than in the other lung cells and that it is restored more slowly after exhaustion due to treatment with NNK (BELINSKY et al. 1988). The high concentration of O^6-methylguanine formed in the Clara cell DNA and the low rate of repair of this adduct may play an important role in pulmonary carcinogenesis induced by NNK. The relatively poor capacity of Clara cells to remove O^6-alkylguanine from their DNA is also supported by studies of the alkylation of hamster lung DNA following treatment with diethylnitrosamine (FONG and RASMUSSEN 1987). In these experiments, O^6-ethylguanine clearly accumulated in these cells to much higher levels than those found in macrophages or alveolar type II cells.

III. Role of Induction of DNA Repair Processes

The AGT activity in rat liver is inducible in response to chronic or even single doses of dialkylnitrosamines (reviewed by MONTESANO 1981; PEGG 1983; SAFFHILL et al. 1985). This induction amounts to as much as a threefold increase, and the increased AGT content clearly contributes to a more rapid removal of O^6-alkylguanine from DNA. Measurements of the hepatic content of O^6-alkylguanine after treatment with nitrosamines showed that there was an increased extent of removal of the adduct from rat liver DNA (MONTESANO 1981; PEGG 1983; SAFFHILL et al. 1985). The inducibility of AGT in rat liver may well contribute to the relative resistance of this organ to carcinogenesis by alkylating agents, but the mechanism of the induction is at present unclear. In the rat, AGT content is increased by a range of other hepatotoxins which are not converted to alkylating agents, by irradiation, and during liver regeneration after partial hepatectomy (PEGG et al. 1981; PEGG 1983; SAFFHILL et al. 1985). In mice, hamsters, and gerbils, hepatic AGT is not induced by either dimethylnitrosamine or by partial hepatectomy (PEGG 1983, SAFFHILL et al. 1985).

Although the AGT content is elevated by as much as sixfold after partial hepatectomy (PEGG et al. 1981), which would be expected to increase resistance to carcinogenesis, it should be noted that such treatment following administration of dimethylnitrosamine or *N*-methyl-*N*-nitrosourea clearly increases the probability of tumor initiation (CRADDOCK 1971; MAGEE et al. 1976; PEGG 1983; PREUSSMANN and STEWART 1984). However, this can be explained by the substantial rise in DNA replication which more than compensates for the increased removal of O^6-methylguanine from DNA in increasing the probability that a segment of DNA containing an O^6-methylguanine will be replicated prior to repair.

A small increase in the activity of the 3-methyladenine DNA glycosylase has also been reported to occur in rat liver cells proliferating after partial hepatectomy (GOMBAR et al. 1981; CRADDOCK and HENDERSON 1982). This increase is much less than the rise in AGT (PEGG et al. 1981) and may be explained by a small variation in the activity of the glycosylase during the cell cycle.

IV. Summary of In Vivo Adduct Persistence Experiments

In summary, the results of these experiments support the concept that the presence of unrepaired O^6-alkylguanine or O^4-alkylthymine may lead to the initiation of tumors if the adduct is present at the time that DNA is replicated. The presence of adequate amounts of the AGT protein can remove O^6-alkylguanine from DNA and minimize the initiation from this adduct. When substantial removal of the O^6-alkylguanine can take place, the accumulation of O^4-alkylthymine may play a major role in tumor initiation. Although formed in smaller amounts, the lack of repair of this adduct can lead to its persisting at levels equal to or exceeding those of O^6-alkylguanine. This is of particular significance with diethylnitrosamine and *N*-ethyl-*N*-nitrosourea since the initial ratio of formation of the thymine adduct to the guanine adduct is only about 1:4, and the O^4-ethylthymine is removed from DNA very slowly. It has by no means been ruled out that other pyrimidine adducts such as O^2-alkylcytosine and O^2-alkylthymine may also accumulate relative to O^6-alkylguanine due to their ineffective removal whilst O^6-alkylguanine is removed by AGT, but these products have simply not been quantitated in a sufficient number of studies for this possibility to be examined.

F. Cell-Specific DNA Repair

The content of AGT protein and the capacity to repair O^6-alkylguanine in DNA differs dramatically not only from one organ to another but also between different cell types in the same organ. This has been demonstrated most clearly in the liver as hepatocytes have a much larger content of AGT than nonparenchymal cells (SWENBERG et al. 1982). This distribution of AGT activity leads to a much greater accumulation of O^6-methylguanine in DNA of the nonparenchymal cells when rats or mice are chronically exposed to 1,2-dimethylhydrazine or dimethylnitrosamine. In rats, this difference was accentuated because chronic treatment with 1,2-dimethylhydrazine leads to a threefold induction of the AGT content in the hepatocytes but did not increase the content of this protein in the nonparenchymal cells (SWENBERG et al. 1982).

Striking differences also exist in the AGT content of the various cell types in the lung. In rabbit lung, the activity of AGT in Clara cells was much less than in alveolar type II cells or in macrophages (DEILHAUG et al. 1985). In the rat, constitutive levels of AGT were at least twofold greater in macrophages and type II cells compared with Clara cells and alveolar small cells. Treatment with the potent lung carcinogen NNK for 4 days did not affect the AGT content of macrophages, reduced the AGT content in the alveolar small cells and type II cells, and completely depleted the activity in Clara cells (BELINSKY et al. 1988). These findings are also in agreement with the persistence of O^6-methylguanine in these cell types when the treatment with NNK was stopped. The loss of this product from macrophages followed first order kinetics with a half-life of about 48 h, whereas there was little detectable loss of O^6-methylguanine from the Clara cell DNA in 8 days (BELINSKY et al. 1988).

These findings may help to explain the sensitivity of certain cells to tumor initiation by alkylating agents, and further study of the distribution of the AGT protein in organs comprised of multiple cell types is clearly very important in evaluating this possibility. Unfortunately, at present no antibodies to the mammalian AGT are available, and studies of its cellular distribution have been limited to investigations of organs from which isolated cell preparations enriched for a particular cell type can be prepared.

G. Fine-Structure Specificity in DNA Repair

I. Repair in Specific Genes and Regions of Chromatin

Although it was for some time assumed that DNA damage and repair sites are uniformly distributed throughout the genome, there is now an overwhelming body of evidence that this is not the case (reviewed by BOHR et al. 1987). It appears that some regions of the chromatin are not easily accessible to DNA repair enzymes, and there is evidence that favors the concepts that many DNA-damaging agents bind more extensively to euchromatin and that preferential repair occurs in more accessible regions of the chromatin structure (BOHR et al. 1987; GALE et al. 1987; ARNOLD et al. 1987). Also, carcinogens may interact with mitochondrial DNA to a different extent from nuclear DNA, and the repair of mitochondrial and nuclear DNA may also vary. In fact, alkylating agents, such as dimethylnitrosamine and *N*-methyl-*N*-nitrosourea, seem to interact more extensively with mitochondrial DNA (WILKINSON et al. 1975; BOHR et al. 1987; MYERS et al. 1988). The repair of alkylated DNA in mitochondria has not been studied in any detail, but it is known that this repair does take place, and AGT has been found in extracts from isolated mitochondria (MYERS et al. 1988).

At an even more specific level of analysis, Hanawalt and colleagues have shown that CHO cells in culture remove UV-induced damage from the DNA of active genes much more efficiently than from sequences not actively involved in transcription (BOHR et al. 1987; MELLON et al. 1986; BOHR and HANAWALT 1987). Similarly, in human cells, the removal of pyrimidine dimers occurs earlier from transcribed than from nontranscribed sequences. Furthermore, it appears that there is even specificity at the level of the DNA strand that is repaired, with a more rapid repair taking place in the transcribed strand (MELLON et al. 1987). Although most of these experiments have been carried out with UV radiation as the DNA damaging agents, it is clear that a similar pattern of preferential repair in actively described genes also applies to bulky adducts (VOS and HANAWALT 1987; BOHR et al. 1987). Equivalent studies have not yet been published for alkylating agents, but it seems likely that alkylation damage will also show selectivity of repair at this level.

II. Sequence Specificity of DNA Alkylation, Repair and Mutagenesis

Many early studies (reviewed in RYAN et al. 1986; BOHR et al. 1987) have shown by nuclease digestion or other chromatin fractionation techniques that the alkylation damage in cells treated with alkylating agents is not randomly distributed. More recently, the presence of selective alkylation has been demonstrated by other, more precise methodology (NEHLS et al. 1984; BRISCOE and COTTER 1985; MATTES et al. 1986; HARTLEY et al. 1986; DOLAN et al. 1988 b; WURDEMAN and GOLD 1988). There is also some evidence that nonrandom repair of methylated purines, with a faster rate of repair occurring in the active chromatin, may contribute to the nonrandom distribution of alkylation damage (RYAN et al. 1986).

Several lines of evidence suggest that both the incidence of alkylation and its repair may be sequence-specific. In fact, direct demonstrations that DNA alkylation is sequence-specific have been provided in studies of the alkylation of oligodeoxynucleotides of defined composition by *N*-methyl-*N*-nitrosourea (DOLAN et al. 1988 b), by alkylation of deoxynucleotide polymers by alkylnitrosoureas (BRISCOE and COTTER 1984, 1985; BRISCOE and DUARTE 1988), and by alkylation of sections of plasmid or viral DNA (MATTES et al. 1986; HARTLEY et al. 1986, 1988; KOHN et al. 1987; WURDEMAN and GOLD 1988). In all cases it appeared that the alkylation of guanine was not random. Preferential formation of 7-methylguanine in regions consisting of runs of contiguous guanines was seen after treatment with chloroethylating agents (HARTLEY et al. 1986), nitrogen mustards (MATTES et al. 1986; KOHN et al. 1987), triazines (HARTLEY et al. 1988), and *N*-methyl-*N*-nitrosourea (WURDEMAN and GOLD 1988). Briscoe and colleagues (BRISCOE and COTTER 1984, 1985; BRISCOE and DUARTE 1988) have also observed sequence specificity in the alkylation of synthetic polydeoxynucleotides by *N*-methyl-*N*-nitrosourea and by 1,3-bis(3-chloroethyl)-1-nitrosourea and found that runs of guanines are preferentially alkylated at the N-7 position.

Analysis of the alkylation of double-stranded dodecadeoxynucleotides by *N*-methyl-*N*-nitrosourea indicated that the formation of both 7-methylguanine and O^6-methylguanine was favored when the alkylated guanine was flanked on the 5′ side by a purine, particularly guanine (DOLAN et al. 1988 b).

The mechanistic basis for the preferential alkylation of guanine in sequences in which it is on the 3′ side of a purine is not yet clear, but several plausible mechanisms have been suggested which would be compatible with this finding. BUCKLEY (1987) has proposed a regioselective mechanism which involves the initial formation from the imidourea derivative of a tetrahydral intermediate at the 5′ dG and the subsequent intramolecular displacement reaction of the 3′ dG with this tetrahydral intermediate to generate the alkylated guanines at this site. Other explanations for neighboring base effects of the specificity of alkylation which involve the negative molecular electrostatic potential induced by the neighboring base have also been advanced (FUROIS-CORBIN and PULLMAN 1985; KOHN et al. 1987).

Evidence for sequence specificity in the repair of DNA by *E. coli* AGT was obtained by Topal and colleagues who synthesized a bacteriophage f1/pBR 322 chimera containing O^6-methylguanine opposite T in the ampicillinase gene. Analysis of the mutations seen in this gene after passage of the DNA through *E. coli* indicated that certain sequences were unmutated, and this was interpreted to mean that these sequences were not repaired by the AGT protein (TOPAL et al. 1986; TOPAL 1988). The consensus sequence from these sites has considerable similarity to the sequence present in the H-*ras* oncogene about the 12th codon, which, as discussed below (Sect. H), is known to be mutated by *N*-methyl-*N*-nitrosourea.

The AGT repair protein is able to act on relatively short oligodeoxynucleotides containing O^6-methylguanine (SCICCHITANO et al. 1986; DOLAN et al. 1988 c). In fact, the repair of double-stranded dodecamers by the AGT protein occurs at rates comparable to the repair of alkylated, high molecular weight DNA. Such dodecamers were therefore used to demonstrate the sequence specificity of DNA repair (DOLAN et al. 1988 b; Dolan and Pegg, unpublished observations). It was found that repair was slowest when the 5′ flanking base was guanine. Combining these results with the studies on the nonrandom alkylation of dodecadeoxynucleotides by *N*-methyl-*N*-nitrosourea described above indicates that O^6-methylguanine is likely to accumulate in certain sequences (DOLAN et al. 1988 b). A crude quantitation of this effect can be obtained by dividing the amount of O^6-methylguanine formed in a particular sequence by its rate of repair. Such a calculation indicates that the relative risk associated with the 5′ flanking base is in the ratio of about 7:2:1:1 for guanine:adenine:thymine:cytosine, respectively. This is of course a very rough calculation for many reasons. These include the fact that only the 5′ preceding base is taken into account, and it is highly probable that other bases close to the guanine considered also play a role.

However, several in vivo studies support this concept. Glickman and colleagues studied the influence of the neighboring base sequence on the mutations caused by *N*-methyl-*N*′-nitro-*N*-nitrosoguanidine, ethyl methanesulfonate, or *N*-ethyl-*N*-nitrosourea in the *lac*I gene of *E. coli* (GLICKMAN et al. 1987; BURNS et al.

1986, 1987, 1988). In all cases, the majority of the mutations were G·C→A·T transitions. With *N*-methyl-*N'*-nitro-*N*-nitrosoguanidine and *N*-ethyl-*N*-nitrosourea as alkylating agents, these mutations occurred predominantly in the sequence 5'-purine-G-3' (Burns et al. 1986, 1988), whereas ethyl methanesulfonate did not show this specificity (Burns et al. 1986). In contrast to the results of treatment with *N*-methyl-*N'*-nitro-*N*-nitrosoguanidine or ethyl methanesulfonate, where virtually all of the mutations (98% and 97%, respectively) were the G·C→A·T transitions, *N*-ethyl-*N*-nitrosourea gave rise to 17% A·T→G·C transitions (Burns et al. 1988). This is consistent with the miscoding potential of the O^4-alkylthymine adduct since *N*-ethyl-*N*-nitrosourea forms a much higher proportion of O^4-alkylthymine than the other agents. Also, these mutations occurred in sequences 5'-purine-T-3', again suggesting that formation and/or repair favors the persistence of the adduct at such sites.

It is also noteworthy that the incidence of mutations induced by *N*-ethyl-*N*-nitrosourea and ethyl methanesulfonate in this system was increased fivefold when the experiments were carried out in a strain unable to repair DNA by excision repair (Burns et al. 1986, 1988). This provides further support for the repair of both O^4-ethylthymine and O^6-ethylguanine by excision repair.

On average, the mutations produced by *N*-methyl-*N'*-nitro-*N*-nitrosoguanidine at guanine residues were 9 and 5 times more likely to occur when the 5' flanking base was guanine or adenine, respectively (Glickman et al. 1987; Burns et al. 1988). Mutations induced by *N*-ethyl-*N*-nitrosourea also showed a bias towards sites in which the 5' flanking base was a purine. In addition, mutations induced by *N*-ethyl-*N*-nitrosourea and ethyl methanesulfonate in the wild-type *E. coli* strain were more prevalent when the site was flanked by G·C base pairs. This bias was not seen in the $UvrB^-$ strain, and the results therefore suggest that neighboring base sequences influence the repairability of the lesions by the excision repair complex (Burns et al. 1988).

Similar results were obtained from studies by Richardson et al. (1987a, b) in which the *E. coli gpt* gene was used as a target. In this system, all of the mutations induced by *N*-methyl-*N*-nitrosourea were G·C→A·T transitions whereas 73% of the point mutations induced by *N*-ethyl-*N*-nitrosourea were G·C→A·T transitions and 21% were A·T→G·C transitions. The majority of the G·C→A·T transitions induced by these agents (82% and 71%, respectively) were present in the middle guanine of the sequence 5'-GG(A,T)-3' (Richardson et al. 1987a). When *N*-methyl-*N'*-nitro-*N*-nitrosoguanidine was used as an alkylating agent in both adapted (induced for high AGT levels) and unadapted *E. coli,* most (more than 92%) of the mutations were G·C→A·T transitions, and these also existed predominantly in the 5'-GG(A,T)-3' sequence (Richardson et al. 1987b).

It is noteworthy that in these experiments (Richardson et al. 1987a, b) and in those of Reed and Hutchinson (1987) on the mutation of bacteriophage λ with *N*-methyl-*N'*-nitro-*N*-nitrosoguanidine, the majority of the mutations occurred in the nontranscribed strand. A possible reason for the strand-specificity effect is that the antisense strand is single-stranded for part of the time and in this state is more resistant to repair by the AGT protein. [Strand specificity in the mutagenesis of M13 plasmids containing O^6-methylguanine was also observed, but this may be accounted for by the difference in infectivity of the plus and minus strands or the mode of M13 replication (BASU and ESSIGMANN 1988).]

H. Correlation of Mutations in Oncogenes with DNA Damage and Repair

Analysis of the activated H-*ras* oncogenes found in rat mammary tumors induced by the administration of *N*-methyl-*N*-nitrosourea indicted that, in all cases examined, a G·C→A·T transition mutation had occurred (SUKUMAR et al. 1983; ZARBL et al. 1985). This is consistent with the lack of repair of an O^6-methylguanine adduct at this site.

All of the single base changes observed as a result of mutations induced by ethyl methanesulfonate in a shuttle vector system containing the *lac*I gene of *E. coli* as a target in human adenovirus-transformed embryonic kidney cells were also found to be G·C→A·T transitions (LEBKOWSKI et al. 1986). This system, which is based on using portions of SV40 in the constructions to allow for replication in human cells, has a high background, and these vectors replicate poorly in human cells. However, similar results were obtained using vectors with the same gene which are autonomously replicating derivatives of Epstein-Barr virus (DUBRIDGE et al. 1987). In this system, virtually all of the mutations induced by *N*-methyl-*N*-nitrosourea were also G·C→A·T transitions (DUBRIDGE et al. 1987). Furthermore, there was an apparent hot spot for mutations at a run of three consecutive guanine residues and, of ten mutations at this site, seven were at the central guanine and the other three at the 3′ guanine. This is in good agreement with the studies on the specificity of DNA alkylation and repair and the sites of mutation seen in bacterial cells as discussed in Sect. G.II above. Unfortunately, the DNA repair status of these cells was not investigated, and it is not known to what extent the target DNA was exposed to the action of the AGT protein. Again, however, these results are consistent with the critical role of unrepaired O^6-alkylguanines in mutagenesis by *N*-methyl-*N*-nitrosourea and ethyl methanesulfonate.

These results are supported by studies of the mutations in a retroviral shuttle vector containing the *E. coli gpt* gene (ASHMAN and DAVIDSON 1987). This vector has the advantage of becoming inserted into the chromosomal DNA. When mouse fibroblasts were infected with it and exposed to ethyl methanesulfonate, all of the mutations involving single base changes were found to be G·C→A·T transitions.

A recent analysis of the mutations induced in human cells by *N*-methyl-*N*-nitrosourea has been carried out using the pHET shuttle vector which contains a 1100-base pair HSV tk gene (ECKERT et al. 1988). After treatment with *N*-ethyl-*N*-nitrosourea, 48% of the mutations detected in this gene were G·C→A·T transitions, 17% were A·T→G·C transitions, 20% were A·T→T·A transversions, and 9% were A·T→C·G transversions. These results were obtained in EBV-transformed human lymphoblastoid cells. The DNA repair capacity of these cells was not examined directly, but it is likely that they contain both AGT and 3-methyladenine DNA glycosylase, which would be expected to reduce the content of O^6-ethylguanine and 3-ethyladenine, respectively. The two transition mutations are likely to be due to the mispairing effects of O^6-ethylguanine and O^4-ethylthymine and are consistent with the previous studies described above. The relative proportion of the G·C→A·T and A·T→G·C transitions is somewhat greater less the ratio of production of O^6-ethylguanine to O^4-ethylthymine, which is consistent with the more extensive repair of the former adduct.

The significant number of transversion mutations at A·T base pairs found in this study may be explained either by the presence of apurinic/apyrimidinic sites in the DNA [which arise through the action of glycosylases or via spontaneous hydrolysis of adducts] or by the inhibition of DNA replication by ethylated adenine residues or thymine residues.

Since the frequency of such A·T transversions was much higher in the experiments with *N*-ethyl-*N*-nitrosourea (ECKERT et al. 1988) than in previous studies with ethyl methanesulfonate or *N*-methyl-*N*-nitrosourea (LEBKOWSKI et al. 1986; DUBRIDGE et al. 1987) and the former agent modifies the O^2 position of thymine to a much greater extent than the latter compounds, it has been proposed that O^2-ethylthymine is a major cause of the transversions (ECKERT et al. 1988). This interesting hypothesis requires more investigation, and the generality of these results is not yet established. Clearly, it cannot be ruled out that O^4-ethylthymine contributes to the transversions observed.

Experiments with shuttle vectors may not be a perfect model for replication of the chromatin in mammalian cells, and the DNA repair status of cells used in these studies was not investigated and may differ between the cell types. However, the mutations observed by ECKERT et al. (1988) are consistent with studies in which *N*-ethyl-*N*-nitrosourea was administered to animals. Such treatment induces A·T→T·A transversion mutations in mouse globin genes (LEWIS et al. 1985), and transplacental administration of *N*-ethyl-*N*-nitrosourea to rats results in the appearance of tumors of the nervous system containing the *neu* oncogene which had been activated by A·T→T·A transversion (BARGMANN et al. 1986; PERANTONI et al. 1987). Analysis of the H-*ras* genes of liver tumors induced by diethylnitrosamine showed that of 14 tumors which were analyzed and had mutations at codon 61, 20% were A·T→G·C transitions (which are consistent with the expected effects of unrepaired O^4-ethylthymine), 50% were G·C→T·A transversions, and 30% were A·T→T·A transversions (STOWERS et al. 1988). The absence of G·C→A·T transitions is consistent with the rapid repair of O^6-ethylguanine in liver cells and emphasizes the probable role of unrepaired ethylated pyrimidines in the initiation of liver tumors by this ethylating agent.

Finally, the possible role of adenine adducts is indicated by the finding that a mouse skin squamous cell carcinoma induced by β-propiolactone contained an activated H-*ras* oncogene with an A·T→T·A transversion mutation at codon 61 (HOCHWALT et al. 1988). This is consistent with the formation of the 1-carboxyethyladenine adduct, which has been found when this agent reacts with DNA, but only one of the six tumors studied contained this mutation.

I. Transfer and Expression of DNA Repair Genes

Several groups have transferred an *E. coli* AGT gene to mammalian cells (BRENNAND and MARGISON 1986a; ISHIZAKI et al. 1986; SAMSON et al. 1986; KATAOKA et al. 1986). The expression of this gene clearly protects against the toxic and mutagenic effects of alkylating agents. This shows definitively that the damage repaired by the products of this gene is of major importance in causing these effects. These experiments have, as yet, only been carried out in cultured cells which are not suitable for definitive experiments in carcinogenesis. The use of transgenic animals expressing the AGT gene in high amounts would provide an important model system in which the role of the AGT in protection against carcinogenesis by alkylating agents could be evaluated. Production of such transgenic mice has been reported (SEARLE et al. 1988), but their properties have not yet been described. It should be noted that the *E. coli* AGT gene which has been used for some expression studies not only repairs O^6-methylguanine but also O^4-methylthymine and alkylphosphate triesters (LINDAHL 1982; YAROSH 1985; BRENNAND and MARGISON 1986a; ISHIZAKI et al. 1986; SAMSON et al. 1986). A truncated form also has been used which lacks the latter activity against alkylphosphate triesters (BRENNAND and MARGISON 1986b; WHITE et al. 1986; ISHIZAKI et al. 1987) or expresses only this activity (KATAOKA et al. 1986). The studies with these constructs suggest that the repair of alkylphosphate triesters does not contribute to the biological effects observed, but it remains possible that the bacterial protein is restoring thymine sites from O^4-alkylthymine, a reaction not carried out by the mammalian AGT (YAROSH 1985; PEGG and DOLAN 1987; DOLAN et al. 1988a).

The use of molecular biological techniques to achieve overproduction of mammalian AGT is also required to verify the importance of O^6-alkylguanine repair. This gene has not yet been isolated, and totally definitive experiments have therefore not been carried out, but it has been possible to transfect CHO cells lacking AGT activity with human DNA and to obtain clones which express the human AGT gene (DING et al. 1985; YAROSH et al. 1986; BARROWS et al. 1987). Such clones are also resistant to the mutagenic and toxic effects of alkylating agents (DING et al. 1985; YAROSH et al. 1986; BIGNAMI et al. 1987; BARROWS et al. 1987). This finding is consistent with the importance of unrepaired O^6-methylguanine in causing these effects, but it is certainly possible that other genes are transferred and expressed in these experiments. Interestingly, the induction of sister chromatid exchanges by alkylating agents is also greatly reduced by expression of the mammalian or bacterial AGT gene (WHITE et al. 1986; ISHIZAKI et al. 1986; SAMSON et al. 1986; BIGAMI et al. 1987). The mechanism by which O^6-methylguanine enhances the production of sister chromatid exchanges is at present unclear.

J. Conclusions

Although many of the different alkylation products formed in DNA may have the potential to cause mutations or chromosome rearrangements and thus lead to the initiation of neoplastic growth, it appears that a major cause of such changes is O^6-alkylguanine. However, this product is subject to very rapid repair by the AGT protein. Such repair reduces the probability that tumors will be initiated. The content of AGT differs substantially according to the species, tissue, and cell type. The relative lack of AGT protein in some cells may increase their susceptibility to alkylating carcinogens, particularly if they turn over sufficiently rapidly for there to be a high probability of DNA replication occurring whilst their DNA contains O^6-alkylguanine. However, the presence of O^6-alkylguanine in replicating DNA is not always sufficient for an increased incidence of tumors, and other factors, which at present are not well understood, may confer resistance to the initiation of neoplastic growth.

In cells that contain substantial amounts of AGT and therefore remove the O^6-alkylguanine from their DNA, other alkylation products such as O^4-alkylthymine (and possibly O^2-alkylthymine), which are formed in much smaller amounts than O^6-alkylguanine, may accumulate to levels comparable with, or greater than, O^6-alkylguanine. Since these pyrimidine adducts are also active in inducing mutations, it appears that they also have the potential to initiate neoplastic growth. Such pyrimidine adducts may be of particular significance in carcinogenesis by ethylating agents such as diethylnitrosamine because the amount of O^4-ethylthymine formed is only about 4 times less than the amount of O^6-ethylguanine initially found. O^4-Ethylthymine therefore accumulates in the DNA of hepatocytes, which contain high levels of AGT, to levels greatly exceeding those of O^6-ethylguanine. It is unclear why there is no rapid DNA repair process to remove O^4-alkylthymine from mammalian DNA since this product appears to pose as serious a threat as O^6-alkylguanine. One possible reason is that the most likely agents to be accounted in the environment are methylating agents and, with them, production of the O^4-methylthymine occurs in very small amounts, less than 1% of the formation of O^6-methylguanine. Therefore, a process of very low capacity, which may not yet have been detected, could be sufficient to reduce the potential effects of O^4-methylthymine.

It is now obvious that there are marked sequence and gene specific effects in both DNA alkylation and DNA repair. Although the AGT protein is capable of acting on O^6-methylguanine in virtually any sequence, it is quite possible that in vivo there are sufficient differences in relative rates of repair that alkylation damage in certain sequences and/or certain genes will remain for a long period. This specificity may depend both on the sequences under consideration and on the status of the gene in which the alkylation damage is present. This specificity of repair could also be magnified by the stoichiometric nature of the AGT reaction since all of the available AGT may be used up in repairing the most favored sites. These considerations, and the preference for alkylation to occur at particular sequences and regions of the chromatin, may lead to the presence of regions at which the chance of alkylation occurring and persisting is much greater than the average for the whole genome. The presence of such sequences in cellular equivalents of oncogenes could play an important role in the potency and specificity of the alkylating carcinogens.

Acknowledgement. Work in my laboratory is supported by grant CA-18137 from the National Cancer Institute.

References

Arnold GE, Dunker AK, Smerdon MJ (1987) Limited nucleosome migration can completely randomize DNA repair patches in intact human cells. J Mol Biol 196:433–436

Ashman CR, Davidson RL (1987) DNA base sequence changes induced by ethyl methanesulfonate in a chromosomally integrated shuttle vector gene in mouse cells. Somatic Cell Mol Genet 13:563–568

Bargmann CI, Hung M, Weinberg RA (1986) Multiple independent activations of the *neu* oncogene by a point mutation altering the transmembrane domain of p185. Cell 45:649–657

Barrows LR, Borchers AH, Paxton MB (1987) Transfectant cells expressing O^6-alkylguanine-DNA-alkyltransferase display increased resistance to DNA damage other than O^6-guanine alkylation. Carcinogenesis 8:1853–1859

Bartsch H, Montesano R (1984) Relevance of nitrosamines to human cancer. Carcinogenesis 5:1381–1383

Bartsch H, O'Neill IK, Schulte-Herman R (1987) The relevance of *N*-nitroso compounds to human cancer: exposures and mechanisms. IARC, Lyon (IARC Sci Publ no 84)

Basu AK, Essigmann JM (1988) Site specifically modified oligodeoxynucleotides as probes for the structural and biological effects of DNA damaging agents. Chem Res Toxicol 1:18

Bedell MA, Lewis JG, Billings KC, Swenberg JA (1982) Cell specificity in hepatocarcinogenesis: preferential accumulation of O^6-methylguanine in target cell DNA during continuous exposure of rats to 1,2-dimethylhydrazine. Cancer Res 42:3079–3083

Belinksy SA, White CM, Boucheron JA, Richardson FC, Swenberg JA, Anderson M (1986) Accumulation and persistence of DNA adducts in respiratory tissue of rats following multiple administrations of the tobacco specific carcinogen 4-(*N*-methyl-*N*-nitrosamine)-1-(3-pyridyl)-1-butanone. Cancer Res 46:1280–1284

Belinsky SA, Walker VE, Maronpot PR, Swenberg JA, Anderson MW (1987a) Molecular dosimetry of DNA adduct formation and cell toxicity in rat nasal mucosa following exposure to the tobacco specific nitrosamine 4-(*N*-methyl-*N*-nitrosamino)-1-(3-pyridyl)-1-butanone and their relationship to induction of neoplasia. Cancer Res 47:6058–6065

Belinsky SA, White CM, Devereux T, Swenberg JA, Anderson MW (1987b) Cell selective alkylation of DNA in rat lung following low dose exposure to the tobacco specific carcinogen 4-(*N*-methyl-*N*-nitrosamino)-1-(3-pyridyl)-1-butanone. Cancer Res 47:1143–1148

Belinsky SA, Dolan ME, White CW, Maronpot RR, Pegg AE, Anderson ME (1988) Cell specific differences in O^6-methylguanine-DNA methyltransferase activity and removal of O^6-methylguanine in rat pulmonary cells. Carcinogenesis 9:2053–2058

Beranek DT, Weis CC, Evans FE, Chetsanga CJ, Kadlubar FF (1983) Identification of *N*5-methyl-*N*5-formyl-2,5,6-triamino-5-hydroxypyrimidine as a major adduct in rat liver DNA after treatment with the carcinogens *N*,*N*-dimethylnitrosamine or 1,2-dimethylhydrazine. Biochem Biophys Res Comm 110:625–631

Bhanot OS, Ray A (1986) The in vivo mutagenic frequency and specificity of O^6-methylguanine in ϕX174 replicative form DNA. Proc Natl Acad Sci USA 83:7348–7352

Bignami M, Terlizzese M, Zijno A, Calcagnile A, Frosina G, Abbondandolo A, Dogliotti E (1987) Cytotoxicity, mutations and SCEs induced by methylating agents are reduced in CHO cells expressing an active mammalian O^6-methylguanine-DNA methyl transferase gene. Carcinogenesis 8:1417–1421

Bohr VA, Hanawalt PC (1987) Enhanced repair of pyrimidine dimers in coding and noncoding genomic sequences in CHO cells expressing a prokaryotic DNA repair gene. Carcinogenesis 8:1333–1336

Bohr VA, Phillips DH, Hanawalt PC (1987) Heterogeneous DNA damage and repair in the mammalian genome. Cancer Res 47:6426–6436

Boiteaux S, Laval J (1983) Imidazole open ring 7-methylguanine: an inhibitor of DNA synthesis. Biochem Biophys Res Comm 110:552–558

Boiteaux S, O'Connor TR, Laval J (1987) Formamidopyrimidine-DNA glycosylase of *Escherichia coli:* cloning and sequencing of the *fpg* structural gene and overproduction of the protein. EMBO J 6:3177–3183

Boucheron JA, Richardson FC, Morgan PH, Swenberg JA (1987) Molecular dosimetry of O^4-ethyldeoxythymidine in rats continuously exposed to diethylnitrosamine. Cancer Res 47:1577–1581

Brennand J, Margison GP (1986a) Reduction of the toxicity and mutagenicity of alkylating agents in mammalian cells harboring the *Escherichia coli* alkyltransferase gene. Proc Natl Acad Sci USA 83:6292–6296

Brennand J, Margison GP (1986b) Expression in mammalian cells of a truncated *Escherichia coli* gene coding for O^6-alkylguanine alkyltransferase reduces the toxic effects of alkylating agents. Carcinogenesis 7:2081–2084

Brent TP, Dolan ME, Fraenkel-Conrat H, Hall J, Karran P, Laval F, Margison GP, Montesano R, Pegg AE, Potter PM, Singer B, Swenberg JA, Yarosh DB (1988) Repair of *O*-alkylpyrimidines in mammalian cells: a present consensus. Proc Natl Acad Sci USA 85:1759–1762

Briscoe WT, Cotter LE (1984) The effect of neighboring bases on *N*-methyl-*N*-nitrosourea alkylation of DNA. Chem Biol Interact 52:103–110

Briscoe WT, Cotter LE (1985) DNA sequence has an effect on the extent and kinds of alkylation of DNA by a potent carcinogen. Chem Biol Interact 56:321–331

Briscoe WT, Duarte SP (1988) Preferential alkylation by 1,3-bis(2-chloroethyl)-1-nitrosourea (BCNU) of guanines with guanines as neighboring bases in DNA. Biochem Pharm 37:1061–1066

Buckley N (1987) A regioselective mechanism for mutagenesis and oncogenesis caused by alkylnitrosourea sequence-specific DNA alkylation. J Am Chem Soc 109:7918–7920

Burns PA, Allen FL, Glickman BW (1986) DNA sequence analysis of mutagenicity and site specificity of ethyl methanesulfonate in Uvr^+ and $UvrB^-$ strains of *Escherichia coli.* Genetics 113:811–819

Burns PA, Gordon AJE, Glickman BW (1987) Influence of neighboring base sequence on *N*-methyl-*N*-nitrosoguanidine mutagenesis in the *lac*I gene of *Escherichia coli.* J Mol Biol 194:385–390

Burns PA, Gordon AJE, Kunsmann K, Glickman BW (1988) Influence of neighboring base sequence on the distribution and repair of *N*-ethyl-*N*-nitrosourea-induced lesions in *Escherichia coli.* Cancer Res 48:4455–4458

Chambers RW, Sledziewska-Gojsda E, Hirani-Hojatti S, Borowy-Borowski H (1985) uvrA and recA mutations inhibit a site specific transition produced by a single O^6-methylguanine in gene G of bacteriophage ϕX174. Proc Natl Acad Sci USA 82:7173–7177

Craddock VM (1971) Liver carcinomas induced in rats by single administration of dimethylnitrosamine after partial hepatectomy. JNCI 47:899–907

Craddock VM, Henderson AR (1982) The activity of 3-methyladenine DNA glycosylase in animals tissues in relation to carcinogenesis. Carcinogenesis 3:747–750

Craddock VM, Henderson AR (1984) Repair and replication of DNA in rat and mouse tissues in relation to cancer induction by *N*-nitroso-*N*-alkylureas. Chem Biol Interact 52:223–231

Deilhaug T, Myrnes B, Aune T, Krokan H, Haugen A (1985) Differential capacities for DNA repair in Clara cells, alveolar type II cells and macrophages of rabbit lung. Carcinogenesis 6:661–663

Devereux TR, Anderson MW, Belinsky SA (1988) Factors regulating activation and DNA alkylation by 4-(*N*-methyl-*N*-nitrosamino)-1-(3-pyridyl)-1-butanone and nitrosodimethylamine in rat lung and isolated lung cells, and the relationship to carcinogenicity. Cancer Res 48:4215–4221

Diaz Gomez MI, Swann PF, Magee PN (1977) The absorption and metabolism in rats of small oral doses of dimethylnitrosamine. Biochem J 164:497–500

Ding R, Ghosh K, Eastman A, Bresnick E (1985) DNA-mediated transfer and expression of a human DNA repair gene that demethylates O^6-methylguanine. Mol Cell Biol 5: 3293–3296

Dolan ME, Oplinger M, Pegg AE (1988 a) Use of a dodecadeoxynucleotide to study repair of the O^4-methylthymine lesion. Mutat Res 193:131–137
Dolan ME, Oplinger M, Pegg AE (1988 b) Sequence specificity of guanine alkylation and repair. Carcinogenesis 9:2139–2143
Dolan ME, Scicchitano D, Pegg AE (1988 c) Use of oligodeoxynucleotides containing O^6-alkylguanine for the assay of O^6-alkylguanine-DNA-alkyltransferase activity. Cancer Res 48:1184–1188
Dubridge RB, Tang P, Hsia HC, Leong P, Miller JH, Calos MP (1987) Analysis of mutation in human cells by using an Epstein-Barr virus shuttle system. Mol Cell Biol 7:379–387
Dyroff MC, Richardson FC, Popp JA, Bedell MA, Swenberg JA (1986) Correlation of O^4-ethyldeoxythymidine accumulation, hepatic initiation and hepatocellular carcinoma induction in rats continuously administered diethylnitrosamine. Carcinogenesis 7:241–246
Eckert KA, Ingle CA, Klinedinst DK, Drinkwater NR (1988) Molecular analysis of mutations induced in human cells by *N*-ethyl-*N*-nitrosourea. Mol Carcinogenesis 1:50–56
Fong AT, Rasmussen RE (1987) Formation and accumulation of O^6-ethylguanine in DNA of enriched populations of Clara cells, alveolar type II cells and macrophages of hamsters exposed to diethylnitrosamine. Toxicology 43:289–299
Furois-Corbin S, Pullman B (1985) Specificity in carcinogen-DNA interaction; a theoretical exploration of the factors involved in the effect of neighboring bases on *N*-methyl-*N*-nitrosourea alkylation of DNA. Chem Biol Interact 54:9–13
Gale JM, Nissen KA, Smerdon MJ (1987) UV-induced formation of pyrimidine dimers in nucleosome core DNA is strongly modulated with a period of 10.3 bases. Proc Natl Acad Sci USA 84:6644–6648
Gallagher P, Brent TP (1984) Further purification and characterization of human 3-methyladenine-DNA glycosylase. Biochim Biophys Acta 782:394–401
Glickman BW, Horsfall MJ, Gordon AJE, Burns PA (1987) Nearest neighbor affects G:C to A:T transitions induced by alkylating agents. Environ Health Perspect 76:29–32
Gombar CT, Katz EJ, Magee PN, Sirover MA (1981) Induction of the DNA repair enzymes uracil DNA glycosylase and 3-methyladenine DNA glycosylase in regenerating rat liver. Carcinogenesis 2:595–599
Goth R, Rajewsky MF (1974) Persistence of O^6-ethylguanine in rat brain DNA. Correlation with nervous system specific carcinogenesis by ethylnitrosourea. Proc Natl Acad Sci USA 71:639–653
Grafstrom RC, Pegg AE, Trump BF, Harris CC (1984) O^6-Alkylguanine-DNA-alkyltransferase activity in normal human tissues and cells. Cancer Res 44:2855–2857
Hartley JA, Gibson NW, Kohn KW, Mattes WB (1986) DNA sequence selectivity of guanine-N7 alkylation by three antitumor chloroethylating agents. Cancer Res 46:1943–1947
Hartley JA, Mattes WB, Vaughan K, Gibson NW (1988) DNA sequence specificity of guanine N7-alkylations for a series of structurally related triazenes. Carcinogenesis 9:669–674
Hecht SS, Hoffmann D (1988) Tobacco-specific nitrosamines, an important group of carcinogens in tobacco and tobacco smoke. Carcinogenesis 9:875–884
Hochwalt AE, Solomon JJ, Garte SJ (1988) Mechanism of H-*ras* oncogene activation in mouse squamous carcinoma induced by an alkylating agent. Cancer Res 48:556–558
Hodgson R, Swann PF, Clothier R, Balls M (1980) The persistence in *Xenopus laevis* DNA of O^6-methylguanine produced by exposure to *N*-methyl-*N*-nitrosourea. Eur J Cancer 16:481–486
Hong J, Pan J, Dong Z, Ning SM, Yang CS (1987) Regulation of *N*-nitrosodimethylamine demethylase in rat liver and kidney. Cancer Res 47:5948–5953
Ishizaki K, Tsujimura T, Yawata H, Fujio C, Nakabeppu Y, Sekiguchi M, Ikenaga M (1986) Transfer of the *E. coli* O^6-methylguanine methyltransferase gene into repair-deficient human cells and restoration of cellular resistance to *N*-methyl-*N'*-nitro-*N*-nitrosoguanidine. Mutat Res 166:135–141

Ishizaki K, Tsujimura T, Fujio C, Yangpei Z, Yawata H, Nakabeppu Y, Sekiguchi M, Ikenaga M (1987) Expression of the truncated *E. coli* O^6-methyltransferase gene in repair-deficient human cells and restoration of cellular resistance to alkylating agents. Mutat Res 184:121–128

Jung G, Ro J, Kim MH, Park G, Paik WK, Magee PN, Kim S (1986) Studies on the distribution of O^6-methylguanine-DNA methyltransferase in rat. Biochem Pharmacol 35:377–384

Kadlubar FF, Beranek DT, Weis CC, Evans FE, Cox R, Irving CC (1984) Characterization of the purine ring-opened 7-methylguanine and its persistence in rat bladder epithelial DNA after treatment with the carcinogen *N*-methylnitrosourea. Carcinogenesis 5:587–592

Kalnik MW, Kouchakdjian M, Li BFL, Swann PF, Patel DJ (1988a) Base pair mismatches and carcinogen-modified bases in DNA: an NMR study of A.C and A.O^4meT pairing in dodecanucleotide duplexes. Biochemistry 27:100–108

Kalnik MW, Kouchakdjian M, Li BFL, Swann PF, Patel DJ (1988b) Base pair mismatches and carcinogen-modified bases in DNA: an NMR study of G.T and G.O^4meT pairing in dodecanucleotide duplexes. Biochemistry 27:108–115

Kataoka H, Hall J, Karran P (1986) Complementation of sensitivity to alkylating agents in *Escherichia coli* and Chinese hamster ovary cells by expression of a cloned bacterial repair gene. EMBO J 5:3195–3200

Kleihues P, Margison GP (1974) Carcinogenicity of *N*-methyl-*N*-nitrosourea. Possible role of excision repair of O^6-methylguanine from DNA. JNCI 53:1839–1842

Kleihues P, Rajewsky MF (1984) Chemical neurooncogenesis: role of structural DNA modifications, DNA repair and neural target cell population. Prog Exp Tumor Res 27:1–16

Kohn K, Hartley JA, Mattes WB (1987) Mechanisms of DNA sequence selective alkylation of guanine-N7 position by nitrogen mustards. Nucleic Acids Res 15:10531–10549

Lawley PD (1976) Carcinogenesis by alkylating agents. In: Searle CE (ed) Chemical carcinogens. American Chemical Society, Washington DC, pp 83–244 (ACS Symp Series 173)

Lebkowski JS, Miller JH, Calos MP (1986) Determination of DNA sequence changes induced by ethyl methanesulfonate in human cells, using a human shuttle vector system. Mol Cell Biol 6:1838–1847

Lewis JG, Swenberg JA (1983) The kinetics of DNA alkylation, repair and replication in hepatocytes, Kupffer cells and sinusoidal endothelial cells in rat liver during continuous exposure to 1,2-dimethylhydrazine. Carcinogenesis 4:529–536

Lewis SE, Johnson FM, Skow LC, Popp DB, Popp RA (1985) A mutation in the β-globin gene detected in the progeny of a female mouse treated with ethylnitrosourea. Proc Natl Acad Sci USA 82:5829–5831

Lijinsky W (1987) Structure-activity relations in carcinogenesis by *N*-nitroso compounds. Cancer Metastasis Rev 6:301–356

Lindahl T (1982) DNA repair enzymes. Annu Rev Biochem 51:61

Lindamood C III, Bedell MA, Billings KC, Swenberg JA (1982) Alkylation and de novo synthesis of liver cell DNA from C3H mice during continuous dimethylnitrosamine exposure. Cancer Res 42:4153–4157

Lindamood C III, Bedell MA, Billings KC, Dyroff MC, Swenberg JA (1984) Dose response for DNA alkylation [^{3}H]thymidine uptake into DNA and O^6-methylguanine-DNA methyltransferase activity in hepatocytes of rats and mice continuously exposed to dimethylnitrosamine. Cancer Res 44:196–200

Loechler EL, Green CL, Essigmann JM (1984) In vivo mutagenesis of O^6-methylguanine built into a unique site in a viral genome. Proc Natl Acad Sci USA 81:6271–6275

Loveless A (1969) Possible relevance of O-6 alkylation of deoxyguanosine to the mutagenicity and carcinogenicity of nitrosamines and nitrosamides. Nature 223:206–207

Magee PN, Montesano R, Preussmann R (1976) *N*-nitroso compounds and related carcinogens. In: Searle CE (ed) Chemical carcinogens. American Chemical Society, Washington DC, pp 491–625 (ACS Symp Series 173)

Male R, Helland DE, Kleppe K (1985) Purification and characterization of 3-methyladenine-DNA glycosylase from calf thymus. J Biol Chem 260:1623–1629
Male R, Haukanes BI, Helland DE, Kleppe K (1987) Substrate specificity of 3-methyladenine-DNA glycosylase from calf thymus. Eur J Biochem 165:13–19
Margison GP, Kleihues P (1975) Chemical carcinogenesis in the nervous system. Preferential accumulation of O^6-methylguanine in rat brain DNA during repetitive administration of *N*-methyl-*N*-nitrosourea. Biochem J 148:521–525
Margison GP, Pegg AE (1981) Enzymatic release of 7-methylguanine from methylated DNA by rodent liver extracts. Proc Natl Acad Sci USA 78:861–865
Mattes WB, Hartley JA, Kohn KW (1986) DNA sequence of guanine-N7 alkylation by nitrogen mustards. Nucleic Acid Res 14:2971–2987
McCarthy JG, Edington BV, Schendel PF (1983) Inducible repair of phosphotriesters in *Escherichia coli.* Proc Natl Acad Sci USA 80:7380–7384
McCarthy TV, Karran P, Lindahl T (1984) Inducible repair of *O*-alkylated pyrimidines in *Escherichia coli.* EMBO J 3:545–550
Mellon I, Bohr VA, Smith CA, Hanawalt PC (1986) Preferential DNA repair of an active gene in human cells. Proc Natl Acad Sci USA 83:8878–8882
Mellon I, Spivak G, Hanawalt PC (1987) Selective removal of transcription-blocking DNA damage from the transcribed strand of the mammalian DHFR gene. Cell 51:241–249
Montesano R (1981) Alkylation of DNA and tissue specificity in nitrosamine carcinogenesis. J Supramolec Struct Cell Biochem 17:259–273
Morimoto K, Dolan ME, Scicchitano D, Pegg AE (1985) Repair of O^6-propylguanine and O^6-butylguanine in DNA by O^6-alkylguanine-DNA alkyltransferases from rat liver and *E. coli.* Carcinogenesis 6:1027–1031
Myers KA, Saffhill R, O'Connor PJ (1988) Repair of alkylated purines in the hepatic DNA of mitochondria and nuclei in the rat. Carcinogenesis 9:285–292
Nehls P, Rajewsky MF, Spiess E, Werner D (1984) Highly sensitive sites for guanine O^6 ethylation in rat brain exposed to *N*-ethyl-*N*-nitrosourea. EMBO J 3:327–332
Nicoll JW, Swann PF, Pegg AE (1975) Effect of dimethylnitrosamine on persistence of methylated guanines in rat liver and kidney DNA. Nature 256:261–262
Patel DJ, Shapiro L, Kozlowski SA, Gaffney BL, Kuzmich S, Jones RA (1985) Covalent carcinogenic lesions in DNA: NMR studies of O^6-methylguanosine containing oligonucleotide duplexes. Biochimie 67:861–886
Patel DJ, Shapiro L, Kozlowski SA, Gaffney BL, Jones RA (1986a) Covalent carcinogenic O^6-methylguanosine lesions in DNA: structural studies of the O^6meG.A and O^6meG.G interactions in dodecanucleotide duplexes. J Mol Biol 188:677–692
Patel DJ, Shapiro L, Kozlowski SA, Gaffney BL, Jones RA (1986b) Structural studies of the O^6meG.T interaction in the d(C-G-T-G-A-A-T-T-C-O^6meG-C-G) duplex. Biochemistry 25:1027–1036
Patel DJ, Shapiro L, Kozlowski SA, Gaffney BL, Jones RA (1986c) Structural studies of the O^6meG.C interaction in the d(C-G-C-G-A-A-T-T-C-O^6meG-C-G) duplex. Biochemsitry 25:1027–1036
Pegg AE (1977a) Formation and metabolism of alkylated nucleosides: possible role in carcinogenesis by nitroso compounds and alkylating agents. Adv Cancer Res 25:195–269
Pegg AE (1977b) Alkylation of rat liver DNA by dimethylnitrosamine: effect of dosage on O^6-methylguanine levels. JNCI 58:681–687
Pegg AE (1978) Dimethylnitrosamine inhibits enzymatic removal of O^6-methylguanine from DNA. Nature 274:182–184
Pegg AE (1983) Alkylation and subsequent repair of DNA after exposure to dimethylnitrosamine and related carcinogens. Rev Biochem Toxicol 5:83–133
Pegg AE (1984) Methylation of the O^6 position of guanine in DNA is the most likely initiating event in carcinogenesis by methylating agents. Cancer Invest 2:223–231
Pegg AE, Dolan ME (1987) Properties and assay of mammalian O^6-alkylguanine-DNA-alkyltransferase. Pharmacol Ther 34:167–179
Pegg AE, Hui G (1978) Formation and subsequent removal of O^6-methylguanine from DNA in rat liver and kidney after small doses of dimethylnitrosamine. Biochem J 173:739–748

Pegg AE, Perry W (1981) Alkylation of nucleic acids and metabolism of small doses of dimethylnitrosamine in the rat. Cancer Res 41:3128–3132

Pegg AE, Perry W, Bennett RA (1981) Partial hepatectmy increases the ability of rat liver extracts to catalyze removal of O^6-methylguanine from alkylated DNA. Biochem J 197:195–201

Pegg AE, Roberfroid M, von Bahr C, Foote RS, Mitra S, Bresil H, Likachev A, Montesano R (1982) Removal of O^6-methylguanine from DNA by human liver fractions. Proc Natl Acad Sci USA 79:5162–5165

Pegg AE, Wiest L, Foote RS, Mitra S, Perry W (1983) Purification and properties of O^6-methylguanine-DNA transmethylase from rat liver. J Biol Chem 258:2327–2333

Pegg AE, Scicchitano D, Dolan ME (1984) Comparison of the rates of repair of O^6-alkylguanines in DNA by rat liver and bacterial O^6-alkylguanine-DNA alkyltransferase. Cancer Res 44:3806–3811

Pegg AE, Dolan ME, Scicchitano D, Morimoto K (1985) Studies of the repair of O^6-alkylguanine and O^4-alkylthymine in DNA by alkyltransferases from mammalian cells and bacteria. Environ Health Perspect 62:109–114

Perantoni AO, Rice JM, Reed CD, Watatani M, Wenk ML (1987) Activated *neu* oncogene sequences in primary tumors of the peripheral nervous system induced in rats by transplacental exposure to ethylnitrosourea. Proc Natl Acad Sci USA 84:6317–6321

Peto R, Gray R, Branton P, Grasso P (1984) Nitrosamine carcinogenesis in 1529 rodents: chronic administration of sixteen different concentrations of NDEA, NDMA, NPYR and NPIP in the water of 4440 inbred rats with parallel studies on NDEA alone on the effect of age of starting (3,6 or 20 weeks) and of species (rats, mice or hamsters). In: O'Neill IK, Von Borstel RC, Miller CT, Long J, Bartsch H (eds) *N*-nitroso compounds: occurrence, biological effects and relevance to human cancer. International Agency for Research on Cancer, Lyon, pp 627–655 (IARC Sci Publ no 57)

Preston BD, Singer B, Loeb LA (1986) Mutagenic potential of O^4-methylthymine in vivo determined by an enzymatic approach to site specific mutagenesis. Proc Natl Acad Sci USA 83:8501–8505

Preston BD, Singer B, Loeb LA (1987) Comparison of the relative mutagenicities of *O*-alkylthymines site-specifically incorporated into ϕX174 DNA. J Biol Chem 262:13821–13827

Preussmann R, Stewart BW (1984) *N*-nitroso carcinogens. In: Searle CE (ed) Chemical carcinogens, 2nd edn. American Chemical Society, Washington DC, pp 643–828 (ACS Monograph 182)

Reed J, Hutchinson F (1987) Effect of the direction of DNA replication on mutagenesis by *N*-methyl-*N'*-nitro-*N*-nitrosoguanidine in adapted cells of *Escherichia coli*. Mol Gen Genet 208:446–449

Richardson FC, Dyroff MC, Boucheron JA, Swenberg JA (1985) Differential repair of O^4-alkylthymidine following exposure to methylating and ethylating hepatocarcinogens. Carcinogenesis 6:625–629

Richardson FC, Boucheron JA, Dyroff MC, Popp JA, Swenberg JA (1986) Biochemical and morphologic studies of heterogeneous lobe responses in hepatocarcinogenesis. Carcinogenesis 7:247–251

Richardson FC, Morgan PH, Boucheron JA, Deal FH, Swenberg JA (1988) Hepatocyte initiation during continuous administration of diethylnitrosamine and 1,2-symdimethylhydrazine. Cancer Res 48:988–992

Richardson KK, Crosby RM, Richardson FC, Slopek TR (1987a) DNA base changes induced following in vivo exposure of unadapted, adapted or Ada-*Escherichia coli* to *N*-methyl-*N'*-nitro-*N*-nitrosoguanidine. Mol Gen Genet 209:526–532

Richardson KK, Richardson FC, Crosby RM, Swenberg JA, Skopek TR (1987b) DNA base changes and alkylation following in vivo exposure of *Escherichia coli* to *N*-methyl-*N*-nitrosourea or *N*-ethyl-*N*-nitrosourea. Proc Natl Acad Sci USA 84:344–348

Ryan AJ, Billett MA, O'Connor PJ (1986) Selective repair of methylated purines in regions of chromatin DNA. Carcinogenesis 7:1497–1503

Saffhill R, Margison GP, O'Connor PJ (1985) Mechanisms of carcinogenesis induced by alkylating agents. Biochim Biophys Acta 823:111–145

Samson N, Derfler B, Waldstein EA (1986) Suppression of human DNA alkylation-repair defects by *Escherichia coli* DNA-repair genes. Proc Natl Acad Sci USA 83:5607–5610
Samson L, Thomale J, Rajewsky MF (1988) Alternative pathways for the in vivo repair of O^6-alkylthymine in *Escherichia coli:* the adaptive response and nucleotide excision repair. EMBO J 7:2261–2267
Sancar A, Sancar GB (1988) DNA repair enzymes. Annu Rev Biochem 57:29–67
Scicchitano D, Jones RA, Kuzmich S, Faffney B, Lasko DD, Essigmann JM, Pegg AE (1986) Repair of oligodeoxynucleotides containing O^6-methylguanine by O^6-alkylguanine-DNA-alkyltransferase. Carcinogenesis 7:1383–1386
Searle P, Tinsley JM, O'Connor PJ, Margison GP (1988) Generation of transgenic mice containing a DNA repair gene from *Escherichia coli.* Proc Am Assoc Cancer Res 29:111
Singer B (1984) Alkylation of the O^6 of guanine is only one of many chemical events that may initiate carcinogenesis. Cancer Invest 2:233–238
Singer B (1986) *O*-Alkyl pyrimidines in mutagenesis and carcinogenesis: occurrence and significance. Cancer Res 46:4879–4885
Singer B, Grunberger D (1983) Molecular biology of mutagenesis and carcinogenesis. Plenum, New York
Singer B, Kusmierek JT (1982) Chemical mutagenesis. Annu Rev Biochem 52:655–693
Snow ET, Mitra S (1987) Do carcinogen-modified deoxynucleotide precursors contribute to cellular mutagenesis? Cancer Invest 5:119–125
Stowers SJ, Wiseman RW, Ward JM, Miller EC, Miller JA, Anderson MW, Eva A (1988) Detection of activated proto-oncogenes in *N*-nitrosodiethylamine-induced liver tumors. Carcinogenesis 9:271–276
Strauss BS (1985) Cellular aspects of DNA repair. Adv Cancer Res 45:45–105
Stumpf R, Margison GP, Montesano R, Pegg AE (1979) Formation and loss of alkylated purines from DNA of hamster liver after administration of dimethylnitrosamine. Cancer Res 39:50–54
Sukumar S, Notario V, Martin-Zanca D, Barbacid M (1983) Induction of mammary carcinomas in rats by nitroso-methylurea involves malignant activation of H-*ras*-1 locus by single point mutations. Nature 306:658–661
Swann PF, Magee PN (1986) Nitrosamine-induced carcinogenesis. The alkylation of nucleic acids of the rat by *N*-methyl-*N*-nitrosourea, dimethylnitrosamine, dimethyl sulphate and methyl methanesulphonate. Biochem J 110:39–47
Swenberg JA, Bedell MA, Billings KC, Umbenhauer DR, Pegg AE (1982) Cell specific differences in O^6-alkylguanine DNA repair activity during continuous carcinogen exposure. Proc Natl Acad Sci USA 79:5499–5502
Swenberg JA, Dyroff MC, Bedell MA, Popp JA, Huh N, Kirstein U, Rajewsky MF (1984) O^4-Ethyldeoxythymidine, but not O^6-ethyldeoxyguanosine, accumulates in hepatocyte DNA of rats exposed continuously to diethylnitrosamine. Proc Natl Acad Sci USA 81:1692–1695
Swenberg JA, Richardson FC, Boucheron JA, Dyroff MC (1985) Relationships between DNA adduct formation and carcinogenesis. Environ Health Perspect 62:177–183
Swenberg JA, Richardson FC, Boucheron JA, Deal FH, Belinsky SA, Charbonneau M, Short BG (1987) High- to low-dose extrapolation: critical determinants involved in the dose response of carcinogenic substances. Environ Health Perspect 76:57–63
Thomas PE, Bandiera S, Maines S, Ryan DE, Levin W (1987) Regulation of cytochrome P-450j, a high-affinity *N*-nitrosodimethylamine demethylase in rat hepatic microsomes. Biochemistry 26:2280–2289
Topal MD (1988) DNA repair, oncogenes and carcinogenesis. Carcinogenesis 9:691–696
Topal MD, Baker MS (1982) DNA precursor pool: a significant target for *N*-methyl-*N*-nitrosourea in C3H/10T1/2 clone 8 cells. Proc Natl Acad Sci USA 79:2211–2215
Topal MD, Eadie JS, Conrad M (1986) O^6-Methylguanine mutation and repair is nonuniform. Selection for DNA most interactive with O^6-methylguanine. J Biol Chem 261:9879–9885
Vos J-H, Hanawalt PC (1987) Processing of psoralen adducts in an active human gene: repair and replication of DNA containing monoadducts and interstrand cross-links. Cell 50:789–799

Washington WJ, Dunn WC, Generoso WM, Mitra S (1988) Tissue-specific variation in repair activity for 3-methyladenine in DNA in two stocks of mice. Mutat Res 207:165–169
Weinfeld M, Drake AF, Saunders JK, Patterson M (1985) Stereospecific removal of methyl phosphotriesters from DNA by an *Escherichia coli* ada$^+$ extract. Nucleic Acids Res 13:7067–7077
White GRM, Ockey CH, Brennand J, Margison GP (1986) Chinese hamster cells harboring the *Escherichia coli* O^6-alkyltransferase gene are less susceptible to sister chromatid exchange and chromosome damage by methylating agents. Carcinogenesis 7:2077–2080
Wiestler O, Kleihues P, Pegg AE (1984) O^6-Alkylguanine-DNA alkyltransferase activity in human brain and brain tumors. Carcinogenesis 5:121–124
Wilkinson R, Hawks A, Pegg AE (1975) Methylation of rat liver mitochondrial DNA by chemical carcinogens and associated alterations in physical properties. Chem Biol Interact 10:157–167
Wurdeman RL, Gold B (1988) The effect of DNA sequence, ionic strength, and cationic DNA affinity binders on the methylation of DNA by *N*-methyl-*N*-nitrosourea. Chem Res Toxicol 1:146–147
Yarosh DB (1985) The role of O^6-methylguanine-DNA methyltransferase in cell survival, mutagenesis and carcinogenesis. Mutat Res 145:1–16
Yarosh DB, Barnes D, Erickson LC (1986) Transfection of DNA from a chloroethylnitrosourea-resistant tumor cell line (Mer$^+$) to a sensitive tumor cell line (Mer$^-$) results in a tumor cell line resistant to MNNG and CNU that has increased O^6-methylguanine-DNA methyltransferase levels and reduced levels of DNA interstrand crosslinking. Carcinogenesis 7:1603–1606
Zarbl H, Sukumar S, Arthur AV, Martin-Zanca D, Barbacid M (1985) Direct mutagenesis in Ha-*ras*-1 oncogenes by *N*-nitroso-*N*-methylurea during initiation of mammary carcinogenesis in rats. Nature 315:382–385

Part II. Modifiers of Chemical Carcinogenesis

CHAPTER 6

Tumour Promotion: Biology and Molecular Mechanisms

A. W. MURRAY, A. M. EDWARDS, and C. S. T. HII

A. Introduction

A promoter can be operationally defined as a compound which enhances the tumour yield in an animal exposed to a low dose of a carcinogen. Promoters act at an early stage in the step-wise induction of malignant cancers, by favouring the cloncal expansion of initiated cells to benign lesions such as papillomas, nodules and polyps. Although mouse skin has been most extensively studied, a number of animal models have been developed in order to study promotion at several other organ sites (SLAGA 1983). The restricted aim of this review is to discuss what is known about promotion by some specific agents in two organs (skin and liver) at both the biological and molecular levels. This summary will not be encyclopedic but will attempt to highlight the important unresolved issues. One characteristic of tumour promotion is that there is frequently tissue specificity associated with a particular promoter (see Table 1), and certainly promotion in skin and liver is achieved by different groups of compounds. Consequently, we believe it would also be useful to consider whether the same, different or overlapping mechanisms are involved.

Table 1. Tissue specificity of tumour promoters

Compound	Active as promoter	Inactive as promoter	References
TPA (12-*O*-tetradecanoyl-phorbol-13-acetate)	Mouse skin, lung and forestomach	Rat liver	BOUTWELL (1964); PAL et al. (1978); GOERTTLER et al. (1979); TATEMATSU et al. (1983a)
TCDD (2,3,7,8-tetrachloro-dibenzo-*p*-dioxin)	Female rat liver Skin, hairless mice	Skin, normal mice	PITOT et al. (1980); POLAND et al. (1982)
Phenobarbital	Rat liver and thyroid	Rat bladder, lung, testis, kidney and nasal cavity	PERAINO et al. (1983); HIASA et al. (1982); NAKANISHI et al. (1982); DIWAN et al. (1985)
Ethinylestradiol	Rat liver and kidney	Rat lung, bladder and thyroid	SHIRAI et al. (1987)
α-Hexachlorocyclohexane	Rat liver	Mouse skin	MUNIR et al. (1984)
Ethoxyquin	Rat bladder	Rat liver	ITO et al. (1984); TSUDA et al. (1984)
Lithocholic acid	Rat colon	Rat liver	MAKINO et al. (1986); REDDY
	Hamster pancreas	Hamster liver	and WATANABE (1979)
Saccharin	Rat bladder	Rat liver, thyroid	TSUDA et al. (1983)

In the first two sections a summary of skin and liver promotion will be presented, and this will be followed by a brief discussion of similarities and differences between the two systems.

B. Tumour Promotion in Mouse Skin

I. Characteristics of Initiation and Promotion

Most of the generally accepted views of tumour promotion as a biological phenomenon have come from extensive studies on two-stage carcinogenesis in mouse skin (SHUBIK 1984; WEINSTEIN 1987; FURSTENBERGER and MARKS 1983; ANGYRIS 1985; KINSELLA 1986). In fact, most work has been done using a polycyclic hydrocarbon as initiator and croton oil or a pure phorbol ester such as 12-*O*-tetradecanoylphorbol-13-acetate (TPA) as the promoting stimulus. Within this obviously limited paradigm the following points appear to be established.

Initiation can be achieved by the application of a single low dose of carcinogen to the dorsal skin. If these animals are then left untreated, very few or no tumours will develop. However, repeated exposure of the animals to a promoter such as TPA results in the development of a variable number of benign papillomas. A small proportion of these will eventually progress into fully malignant carcinomas, a process which appears to be independent of the promoting stimulus and may involve additional genetic events (HENNINGS et al. 1983).

Initiation results in a permanent genetic change in the epidermal cells. Thus, animals can be left for up to 1 year before beginning promoter treatment with little change in the lag period before the appearance of tumours (VAN DUUREN et al. 1975; LOEHRKE et al. 1983). A discussion of the key genetic events occurring during initiation is beyond the scope of this review. However, it is noteworthy that initiation with polycyclic hydrocarbons frequently results in a specific mutation of the H-*ras* gene (QUINTANILLA et al. 1986; BROWN et al. 1986). In addition, introduction of Harvey murine sarcoma virus *ras* genes into epidermal cells by the direct application of retroviruses to mouse skin successfully initiated the cells (BROWN et al. 1986). Subsequent repeated exposure to TPA resulted in papilloma formation. These results strongly implicate activated *ras* genes in the initiation phase of at least some chemically induced tumours.

In contrast with initiation, promotion is reversible, at least in its early stages (BOUTWELL 1974). In theory at least, this makes the encouraging prediction that the incidence of some human cancers could be reduced if relevant promoters could be identified and exposure restricted. Although this may be possible in some cases, its general applicability will depend on the proportion of important promoters that are exogenous environmental factors and those that are endogenously produced by the host.

II. Models of Tumour Promotion

1. Clonal Expansion

The simplest model proposes that each papilloma arises from a single initiated cell, i.e. that promotion involves the clonal expansion of initiated cells. This model is largely a result of the work of Yuspa and his colleagues and is based on extensive studies using primary keratinocyte culture (YUSPA and MORGAN 1981; HENNINGS et al. 1987b; YUSPA et al. 1983). In essence, the group has shown that, in primary cultures from initiated epidermis, a small proportion of the cells are resistant to calcium-induced terminal differentiation and continue to proliferate. It is presumed that the differentiation-resistant cells are "initiated" cells which have been genetically modified. These cells are also resistant to TPA-induced differentiation, whereas TPA induces the terminal differentiation of normal keratinocytes. The model therefore proposes that initiated cells continue to proliferate in the presence of TPA while normal keratinocytes are induced to differentiate. The consequence is selection for the initiated cell population and the eventual appearance of a visible tumour.

2. Two-Stage Model for Promotion

The clonal expansion model described above assumes that the promoter itself does not induce any phenotypic change in the initiated cell but simply provides a selection pressure favouring these cells. A number of reports have indicated, however, that promotion can be divided into at least two stages. The existence of stages in promotion was first suggested by BOUTWELL (1964) and subsequently supported by results from other laboratories (SLAGA et al. 1980; FURSTENBERGER et al. 1981; KINZEL et al. 1986).

In the standard protocol mice are first initiated and then treated with one or a few applications of TPA such that no papilloma formation occurs. This first stage of promotion, by agents such as TPA, calcium ionophore and hydrogen peroxide, is considered to involve a relatively stable change in the differentiated state of initiated cells which decays slowly over a period of about 10 weeks (FURSTENBERGER et al. 1983).

Subsequent repeated treatment with a second-stage promoter will then result in the selective growth of these cells and papilloma formation. Second-stage promoters include TPA as well as compounds which are effective as hyperplastigenic agents but are weak as complete promoters such as mezerein, 12-*O*-retinoyl-phorbol-13-acetate and turpentine. A number of inhibitors have also been reported to exert differential effects on first- and second-stage promotion (KINSELLA 1986).

3. Assessment of One- and Two-Stage Models

The key question is whether first-stage promotion involves the conversion of initiated cells to benign tumour cells which are then selected for during second-stage promotion. As recently discussed (HENNINGS and YUSPA 1985; HENNINGS 1987), the alternative is to propose that there are no independent stages and that both "first" and "second" stage promoters induce the clonal expansion of initiated cells. However, second-stage promoters are ineffective when applied as the only stimulus because of their greater toxicity, which is more manifest when there are only a small number of initiated cells present. In support of this alternative proposal, Hennings and Yuspa cite results indicating that the latent period for papilloma formation could be shortened by increasing the time interval between initiation and the start of promoter treatment. One interpretation of this result is that initiated cells have some inherent growth advantage resulting in an increase in clone size even in the absence of a promoting substance. Application of a promoter then induces the more rapid appearance of a visible tumour. In fact HENNINGS (1987) has reported that mezerein, a second-stage promoter, is an effective complete promoter if a 10-week period is left between initiation and the start of mezerein treatments. Other studies have shown that mezerein is toxic to initiated cells (MUFSON et al. 1979). This suggests that mezerein may not become effective as a promoter until a critical mass of initiated cells is achieved, either by exposure to a "first-stage" promoter or by proliferation of initiated cells in the absence of applied promoter. It has been reported that first-stage promotion can be achieved by a single treatment with TPA given 3 weeks before initiation (KINZEL et al. 1986). This observation is more difficult to explain within the clonal expansion model, although one possibility is that the TPA-induced hyperplastic epidermis remains less susceptible to toxic effects of "second-stage" promoters for several weeks. In support of this it has been reported that the first application of TPA is more toxic than subsequent applications (RAIK et al. 1972).

Although we favour the clonal expansion model of promotion, the current data do not allow an unequivocal decision between this and the two-stage model. The issue is, however, of crucial importance to a consideration of promoter mechanism. If functionally different stages exist, this implies that different biochemical changes are induced during each stage. It follows that complete promoters like TPA must be able to induce changes that second-stage promoters cannot. The significance of this possibility will be discussed further below.

III. Molecular Mechanism of Phorbol Ester-Induced Tumour Promotion

1. Specific Binding Sites Mediate Many Phorbol Ester Actions

Most of the information on mechanisms has come from studies in which phorbol esters have been added to a variety of cultured cell systems, and clearly the relevance of much of this information to in vivo promotion is unknown. However, it is certain that phorbol esters induce an enormous array of cellular and biochemical changes (see Table 2). In addition, there is frequently a reasonable correlation between the relative potency of a series of esters to induce a particular response and to promote tumours in vivo (BLUMBERG 1981). Although many of these responses are relatively long term, others occur within minutes and involve events at the cell surface. This, together with the demonstration of clear structure-activity responses and effects at nanomolar concentrations, suggested that the phorbol esters had a "hormone-like" function. This analogy was given further credence by the demonstration that particulate preparations from many cell lines and tissues, including skin, contained specific, high-affinity binding sites for phorbol esters (DRIEDGER and BLUMBERG 1980; BLUMBERG et al. 1984). Further, the affinity of binding of a series of phorbol esters to tissue preparations largely paralleled their in vivo potency as promoters (BLUMBERG et al. 1984). It is therefore widely believed that the phorbol ester promoters functioned by interacting with the receptor for some naturally occurring regulatory molecule.

Table 2. Biological effects of phorbol esters on cultured mammalian cells[a]

Biological response[b]	References[c]
Stimulation and inhibition of terminal differentiation	KINSELLA (1986)
Modulation of growth factor/hormone interactions with cells	
*Adrenergic receptor/adenylate cyclase	YOSHIMASA et al. (1987)
*EGF receptor	KING and COOPER (1986)
*Insulin and insulin-like growth factor receptor	JACOBS and CUATRECASAS (1986)
Interleukin 2 receptor	TAGUCHI et al. (1986)
Muscarinic receptor	SERRA et al. (1986)
*Mitogenesis	PASTI et al. (1986); HUANG and IVES (1987)
Enhanced gene expression	CARTER (1987)
*Ornithine decarboxylase	VERMA et al. (1986)
Plasminogen activator	CRUTCHLEY et al. (1980)
Glycophorin	SIEBERT and FUKUDA (1985)
Calcitonin	DE BUSTROS et al. (1985)
*Proto-oncogenes	ANGEL et al. (1987)
	Verma and SASSONE-CORSI (1987)
*Inhibition of cell-cell communication	YAMASAKI and ENOMOTO (1985); GAINER and MURRAY (1985)
*Enhanced hydrolysis of phosphatidylcholine	BESTERMAN et al. (1986); MUIR and MURRAY (1987)

[a] The examples listed are not comprehensive and have been chosen to illustrate the range of responses to phorbol esters.
[b] There is direct evidence for protein kinase C involvement in those responses marked with an asterisk.
[c] References have been chosen to give maximum access to the literature.

2. Protein Kinase C is a Major Cellular Target for Phorbol Esters

It is now generally accepted that many of the biological responses to phorbol esters result from activation of a calcium- and phospholipid-dependent protein kinase (protein kinase C; NISHIZUKA 1986; BERRIDGE 1987). Protein kinase C is an important component of transmembrane signalling systems which are associated with the turnover of inositol phospholipids (BERRIDGE 1987). In these systems, interaction with a surface receptor leads to activation of a specific phospholipase C, which hydrolyses phosphatidylinositol-4,5-bisphosphate, producing inositol trisphosphate and diacylglycerol (DG). Inositol trisphosphate is a water-soluble metabolite which increases cytosolic calcium concentrations by enhancing mobilisation from intracellular calcium stores. The highly lipophilic molecule DG remains associated with the membrane, where it forms a complex consisting of protein kinase C, calcium, DG and phospholipid. Experimentally, this can be detected by measuring a translocation of protein kinase C from the soluble fraction to the particulate fraction of cells. Phorbol esters have been shown to act as structural analogues of DG, which bind to and activate protein kinase C (CASTAGNA et al. 1982; NISHIZUKA 1986). A major consequence of incubating mammalian cells with phorbol esters is therefore a massive and sustained translocation of protein kinase C to the membrane, resulting in the phosphorylation of specific protein substrates. An additional complication is that association of protein kinase C with the membrane also facilitates its proteolysis into two fragments of molecular weights 50 K and 30–35 K (MURRAY et al. 1987). The 50 K fragment retains constitutive protein kinase activity which is now independent of calcium and phospholipid, whereas the 30–35 K fragment (regulatory domain) contains the binding sites for calcium, phospholipid and DG/phorbol ester. Both fragments have been shown to accumulate in cells exposed to TPA (MURRAY et al. 1987; FOURNIER and MURRAY 1987), and their functions, if any, are quite unknown. Finally, it should be mentioned that protein kinase C comprises a complex family of closely related proteins (see ONO et al. 1987 for references) which are differentially expressed in different tissues (OHNO et al. 1987; BRANDT et al. 1987) and which may have different substrate specificities. The biological functions of these protein kinase C isozymes are also unknown.

The demonstration that phorbol esters activate protein kinase C has revolutionised understanding of the biological activities of these compounds. It seems clear that protein kinase C has a central role in transmitting external signals to the interior of cells and in modulating the interactions of other regulatory factors with cells (see Table 2). Consequently, it is easy to theorise that phorbol esters could modulate the receipt of signals by initiated cells, resulting in their selective proliferation. Two questions emerge from these observations. Firstly, how firm is the evidence that protein kinase C is the only molecular target of phorbol esters, and secondly, do non-phorbol ester tumour promoters also activate protein kinase C?

3. Are All Phorbol Ester Effects Mediated via Protein Kinase C?

There is no doubt that protein kinase C is a major cellular target of phorbol esters. In fact, many authors are prepared to assume that demonstration of a phorbol ester response provides good evidence of a protein kinase C involvement in that response. If protein kinase C is the only cellular target for phorbol esters, it follows that prolonged activation of this enzyme should be sufficient to promote tumour formation. This question is of crucial importance when considering the possible existence of multiple stages in promotion. At least some of the second-stage promoters (e.g. mezerein, 12-*O*-retinoylphorbol-13-acetate) are potent activators of protein kinase C (GSCHWENDT et al. 1983). It therefore follows that if these compounds are functionally different from "complete" promoters such as TPA, then the latter must induce biochemical changes independent of protein kinase C. There is some indirect evidence that this may be the case, although the issue is far from resolved. For example, incubation of HL-60 cells with cell-permeable synthetic DGs and TPA stimulated the phosphorylation of common proteins, presumably indicating that both classes of compounds activated protein kinase C in intact cells (MORIN et al. 1987). However, only TPA and not the DGs was effective as an inducer of cell differentiation. However it should be noted that, although the profiles of protein phosphorylation induced by TPA and the DGs overlapped, they were not identical. It is therefore possible, for example, that the two compounds differentially activate individual protein kinase C isozymes. This study did not support a functional distinction between first- and second-stage promoters, however, as both TPA and mezerein were effective inducers of differentiation. In a separate study with C62B glioma cells (Brooks et al. 1987), it was shown that the ability of phorbol-12,13-dibutyrate to inhibit inositol phospholipid hydrolysis was not blocked by the protein kinase C inhibitor sphingosine, whereas the protein kinase C-mediated activation of inositol trisphosphate phosphatase was. Again this result implies that some phorbol ester effects may be independent of protein kinase C.

Other studies have further complicated the notion of a simple relationship between protein kinase C activation and tumour promotion. For example, the macrocyclic lactone bryostatin 1 binds to and activates protein kinase C in vitro and mimics many phorbol ester responses in cultured cells (Smith et al. 1985; Berkow and Kraft 1985). However, bryostatin 1 is not a complete promoter in mouse skin and is, in fact, an effective inhibitor of promotion by TPA (Hennings et al. 1987a). Results of this sort could be explained by differential effects of TPA and bryostatin on different protein kinase C isozymes. The results emphasise, however, that the role of protein kinase C in tumour promotion is complex and stress the urgent need to understand the biological roles of the different forms of protein kinase C.

4. Do Non-Phorbol Ester Tumour Promoters Activate Protein Kinase C?

It is well-established that several non-phorbol tumour promoters such as aplysiatoxin, lyngbatoxin, teleocidin B and mezerein can activate protein kinase C in vitro (Fujiki et al. 1984; Miyaka et al. 1984). These compounds all have structural similarities to the phorbol esters (Jeffrey and Liskamp 1986) and probably act in a similar way as analogues of DG. More recently, however, several other non-related compounds known or suspected of being tumour promoters have also been shown to activate protein kinase C. These include benzene (Roghani et al. 1987a), chloroform (Roghani et al. 1987b), *cis*-fatty acids (Murakami et al. 1986) and saccharin (Kleine et al. 1986). There is also indirect evidence that bile salts, which are promoters of colon carcinogenesis, can activate protein kinase C in Swiss 3T3 cells (Fitzer et al. 1987) and that the first-stage promoter benzoyl peroxide enhances the translocation of protein kinase C to the particulate fraction of keratinocytes when applied topically to mouse skin (Donnelly et al. 1987a). Finally, maintenance of mice for 4–6 weeks on a high-fat diet resulted in an elevated protein kinase C activity in the particulate fraction of basal epidermal cells (Donnelly et al. 1987b). In laboratory animals high-fat diets have been associated with promotion of tumours of the skin, breast, pancreas and colon (see Donnelly et al. 1987b for refs.). It is far too early to conclude that the diverse range of tumour-promoting compounds all act by stimulating protein kinase C. However, this remains a possibility, as it seems highly likely that the enzyme can be activated by a variety of mechanisms other than the highly specific interactions associated with the phorbol esters and some other classes of promoters. As has been noted before (Jones and Murray 1986), manipulation of kinase activity by perturbation of membrane structure would seem a particularly profitable future area of study.

C. Tumour Promotion in Liver

I. Outline of Chemical Carcinogenesis

Steps in the chemical induction of liver cancers have been characterised using rodents as experimental models. In the models most extensively studied, fully differentiated hepatocytes are considered to be the targets for initiation. A single exposure to an active carcinogen is sufficient for initiation. This exposure must be coupled with a proliferative stimulus to ensure that the carcinogen-induced changes, presumed to be in DNA, are fixed by a round of replication. A variety of promoting regimes have been used to induce the expansion of initiated cells to altered hepatic foci (AHF), some of which may expand into grossly visible hyperplastic nodules. Although the majority of the foci and nodules regress, some represent populations with an increased probability of undergoing the further genetic changes necessary for progression to malignant hepatocellular carcinoma.

Various experimental models used in studies of hepatocarcinogenesis have been extensively reviewed (PITOT and SIRICA 1980; PERAINO et al. 1983; MOORE and KITAGAWA 1986; GOLDSWORTHY et al. 1986; FARBER and SARMA 1987). In the following discussion emphasis will be placed on tumour promotion by phenobarbital (PB) and some related agents.

II. Characteristics of Initiated Hepatocytes and Their Progeny

Since initiation is believed to be a rare event with estimates ranging from 1–100 initiated cells/10^6 hepatocytes (FARBER 1980; PITOT et al. 1978), conclusions about the nature of initiated cells come largely from their progeny, AHF and nodules, which have been shown to arise by clonal proliferation from single initiated cells (WEINBERG et al. 1987). These putative pre-neoplastic lesions are distinguished from the surrounding cells by their altered histology and pattern of gene expression (PERAINO et al. 1983; MOORE and KITAGAWA 1986). Despite diversity in the phenotypic expression of individual lesions, there is a relatively consistent altered biochemical pattern common to most AHF and nodules and many of the hepatocellular carcinomas produced by a variety of initiation and promotion regimes (FARBER 1984). This "pre-neoplastic" phenotype includes changes in multiple enzymes involved in phase I and phase II xenobiotic metabolism (ERIKSSON et al. 1987) as well as changes in enzymes of carbohydrate metabolism (reviewed by SCHULTE-HERMANN 1985). Among the more widely used markers, elevated levels of γ-glutamyl transpeptidase or the placental form of glutathione-*S*-transferase identify a high proportion of AHF detectable by any marker (PITOT et al. 1985). Although there may be variations in detail between the phenotypes of PB-promoted lesions and those produced by other regimes (e.g. SCHULTE-HERMANN et al. 1986; CARR 1987), the phenotypic features common to all suggest that the nature of the pre-neoplastic phenotype may provide important pointers to the change produced in cells either at initiation, or as a result of promoter action on initiated cells.

It should be noted that there may be multiple pathways for liver cancer induction. There is evidence for the existence of liver stem cells which may give rise to hepatocytes, biliary epithelial cells and various transitional cells (FAUSTO et al.

1987), any of which may be targets for initiation. In particular "oval cells" may be important in some experimental models (FAUSTO et al. 1987; SELL and LEFFERT 1982). In the models discussed below, initiation and promotion appear to involve "neo-differentiation" of mature hepatocytes. However, given the fragmentary knowledge of normal liver cell lineages, it is difficult to exclude the possibility that key events leading to eventual tumour formation involve a population of cells in which initiation has arrested the potential for terminal differentiation of a hepatocyte-precursor cell. This situation would be analogous to that proposed for initiation in skin (see Sect. B.I).

III. Characteristics of Promotion by Phenobarbital

PERAINO et al. (1971, 1977) and PITOT et al. (1978) were among the first to show promotion by PB in rat liver. In the Pitot model, initiation by a single low dose of diethylnitrosamine given after partial hepatectomy resulted in the appearance of some AHF but no carcinomas. Feeding PB alone for 6 months produced no AHF, but after initiation, it increased the number of detectable AHF fivefold and led to development of small numbers of carcinomas. In this model, 3–4 months of PB feeding was required for maximal enhancement of the number and volume of AHF. No-effect thresholds were observed, above which PB caused a dose-dependent increase up to a maximum effective dose (GOLDSWORTHY et al. 1984; PITOT et al. 1987). PB increased AHF when given after initiation but had no effect prior to initiation (SCHWARZ et al. 1983). During promotion, detectable AHF emerge progressively with time, and a major effect of PB is to favour the earlier appearance of detectable foci (KAUFMANN et al. 1987) as well as an increase in the total AHF detectable after longer periods of treatment (PITOT et al. 1987). MOORE et al. (1983) observed a PB-induced shift towards a higher proportion of larger AHF, and PB also causes a reduction in the latent period before the appearance of tumours (PITOT et al. 1978; MOORE et al. 1983; KUNZ et al. 1983). By contrast, PERAINO et al. (1980) reported that PB increased the number of nodules but did not change their size distribution or time of appearance. On balance, it seems that during the genesis of AHF, initiated cells require a direct or indirect proliferative stimulus from the promoter to allow clonal expansion to a focus of detectable size. In the absence of promoter, the growth of cells in early foci or nodules appears to be similar to that of normal hepatocytes (ROTSTEIN et al. 1986), as is their expression of some growth-related proto-oncogenes (BEER et al. 1986). The fraction of cells active in DNA synthesis is higher in AHF than in normal cells when measured during chronic carcinogen exposure (RABES and SZYMKOWIAK 1979) and further increases after exposure to PB (SCHULTE-HERMANN et al. 1981, 1986).

The altered enzyme markers expressed by different AHF provide the basis for recognising these lesions and, as elements of a common liver pre-neoplastic phenotype, may have relevance to cancer induction. Thus, a number of studies have followed changes in the expression of various markers as an indication of cellular changes occurring during promotion. Because the rare initiated cells are difficult to detect, the extent to which expression of an altered phenotype is dependent on genetic changes at initiation alone or on subsequent exposure to promoter is unclear. Small groups of γ-glutamyl transpeptidase-positive hepatocytes were detectable 7 days after initiation (OGAWA et al. 1980), and single hepatocytes or small groups of cells positive for placental glutathione-*S*-transferase were observed 48 h after carcinogen treatment (MOORE et al. 1987), suggesting that some markers are expressed at early times after initiation without exposure to promoters. This is consistent with the view (FARBER and SARMA 1987) that at least a sub-population of initiated cells constitutively express the altered phenotype observed in AHF or nodules. During promotion by PB, however, evidence suggests that expression of markers of the preneoplastic phenotype is dependent, at least in part, on the presence of promoter. In the Pitot model, in addition to the appearance of many new AHF during PB promotion, the pattern of markers expressed in foci changed. For example, the proportion of AHF positive for γ-glutamyl transpeptidase was greater after PB than after initiation alone (PITOT et al. 1978; SCHULTE-HERMANN et al. 1986). This apparently reflects a reversible effect of promoter on expression of this enzyme, a view supported by findings that most liver tumour promoters cause a modest induction of the enzyme in normal hepatocytes (EDWARDS and LUCAS 1985a), that PB further increases enzyme activity in existing AHF (SIRICA et al. 1984) and that PB reversibly induces high levels of activity in spontaneous liver tumours in mice (WILLIAMS et al. 1980). Similarly, initiation alone resulted in AHF with increased UDP-glucuronyl transferase activity which was further reversibly increased in foci by exposure to PB (FISCHER et al. 1985).

When PB was withdrawn after a period of promotion and rats fed a purified diet, AHF rapidly regressed, and 50% of foci detected by enzyme markers were lost in 10 days (GLAUERT et al. 1986). Similarly, 95%–98% of nodules produced by the Solt/Farber protocol (see below) regressed, although apparently more slowly than after PB withdrawal. The disappearance of nodules was not the result of cell death or replacement with hepatocytes from the surrounding liver but from "redifferentiation" to an adult liver phenotype (TATEMATSU et al. 1983b). In the Pitot model, foci regressed when PB was withdrawn, but even after 90 days, readministration of PB for 30 days caused rapid reappearance of foci occupying a greater volume than after the original 4-month period of promotion (HENDRICH et al. 1986). It therefore seems possible that after PB withdrawal, the expanded populations of initiated cells remain, but many are not detected because they do not express the pre-neoplastic phenotype in the absence of promoter. Thus, PB appears to exert reversible effects on the extent to which at least some elements of the pre-neoplastic phenotype are expressed.

PERAINO et al. (1984) have argued that different initiation events determine the potential of individual AHF for expressing different sets of markers. There is evidence that a variety of different genetic changes can lead to AHF in the presence of promoter, although only a subset of such changes yield AHF with enhanced probability of further progression to carcinomas. Thus the ratio of AHF to tumours observed is different when different carcinogens or γ-radiation are employed as an initiating stimulus prior to PB promotion (PERAINO et al. 1987; KAUFMANN et al. 1986a). Those AHF expressing more altered enzyme markers (greater "phenotypic complexity") are the most active in DNA synthesis and proliferate most rapidly (PUGH and GOLDFARB 1978; OGAWA et al. 1980; MOORE et al. 1983). These may represent the fraction of AHF most significant as precursors of tumours.

IV. Alternative Regimes for Promotion

In addition to PB, a variety of other regimes have been used to promote tumours in liver. Xenobiotics such as DDT, hexachlorocyclohexane, some polychlorinated and polybrominated biphenyls promote and induce adaptive changes similar to PB. Like PB, 2,3,7,8-tetrachlorodibenzo-*p*-dioxin (TCDD) and some progestins promote but cause differing sets of adaptive changes (SCHULTE-HERMANN 1985). This is also true for estrogens such as ethinyl estradiol (YAGER et al. 1986) and "peroxisomal proliferators" such as clofibrate (RAO and REDDY 1987). This whole group, whose members share the ability to cause liver hypertrophy and hyperplasia, has been tentatively linked under the heading "stimulators of additive hyperplasia" by LOURY et al. (1987). Other protocols for liver promotion include feeding choline-deficient diets (CHANDAR et al. 1987), orotic acid (SARMA et al. 1986) and regimes such as a high fat diet or portacaval anastomosis (DE GERLACHE et al. 1987). It is therefore clear that quite diverse conditions which provoke adaptive changes in liver may promote.

Farber and his colleagues have made extensive use of a "resistant hepatocyte" model, in which initiation is considered to result in constitutive expression of the pre-neoplastic phenotype. The altered expression of xenobiotic-metabolising enzymes in this model is considered central to clonal expansion in that initiated cells are selectively resistant to toxins such as 2-acetylaminofluorene. After initiation, livers are given a proliferative stimulus (partial hepatectomy) during a period of 2-acetylaminofluorene feeding (Solt/Farber regime). The growth of normal cells is inhibited, but the resistant, initiated cells rapidly proliferate to produce nodules (FARBER and SARMA 1987). This protocol yields many fewer lesions than are observed in the Pitot model (HENDRICH et al. 1987). PB treatment following the Solt/Farber regime increases the tumour yield and decreases the latent period before tumours are observed (LANS et al. 1983; PREAT et al. 1987), suggesting that PB promotion and Solt/Farber selection differ in mechanism. The extent to which this reflects complementary effects on the same group of initiated cells or effects on different subsets of initiated cells is unclear.

V. Mechanism of Promotion by Phenobarbital and Related Xenobiotics

1. Mechanisms at the Cellular Level

As indicated above, PB is known to induce a broad adaptive response in normal liver, including hyperplasia and induction of a variety of enzymes involved in xenobiotic metabolism. SCHULTE-HERMANN (1985) has noted the similarities between this response and the pre-neoplastic phenotype. Most of the observations above are consistent with the idea that initiation results in irreversible changes to major regulatory gene(s) involved in adaptive responses to promoter or further carcinogen exposure (SCHULTE-HERMANN 1985; MOORE and KITAGAWA 1986; FARBER and SARMA 1987). This change may facilitate expression of some genes and depress others, favouring a new state of differentiation in hepatocytes. In the case of promotion by PB, SCHULTE-HERMANN (1985) has suggested that this potential becomes fully manifest when initiated cells exhibit an excessive, or aberrant, response to the adaptive programme favoured in all hepatocytes by PB. Presumably, some elements in the altered response of initiated cells to PB contribute directly to their clonal expansion to AHF, while others (which may still be useful as markers) are not directly relevant to the genesis of AHF.

Since the key event in promotion is proliferation of a single initiated cell to an expanded focus of cells, any mechanism must provide the basis for the selective growth advantage of these cells. PB may be representative of the group of non-genotoxic xenobiotics and steroids listed above ("inducers of additive hyperplasia") which share an ability to stimulate liver hypertrophy and hyperplasia (SCHULTE-HERMANN 1985; LOURY et al. 1987). Many of the compounds stimulate DNA synthesis in normal cultured hepatocytes (EDWARDS and LUCAS 1985b). For several agents in this group, the relative abilities of various compounds to promote in rodents of different species or sex paralleled their ability to stimulate DNA synthesis in vivo (BUSSER and LUTZ 1987). SCHULTE-HERMANN et al. (1981, 1986) have shown that cells in AHF exhibit a greater proliferative response to PB than normal hepatocytes. Such a selective mitogenic effect of PB and related promoters on initiated cells would eventually result in the formation of foci. Although this model is attractive, it has not yet been conclusively shown that the mitogenic responses to PB-like promoters are sufficient to explain promotion. Thus, the response of normal cells (BARBASON et al. 1983) and cells in AHF (SCHULTE-HERMANN et at. 1981, 1986) to xenobiotics is relatively short-lived. Indeed, there is evidence that longer-term treatment with PB may inhibit proliferation in normal hepatocytes (BARBASON et al. 1983), possibly by depression of epidermal growth factor (EGF) and insulin receptor levels (BETSCHART et al. 1987; GUPTA et al. 1987). Cells in AHF may be less susceptible to this inhibitory component of PB action, since DNA synthesis was inhibited less by PB in AHF than in normal cells (ECKL et al. 1987). Direct evidence that initiation may yield cells which are more sensitive to the mitogenic action of PB is provided by the isolation from initiated (but not normal) liver of small numbers of hepatocytes which grow slowly to yield colonies in the presence of PB but not in its absence. After some months in culture the colonies acquired anchorage independence and produced hepatocellular carcinomas in rats (KITAGAWA et al. 1980; KAUFMANN et al. 1986b, 1987). However, a proliferative stimulus per se cannot account for promotion since partial hepatectomy had little effect alone or in addition to PB on the appearance of AHF after initiation (HASEGAWA et al. 1986), and even repeated partial hepatectomy only promoted weakly (POUND and MCGUIRE 1978). The ability of initiated cells to respond *selectively* to the mitogenic effects of PB may be important for promotion.

The significance for promotion of the effects of PB that are less directly related to mitogenesis is unclear. Differential effects of PB in inhibiting cell death (or apoptosis; BURSCH et al. 1984) or inhibition of intercellular communication (KLAUNIG and RUCH 1987) may contribute to the net growth of AHF. While, in the resistant hepatocyte model, expression of the pre-neoplastic phenotype is fundamental to selection of initiated cells (Farber and Sarma 1987), the basis for promotion by PB does not appear to involve toxicity to cells (Hayes et al. 1987), and it is currently not clear how the altered pattern of xenobiotic metabolising enzymes in PB-induced AHF contributes to promotion.

2. Mechanisms at the Molecular Level

It has been widely suggested that PB causes an adaptive response by co-ordinately regulating the expression of multiple genes. PB-stimulated changes that may be mediated by pleiotropic control include induction of specific isozymes of cytochrome P450, glucuronyl transferase, glutathione-*S*-transferase and epoxide hydrolase, induction of NADPH-cytochrome P450 reductase and decreases in RNase and glucose 6-phosphatase (Schulte-Hermann 1985; Nims et al. 1987; Lechner et al. 1987; Poland et al. 1981). PB has been shown to cause relatively synchronous increases or decreases in several mRNA species by increasing transcription (Hardwick et al. 1983; Lechner et al. 1987), although post-transcriptional control may also occur (Lechner et al. 1987). Within a group of barbiturates, promoter potency correlated with the ability to induce cytochrome P450-dependent drug metabolism, epoxide hydrolase and liver weight gain (Nims et al. 1987) consistent with a linkage of these changes in a single adaptive programme. However, the extent to which mitogenic and other responses to BP are co-ordinated via a single mechanism is unclear.

Little is known about the primary effects which precede the PB-induced changes in transcription. TCDD is known to induce a variety of xenobiotic metabolising enzymes via binding to a specific receptor which then interacts with regulatory elements in DNA (Whitlock 1987). A related mode of action might be expected for BP, but, because most PB-like compounds act only at relatively high concentrations, it has not yet been possible to identify a specific receptor (Fonne and Meyer 1987). The finding that 1,4-bis[2-(3,5-dichloropyridyloxy)]-benzene is a potent promoter in mouse liver (DellaPorta et al. 1987) and produces a PB-like spectrum of adaptive changes (Poland et al. 1981) may help the search for specific cellular binding sites.

In the absence of definitive evidence for one or more specific receptors mediating liver tumour promotion, alternative mechanisms have been sought. Most of the promoters are lipophilic and known to penetrate into membranes, increasing membrane fluidity. It has been suggested that resultant changes in ion channels or membrane-bound enzyme activities may account for some of the actions of barbiturates in the nervous system (Ho and Harris 1981). Such direct effects on membranes could result in their mimicking, or modulating, the action of hormones which bind to surface receptors and alter cellular function via signal transducers such as cytoplasmic calcium or various protein kinase enzymes. In this context, the question as to whether or not liver promoters activate protein kinase C is of particular interest. Chloroform and some bile acids, which are active as liver promoters, did activate the kinase, possibly by modifying the lipid environment and/or stimulating DG formation, but other agents, including PB and TCDD, did not alter protein kinase C activity in a reconstituted cell-free system (reviewed by Castagna 1987). In isolated hepatocytes TPA, but not PB, caused a rapid loss of the ability of α_1-adrenergic agonists to activate glycogen phosphorylase (Blight and Edwards, unpublished data), suggesting that PB has no rapid effect on protein kinase C activity. Longer-term effects during in vivo promotion are still possible. In cultured hepatocytes both TPA and PB stimulate DNA synthesis but appear to act by different mechanisms (EDWARDS and YUSOF 1986; SAWADA et al. 1987). PB has been shown to influence the calcium dependence of growth of various liver-derived cells (ARMATO et al. 1985; BOYNTON and WHITFIELD 1980; ECKL et al. 1987) consistent with some effect of PB, direct or indirect, at the level of signal transmission mechanisms.

In summary, there is currently no clear-cut evidence as to whether the various adaptive responses to PB-like promoters involve direct interaction of receptors with chromatin or effects on signalling mechanisms via second messengers or protein kinases. However, rapid progress in delineating mechanisms in signal transduction and changes in gene expression controlling the cell cycle in various tissues including liver (SOBCZAK and DUGUET 1986; MCGOWAN 1986) should facilitate studies on how PB influences these events.

D. Concluding Remarks

The unifying feature of tumour promotion in skin and liver (and probably in other tissues) is the provision of a selective growth advantage to genetically modified ("initiated") cells. In addition, promoters like phorbol esters and PB each induce pleiotropic responses in target cells, suggesting that both compounds modulate some central regulatory mechanism(s). It cannot, of course, be assumed that all responses to either compound are related to promotion, and identification of the key events remains a major challenge. Given that proliferation is regulated by the interaction of cells with a large number of growth factors/hormones, it seems likely that this process could be modulated by promoter interactions at a number of different points. Consequently, there is unlikely to be a single "mechanism" for tumour promotion.

In the context of the present discussion, the three most obvious models for key events in promotion are as follows:

1. As a result of initiation, cells express an abnormal phenotype which enables them to proliferate in the presence of promoters.
2. The initiated phenotype is not fully expressed except in the presence of the promoter. Full expression of this phenotype is sufficient to confer a growth advantage on the initiated cells.
3. Promoters both enhance the expression of the initiated cell phenotype and provide a direct growth stimulus to cells expressing this phenotype.

The requirements for generating a growth advantage in initiated cells may be quite different in skin and liver, the specific models discussed in this review. The epidermis is a constantly regenerating tissue, and the differentiation block imposed by initiation (Yuspa model) therefore provides for the "immortality" of the genetically modified cells. The permanence of the initiated cell in a regenerating tissue implies that its abnormal phenotype is expressed even in the absence of the promoter. In the presence of the promoter these cells are then stimulated to proliferate whereas normal cells differentiate and die. This sequence is consistent with model 1 above. The alternative view that skin promotion occurs in two stages would be consistent with model 3. On the other hand, liver is a fully differentiated tissue containing predominately quiescent cells. Consequently, in contrast to the "differentiation block" in epidermis, initiation in liver can lead to "neo-differentiated" cells with enhanced growth potential. Apparently, promotion by PB enhances expression of the initiated cell phenotype, which may itself enable proliferation (model 2) or which may increase sensitivity to growth stimulation by the promoter (model 3).

In addition, different mechanisms for the selection of initiated cells are possible within a given tissue. For example, in the Solt/Farber regime, the initiated liver cells are believed to express the initiated cell phenotype (model 1), which confers resistance to cytotoxins and enables them to respond to a proliferative stimulus in a toxic environment. Similarly, in mouse skin the promoter benzoyl peroxide is less toxic to initiated cells than to normal cells. This increased resistance may enable the selective growth of initiated cells following the regenerative hyperplasia induced by benzoyl peroxide (Hartley et al. 1987).

Several factors may be relevant to the apparent tissue specificity noted in the actions of many tumour promoters (Table 1). Differences in effective concentrations of active promoters in various tissues in vivo, influenced by delivery mode and metabolism, must affect promoting potency at different sites. For example, active esterases in liver may degrade TPA (Mentline 1986), obscuring a potential promoting action. In models 2 or 3, discussed above, promotion results from an interaction between genetic change at initiation and modulation of the pattern of differentiated gene expression by promoters. Since the sets of genes with potential for expression vary from one tissue to another, it is likely that agents which can modulate the pattern of expression (in a manner contributing to clonal expansion) will also vary. Thus, PB can evoke adaptive changes involving many genes in normal liver, whereas little change occurs in skin. In situations in which promotion involves model 1, it may be that agents directly affecting growth-controlling mechanisms common to many tissues (e.g. protein kinase C) will exhibit promoter activity in a wider range of tissues than for models 2 or 3.

References

Angel P, Imagawa M, Chiu R, Stein B, Imbra RJ, Rahmsdorf HJ, Jonat C, Herrlich P, Karin M (1987) Phorbol ester-inducible genes contain a common *cis* element recognised by a TPA-modulated *trans*-acting factor. Cell 49:729–739

Argyris TS (1985) Regeneration and the mechanism of epidermal tumor promotion. CRC Crit Rev Toxicol 14:211–257

Armato U, Andreis PG, Romano F (1985) The stimulation by the tumor promoters 12-*O*-tetradecanoylphorbol-13-acetate and phenobarbital of the growth of primary neonatal rat hepatocytes. Carcinogenesis 6:811–821

Barbason H, Rassenfosse C, Betz EH (1983) Promotion mechanism of phenobarbital and partial hepatectomy in DNEA hepatocarcinogenesis cell kinetics effect. Br J Cancer 47:517–525

Beer DG, Schwarz M, Sawada N, Pitot HC (1986) Expression of H-*ras* and c-*myc* protooncogenes in isolated γ-glutamyl transpeptidase-positive rat hepatocytes and in hepatocellular carcinomas induced by diethylnitrosamine. Cancer Res 46:2435–2441

Berkow RL, Kraft AS (1985) Bryostatin, a non-phorbol macrocyclic lactone, activates intact human polymorphonuclear leukocytes and binds to the phorbol ester receptor. Biochem Biophys Res Commun 131:1109–1116

Berridge MJ (1987) Inositol trisphosphate and diacylglycerol: two interacting second messengers. Ann Rev Biochem 56:159–193

Besterman JM, Duronio V, Cuatrecasas P (1986) Rapid formation of diacylglycerol from phosphatidylcholine: a pathway for generation of a second messenger. Proc Natl Acad Sci USA 83:6785–6789

Betschart JM, Gupta C, Shinozuka H, Virji MA (1987) Modulation of hepatocyte insulin and glucagon receptors in rats fed phenobarbital, a tumor promoter. Proc Am Assoc Cancer Res 28:165

Blumberg PM (1981) In vitro studies on the mode of action of the phorbol esters, potent tumor promoters, part 2. CRC Crit Rev Toxicol 8:199–234

Blumberg PM, Dunn JA, Jaken S, Jeng AY, Leach KL, Sharkey NA, Yeh E (1984) Specific receptors for phorbol ester tumor promoters and their involvement in biological responses. In: Slaga TJ (ed) Mechanisms of tumor promotion, vol 3. Tumor promotion and carcinogenesis in vitro. CRC Press, Baton Rouge, p 143

Boutwell RK (1964) Some biological aspects of skin carcinogenesis. Prog Exp Tumor Res 4:207–250

Boutwell RK (1974) The function and mechanism of promoters of carcinogenesis. CRC Crit Rev Toxicol 2:419–443

Boynton AL, Whitfield JF (1980) Stimulation of DNA synthesis in calcium-deprived T51B liver cells by the tumor promoters phenobarbital, saccharin and 12-*O*-tetradecanoylphorbol-13-acetate. Cancer Res 40:4541–4545

Brandt SJ, Niedel JE, Bell RM, Young WS (1987) Distinct patterns of different protein kinase C mRNAs in rat tissues. Cell 49:57–63

Brooks RC, Morell P, DeGeorge JJ, McCarthy KD, Lapetina EG (1987) Differential effects of phorbol ester and diacylglycerols on inositol phosphate formation in C62B glioma cells. Biochem Biophys Res Commun 148:701–708

Brown K, Quintanilla M, Ramsden M, Kerr IB, Young S, Balmain A (1986) v-*ras* genes from Harvey and BALB murine sarcoma viruses can act as initiators of two-stage mouse skin carcinogenesis. Cell 46:447–456

Bursch W, Lauer B, Timmermann-Trosiener I, Barthel G, Schuppler J, Schulte-Hermann R (1984) Controlled death (apoptosis) of normal and putative preneoplastic cells in rat liver following withdrawal of tumor promoters. Carcinogenesis 5:453–458

Busser M-T, Lutz WK (1987) Stimulation of DNA synthesis in rat and mouse liver by various tumor promoters. Carcinogenesis 8:1433–1437

Carr BI (1987) Pleiotropic drug resistance in hepatocytes induced by carcinogenes administered to rats. Cancer Res 47:5577–5583

Carter TH (1987) The regulation of gene expression by tumor promoters. In: Barrett JC (ed) Mechanisms of environmental carcinogenesis, vol 1. CRC Press, Baton Rouge, p 47

Castagna M (1987) Phorbol esters as signal transducers and tumor promoters. Biol Cell 59:3–13

Castagna M, Takai Y, Kaibuchi K, Sano K, Kikkawa U, Nishizuka Y (1982) Direct activation of calcium-activated phospholipid-dependent protein kinase by tumor-promoting phorbol esters. J Biol Chem 257:7847–7851

Chandar N, Amenta J, Kandala JC, Lombardi B (1987) Liver cell turnover in rats fed a choline-devoid diet. Carcinogenesis 8:669–673

Crutchley DY, Conan LB, Maynard JR (1980) Induction of plasminogen activator and prostaglandin biosynthesis in HeLa cells by 12-*O*-tetradecanoylphorbol-13-acetate. Cancer Res 40:849–852

de Bustros A, Baylin SB, Berger CL, Roos BA, Leong SS, Nelkin BD (1985) Phorbol esters increase calcitonin gene transcription and decrease c-*myc* mRNA levels in cultured human medullary thyroid carcinoma. J Biol Chem 260:98–104 de Gerlache J, Taper HS, Lans M, Preat V, Roberfroid M (1987) Dietary modulation of rat liver carcinogenesis. Carcinogenesis 8:337–340

DellaPorta G, Dragani TA, Maneti G (1987) Two-stage liver carcinogenesis in the mouse. Toxicol Pathol 15:229–233

Diwan BA, Rice JM, Ohshima M, Ward JM, Dove LF (1985) Comparative tumor-promoting activities of phenobarbital, amobarbital, barbital sodium, and barbituric acid on livers and other organs of male F344/NCr rats following initiation with *N*-nitrosodiethylamine. JNCI 74:509–516

Donnelly TE, Pelling JC, Anderson CL, Dalby D (1987a) Benzoyl peroxide activation of protein kinase C activity in epidermal cell membranes. Carcinogenesis 8:1871–1874

Donnelly TE, Birt DF, Sittler R, Anderson CL, Choe M, Julius A (1987b) Dietary fat regulation of the association of protein kinase C activity with epidermal cell membranes. Carcinogenesis 8:1867–1870

Driedger PE, Blumberg PM (1980) Specific binding of phorbol ester tumor promoters. Proc Natl Acad Sci USA 77:567–571

Eckl PM, Whitcomb WR, Meyer SA, Jirtle RL (1987) Effects of phenobarbital and calcium on the growth of normal and preneoplastic hepatocytes. Proc Am Assoc Cancer Res 28:171

Edwards AM, Lucas CM (1985a) Induction of γ-glutamyl transpeptidase in primary cultures of normal rat hepatocytes by liver tumor promoters and structurally related compounds. Carcinogenesis 6(5):733–739

Edwards AM, Lucas CM (1985b) Phenobarbital and some other liver tumor promoters stimulate DNA synthesis in cultured rat hepatocytes. Biochem Biophys Res Commun 131:103–108

Edwards AM, Yusof YAM (1986) Stimulation of hepatocyte DNA synthesis by phorbol esters and mezerein: comparison with effects of xenobiotic liver tumour promoters. Proc Aust Biochem Soc 19:15

Eriksson LC, Blanck A, Bock KW, Mannervik B (1987) Metabolism of xenobiotics in hepatocyte nodules. Toxicol Pathol 15:27–42

Farber E (1980) The sequential analysis of liver cancer induction. Biochim Biophys Acta 605:149–166

Farber E (1984) The biochemistry of preneoplastic liver: a common metabolic pattern in hepatocyte nodules. Can J Biochem Cell Biol 62:486–494

Farber E, Sarma DSR (1987) Hepatocarcinogenesis: a dynamic cellular perspective. Lab Invest 56:4–22

Fausto N, Thompson NL, Braun L (1987) Purification and culture of oval cells from rat liver. In: Pretlow TG, Pretlow TP (eds) Cell separation. Methods and selected applications. Academic, Orlando, p 45

Fischer G, Ullrich D, Bock KW (1985) Effects of *N*-nitrosomorpholine and phenobarbital on UDP-glucuronyl-transferase in putative preneoplastic foci of rat liver. Carcinogenesis 6:605–609

Fitzer CJ, O'Brian CA, Guillem JG, Weinstein IB (1987) The regulation of protein kinase C by chenodeoxycholate, deoxycholate and several structurally related bile acids. Carcinogenesis 8:217–220

Fonne R, Meyer UA (1987) Mechanisms of phenobarbital-type induction of cytochrome P-450 isozymes. Pharmacol Ther 33:19–22

Fournier A, Murray AW (1987) Application of phorbol ester to mouse skin causes a rapid and sustained loss of protein kinase C. Nature 330:767–769

Fujiki H, Tanaka Y, Miyaka R, Kikkawa U, Nishizuka Y, Sugimura T (1984) Activation of protein kinase C by new classes of tumor promoters; teleocidin and debromaplysiatoxin. Biochem Biophys Res Commun 120:339–343

Furstenberger G, Marks F (1983) Growth stimulation and tumor promotion in skin. J Invest Dermatol 81:157s–161s

Furstenberger G, Berry DL, Sorg B, Marks F (1981) Skin tumor promotion by phorbol esters is a two-stage process. Proc Natl Acad Sci USA 78:7722–7726

Furstenberger G, Sorg B, Marks F (1983) Tumor promotion by phorbol esters in skin: evidence for a memory effect. Science 220:89–92

Gainer HStC, Murray AW (1985) Diacylglycerol inhibits gap junctional communication in cultured epidermal cells: evidence for a role of protein kinase C. Biochem Biophys Res Commun 126:1109–1113

Glauert HP, Schwarz M, Pitot HC (1986) The phenotypic stability of altered hepatic foci: effect of the short-term withdrawal of phenobarbital and of the long-term feeding of purified diets after the withdrawal of phenobarbital. Carcinogenesis 7:117–121

Goerttler K, Loehrke H, Schweizer J, Hesse B (1979) Systemic two-stage carcinogenesis in the epithelium of the forestomach of mice using 7,12-dimethylbenz(a)anthracene as initiator and the phorbol ester 12-*O*-tetradecanoylphorbol-13-acetate as promoter. Cancer Res 39:1293–1299

Goldsworthy T, Campbell HA, Pitot HC (1984) The natural history and dose response characteristics of enzyme-altered foci in rat liver following phenobarbital and diethylnitrosamine administration. Carcinogenesis 5:67–71

Goldsworthy TL, Hanigan MH, Pitot HC (1986) Models of hepatocarcinogenesis in the rat – contrasts and comparisons. CRC Crit Rev Toxicol 17:61–89

Gschwendt M, Horn F, Kittstein W, Furstenberger G, Marks F (1983) Soluble phorbol ester binding sites and phospholipid- and calcium-dependent protein kinase activiy in cytosol of chick oviduct. FEBS Lett 162:147–150

Gupta C, Hattori A, Betschart JM, Virji MA, Shinozuka H (1987) Inhibition of EGF binding in rat hepatocytes by liver tumor promoters. Proc Am Assoc Cancer Res 28:113

Hardwick JP, Gonzalez FJ, Kasper CB (1983) Transcriptional regulation of rat liver epoxide hydratase, NADPH-cytochrome P-450 oxidoreductase and cytochrome P-450_b genes by phenobarbital. J Biol Chem 258:8081–8085

Hartley JA, Gibson NW, Kilkenny A, Yuspa S (1987) Mouse keratinocytes derived from initiated skin or papillomas are resistant to DNA strand breakage by benzoyl peroxide: a possible mechanism for tumor promotion mediated by benzoyl peroxide. Carcinogenesis 8:1827–1830

Hasegawa R, Tsuda H, Shirai T, Kurata Y, Masuda A, Ito N (1986) Effect of timing of partial hepatectomy on the induction of preneoplastic liver foci in rats given hepatocarcinogens. Cancer Lett 32:15–23

Hayes MA, Lee G, Tatematsu M, Farber E (1987) Influences of diethylnitrosamine on longevity of surrounding hepatocytes and progression of transplanted persistent nodules during phenobarbital promotion of hepatocarcinogenesis. Int J Cancer 40:58–63

Hendrich S, Glauert HP, Pitot HC (1986) The phenotypic stability of altered hepatic foci: effects of withdrawal and subsequent readministration of phenobarbital. Carcinogenesis 7:2041–2045

Hendrich S, Campbell HA, Pitot HC (1987) Quantitative stereological evaluation of four histochemical markers of altered foci in multistage hepatocarcinogenesis in the rat. Carcinogenesis 8:1245–1250

Hennings H (1987) Tumor promotion and progression in mouse skin. In: Barrett JC (ed) Mechanisms of environmental carcinogenesis, vol 2. CRC Press, Baton Rouge, p 59

Hennings H, Yuspa SH (1985) Two-stage tumor promotion in mouse skin: an alternative interpretation. JNCI 74:735–740

Hennings H, Shores R, Wenk ML, Spangler EF, Tarone R, Yuspa SH (1983) Malignant conversion of mouse skin tumors is increased by tumor initiators and unaffected by tumor promoters. Nature 304:67–69

Hennings H, Blumberg PM, Pettit GR, Herald CL, Shores R, Yuspa SH (1987a) Bryostatin 1, an activator of protein kinase C, inhibits tumor promotion by phorbol esters in SENCAR mouse skin. Carcinogenesis 8:1343–1346

Hennings H, Michael D, Lichti U, Yuspa SH (1987b) Response of carcinogen-altered mouse epidermal cells to phorbol esters and calcium. J Invest Dermatol 88:60–65

Hiasa Y, Kitahori Y, Ohshima M, Fujita T, Yuasa T, Konishi H, Miyashiro A (1982) Promoting effects of phenobarbital and barbital on development of thyroid tumors in rats treated with *N*-bis(2-hydroxypropyl)-nitrosamine. Carcinogenesis 3:1187–1190

Ho IK, Harris RA (1981) Mechanism of action of barbiturates. Annu Rev Pharmacol Toxicol 21:83–111

Huang CL, Ives HE (1987) Growth inhibition by protein kinase C late in mitogenesis. Nature 329:849–850

Ito N, Fukushima S, Tsuda H (1984) Carcinogenicity and modification of carcinogenic response by BHA, BHT and other antioxidants. CRC Crit Rev Toxicol 15:109–150

Jacobs S, Cuatrecasas P (1986) Phosphorylation of receptors for insulin and insulin-like growth factor. 1. Effects of hormones and phorbol esters. J Biol Chem 261:934–939

Jeffrey AM, Liskamp RMJ (1986) Computer-assisted molecular modelling of tumor promoters: rationale for the activity of phorbol esters, teleocidin B, and aplysiatoxin. Proc Natl Acad Sci USA 83:241–245

Jones MJ, Murray AW (1986) Effect of membrane perturbation on protein kinase C activation: treatment with exogenous phospholipase C decreases translocation of enzyme to cellular membranes. Biochem Biophys Res Commun 136:1083–1089

Kaufmann WK, MacKenzie SA, Rahija RJ, Kaufman DG (1986a) Quantitative relationship between initiation of hepatocarcinogenesis and induction of altered cell islands. J Cell Biochem 30:1–9

Kaufmann WK, Tsao M-S, Novicki DL (1986b) In vitro colonization ability appears soon after initiation of hepatocarcinogenesis in the rat. Carcinogenesis 7:669–671

Kaufmann W, Rahija R, Kaufman D, Goyette M, Dolan M, Shank P, Fausto N (1987) Cellular and genetic alterations during hepatocarcinogenesis in the rat. Fed Proc 46:972

King CS, Cooper JA (1986) Effects of protein kinase C activation after epidermal growth factor binding on epidermal growth factor receptor phosphorylation. J Biol Chem 261:10073–10078

Kinsella AR (1986) Multistage carcinogenesis and the biological effects of tumor promoters. In: Evans FJ (ed) Naturally occurring phorbol esters. CRC Press, Baton Rouge, p 33

Kinzel V, Furstenberger G, Loehrke H, Marks F (1986) Three-stage tumorigenesis in mouse skin: DNA synthesis as a prerequisite for the conversion stage induced by TPA prior to initiation. Carcinogenesis 7:779–782

Kitagawa T, Watanabe R, Kayano T, Sugano H (1980) In vitro carcinogenesis of hepatocytes obtained from acetylaminofluorene-treated rat liver and promotion of their growth by phenobarbital. Gann 71:747–754

Klaunig JE, Ruch RJ (1987) Strain and species effects on the inhibition of hepatocyte intercellular communication by liver tumor promoters. Cancer Lett 36:161–168

Kleine LP, Whitfield JF, Boynton AL (1986) The glucocorticoid dexamethasone and the tumor promoting artifical sweetener saccharin stimulate protein kinase C from T51B rat liver cells. Biochem Biophys Res Commun 135:33–40

Kunz HW, Tennekes HA, Port RE, Schwartz M, Lorke D, Schaude G (1983) Quantitative aspects of chemical carcinogenesis and tumor promotion in liver. Environ Health Perspect 50:113–122

Lans M, De Gerlache J, Taper H, Preat V, Roberfroid MB (1983) Phenobarbital as a promoter in the initiation/selection process of experimental rat hepatocarcinogenesis. Carcinogenesis 4:141–144

Lechner MC, Sinogas C, Osorio-Almeida ML, Freire MT, Chaumet-Riffaud P, Frain M, Sala-Trepat JM (1987) Phenobarbital-mediated modulation of gene expression in rat liver. Analysis of cDNA clones. Eur J Biochem 163:231–238

Loehrke H, Schweizer J, Dederer E, Hesse B, Rosenkranz G, Goerttler K (1983) On the persistence of tumor initiation in two-stage carcinogenesis on mouse skin. Carcinogenesis 4:771–775

Loury DJ, Goldsworthy TL, Butterworth BE (1987) The value of measuring cell replication as a predictive index of tissue-specific tumorigenic potential. In: Butterworth BE (ed) Nongenotoxic mechanisms in carcinogenesis. Cold Spring Harbor Laboratory, Cold Spring Harbor, NY, USA, pp 119–136 (Banbury report 25)

Makino T, Obara T, Ura H, Kinugasa T, Kobayashi H, Takahashi S, Konishi Y (1986) Effects of phenobarbital and secondary bile acids on liver, gallbladder and pancreas carcinogenesis initiated by *N*-nitroso-bis(2-hydroxypropyl)amine in hamsters. JNCI 76:967–975

McGowan JA (1986) Hepatocyte proliferation in culture. In: Guillouzo A, Guguen-Guillouzo C (eds) Research in isolated and cultured hepatocytes. Libbey Eurotext, London, p 14

Mentline R (1986) The tumor promoter 12-*O*-tetradecanoyl phorbol 13-acetate and regulatory diacylglycerols are substrates for the same carboxylesterase. J Biol Chem 261:7816–7818

Miyaka R, Tanaka Y, Tsuda T, Kaibuchi K, Kikkawa U, Nishizuka Y (1984) Activation of protein kinase C by non-phorbol tumor promoter mezerein. Biochem Biophys Res Commun 121:649–656

Moore MA, Kitagawa T (1986) Hepatocarcinogenesis in the rat: the effect of promoters and carcinogens in vivo and in vitro. Int Rev Cytol 101:125–173

Moore MA, Hacker H-J, Kunz HW, Bannasch P (1983) Enhancement of NNM-induced carcinogenesis in the rat liver by phenobarbital: a combined morphological and enzyme histochemical approach. Carcinogenesis 4:473–479

Moore MA, Nakagawa K, Satoh K, Ishikawa T, Sato K (1987) Single GST-P positive liver cells – putative initiated hepatocytes. Carcinogenesis 8:483–486

Morin MJ, Kreutter D, Rasmussen H, Sartorelli AC (1987) Disparate effects of activators of protein kinase C on HL-60 promyelocytic leukemia cell differentiation. J Biol Chem 262:11758–11763

Mufson RA, Fischer SM, Verma AK, Gleason GL, Slaga TJ, Boutwell RK (1979) Effects of 12-*O*-tetradecanoylphorbol-13-acetate and mezerein on epidermal ornithine decarboxylase activity, isoproterenol-stimulated levels of cyclic adenosine 3′:5′-monophosphate, and induction of mouse skin tumors in vivo. Cancer Res 39:1979–1984

Muir JG, Murray AW (1987) Bombesin and phorbol ester stimulate phosphatidylcholine hydrolysis by phospholipase C: evidence for a role of protein kinase C. J Cell Physiol 130:382–391

Munir KM, Rao KV, Bhide SV (1984) Effect of hexachlorocyclohexane on diethylnitrosamine-induced hepatocarcinogenesis in rat and its failure to promote skin tumors on dimethylbenz(a)anthracene initiation in mouse. Carcinogenesis 5:479–481

Murakami K, Chan SY, Routtenberg A (1986) Protein kinase C activation by *cis*-fatty acid in the absence of Ca^{2+} and phospholipids. J Biol Chem 261:15424–15429

Murray AW, Fournier A, Hardy SJ (1987) Proteolytic activation of protein kinase C: a physiological reaction? Trend Biochem Sci 12:53–54

Nakanishi K, Fukushima S, Hagiwara A, Tamamo S, Ito N (1982) Organ-specific promoting effects of phenobarbital sodium and sodium saccharin in the induction of liver and urinary bladder tumors in male F344 rats. JNCI 68:497–500

Nims RW, Devor DE, Henneman JR, Lubet RA (1987) Induction of alkoxyresorufin O-dealkylases, epoxide hydrolase and liver weight gain: correlation with liver tumor-promoting potential in a series of barbiturates. Carcinogenesis 8:67–71

Nishizuka Y (1986) Studies and perspectives of protein kinase C. Science 233:305–311

Ogawa K, Solt DB, Farber E (1980) Phenotypic diversity as an early property of putative preneoplastic hepatocyte populations in liver carcinogenesis. Cancer Res 40:725–733

Ohno S, Kawasaki H, Imajoh S, Inagaki M, Yokokura H, Sakoh T, Hidaka H (1987) Tissue-specific expression of three distinct types of rabbit protein kinase C. Nature 325:161–166

Ono Y, Fujii T, Ogita K, Kikkawa U, Igarashi K, Nishizuka Y (1987) Identification of three additional members of rat protein kinase C family: δ-, ε-, and ζ-subspecies. FEBS Lett 226:125–128

Pal BC, Topping DC, Griesemer RA, Nelson FP, Nettesheim P (1978) Development of a system for controlled release of benzo(a)pyrene,7,12-dimethylbenz(a)anthracene and phorbol ester for tumor induction in heterotropic tracheal grafts. Cancer Res 38:1376–1383

Pasti G, Lacal JC, Warren BS, Aaronson SA, Blumberg PM (1986) Loss of mouse fibroblast cell response to phorbol esters restored by microinjected protein kinase C. Nature 324:375–377

Peraino C, Fry RJ, Staffeldt E (1971) Reduction and enhancement by phenobarbital of hepatocarcinogenesis induced in the rat by 2-acetylaminofluorene. Cancer Res 31:1506–1513
Peraino C, Fry RJM, Staffeldt E (1977) Effects of varying the onset and duration of exposure to phenobarbital on its enhancement of 2-acetylaminofluorene-induced hepatic tumorigenesis. Cancer Res 37:3623–3627
Peraino C, Staffeldt EF, Haugen DA, Lombard LS, Stevens FJ, Fry RJM (1980) Effects of varying the dietary concentration of phenobarbital on its enhancement of 2-acetylaminofluorene-induced hepatic tumorigenesis. Cancer Res 40:3268–3273
Peraino C, Richards WL, Stevens FJ (1983) Multistage hepatocarcinogenesis. In: Slaga TJ (ed) Mechanisms of tumor promotion, vol 1. CRC Press, Baton Rouge, p 1
Peraino C, Staffeldt EF, Carnes BA, Ludeman VA, Blomquist JA, Vesselinovitch SD (1984) Characterization of histochemically detectable altered hepatocyte foci and their relationship to hepatic tumorigenesis in rats treated once with diethylnitrosamine or benzo(a)pyrene within one day after birth. Cancer Res 44:3340–3347
Peraino C, Grdina DJ, Staffeldt EF, Russell JF, Prapuolenis A, Carnes BA (1987) Effects of separate and combined treatments with gamma radiation and diethylnitrosamine in neonatal rats on the induction of altered hepatocyte foci and hepatic tumors. Carcinogenesis 8:599–600
Pitot HC, Sirica AE (1980) The stages of initiation and promotion in hepatocarcinogenesis. Biochim Biophys Acta 605:191–215
Pitot HC, Barsness L, Goldsworthy T, Kitagawa T (1978) Biochemical characterisation of stages of hepatocarcinogenesis after a single dose of diethylnitrosamine. Nature 271:456–458
Pitot HC, Goldsworthy T, Campbell HA, Poland A (1980) Quantitative evaluation of the promotion by 2,3,7,8-tetrachlorodibenzo-*p*-dioxin of hepatocarcinogenesis from diethylnitrosamine. Cancer Res 40:3616–3620
Pitot HC, Glauert HP, Hanigan M (1985) The significance of selected biochemical markers in the characterization of putative initiated cell populations in rodent liver. Cancer Lett 29:1–14
Pitot HC, Goldsworthy TL, Moran S, Kennan W, Glauert HP, Maronpot RR, Campbell HA (1987) A method to quantitate the relative initiating and promoting potencies of hepatocarcinogenic agents in their dose-response relationships to altered hepatic foci. Carcinogenesis 8:1491–1499
Poland A, Mak I, Glover E (1981) Species differences in the responsiveness to 1,4-bis[2-(3,5-dichloropyridyloxy)]-benzene, a potent phenobarbital-like inducer of microsome monooxygenase activity. Mol Pharmacol 20:442–450
Poland A, Palen D, Glover E (1982) Tumor promotion by TCDD in skin of HRS/J hairless mice. Nature 300:271–273
Pound AW, McGuire LJ (1978) Repeated partial hepatectomy as a promoting stimulus for carcinogenic response of liver to nitrosamines in rats. Br J Cancer 37:585–594
Preat V, Lans M, de Gerlache J, Taper H, Roberfroid M (1987) Influence of the duration and delay of administration of phenobarbital on its modulating effect on rat hepatocarcinogenesis. Carcinogenesis 8:333–335
Pugh TD, Goldfarb S (1978) Quantitative histochemical and autoradiographic studies of hepatocarcinogenesis in rats fed 2-acetylaminofluorene followed by phenobarbital. Cancer Res 38:4450–4457
Quintanilla M, Brown K, Ramsden M, Balmain A (1986) Carcinogen-specific mutation and amplification of Ha-*ras* during mouse skin carcinogenesis. Nature 322:78–80
Rabes HM, Szymkowiak R (1979) Cell kinetics of hepatocytes during the preneoplastic period of diethylnitrosamine-induced liver carcinogenesis. Cancer Res 39:1298–1304
Raik AN, Thum K, Chivers BR (1972) Early effects of 12-*O*-tetradecanoylphorbol-13-acetate on the incorporation of tritiated precursor into DNA and the thickness of the interfollicular epidermis, and their relation to tumor promotion in mouse skin. Cancer Res 32:1562–1568
Rao MS, Reddy JK (1987) Peroxisome proliferation and hepatocarcinogenesis. Carcinogenesis 8:631–636

Reddy BS, Watanabe K (1979) Effects of cholesterol metabolites and promoting effect of lithocholic acid in colon carcinogenesis in germ-free and conventional F344 rats. Cancer Res 39:1521–1524

Roghani M, DaSilva C, Guvelli D, Castagna M (1987a) Benzene and toluene activate protein kinase C. Carcinogenesis 8:1105–1107

Roghani M, DaSilva C, Castagna M (1987b) Tumour promoter chloroform is a potent protein kinase C activator. Biochem Biophys Res Commun 142:738–744

Rotstein J, Sarma DSR, Farber E (1986) Sequential alterations in growth control and cell dynamics of rat hepatocytes in early precancerous steps in hepatocarcinogenesis. Cancer Res 46:2377–2385

Sarma DSR, Rao PM, Rajalakshmi S (1986) Liver tumor promotion by chemicals: models and mechanisms. Cancer Surv 5:781–798

Sawada N, Staecker JL, Pitot HC (1987) Effects of tumor-promoting agents 12-O-tetradecanoylphorbol-13-acetate and phenobarbital on DNA synthesis of rat hepatocytes in primary culture. Cancer Res 47:5665–5671

Schulte-Hermann R (1985) Tumor promotion in the liver. Arch Toxicol 57:147–158

Schulte-Hermann R, Ohde G, Schuppler J, Timmermann-Trosiener I (1981) Enhanced proliferation of putative preneoplastic cells in rat liver following treatment with the tumor promoters phenobarbital, hexachlorocyclohexane, steroid compounds and nafenopin. Cancer Res 41:2556–2562

Schulte-Hermann R, Timmermann-Trosiener I, Schuppler J (1986) Facilitated expression of adaptive responses to phenobarbital in putative pre-stages of liver cancer. Carcinogenesis 7:1651–1655

Schwarz M, Bannasch P, Kunz W (1983) The effect of pre- and post-treatment with phenobarbital on the extent of γ-glutamyl transpeptidase positive foci induced in rat liver by *N*-nitrosomorpholine. Cancer Lett 21:17–21

Sell S, Leffert HL (1982) An evaluation of cellular lineages in the pathogenesis of experimental hepatocellular carcinoma. Hepatology 2:77–86

Serra M, Smith TL, Yamamura HI (1986) Phorbol esters alter muscarinic receptor binding and inhibit polyphosphoinositide breakdown in human neuroblastoma (SH-5Y5Y) cells. Biochem Biophys Res Commun 140:160–166

Shirai T, Tsuda H, Ogiso T, Hirose M, Ito N (1987) Organ specific modifying potential of ethinyl estradiol on carcinogenesis initiated with different carcinogens. Carcinogenesis 8:115–119

Shubik P (1984) Progression and promotion. JNCI 73:1005–1011

Siebert PD, Fukuda M (1985) Regulation of glycophorin gene expression by a tumor promoting phorbol ester in human leukemic K562 cells. J Biol Chem 260:640–645

Sirica AE, Jicinsky JK, Heyer EF (1984) Effect of chronic phenobarbital administration on the gamma-glutamyl transpeptidase activity of hyperplastic liver lesions induced in rats by the Solt/Farber initiation selection process of hepatocarcinogenesis. Carcinogenesis 5:1737–1740

Slaga TJ (1983) Mechanisms of tumor promotion, vol 1. Tumor promotion in internal organs. CRC Press, Baton Rouge

Slaga TJ, Fischer SM, Nelson K, Gleason GL (1980) Studies on the mechanism of skin tumor promotion: evidence for several stages in promotion. Proc Natl Acad Sci USA 77:3659–3663

Smith JB, Smith L, Pettit GR (1985) Bryostatins: potent new mitogens that mimic phorbol ester tumor promoters. Biochem Biophys Res Commun 132:939–945

Sobczak J, Duguet M (1986) Molecular biology of liver regeneration. Biochimie 68:957–967

Taguchi M, Thomas TP, Anderson WB, Farrar WL (1986) Direct phosphorylation of the interleukin 2 receptor TCA antigen epitope by protein kinase C. Biochem Biophys Res Commun 135:239–247

Tatematsu M, Hasegawa R, Imaida K, Tsuda H, Ito N (1983a) Survey of various chemicals from initiating and promoting activities in a short-term in vivo system based on generation of hyperplastic liver nodules in rats. Carcinogenesis 4:381–386

Tatematsu M, Nagamine Y, Farber E (1983 b) Redifferentiation as a basis for remodelling of carcinogen-induced hepatocyte nodules to normal appearing liver. Cancer Res 43:5049–5058

Tsuda H, Fukushima S, Imaida K, Kurata Y, Ito N (1983) Organ specific promoting effects of phenobarbital and saccharin in induction of thyroid, liver and urinary bladder tumors in rats after initiation with *N*-nitrosodiethylurea. Cancer Res 43:3292–3296

Tsuda H, Hasegawa R, Imaida K, Masui T, Moore MA, Ito N (1984) Modifying potential of thirty-one chemicals on the short-term development of γ-glutamyl transpeptidase-positive foci in diethylnitrosamine-initiated rat liver. Gann 75:876–883

Van Duuren BL, Sivak A, Katz C, Siedman I, Melchionne S (1975) The effect of ageing and interval between primary and secondary treatment in two-stage carcinogenesis in mouse skin. Cancer Res 35:502–505

Verma IM, Sassone-Corsi P (1987) Proto-oncogene *fos:* complex but versatile regulation. Cell 51:513–514

Verma AK, Pong RC, Erickson D (1986) Involvement of protein kinase C activation in ornithine decarboxylase gene expression in primary culture of newborn mouse epidermal cells and in skin tumor promotion by 12-*O*-tetradecanoylphorbol-13-acetate. Cancer Res 46:6149–6155

Weinberg WC, Berkwits L, Iannaccone PM (1987) The clonal nature of carcinogen-induced altered foci of γ-glutamyl transpeptidase expression in rat liver. Carcinogenesis 8:565–570

Weinstein IB (1987) Growth factors, oncogenes and multistage carcinogenesis. J Cell Biochem 33:213–224

Whitlock JP (1987) The regulation of gene expression by 2,3,7,8-tetrachlorodibenzo-*p*-dioxin. Pharmacol Rev 39:147–161

Williams GM, Ohmori T, Katayama S, Rice JM (1980) Alteration by phenobarbital of membrane associated enzymes including gamma-glutamyl transpeptidase in mouse liver neoplasma. Carcinogenesis 1:813–818

Yager JD, Roebuck BD, Paluszcyk TL, Memoli VA (1986) Effects of ethinyl estradiol and tamoxifen on liver DNA turnover and new synthesis and appearance of gamma glutamyl transpeptidase-positive foci in female rats. Carcinogenesis 7:2007–2014

Yamasaki H, Enomoto T (1985) Role of intercellular communication in BALB/c 3T3 cell transformation. In: Barrett JC, Tennant RW (eds) Carcinogenesis, vol 9. Raven, New York, p 179

Yoshimasa T, Sibley DR, Souvier M, Lefkowitz RJ, Caron MG (1987) Cross-talk between cellular signalling pathways suggested by phorbol-ester-induced adenylate cyclase phosphorylation. Nature 327:67–70

Yuspa SH, Morgan DL (1981) Mouse skin cells resistant to terminal differentiation associated with initiation of carcinogenesis. Nature 293:72–74

Yuspa SH, Kulesz-Martin M, Ben T, Hennings H (1983) Transformation of epidermal cells in culture. J Invest Dermatol 81:162s–168s

CHAPTER 7

Inhibition of Chemical Carcinogenesis

J. DiGiovanni

A. Introduction

Many compounds are known to modify (inhibit or enhance) the development of tumors in response to various carcinogenic agents in experimental animals (reviewed in Slaga 1980). Epidemiological studies have suggested that specific, pharmcologically active agents present in the diet may reduce the relative risk of exposure to carcinogens. For example, an inverse correlation between dietary selenium levels and human cancer mortality has been noted (Shamberger and Willis 1971; Shamberger et al. 1976). Also, Graham et al. (1978) have reported that there is an inverse correlation between the intake of cabbage and human colon cancer. An important consideration in cancer research today is that exposure to pharmacologically active "modifying" chemicals may play an important role in reducing or, in some instances, increasing the relative risks resulting from exposure to carcinogenic chemicals. This chapter focuses on inhibitors of chemical carcinogenesis in experimental animals. Interest in chemical inhibitors of carcinogenesis is twofold: (i) they provide valuable tools for studying the biochemical and molecular mechanism(s) of carcinogenesis; and (ii) understanding their chemical nature, mechanism(s), and specificity may lead to a rational approach for the chemoprevention of chemically induced tumors in humans.

Chemical carcinogenesis in experimental animals can be characterized as either multistage or complete (Boutwell 1964, 1974). *Complete* carcinogenesis experimental protocols involve the administration of a single large dose or repeated applications of smaller doses of a carcinogen to an experimental animal. For example, multiple papillomas and carcinomas can be produced on the backs of mice following a single application of as little as 600–800 nmol of 7,12-dimethylbenz[*a*]anthracene (DMBA) (Teracini et al. 1960; Turusov et al. 1971). Presumably, both the initiating and promoting components are present under these experimental conditions. The induction of mouse skin tumors can also be described by using a multistage model that involves the processes defined operationally and mechanistically as initiation and promotion (Boutwell 1964, 1974, and references therein). Initiation is accomplished by topical application of a single small dose of a skin carcinogen, such as DMBA, and is essentially irreversible. An initiating dose of a carcinogen per se will not lead to the development of visible tumors. Visible tumors will result only following prolonged and repeated applications of a tumor promoter, such as 12-*O*-tetradecanoylphorbol-13-acetate (TPA), to the initiated skin (Boutwell 1964, 1974). Although the multistage model of chemical carcinogenesis has been best documented in mouse skin, the system has been extended to include a variety of other tissues including liver, lung, bladder, colon, esophagus, mammary gland, and stomach, as well as various cells in culture (Farber and Solt 1978; Fukushima et al. 1984; Heidelberger et al. 1978; Hicks et al. 1978; Huggins et al. 1962; Newberne 1976; Peraino et al. 1978; Pitot et al. 1978; Reddy et al. 1978; Witschi and Lock 1978). An important advantage of multistage models is that one can study the specific stage of the process at which a given modifier or inhibitor acts. Mouse skin has been particularly useful in this regard an is the system with which I am most familiar. Therefore, further description of the multistage system of mouse skin tumorigenesis will facilitate an understanding of many of the potentially useful inhibitors to be described in this chapter.

B. Multistage Carcinogenesis in Mouse Skin

The protocol used in most multistage carcinogenesis studies in mouse skin was first described by Mottram (1944) and forms the basis of the operational criteria defining initiators and promoters. Mottram elicited skin tumors by treating the backs of mice with a single, subcarcinogenic dose of benzo[*a*]pyrene (B[a]P) and followed this with repeated applications of croton oil, obtained from the seeds of *Croton tiglium*. Initiators such as the polycyclic aromatic hydrocarbon (PAH), B[a]P, require metabolic activation to highly reactive diolepoxide intermediates for their cytotoxic, mutagenic, and carcinogenic activities (Heidelberger 1975; Dipple et al. 1984; Miller 1978; Gelboin and Ts'o 1978; Phillips and Sims 1979; Sims 1980). These reactive intermediates are capable of interacting covalently with a wide variety of cellular macromolecules (Heidelberger 1975; Dipple et al. 1984; Miller 1978; Gelboin and Ts'o 1978; Phillips and Sims 1979; Sims 1980). Interactions with DNA as a target for activated hydrocarbon intermediates have received considerable attention in recent years and are presently considered as one of the critical cellular events involved in the initiation of chemical carcinogenesis in mouse skin as well as in other tissues (Heidelberger 1975; Dipple et al. 1984; Miller 1978; Gelboin and Ts'o 1978; Phillips and Sims 1979; Sims 1980). The initiation stage of mouse skin carcinogenesis has been postulated to arise through genetic change(s) in a gene(s) controlling epidermal differentiation (Yuspa et al. 1982; Balmain et al. 1984).

PAHs are metabolized to a wide variety of other organic solvent-soluble metabolites, including epoxides, dihydrodiols, quinones, and phenols (including their sulfate esters), as well as to the more water soluble glutathione and glucuronide conjugates (reviewed in Dipple et al. 1984; Gelboin and Ts'o 1978; Selkirk 1985). Potential inhibitors of tumor initiation may affect various aspects of the metabolism of PAHs as well as that of other carcinogens. Examples of compounds affecting the metabolic activation and/or detoxification of chemical carcinogens, including PAHs, are shown in Fig. 1. Figure 1 summarizes our current understanding of multistage carcinogenesis in SENCAR mouse skin.

The most active tumor-promoting component of croton oil was identified as the phorbol diester, TPA (Hecker 1968, 1971; Van Duuren and Orris 1965;

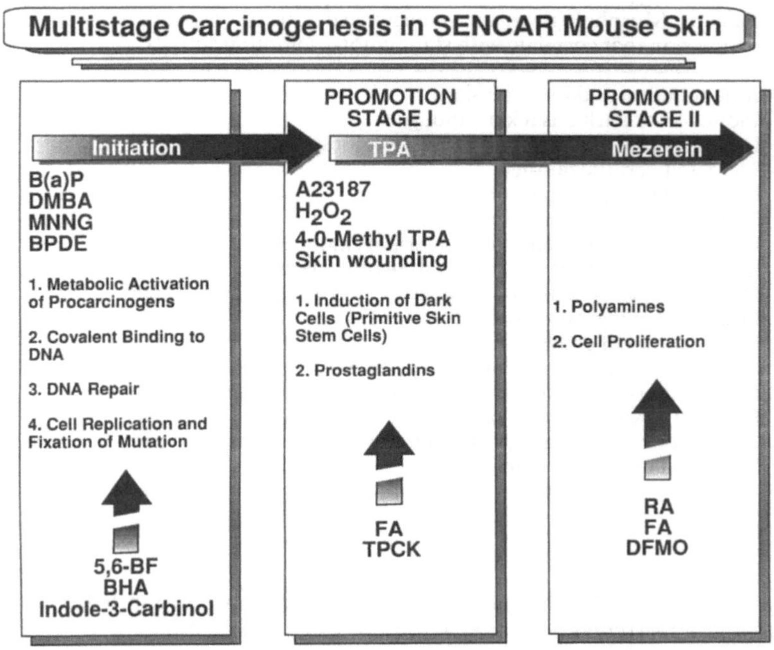

Fig. 1. Current concept of multistage carcinogenesis in SENCAR mouse skin. Initiation can be effected by a variety of agents including: benzo[*a*]pyrene [*B(a)P*]; 7,12-dimethylbenz[*a*]anthracene (*DMBA*); *N*-methyl-*N'*-nitro-*N*-nitrosoguanidine (*MNNG*); or (±)7β-8α-dihydroxy-9α,10α-epoxy-7,8,9,10-tetrahydrobenzo[*a*]pyrene (*BPDE*). Several representative inhibitors of initiation, which affect the metabolic activation and subsequent covalent DNA binding of initiators include: 5,6-benzoflavone (*5,6-BF*), butylated hydroxyanisole (*BHA*), and indole-3-carbinol. Following initiation of mice, stage I of promotion can be effected by limited treatments with *TPA, A23187*, H_2O_2, *4-O-methyl-TPA*, or skin wounding. Subsequently, stage II of promotion can be effected by multiple, continuous applications of the weak complete promoter mezerein. Through the use of the specific inhibitors indicated, it has been hypothesized that the induction of epidermal dark cells and increased prostaglandin biosynthesis which occur after TPA treatment are important for stage I, whereas increased epidermal polyamines and cell proliferation are important for stage II of promotion
Abbreviations used are: B(a)P, benzo(a)pyrene; DMBA, 7,12-dimethylbenz[*a*]anthracene; MNNG, *N*-methyl-*N'*-nitro-*N*-nitrosoguanidine; BPDE, (±)7β-8α-dihydroxy-9α,10α-epoxy-7,8,9,10-tetrahydrobenzo[*a*]pyrene; 5,6-BF, 5,6-benzoflavone; BHA, butylated hydroxyanisole; FA, fluocinolone acetonide; TPCK, tosyl phenylalanine chloromethyl ketone; RA, retinoic acid; DFMO, α-difluoromethylornithine

Van Duuren 1969), and today it is the most frequently used promoter for mechanistic studies. A summary of the morphological and biochemical effects produced in mouse skin by phorbol esters which correlate with their skin tumor-promoting ability is given in Table 1. Of these many effects, those that correlate most closely with the promoting activity of phorbol esters include: the induction of epidermal ornithine decarboxylase (ODC, EC 4.1.1.17) followed by increased levels of polyamines, the induction of dark basal keratinocytes (DCs), and the

Table 1. Morphological and biochemical effects of phorbol esters in mouse skin that correlate with skin tumor-promoting activity

Inflammation and hyperplasia
Induction of dark basal keratinocytes
Increased DNA, RNA, and protein synthesis
Increased keratinization
Increased phospholipid turnover
Increased prostaglandin synthesis
Decreased responsiveness to epidermal chalones
Decreased responsiveness to β-adrenergic agonists
Decreased basal activities of superoxide dismutase and catalase
Decreased epidermal histidase
Modification of epidermal keratins
Increased synthesis and phosphorylation of histones
Increased ornithine decarboxylase and accumulation of polyamines
Interaction with specific, saturable receptors
Activation of epidermal protein kinase C
Induction of active oxygen species

ability to induce a sustained epidermal hyperplasia (reviewed in Boutwell 1964; Hennings and Boutwell 1970; Slaga et al. 1982; Slaga 1984b).

Tumor promotion in mouse skin can itself be divided into at least two steps that can each be mediated by TPA. Boutwell (1964) was the first to recognize an early step involving the "conversion" of an initiated cell into a dormant or latent tumor cell and a second step in which the "propagation" of the cell is promoted. In these early studies, Boutwell used a limited number of croton oil treatments followed by repetitive applications of turpentine or skin wounding (Hennings and Boutwell 1970) on STS female mice (the progenitors of the current outbred SENCAR mouse). These steps or stages were further defined by Slaga and coworkers (1982; Slaga 1984b). They found that mezerein, a diterpene ester similar in structure to TPA but with weak, complete, papilloma-promoting activity, acts as a potent second stage promoter if applied to initiated skin that had already been treated with TPA, even if with only a single application (Slaga et al. 1982; Slaga 1984b). Furstenberger et al. (1981), using an outbred mouse line called NMRI, showed that 12-*O*-retinoylphorbol-13-acetate (RPA) was inactive as a complete promoter but was quite effective as a second stage promoter in this mouse line. Using TPA and mezerein in a two-stage promotion protocol, Slaga and coworkers (1982; Slaga 1984b) demonstrated that compounds known to inhibit phorbol ester skin tumor promotion affect specific stages of promotion, presumably by inhibiting events that are important for those stages (see Fig. 1). For example, the protease inhibitor tosyl phenylalanine chloromethyl ketone (TPCK) is a potent inhibitor specifically of stage I of promotion, whereas retinoic acid (RA), an inhibitor of promoter-induced ODC activity, specifically inhibited stage II. Fluocinolone acetonide (FA), on the other hand, inhibited both stages of tumor promotion. Thus, Slaga and coworkers (1984b) have proposed that the induction of DCs and prostaglandin biosynthesis which occurs following TPA application is specific for stage I of promotion, while increased polyamine levels and cell proliferation are important for stage II.

In this chapter, we will discuss inhibitors of the initiation phase of chemical carcinogenesis first, followed by inhibitors of the tumor promotion phase. Emphasis will be placed on the mechanism(s) of inhibitory action wherever possible, although, for many inhibitors, the mechanism is not completely understood. It should be readily apparent from reading this chapter that certain classes of compounds are capable of inhibiting both stages of chemical carcinogenesis. In addition, because the process of tumor promotion is reversible and generally takes place over an extended period of time, inhibitors of tumor promotion may prove to be more practical in terms of chemoprevention in humans.

C. Inhibitors of Tumor Initiation

Inhibitors of tumor initiation may act by a variety of mechanisms which are outlined in the scheme shown in Fig. 2. Some potential mechanisms such as stimulation or alteration of DNA repair (mechanism E) have not so far been associated with chemoprevention by a significant number of agents. Nevertheless, this mechanism(s) remains open for the action(s) of certain chemicals. Mechanisms

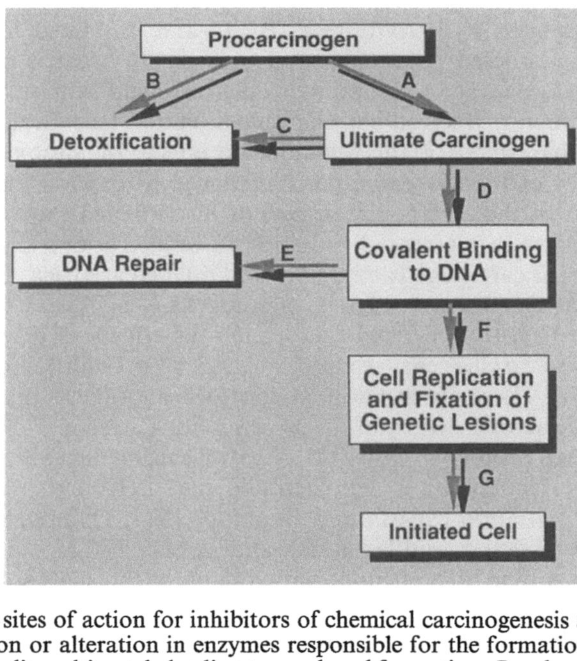

Fig. 2. Potential sites of action for inhibitors of chemical carcinogenesis and tumor initiation. *A*, inhibition or alteration in enzymes responsible for the formation of reactive carcinogenic metabolites ultimately leading to a reduced formation; *B*, enhancement or induction of enzyme pathways which produce products of lower carcinogenic potential; *C*, enhanced detoxification of ultimate carcinogenic metabolites through increased or altered metabolism; *D*, scavenging of reactive, carcinogenic intermediates through direct chemical interaction; *E*, inhibition or enhancement of DNA repair mechanisms; *F*, inhibition of cell proliferation and DNA synthesis; *G*, direct toxicity to initiated cells

A, B, and C involve alterations in the metabolic activation or detoxification of the parent carcinogen or its metabolites. They usually include the stimulation (induction) or inhibition of enzymatic pathways involved in the detoxification or activation of carcinogens, respectively. Mechanism D involves chemicals which are capable of scavenging reactive electrophilic intermediates and forming less toxic, more excretable products. Mechanisms E, F, and G, although less commonly implicated, appear to be involved in some cases. Mechanism E has already been alluded to. Mechanism F involves chemicals that may inhibit cell replication, whereas mechanism G involves the killing of initiated cells. This last mechanism deserves further mention because selective toxicity of a potential inhibitor to initiated cells should always be ruled out as a possible explanation for any observed decrease in tumor response. Since many inhibitors of initiation may function by more than one mechanism, they will be listed by chemical class or chemical property rather than by their potential mechanism of action.

I. Antioxidants

A wide variety of chemical compounds possessing antioxidant properties have been shown to inhibit the formation of chemically induced tumors in experimental animals, and some of the more widely studied antioxidants and the systems in which they produce antagonistic actions are listed in Table 2. The phenolic antioxidants, butylated hydroxytoluene (BHT) and butylated hydroxyanisole (BHA), have been studied most extensively, primarily because of their use as food preservatives. However, compounds such as disulfiram, ethoxyquin, and selenium (Se) have also received considerable attention (see Table 2). Although the detailed mechanism(s) of the antagonistic effects of antioxidants are not known, a number of theories have been advanced to explain their effects: (a) direct interaction of the carcinogen or one of its activated metabolic products with the antioxidant; (b) decreased activities of, or alterations to, enzyme pathways responsible for carcinogen activation; and (c) increased activities of enzyme pathways responsible for detoxifying carcinogens. As illustrated in Table 2, antioxidnts afford protection from the deleterious effects of a wide variety of structurally diverse carcinogens (see also Wattenberg 1980a), including direct-acting carcinogens. These observations support the hypothesis that antioxidants may have multiple actions. Many of the investigations listed in Table 2 used complete carcinogenesis protocols. Other investigations using the two-stage system of mouse skin tumorigenesis have demonstrated that BHA and BHT effectively antagonize both initiation (Slaga and Bracken 1977; DiGiovanni 1978) and promotion stages (Kensler and Jaffe 1986 and see Sect. D).

Studies to determine how phenolic antioxidants such as BHA antagonize the tumor-initiating action of PAHs such as B[a]P have indicated that BHA administration results in an altered metabolism of the hydrocarbon by mouse liver microsomes (Lam and Wattenberg 1977; Speier and Wattenberg 1975). For example, the hepatic aryl hydrocarbon hydroxylase (AHH, EC 1.14.14.2) activity of BHA-fed mice was more sensitive to the inhibitory effects of 7,8-benzoflavone (7,8-BF) than AHH from control animals (Speier and Wattenberg 1975). Alterations seen in the ethyl isocyanide difference spectra for hepatic cytochrome

Table 2. Inhibition of chemical carcinogenesis and tumor initiation by antioxidants

Antioxidant	Carcinogen	Species	Tissue	References
BHA	DMBA	Mouse	Forestomach	WATTENBERG (1972)
			Skin, lung	SLAGA and BRACKEN (1977); WATTENBERG (1973)
	DMBA	Rat	Mammary gland	WATTENBERG (1972)
	B[a]P	Mouse	Lung	WATTENBERG (1973)
			Forestomach	WATTENBERG (1972)
	DB[a,h]A	Mouse	Lung	WATTENBERG (1973)
	7-OHM-12-MBA	Mouse	Lung	WATTENBERG (1973)
	Urethan	Mouse	Lung	WATTENBERG (1973)
	Uracil mustard	Mouse	Lung	WATTENBERG (1973)
	Bracken fern	Rat	Gastrointestinal	PAMUKCU et al. (1977)
	DEN	Mouse	Lung	WATTENBERG (1971)
	4-NQO	Mouse	Lung	WATTENBERG (1971)
	AFB_1	Rat	Liver	WILLIAMS et al. (1986)
	NENHE	Rat	Liver	TSUDA et al. (1984)
	DMBA	Rat	Mammary gland	HIROSE et al. (1986)
	DMBA	Rat	Ear duct	HIROSE et al. (1986)
	NDMA	Mouse	Lung	CHUNG et al. (1986)
BHT	DMBA	Mouse	Skin	DIGIOVANNI (1978); SLAGA and BRACKEN (1977)
	DMBA	Rat	Mammary gland	WATTENBERG (1972)
	B[a]P	Mouse	Forestomach	WATTENBERG (1972)
	Azoxymethane	Rat	Gastrointestinal	WEISBURGER et al. (1977)
	AAF	Rat	Liver	ULLAND et al. (1973)
	N-OH-AAF	Rat	Liver	ULLAND et al. (1973)
	DMH	Mouse	Colon	CLAPP et al.(1979)
	AFB_1	Rat	Liver	WILLIAMS et al.(1986)
Ethoxyquin	DMBA	Rat	Mammary gland	WATTENBERG (1972)
	DMBA	Mouse	Forestomach	WATTENBERG (1972)
	B[a]P	Mouse	Forestomach	WATTENBERG (1972)
	DEN	Mouse	Lung	WATTENBERG (1971)
	4-NQO	Mouse	Lung	WATTENBERG (1971)
	NENHE	Rat	Liver	TSUDA et al.(1984)
Disulfiram	DMBA	Rat	Mammary gland	WATTENBERG (1974)
	b[a]P	Mouse	Forestomach	WATTENBERG (1974)
	Bracken fern	Rat	Gastrointestinal	PAMUKCU et al. (1977)
	DMH	Mouse	Large bowel	WATTENBERG (1975); WATTENBERG et al. (1977)
	Azoxymethane	Mouse	Large intestine	WATTENBERG (1977)
Selenium	DMBA	Mouse	Skin	SHAMBURGER (1970)
	DEN	Rat	Liver	LEBOEUF et al. (1985a)
	AAF	Rat	Mammary gland	HARR et al. (1972)
			Liver	MARSHALL et al. (1979)
	3′-MeDAB	Rat	Liver	GRIFFIN and JACOBS (1977)
	DMH	Rat	Colon	JACOBS et al. (1977)
	Methylazoxymethanol	Rat	Colon	JACOBS et al. (1977)
	B[a]P	Mouse	Forestomach	EL-BAYOUMY (1985)
	DMBA	Rat	Mammary gland	IP (1981a, b); IP and DANIEL (1985); LANE and MEDINA (1985); MEDINA and LANE (1983); THOMPSON et al. (1982)
	AAF	Rat	Liver	LEBOEUF et al. (1985a)
	BOP	Rat	Colon	BIRT et al. (1982)

Table 2. (continued)

Antioxidant	Carcinogen	Species	Tissue	References
Vitamin C	DMBA	Mouse	Skin	Shamberger (1972); Slaga and Bracken (1977)
Vitamin E	DMBA	Mouse	Skin	Shamberger and Rudolph (1966); Slaga and Bracken (1977)
Cu^{2+}Dips	DMBA	Mouse	Skin	Solanki et al. (1984)

7-OHM-12-MBA, 7-hydroxymethyl-12-methylbenz[*a*]anthracene; DMBA, 7, 12-dimethylbenz[*a*]anthracene; B[*a*]P, benzo[*a*]pyrene; DB[*a,h*]A, dibenz[*a,h*]anthracene; DEN, diethylnitrosamine; 4-NQO, 4-nitroquinoline-*N*-oxide; AFB_1, aflatoxin B_1; NENHE, *N*-ethyl-*N*-hydroxyethylnitrosamine; NDMA, *N*-nitrosodimethylamine; AAF, 2-acetylaminofluorene; N-OH-AAF, *N*-hydroxy-2-acetylaminofluorene; DMH, dimethylhydrazine; 3′-Me-DAB, 3′-methyl-4-dimethylaminoazobenzene; BOP, bis(2-oxopropyl)-nitrosamine; BHA, butylated hydroxyanisole; BHT, butylated hydroxytoluene; Cu^{2+}DIPS, copper(II)bis(diisopropylsalicylate).

P-450 from BHA-fed mice also suggest that BHA might produce changes in the metabolite profile of B[a]P that could lead to a decreased activation or to an increased detoxification of the molecule. Further investigation demonstrated that hepatic microsomes from BHA-fed mice produced lower amounts of epoxides from B[a]P (primarily B[a]P 4,5-oxide) and an overall increase in formation of 3-hydroxybenzo[*a*]pyrene (3-OH-B[a]P) (Lam and Wattenberg 1977; Speier et al. 1978). In addition to these changes, a marked decrease in the formation of polar metabolites of B[a]P, presumed to be triols and tetraols, was observed 4 h after BHA treatment (Lam et al. 1980). Speier et al. (1978) showed that administration of BHA by oral intubation 4 h before challenge with B[a]P antagonized the formation of pulmonary adenomas in A/HeJ mice. In addition, incubation of B[a]P with hepatic microsomes from mice that received BHA 4 h prior to killing resulted in reduced binding of the hydrocarbon to DNA compared with control incubations. Similar results were obtained by Slaga and Bracken (1977) in mouse skin. In their study, BHT or BHA applied 5 min prior to initiation with DMBA effectively reduced skin tumor formation. Also, the in vitro epidermal homogenate-mediated binding of radiolabeled B[a]P and DMBA to calf thymus DNA was inhibited as early as 3 h after topical application of BHA or BHT (Slaga and Bracken 1977). All of these results indicate that the phenolic antioxidants produce rapid changes in the metabolism of B[a]P that could account, in part, for their anticarcinogenic activity.

Dietary administration of BHA or its individual 2- or 3-tert-butyl-4-hydroxyanisole isomers (2- or 3-t-BHA) have indicated that 2-t-BHA was more effective at inhibiting B[a]P-induced forestomach tumors (Lam et al. 1979; Wattenberg et al. 1980). 2-t-BHA is the minor isomer of commercial BHA (Lam et al. 1979). More recent metabolism studies using dietary administration of BHA have confirmed significant alterations in the oxidative metabolism of B[a]P. Dietary BHA reduced the formation of 9-OH-B[a]P from B[a]P and inhibited the further metabolism of B[a]P-7,8-diol, despite increasing overall metabolism of the parent hydrocarbon (Dock et al. 1982; Sydor et al. 1983). Furthermore, dietary administration of BHA to mice for 4 h or 4 days led to significant reductions in the in vivo covalent binding of B[a]P to the DNA of forestomach, lung, and liver (ANDERSON et al. 1981). BHA has been shown to act as a non-competitive inhibitor of the metabolism of B[a]P-7,8-diol (SYDOR et al. 1983), suggesting a possible mechanism for the observed effects on oxidative metabolism and DNA-binding of B[a]P. SAWICKI and DIPPLE (1983) also reported that BHA reduced the proportion of DMBA *syn* diol-epoxide DNA adducts formed in mouse embryo cells exposed in culture while not altering total DNA binding.

In addition to the alterations in oxidative metabolism, the phenolic antioxidants are known to alter markedly other enzymes important in carcinogen activation and/or detoxication. CHA and BUEDING (1979) reported the effects of BHA administration on a number of hepatic microsomal and cytosolic enzyme activities in CD-1 mice. The microsomal enzymes aniline hydroxylase (EC 1.14.14.1) and uridine 5′-diphosphoglucuronyltransferase (UDPGT, EC 2.4.1.17) were significantly elevated above control levels. In addition, several cytosolic enzyme activities, including glucose-6-phosphate dehydrogenase (EC 1.1.1.49) and UDP-glucose dehydrogenase (EC 1.1.1.22), also were elevated. BENSON et al. (1978, 1979) demonstrated that BHA added to the diets of mice and rats significantly elevated glutathione-*S*-transferase (GST, EC 2.5.1.18) and epoxide hydratase (EH, EC 4.2.1.63) activities in hepatic and extrahepatic tissues. GRANTHAM et al. (1973) studied the effect of BHT on the metabolism of 2-acetylaminofluorene (AAF) and N-OH-AAF in male and female rats. Administration of BHT in the diet for 4 weeks increased the excretion of glucuronic acid conjugates from both carcinogens.

Subsequent to these studies, dietary administration of 3-t-BHA (the major isomer of commercial BHA) to rats was shown to increase total hepatic GST and UDPGT activities with preferential induction of form A of GST and the late fetal form of UDPGT (SATO et al. 1984). DOCK et al. (1984) have also shown that dietary administration of BHA to rats led to increased cytosolic GST activity towards (±)7β,8α-dihydroxy-9α,10α-epoxy-7,8,9,10-tetrahydrobenzo[*a*]pyrene (*anti*-BPDE) but with a concomitant decreased affinity for this substrate. These authors concluded that BHA-induced GST activity toward BPDE may be of limited value in the detoxification of B[a]P. However, one must consider that elevated GST activity may be more important for detoxifying primary epoxides such as the 7,8-epoxide of B[a]P rather than the diol-epoxides themselves. Further work in these areas is necessary to fully establish the role of elevated GST in the anticarcinogenetic effects of BHA. Other investigators have shown that BHA treatment produces large reductions in aflatoxin B_1 (AFB_1) binding to DNA in the rat. LOTLIKAR et al. (1984) reported a 50% reduction in AFB_1 binding to hepatic nuclear DNA in vivo when rats were fed 7500 ppm BHA for 2 weeks. Similarly, KENSLER et al. (1985) reported a 65% reduction in AFB_1 binding to both hepatic and renal DNA in vivo when rats were fed 4500 ppm BHA for the same period of time. CHANG and BJELDANES (1987) also showed that dietary BHA (7500 ppm) increased hepatic GST activity, reduced AFB_1-DNA binding and increased biliary excretion of AFB_1-thiol conjugates.

Ethoxyquin has been shown to inhibit neoplasia induced by a variety of agents some of which are listed in Table 2. It is known to induce specific forms of cytochrome(s) P-450 (Kahl and Netter 1977; Mandel et al. 1987; Parke et al. 1974) as well as the activity of GST in mouse and rat liver (Benson et al. 1978; Kensler et al. 1986; Miranda et al. 1981). Ethoxyquin, when fed at a level of 0.5% in the diet, induced one or more cytochrome P-450 species with elevated levels of AFB_1-hydroxylating activity, resulting in the increased formation of both AFM_1 and AFQ_1 (Mandel et al. 1987). Ethoxyquin administration also led to a marked induction of hepatic GST (Kensler et al. 1986; Mandel et al. 1987). The result of increased AFB_1 detoxification pathways induced by ethoxyquin was the decreased covalent binding of AFB_1 metabolites to hepatic DNA (Kensler et al. 1986; Mandel et al. 1987).

Disulfiram and diethyl dithiocarbamate added to the diets of CF_1 mice effectively antagonized large bowel neoplasia induced by dimethylhydrazine (DMH) (Wattenberg 1975; Wattenberg et al. 1977). The effects of disulfiram on the metabolism of DMH were investigated by Fiala et al. (1977). They concluded that carbon disulfide, a metabolite of disulfiram, was most likely responsible for the antagonistic action of disulfiram and related compounds. Furthermore, they suggested that the mechanism was due to inhibition of N-oxidation of azomethane to azoxymethane, which is an important step in the metabolic activation sequence for DMH. Disulfiram when added to the diet is also a potent inhibitor of B[a]P-induced neoplasia of the mouse forestomach; it significantly reduced the covalent binding of B[a]P to the RNA, protein, and, to a lesser extent, DNA of this tissue (Borchert and Wattenberg 1976). Disulfiram has also been extensively studied for its protective effects against the carcinogenic action of several nitrosamines including diethylnitrosamine (Schmahl and Kruger 1972; Schmahl et al. 1976) and *N*-butyl-*N*-(4-hydroxybutyl)nitrosamine (Irving et al. 1979). Mechanistic studies have suggested that disulfiram may inhibit chemically induced carcinogenesis by a number of pathways including: (a) Inhibition of oxidative metabolism (Deitrich and Erwin 1970; Dubois et al. 1961; Frank et al. 1984; Freundt 1978; Honjo and Netter 1969; Phillips et al. 1977; Schurr et al. 1978; Stripp et al. 1969; Tottmar and Marchner 1976; Zemaitis and Greene 1976), (b) direct interaction with reactive electrophiles (Bertram et al. 1982), and (c) induction of GST and elevation of cellular thiol levels (Bertram et al. 1985; Frei et al. 1985).

Selenium when added to the diet or drinking water antagonizes the formation of chemically induced tumors in a wide variety of systems (Griffin and Jacobs 1977; Harr et al. 1972; Jacobs et al. 1977; Marshall et al. 1979; Shamberger 1970; and see other examples in Table 2). Marshall et al. (1979) found that oral administration of Se led to an increase in ring hydroxylation and a decrease in N-hydroxylation of AAF. Further studies by Capel et al. (1980) demonstrated that Se deficiency led to decreased levels of AHH and UDPGT activities and elevated levels of GST activity. These latter changes in enzyme activity were not, however, accompanied by changes in the level of [^{3}H]B[a]P bound covalently to rat liver DNA. Similar results were obtained by Wortzman et al. (1980) with AAF. In their study, no difference was detected in the total amount of AAF bound covalently to liver DNA in vivo between Se-deficient and Se-supplemented rats. However, single-strand breaks in DNA induced by AAF in the livers of Se-deficient rats were absent in the livers of Se-supplemented rats. Milner et al.

(1985) have reported that sodium selenite inhibited the covalent binding of DMBA to DNA of tertiary mouse embryo cells in culture. In addition, these authors provided evidence that Se prevents the induction of an enzyme responsible for producing DMBA *anti*-diol-epoxides and the subsequent formation of DNA adducts from this metabolite in mouse embryo cells. On the other hand, IP and DANIEL (1985) reported no effect of Se given in the diet on DMBA DNA-adduct formation in rat mammary gland.

The possibility that Se supplementation might enhance Se-dependent glutathione peroxidase and thus lower the level of potentially damaging peroxide radicals that might be generated during the metabolism of various carcinogenic chemicals has been explored. To date the evidence does not favor this mechanism (LANE and MEDINA 1985, and references therein). Studies by SHAMBERGER (1970) demonstrated an antagonistic effect of Se on the promotion phase of two-stage carcinogenesis in mouse skin, while IP and DANIEL (1985) and MEDINA and LANE (1983) have suggested that Se primarily affects the promotion stage of DMBA-induced mammary tumorigenesis in rats. Although Se has produced alterations in enzyme pathways responsible for carcinogen activation and detoxication, its major effect may be on the promotion phase of carcinogenesis. In particular, Se appears to possess antiproliferative effects which could explain its reported ability to inhibit both the initiation and promotion stages of chemical carcinogenesis (LEBOEUF et al. 1985a, b; MEDINA and LANE 1983; MEDINA and OSBORN 1984). This action of Se has been suggested from studies of its effects on cellular GSH levels (LEBOEUF et al. 1985a).

In summary, the antioxidants may protect against the initiation of chemically induced tumors in a variety of tissues primarily by altering the enzymes responsible for both the activation and detoxification of carcinogens. The information available suggests that the phenolic antioxidants disulfiram and ethoxyquin may act primarily by this mechanism. The overall result is a decrease in the amount of active carcinogen that reaches critical cellular targets, such as DNA. The mechanism by which Se inhibits chemically induced tumors in experimental animals may be different from the other antioxidants. As noted above, Se has been shown to inhibit DNA synthesis in target organs (LEBOEUF et al. 1985a, b; MEDINA and LANE 1983; MEDINA and OSBORN 1984), and this may explain its effects on both tumor initiation and promotion. One of the phenolic antioxidants, BHT, is capable of enhancing urethan tumorigenesis in mouse lung (WITSCHI et al. 1977; WITSCHI and LOCK 1978, 1979; YAMAMOTO et al. 1971) when administered subsequent to carcinogen exposure. Thus, this compound can act as a tumor promoter under certain circumstances. The temporal and spatial relationships between exposure to the modifier and exposure to the carcinogen are extremely critical; different effects – anticarcinogenic vs cocarcinogenic activity – may be observed depending on the time of application of the antioxidant. In addition, BHA and ethoxyquin, although capable of inhibiting chemical carcinogenesis in a specific tissue, may also enhance carcinogenesis in other tissues of the same animal (FUKUSHIMA et al. 1987; MANSON et al. 1987; TSUDA et al. 1984). For example, both BHA and ethoxyquin inhibited *N*-ethyl-*N*-hydroxyethylnitrosamine-induced heptocarcinoma formation while increasing the incidence of renal tumors in rats (TSUDA et al. 1984). Thus, many of these antioxidants have undesirable side effects which may well preclude their utility as chemopreventive agents in humans.

II. Flavones and Polyhydroxylated Compounds

The flavones are a large group of oxygenated, cyclic compounds that are prevalent in a variety of plants (Harborne 1976; Willaman 1955). Early reports indicated that flavonoids possessed diverse biological properties, including enzyme-inducing as well as enzyme-inhibiting properties (Willaman 1955; Wattenberg et al. 1968; Wiebel et al. 1971). Subsequently, a number of flavones, both synthetic and naturally occurring, where shown to antagonize markedly the formation of chemically induced tumors in experimental animals (for previous reviews see Wattenberg 1978; Wiebel 1980). The majority of this work involved the effects of two synthetic flavone derivatives, 7,8- and 5,6-benzoflavone (7,8- and 5,6-BF), on PAH carcinogenesis. Less effort has been made in determining the effects of flavones on carcinogenesis induced by other classes of chemical carcinogens. Table 3 lists some of the flavones and polyphenols, carcinogens, and animal model systems for which antagonism has been consistently observed. The mouse skin tumorigenesis system has been used extensively for these studies.

Wattenberg and Leong (1968, 1970) proposed that certain flavones, by virtue of their ability to induce AHH, inhibited PAH carcinogenesis. For example, intraperitoneal injections of 5,6-BF inhibited the formation of pulmonary adenomas in mice and mammary tumors in rats after exposure to DMBA (Wattenberg 1968). 5,6-BF effectively induced AHH in the liver, lung, and small intestine of the mice and in the liver and lung of the rats used for these studies. Diamond et al. (1972) demonstrated that both 5,6-BF and 7,8-BF inhibited pulmonary adenoma formation by DMBA when administered orally to A/HeJ mice. Lesca (1981) demonstrated that 5,6-BF effectively antagonized papilloma formation when applied 24 h prior to each DMBA application to the skins of NMRI mice. However, when CF_1 Swiss mice were substituted into the same experimental protocol, 5,6-BF potentiated skin papilloma formation (Lesca 1981).

When given topically to mice 5 min prior to initiation with DMBA, 7,8-BF effectively antagonized papilloma induction (Bowden et al. 1974; Gelboin et al. 1970; Kinoshita and Gelboin 1972a; Slaga et al. 1977). On the other hand, 7,8-BF was found to potentiate tumor initiation slightly by B[a]P and dibenz[*a,h*]anthracene (DB[a,h]A) (Bowden et al. 1974; Gelboin et al. 1970; Kinoshita and Gelboin 1972a; Slaga et al. 1977) and have little or no effect on 7-methyl-DB[a,h]A or 7,14-dimethyl-DB[a,h]A skin tumor initiation (Alworth and Slaga 1985). Further studies using the two-stage initiation-promotion system of mouse skin carcinogenesis have shown that 7,8-BF inhibits tumor initiation with 3-methylcholanthrene (MCA) (Slaga et al. 1976) and with various 7- and 12-substituted derivatives of DMBA (DiGiovanni et al. 1978). When applied topically to mice 5 min prior to initiation with either DMBA or DB[a,h]A, 5,6-BF antagonized both tumor initiation and covalent binding of these hydrocarbons to macromolecules (Bowden et al. 1974).

Table 3. Inhibition of chemical carcinogenesis and tumor initiation by flavonoids polyphenols

Flavone	Carcinogen	Species	Tissue	References
7,8-Benzo-flavone	DMBA	Mouse	Skin	BOWDEN et al. (1974); GELBOIN et al. (1970); KINOSHITA and GELBOIN (1972a, b); SLAGA et al. (1977)
			Lung	DIAMOND et al. (1972)
	MCA	Mouse	Skin	SLAGA et al. (1976)
	7-CHO-12-MBA	Mouse	Skin	DIGIOVANNI et al. (1978)
	7-BRMe-12-MBA	Mouse	Skin	DIGIOVANNI et al. (1978)
	12-BRMe-7-MBA	Mouse	Skin	DIGIOVANNI et al. (1978)
	7-BRMe-BA	Mouse	Skin	DIGIOVANNI et al. (1978)
	15,16-DHCP	Mouse	Skin	COOMBS et al. (1975)
5,6-Benzo-flavone	DMBA	Mouse	Skin	BOWDEN et al. (1974); KINOSHITA and GELBOIN (1972b); LESCA (1981)
			Lung	DIAMOND et al. (1972); WATTENBERG and LEONG (1968)
		Rat	Mammary gland	WATTENBERG and LEONG (1968)
	B[a]P	Mouse	Skin	WATTENBERG and LEONG (1970)
	DB[a,h]A	Mouse	Skin	BOWDEN et al. (1977)
	MCA	Mouse	Skin	SLAGA et al. (1976)
	MCA	Mouse	Forestomach, skin and other tissues	ANDERSON and SEETHARAM (1985)
	AFB_1	Trout	Liver	NIXON et al. (1984)
	AFB_1	Rat	Liver	GURTOO et al. (1985)
Quercetin	DMBA	Mouse	Skin	DIGIOVANNI et al. (1978)
	B[a]P	Mouse	Skin	SLAGA et al. (1978a)
Quercetin Pentamethyl ether	B[a]P	Mouse	Lung	WATTENBERG and LEONG (1970)
Ellagic acid	B[a]P	Mouse	Lung	LESCA (1983)
	DMBA	Mouse	Skin	LESCA (1983)
	MCA	Mouse	Skin	MUKHTAR et al. (1986)

7-CHO-12-MBA, 7-formyl-12-methylbenz[*a*]anthracene; 7-BRMe-12-MBA, 7-bromomethyl-12-methylbenz[*a*]anthracene; 12-BRMe-7-MBA, 12-bromomethyl-7-methylbenz-[*a*]anthracene; 7-BRME-BA, 7-bromomethylbenz[*a*]anthracene; 15,16-DHCP, 15,16-dihydro-11-methylcyclopenta[*a*]phenanthrene-17-one; MCA, 3-methylcholanthrene; B[*a*]P, benzo[*a*]apyrene; DMBA, 7,12-dimethylbenz[*a*]anthracene; DB[*a,h*]A, dibenz[*a,h*]anthracene; AFB_1, aflatoxin B_1.

In addition to the synthetic flavones, several naturally occurring flavones have been tested for their effects on PAH carcinogenesis and tumor initiation. Separate feeding of three flavones with various abilities to induce AHH in the liver and lungs of rats (WATTENBERG et al. 1968) inhibited the induction of pulmonary adenomas by B[a]P in A/HeJ mice (WATTENBERG and LEONG 1970). The order of potency for these effects was 5,6-BF > quercetin pentamethyl ether > rutin. 3,3′,4′,5,7-Pentahydroxyflavone (quercetin) was the only other naturally occurring flavone that was adequately tested and that demonstrated reproducible inhibition of skin tumor initiation by both B[a]P (Slaga et al. 1978a) and DMBA (DiGiovanni et al. 1978). Other flavones studied, including 3,3′,4′,5,5′,7-hexahydroxyflavone (myricetin), 4′,5,7-trihydroxyflavanone (naringenin), and 4′,5,7-trihydroxyflavone (apigenin), either had no effect or potentiated skin tumor initiation by DMBA (DiGiovanni et al. 1978; Kinoshita and Gelboin 1972b). It should be noted that quercetin, myricetin and naringenin were found to inhibit the epidermal homogenate-mediated metabolism of DMBA (DiGiovanni and Juchau 1980). Thus, there is no correlation among the abilities of these compounds to inhibit metabolism in vitro and their effects on skin tumor initiation in vivo.

The detailed mechanism(s) by which flavones inhibit PAH carcinogenesis are unknown, but several seem to be operating. When test animals are exposed to inducing doses of flavones (e.g., 5,6-BF) prior to challenge with the carcinogen, inhibition correlates with increased AHH activity in the target tissue. The exact mechanism(s) by which increased AHH activity could bring about a reduction in carcinogen response is not fully understood. Studies have provided direct evidence for the existence of multiple forms of cytochrome(s) P-450, each of which has a different but overlapping substrate specificity (Coon and Vatsis 1978; Lu and West 1980). The relative ratios of these different forms of cytochrome(s) P-450 in a particular tissue depend in part on the prior exposure of test animals to different classes of enzyme-inducing agents. Holder et al. (1974) and Rasmussen and Wang (1974) have shown that hepatic microsomes from phenobarbital (PB)-pretreated, MCA-pretreated, and control rats have different specificities for metabolizing B[a]P. By using different forms of purified cytochrome(s) P-450, several laboratories have demonstrated distinct differences in the position and stereoselectivity of hydroxylation of B[a]P, (−)*trans*-7,8-dihydro-7,8-dihydroxybenzo[*a*]pyrene (B[a]P-7,8-diol), and DMBA (Christou et al. 1986; Deutsch et al. 1978, 1979; McCord et al. 1988; Thakker et al. 1977; Wiebel et al. 1975; Wilson et al. 1984). Because the mixed function oxidases both activate and detoxify carcinogens, differences in the relative distribution of cytochrome(s) P-450 may play a critical part in determining susceptibility to carcinogenic agents. This mechanism may also apply to the compounds discussed in the next section (i.e., halogenated hydrocarbons and other enzyme inducers).

In addition to the mechanism(s) discussed above, flavones that possess inducing activity could reduce carcinogenicity by inducing metabolism of the carcinogen in nontarget tissues, such as the liver. Such activity could result in greater metabolism at these sites and a reduction in the effective concentration of carcinogen in the target tissue. Lesca (1981) demonstrated that pretreatment of the skin of NMRI mice with 5,6-BF prior to application of [^{3}H]DMBA increased the rate of disappearance from the skin and altered the time course of binding to epidermal DNA for this carcinogen. The total binding of [^{3}H]DMBA was reduced in those mice that received pretreatment with 5,6-BF. This finding suggested that the carcinogen is cleared from the skin more rapidly in the pretreated animals. In contrast, when CF_1 mice were used, 5,6-BF pretreatment increased the disappearance of [^{3}H]DMBA but significantly enhanced its binding to epidermal DNA, without altering the time course. The increase in DNA binding caused by 5,6-BF pretreatment correlated with the potentiation of tumorigenesis with DMBA by this flavone in CF_1 mice. On the other hand, ANDERSON and SEETHARAM (1985), in a much more extensive study of aryl hydrocarbon (Ah) responsive and nonresponsive mouse strains, found significant inhibition of MCA-induced tumorigenesis in all strains given i.p. doses of 5,6-BF. No enhancement of MCA-induced tumorigenesis was observed in any of the mouse strains examined in contrast to the study of LESCA (1981). It is not clear at present whether 5,6-BF enhancement of PAH carcinogenesis in Ah-nonresponsive mice is a general phenomenon. It should be pointed out that enzyme inducers such as 5,6-BF affect other enzymes besides the AHH complex and associated cytochrome(s) P-450. For example, 5,6-BF is an effective inducer of UDPGT (OWENS 1977). Inhibition of carcinogenesis by enzyme-inducing flavones, therefore, may be the result of a coordinated response involving several enzymes.

The mechanism by which 7,8- and 5,6-BF inhibit tumorigenesis when they are given 5 min before initiation is qualitatively different from that of 5,6-BF given at pretreatment times that allow for enzyme induction. 7,8-BF is a potent inhibitor of hepatic AHH from MCA-pretreated but not control rats (GELBOIN et al. 1972; WIEBEL et al. 1971; WIEBEL 1980), in homogenates of hamster embryo cells (DIAMOND et al. 1972), and in mouse epidermal tissue preparations from untreated or inducer-pretreated animals (BOWDEN et al. 1974; GELBOIN et al. 1970; SLAGA et al. 1977). 5,6-BF, like 7,8-BF, inhibited mouse epidermal AHH in vitro, but it was less potent (BOWDEN et al. 1974; GELBOIN et al. 1970; SLAGA et al. 1977). Interestingly, 7,8-BF has been used to determine the presence of multiple forms of cytochrome(s) P-450 (GELBOIN et al. 1972). For example, 7,8-BF markedly inhibited the hydroxylation of B[a]P by hepatic microsomes from MCA-pretreated rats or by a cytochrome P_1-450-reconstituted enzyme system (LU and WEST 1978; WIEBEL et al. 1971). In contrast, this flavone slightly stimulated the hydroxylation of B[a]P by hepatic microsomes from control and PB-pretreated rats or by a cytochrome P-450-reconstituted system (LU and WEST 1978; WIEBEL et al. 1971; WIEBEL 1980). Studies using homogenates of human liver revealed that 7,8-BF stimulated the rates of hydroxylation of several substrates including B[a]P (KAPITULNIK et al. 1977).

7,8-BF (BERRY et al. 1977; DIGIOVANNI et al. 1977; SELKIRK et al. 1974) and 5,6-BF (BERRY et al. 1977; DIGIOVANNI et al. 1977) have been analyzed for their effects on the formation of metabolites in vitro from B[a]P and DMBA in adult rat liver and mouse skin. These flavones appear to inhibit the initial oxidative attack by monooxygenases present in the tissues. 7,8-BF also inhibited the formation in vitro of electrophilic intermediates capable of binding covalently to calf thymus DNA (SELKIRK et al. 1974; SLAGA et al. 1977). Furthermore, 7,8-BF inhibited the binding of DMBA and B[a]P to the DNA of hamster embryo cells pretreated with the flavone (BAIRD and DIAMOND 1976, 1978). Interestingly, COOMBS et al. (1975, 1979) found that 7,8-BF markedly inhibited skin tumor induction by repeated applications of 15,16-dihydro-11-methyl-cyclopenta-[*a*]phenanthren-17-one (15,16-DHCP). Ring hydroxylation of this potent carcinogen was inhibited by 7,8-BF, but hydroxylation of the methyl and methylene groups was not (Coombs et al. 1976). Studies have indicated that 7,8-BF is metabolized by hepatic microsomes from MCA-pretreated rats to several metabolites; *trans*-9,10-dihydro-9,10-dihydroxy-7,8-BF is the major metabolite formed (Coombs et al. 1981; Nesnow et al. 1980). Several of these metabolites, like the parent 7,8-BF, inhibited the metabolism of 15,16-DHCP (Coombs et al. 1981).

Buening et al. (1981) reported the effects of several naturally occurring and synthetic flavonoids on the metabolism of B[a]P and AFB_1 in human liver microsomes. Of the compounds tested, flavone, nobiletin, tangeretin, and 7,8-BF stimulated the metabolism of these carcinogens. All of the other naturally occurring flavones, which contain hydroxyl groups, inhibited the hydroxylation of B[a]P. The ability of 7,8-BF to activate the metabolism of B[a]P by hepatic microsomes is highly dependent on the animal species (Huang et al. 1980; Kapitulnik et al. 1977; Thakker et al. 1981). A detailed study by Thakker et al. (1981) investigated this interesting phenomenon. In their study, 7,8-BF stimulated the metabolism of benzo[*e*]pyrene (B[e]P) in hepatic microsomes in a species-dependent manner. 7,8-BF enhanced the conversion of 9,10-dihydro-9,10-dihydroxybenzo[*e*]pyrene (B[e]P-9,10-diol) to B[e]P-9,10-diol-11,12-epoxides most significantly by hepatic microsomes from humans, rabbits, and mice. In contrast, in hepatic microsomes from hamsters, formation of diol-epoxide derivatives from B[e]P was decreased. In hepatic microsomes from rats, 7,8-BF had little effect on diol-epoxide formation, although the overall metabolism was increased. These results correlated with the ability of 7,8-BF to enhance the metabolic activation of B[e]P-9,10-diol to mutagens in tests with *Salmonella typhimurium* TA 100.

The ability of 7,8-BF to enhance the metabolic activation of certain carcinogens in a particular species is of considerable interest. An important question that remains unanswered is whether 7,8-BF, and possibly other flavones, can stimulate the metabolic activation of carcinogens in vivo. The differential effects of 7,8-BF on mouse skin tumor initiation by various PAHs (e.g., B[a]P vs DMBA) may be explained by the ability of this flavone to enhance the metabolic activation of certain types of PAH. Further work will be necessary to prove or disprove this interesting hypothesis.

Ellagic acid (EA), a widely distributed plant phenol has been shown to inhibit PAH-induced lung (Lesca 1983) and skin tumorigenesis (Lesca 1983; Mukhtar et al. 1984a, 1986). The exact mechanism(s) of inhibition by EA is not known, but several have been postulated based on the results of various studies. In aqueous solution, EA has been shown to form both *cis* and *trans* adducts with *anti*-BPDE (Sayer et al. 1983). These adducts subsequently undergo hydrolysis to inactive B[a]P tetraols (Kapitulnik et al. 1978). If this reaction occurs in tissues and cells, then EA could act as a scavenger of the reactive metabolites of carcinogens. Other studies have shown that EA inhibits epidermal microsomal AHH and enzyme-mediated binding of B[a]P to calf thymus DNA in vitro and epidermal DNA in vivo (Del Tito et al. 1983; Dixit et al. 1986). In addition, EA was reported to be an inhibitor of B[a]P metabolism and its subsequent glucuronidation, sulfation, or covalent binding to DNA in cultured mouse keratinocytes (Mukhtar et al. 1984b). Acute and/or chronic administration of

EA in vivo produced a significant decrease of hepatic and pulmonary cytochrome P-450 levels and of AHH and 7-ethoxycoumarin-*O*-deethylase activities (Das et al. 1985). EA has also been reported to induce the activity of detoxificating enzymes, especially GST, in mouse liver (Das et al. 1985). Previous studies have shown that EA inhibits the metabolic activation and binding of B[a]P and B[a]P-7,8-dihydrodiol to DNA in cultured mouse lung explants and human bronchial tissues (Dixit et al. 1985; Teel et al. 1985, 1986). However, EA was significantly more inhibitory towards the dihydrodiol than towards B[a]P itself (Dixit et al. 1985). Collectively, these studies suggest that EA may exert its inhibitory effects by (i) inhibiting the metabolism of B[a]P, (ii) promoting the detoxification of B[a]P, and/or (iii) serving as a scavenger of the reactive metabolites of B[a]P. It should be stressed that several studies have been unable to demonstrate an effect of EA (or several of its more lipid-soluble analogues) on B[a]P or MCA-induced tumorigenesis (Smart et al. 1986). The reasons for these conflicting results are unknown at present. In addition, at least one study has reported toxicity associated with the oral administration EA (Lesca 1983).

In summary, the mechanism of inhibition of carcinogenesis by flavones apparently depends on many factors, including the test system, time of exposure, and the chemical nature and dose of the compounds involved. Investigations with the two-stage system of mouse skin tumorigenesis suggest that certain flavones (e.g., 7,8-BF, 5,6-BF, and quercetin) applied simultaneously with several PAHs inhibit tumor formation; this ability results, at least partially, from inhibition of the enzyme system(s) responsible for generating the electrophilic intermediates that bind to DNA. On the other hand, prior exposure to flavones that possess enzyme-inducing activity (e.g., 5,6-BF) seems to reduce carcinogenic activity via enhanced detoxification. Other mechanisms also may be operating and remain to be explored. One area that has received little attention is the effect of flavonoids on nonoxidative enzyme pathways [e.g., GST, UDPGT, and sulfotransferase (ST, EC 2.8.2.1)]. Analysis of these effects may shed further light on the mechanism of inhibition of carcinogenesis by flavones. Other plant phenols such as EA also appear to involve complex mechanism(s), and further work is necessary to determine whether these compounds can be effective inhibitors of carcinogenesis.

III. Halogenated Hydrocarbons and Other Enzyme Inducers

Many compounds of diverse chemical structure possess the ability to induce microsomal monooxygenase enzymes (Conney 1967; Remmer 1972). Microsomal and nonmicrosomal enzyme induction has been postulated as a mechanism for anticarcinogenesis by a number of compounds, some of which are listed in Table 4. In addition, this mechanism seems to apply, in part, to certain flavonoids (e.g., 5,6-BF) as discussed in the previous section. The most potent member and prototype of this group of inhibitors is 2,3,7,8-tetrachloro-dibenzo-*p*-dioxin (TCDD). TCDD is a highly toxic and widely studied chemical that is found as an environmental contaminant (see Poland and Kende 1976 for a review). Like many of the compounds listed in Table 4, TCDD is known to influence markedly enzyme pathways responsible for both the activation and

Table 4. Inhibition of chemical carcinogenesis and tumor initiation by halogenated hydrocarbons and other enzyme inducers

Inducer	Carcinogen	Species	Tissue	References
Aroclor	DMBA	Mouse	Skin	BERRY et al. (1979); DIGIOVANNI et al. (1979b)
	AFB_1	Trout	Liver	HENDRICKS et al. (1977)
Kanechlor 500	3′-Me-DAB	Rat	Liver	MAKIURA et al. (1974)
	AAF	Rat	Liver	MAKIURA et al. (1974)
	DEN	Rat	Liver	MAKIURA et al. (1974)
Kanechlor 400	3′-Me-DAB	Rat	Liver	KIMURA et al. (1976)
3,4,3′4′-TCB	DMBA	Mouse	Skin	DIGIOVANNI et al. (1979b, 1980a)
2,3,7,8-TCDD	DMBA, MCA, B[a]P, *anti*-BPDE, 7-MBA, 12-MBA, 5-MeC, DB[a,h]A	Mouse	Skin	BERRY et al. (1979); DIGIOVANNI et al. (1979b, 1980a)
DDT	DMBA	Rat	Mammary gland	OKEY (1972); SILINKAS and OKEY (1975)
Coumarin	DMBA	Rat	Mammary gland	WATTENBERG et al. (1979)
	B[a]P	Mouse	Forestomach	WATTENBERG et al. (1979)
4-Methylcoumarin	DMBA	Rat	Mammary gland	FEUER and KELLEN (1974)
MCA	3′-Me-DAB	Rat	Liver	MILLER et al. (1958); MEECHAN et al. (1953); RICHARDSON et al. (1952)
	4′-Fl-DAB	Rat	Liver	MILLER et al. (1958)
	2′-4′-diFl-DAB	Rat	Liver	MILLER et al. (1958)
	AAF	Rat	Liver	MILLER et al. (1958)
	7-Fl-AAF	Rat	Liver	MILLER et al. (1958)
PB	AFB_1	Rat	Liver	MCLEAN and MARSHALL (1971)
	AAF	Rat	Liver	PERAINO et al. (1971)
	Urethan	Mouse	Lung	YAMAMOTO et al. (1971)
Indole-3-carbinol	DMBA	Rat	Mammary gland	WATTENBERG and LOUB (1978)
	B[a]P	Mouse	Forestomach	WATTENBERG and LOUB (1978)
	AFB_1	Trout	Liver	NIXON et al. (1984)
3′,3-Diindolylmethane	DMBA	Rat	Mammary gland	WATTENBERG and LOUB (1978)
	B[a]P	Mouse	Forestomach	WATTENBERG and LOUB (1978)
Indole-3-acetonitrile	B[a]P	Mouse	Forestomach	WATTENBERG and LOUB (1978)
Diallyl sulfide	DMH	Mouse	Colon	WARGOVICH (1987)
	B[a]P	Mouse	Forestomach	SPARNINS et al. (1988)
	B[a]P	Mouse	Lung	SPARNINS et al. (1988)
Allyl methyl trisulfide	B[a]P	Mouse	Forestomach	SPARNINS et al. (1988)
Allyl methyl disulfide	B[a]P	Mouse	Lung	SPARNINS et al. (1988)
Diallyl trisulfide	B[a]P	Mouse	Forestomach	SPARNINS et al. (1988)
[5-(2-pyrazinyl)-4-methyl-1,2-dithiol-3-thione]	B[a]P	Mouse	Lung	WATTENBERG and BUEDING (1986)
	B[a]P	Mouse	Forestomach	WATTENBERG and BUEDING (1986)
	AFB_1	Rat	Liver	KENSLER et al. (1987)

Table 4. (continued)

Inducer	Carcinogen	Species	Tissue	References
Phenothiazine	DMBA	Rat	Mammary gland	WATTENBERG and LEONG (1967)
Chlorpromazine	DMBA	Rat	Mammary gland	WATTENBERG and LEONG (1967)
	DMBA	Hamster	Cheek pouch	LEVIJ and POLLIACK (1970)

DDT, 1,1,1-trichloro-2,2-bis(p-chlorophenyl)ethane; 4′-Fl-DAB, 4′-fluoro-4-dimethylaminoazobenzene; 2′,4′-diFl-DAB, 2′,4′-difluoro-4-dimethylaminoazobenzene; 7-Fl-AAF, 7-fluoroacetylaminofluorene; DMBA, 7,12-dimethylbenz(a)anthracene; AFB_1, aflatoxin B_1; 3′-Me-DAB, 3′-methyl-4-dimethylaminoazobenzene; AAF, acetyaminofluorene; DEN, diethylnitrosamine; MCA, 3-methylcholanthrene; B[a]P, benzo[a]pyrene; *anti*-BPDE, (±)-7β-8α-dihydroxy-9α,10α-epoxy-7,8,9,10-tetrahydrobenzo(a)pyrene; DMH, dimethylhydrazine; DB[a,h]A, dibenz[*a,h*]anthracene; 2,3,7,8-TCDD, 2,3,7,8-tetrachlorodibenzo-*p*-dioxin; 3,4,3′,4′-TCB, 3,4,3′,4′-tetrachlorobiphenyl; 5-MeC, 5-methylchrysene; PB, phenobarbital; Fl-DAB, 4-fluoro-DAB.

detoxification of PAHs. TCDD is an extremely potent inducer of hepatic microsomal monooxygenase activity with properties similar to those of MCA (POLAND and KENDE 1976). Microsomal enzyme induction by MCA is characterized by an increase in the spectrally distinct cytochrome(s) P_1-450 and a shift to a narrow substrate specificity (CONNEY 1967, 1972; SLADEK and MANNERING 1966). TCDD has the added distinction of being 30000 times more potent than MCA, and it also markedly affects other enzyme-catalyzed pathways (BAARS et al. 1978; FOWLER et al. 1977; LUCIER et al. 1975; OWENS 1977; POLAND and GLOVER 1974). When mice are treated with TCDD, the time course and magnitude of inhibition of skin tumor initiation by various PAHs closely follow the time course and magnitude of the induction of AHH and UDPGT in skin (BERRY et al. 1979; COHEN et al. 1979; DIGIOVANNI et al. 1980a, b). Studies by DIGIOVANNI et al. (1979a, b) have shown a marked reduction in the binding of [^{3}H]DMBA to epidermal DNA and RNA in mice pretreated with TCDD. In contrast, the total binding of [^{3}H]B[a]P to epidermal DNA was increased by pretreatment with TCDD (COHEN et al. 1979). However, when the isolated DNA samples were hydrolyzed to deoxyribonucleosides and analyzed by Sephadex LH-20 chromatography, B[a]P-DNA adducts were absent (COHEN et al. 1979). These results suggested that, when exposed to TCDD, epidermal cells were programmed to inactivate PAH carcinogens more efficiently, thus reducing the covalent DNA binding of critical reactive metabolites.

Several polychlorinated biphenyl (PCB) mixtures and pure isomers have been tested for their antagonistic effects on chemical carcinogenesis. In the two-stage system of mouse skin tumorigenesis, Aroclor 1254 (a PCB mixture) and 3,4,3′,4′-tetrachlorobiphenyl (TCB) inhibited tumor initiation by DMBA (DIGIOVANNI et al. 1979b, 1980a). TCB is a pure PCB isomer possessing MCA-type inducing properties (GOLDSTEIN et al. 1977; POLAND and GLOVER 1977). When 2,4,5,2′,4′,5′-hexachlorobiphenyl (HCB) was tested, it had no effect on tumor initiation by DMBA (DIGIOVANNI et al. 1979b, 1980a). HCB is another pure PCB isomer possessing phenobarbital PB-type inducing properties (Goldstein et al. 1977; Poland and Glover 1977). Induction of microsomal monooxygenase enzymes with PB is characterized by increased concentrations of cytochrome(s) P-450. Furthermore, PB enhances the metabolism of a wide variety of substrates (Conney 1967, 1972; Poland and Glover 1974; Poland and Kende 1976; Remmer 1972). Apparently, MCA-type inducing activity is a prerequisite for inhibiting tumor initiation by PAH in mouse skin. This prerequisite does not seem to hold, however, for the inhibition of hepatic tumors induced by other classes of carcinogens (McLean and Marshall 1971; Peraino et al. 1971). For example, Kanechlor 500 effectively inhibited the formation of hepatic tumors induced by 3′-Me-DAB, AAF, and DEN (Makiura et al. 1974). Kimura et al. (1976) reported that inclusion of Kanechlor 400 in the diets of rats 4 months prior to and for 2 months during 3′-Me-DAB administration completely prevented hepatocarcinomas usually induced by this carcinogen. Kanechlor 500 and Kanechlor 400 are PCB mixtures containing isomers with both MCA- and PB-type inducing properties (Goldstein et al. 1977; Ito et al. 1973; Poland and Glover 1977; Yoshimura et al. 1978). MCA (Miller et al. 1958) and PB (Peraino et al. 1971) both effectively inhibited the induction of hepatic tumors by AAF. With regard to detailed mechanism(s), Shelton et al. (1986) have shown that Aroclor 1254 produces dramatic initial changes in the metabolism of AFB_1, resulting in reduced covalent DNA binding to trout hepatic DNA. Thus, PCBs and PB appear to bring about their tumor-inhibitory effects by a mechanism similar to that of TCDD.

A number of organosulfur compounds have recently been shown to be capable of inhibiting chemically induced cancer in several model systems. Many of these compounds occur naturally in the human diet and have recently received considerable attention (Kensler et al. 1987; Nixon et al. 1984; Sparnins et al. 1986, 1988; Wargovich 1987; Wattenberg and Loub 1978). These compounds include indoles and sulfides. Although less is known about their mechanism(s) of action, these compounds appear to possess enzyme-inducing properties similar to the antioxidants. In this regard, all of these organosulfur compounds appear to possess the ability to induce GST in target organs (Sparnins et al. 1982, 1988). Recent studies (Goeger et al. 1986) with indole-3-carbinol have demonstrated that its dietary feeding to trout markedly altered the pharmacokinetics of AFB_1 metabolism, with a dramatic increase in glucuronide metabolites excreted in the bile. These changes in AFB_1 metabolism ultimately led to reduced DNA-adduct formation in hepatic DNA (Dashwood et al. 1988).

The time of treatment and sequence of exposure to many of these enzyme-inducing modifiers and carcinogens are extremely critical in some of these studies. For example, when Kanechlor 400 was administered subsequent to 3′-Me-DAB, the incidence of hepatocarcinomas was significantly increased (Kimura et al. 1976). This finding, as well as other reports, suggests that PCBs are promoters of hepatic tumors when applied after carcinogen exposure (Ito et al. 1973; Nishizumi 1976; Preston et al. 1981). PB is known to enhance the development of hepatocarcinomas induced in the livers of rats after sequential partial hepatectomy and AAF administration (Peraino et al. 1971, 1978) or following 3′-Me-DAB administration (Kitagawa et al. 1984). Studies by Kouri et al. (KOURI 1976; KOURI et al. 1977) on the effects of subcutaneous coinjection of TCDD on MCA-induced fibrosarcomas suggested a potential cocarcinogenic effect in certain mouse strains, for example in DBA/2 mice. Furthermore, PITOT et al. (1980) demonstrated that TCDD was a potent promoting agent for hepatocarcinogenesis in rats exposed to DEN. Finally, indole-3-carbinol has been shown under certain conditions to enhance (or promote) AFB_1-induced hepatocellular carcinomas in trout (BAILEY et al. 1987) and DMH-induced colon carcinogenesis in rats (PENCE et al. 1986).

In summary, the chlorinated hydrocarbons and other related enzyme inducers can be likened to a double-edged sword. Exposure to these agents prior to a carcinogen results, most often, in an anticarcinogenic effect. The mechanistic studies that have been conducted indicate that inhibition of tumorigenesis correlates with increased enzyme activities, both oxidative and conjugative, and reduced formation of DNA adducts in the target tissue. The formation of critical electrophilic intermediates from various carcinogens depends on a balance between their rate of formation and rate of removal. Under conditions of prior exposure to enzyme-inducing agents, such as TCDD, PCBs, and PB, this balance is apparently shifted toward detoxification. Organosulfur compounds appear to work primarily by affecting nonoxidative metabolism of carcinogens, although further work is necessary to substantiate this hypothesis fully. On the other hand, under certain conditions, prior exposure to enzyme inducers may enhance tumor formation. In certain mouse strains, basal levels of AHH are low (NEBERT et al. 1975; ROBINSON et al. 1974). Several reports suggest that enzyme inducers, including 5,6-BF and TCDD, enhance PAH carcinogenesis in these strains by elevating AHH levels and by shifting the balance of metabolism toward greater activation (KOURI 1976; KOURI et al. 1977; LESCA 1981) although, as previously noted, at least one study failed to observe this effect (ANDERSON and SEETHARAM 1985). In addition, many of the compounds listed in Table 4 appear to possess tumor-promoting activity when administered after exposure to carcinogens. Thus, although inducers can provide protection under particular experimental protocols, with other protocols they can have quite the opposite effect.

IV. Weakly Carcinogenic or Noncarcinogenic Polycyclic Aromatic Hydrocarbons

Human exposure to chemical carcinogens probably involves repeated exposure to low doses over an extended period of time. In addition, this exposure may occur in the presence of other noncarcinogenic or weakly carcinogenic chemicals that may modify the carcinogenic process. Combustion processes generate mixtures of PAHs with a wide range of carcinogenic activity (FREUDENTHAL and JONES 1976; GELBOIN and TS'O 1978). Exposure to these mixtures may produce variable effects depending on the amount and degree of carcinogenicity of the constituents. The early literature is replete with studies to determine the outcome of exposure to mixtures of PAHs and has been reviewed (DIGIOVANNI and SLAGA 1981a).

Investigations using the two-stage initiation-promotion system of mouse skin tumorigenesis have aided in an understanding of the combined effects of various

Table 5. Inhibitory effects of weakly or noncarcinogenic polycyclic aromatic hydrocarbons on the mouse skin tumor-initiating activity of other hydrocarbons [a]

Modifying PAH [a]	Initiator	References
Phenanthrene	DMBA	Huh and McCarter (1960)
DB[a,c]A	DMBA	DiGiovanni et al. (1980c, 1982), DiGiovanni (1981); DiGiovanni and Slaga (1981b); Slaga and Boutwell (1977); Slaga et al. (1978b)
	MCA	DiGiovanni (1981); DiGiovanni and Slaga (1981b); DiGiovanni et al. (1982)
	DB[a,h]A	DiGiovanni et al. (1982)
B[e]P	DMBA	DiGiovanni et al. (1980, 1982); DiGiovanni (1981); DiGiovanni and Slaga (1981b); Slaga et al. (1979);
	DB[a,h]A	DiGiovanni et al. (1982)
Pyrene	DMBA	DiGiovanni et al. (1980c); Slaga et al. (1979)
Fluoranthene	DMBA	DiGiovanni et al. (1980c); Slaga et al. (1979)

DMBA, 7,12-dimethylbenz[a]anthracene; MCA, 3-methylcholanthrene; DB[a,c]A, dibenz[*a,c*]anthracene; DB[a,h]A, dibenz[*a,h*]anthracene; B[e]P, benzo[e]pyrene.
[a] The modifying PAH was applied at the same time or just prior to application of the initiator.

PAH mixtures. Table 5 lists a number of weakly carcinogenic or noncarcinogenic PAHs consistently inhibiting skin tumor-initiating activity of several more potent hydrocarbons. To date, little work has been done to determine the effect of weakly or noncarcinogenic PAH on the carcinogenic action of other classes of chemical carcinogens. Huh and McCarter (1960) determined the effect of phenanthrene on tumorigenesis by DMBA. Their results suggested that the modifier was inhibiting the initiation phase of the tumorigenic process in mouse skin. Dibenz[*a,c*]anthracene (DB[a,c]A) markedly inhibited tumor formation when applied topically at doses of $\geqq 20$ nmol/mouse 5 min prior to initiation with DMBA (Slaga and Boutwell 1977; Slaga et al. 1978b). Several other weakly or noncarcinogenic PAHs have been tested for their effects on skin tumor initiation by DMBA and B[a]P (DiGiovanni et al. 1980c; Slaga et al. 1979). B[e]P, pyrene, and fluoranthene inhibited tumor formation when applied topically 5 min before initiation with DMBA. In contrast, none of these compounds inhibited tumor initiation with B[a]P. With B[a]P as the initiator, the tumor response was in some cases mildly potentiated. Under similar experimental conditions, B[e]P and DB[a,c]A effectively inhibited tumor initiation with DB[a,h]A, but only DB[a,c]A antagonized MCA initiation (DiGiovanni 1981; DiGiovanni and Slaga 1981b; DiGiovanni et al. 1982). The tumor-initiating activities of 7-methylbenz[*a*]anthracene (7-MBA), 12-methylbenz[*a*]anthracene (12-MBA), and 5-methylchrysene (5-MeC) were not altered by either modifier (DiGiovanni et al. 1982).

Several possible mechanisms have been proposed to explain the antagonistic effect of one hydrocarbon on the tumor-initiating activity of another. The first possibility is that the modifying hydrocarbon induces enzymes such that the balance of metabolism for the initiator is shifted toward detoxification. This mechanism would be analogous to that postulated for certain flavones, halogenated hydrocarbons, and related compounds discussed in previous sections. This mechanism cannot explain, however, why all the modifiers save one (DB[a,c]A) cannot induce microsomal monooxygenase enzymes (ARCOS et al. 1961; WIEBEL 1980). Furthermore, a study by DIGIOVANNI and SLAGA (1981 b) indicated that DB[a,c]A was an effective inhibitor of skin tumor initiation by B[a]P, but only with pretreatment times that corresponded to maximum induction of epidermal AHH. These results further strengthen the notion that enzyme induction is not involved in the inhibitory effects when weakly or noncarcinogenic hydrocarbons are applied 5 min before the initiator.

A number of studies have demonstrated that other PAH will inhibit the hydroxylation of B[a]P in vitro (SLAGA and BOUTWELL 1977; THOMAS et al. 1979; WIEBEL et al. 1974). For example, the addition of equimolar concentrations of DMBA, phenanthrene, and anthracene inhibited hydroxylation of B[a]P in hepatic microsomes from control rats (WILLIAMS et al. 1971). On the other hand, the addition of DMBA, DB[a,c]A, and MCA markedly inhibited B[a]P hydroxylation in hepatic microsomes from MCA-pretreated rats. MCA and DB[a,c]A had little or no effect when control microsomes were used. Several hydrocarbons including DB[a,c]A, DB[a,h]A, and DMBA inhibited AHH activity to varying degrees in epidermal homogenates from benz[*a*]anthracene (BA)-pretreated mice (SLAGA and BOUTWELL 1977). Interestingly, the differences in specificity toward the control vs induced forms of the hydroxylase for DB[a,c]A are surprisingly similar to those found with 7,8-BF (see Sect. C.II).

The effects of B[e]P on the metabolism of both DMBA and B[a]P have been analyzed (BAIRD and DIAMOND 1981; DIGIOVANNI and SLAGA 1981 b; J. DiGiovanni, unpublished data). B[e]P produced a concentration-dependent inhibition of DMBA metabolite formation in mouse epidermal homogenates (DIGIOVANNI and SLAGA 1981 b; J. DiGiovanni, unpublished data). However, relatively greater decreases in the formation of ring hydroxylated metabolites (e.g., phenols and diols) were observed compared with hydroxymethyl derivatives. With B[a]P as the substrate for epidermal monooxygenase, equimolar concentrations of B[e]P inhibited total metabolism by approximately 37% (J. DiGiovanni, unpublished data). Further increases in B[e]P concentration did not result in greater inhibition of B[a]P metabolism. In the presence of a 10-fold excess of B[e]P, the formation of 3-OH-B[a]P was reduced to a greater extent than that of B[a]P-7,8-diol. BAIRD and DIAMOND (1981) reported that, at doses equimolar or fourfold higher than B[a]P, B[e]P had little or no effect on the total amount of B[a]P metabolized by secondary cultures of hamster embryo cells. However, the proportion of B[a]P metabolites recovered as the 9,10- and 7,8-diols was greater in the cells treated with both B[e]P and B[a]P. In order to define better the complex mechanisms by which B[e]P alters the metabolic activation of B[a]P and DMBA, the effects of coadministration of B[e]P on the DNA binding of B[a]P and DMBA were examined in hamster embryo cells in culture (BAIRD et al. 1984; SMOLAREK and BAIRD 1984, 1986). B[e]P treatment resulted in a dose-dependent decrease in the binding of B[a]P to DNA. Analysis of the DNA adducts indicated that those formed from both *syn*- and *anti*-diol-epoxides of B[a]P decreased; however, the larger decrease in the proportion of *anti*-BPDE adducts indicated that B[e]P affected the stereochemical selectivity of the metabolism of B[a]P to diol-epoxides (Smolarek and Baird 1984). B[e]P:DMBA dose ratios comparable to those used in mouse skin initiation experiments greatly inhibited the binding of DMBA to DNA but did not alter the ratio of *anti*- to *syn*-DMBA diol-epoxide adducts (Smolarek and Baird 1986). In contrast, lower B[e]P:DMBA dose ratios caused a decrease in the *syn*-DMBA diol-epoxide DNA adducts and increased the amount of *anti*-DMBA diol-epoxide DNA adducts present (Smolarek and Baird 1986). Thus, B[e]P altered both the level of hydrocarbon metabolism in the cell cultures and the proportions of specific metabolites formed.

Studies on the binding of DMBA and B[a]P to DNA in vitro and mouse skin in vivo have been conducted. Under conditions identical to those used in tumor initiation experiments, DB[a,c]A, B[e]P, pyrene, and fluoranthene significantly inhibited the covalent binding of DMBA to calf thymus DNA in vitro when using mouse epidermal homogenates as the enzyme source (Slaga and Boutwell 1977; Slaga et al. 1978b, 1979). The same modifiers (except DB[a,c]A, which was not tested) slightly enhanced the in vitro covalent binding of B[a]P to calf thymus DNA (Slaga et al. 1979). B[e]P produced a marked inhibition of total binding of DMBA to mouse epidermal DNA in vivo but had no effect on the level of B[a]P binding (DiGiovanni 1981; DiGiovanni and Slaga 1981b; J. DiGiovanni, unpublished data). Rice et al. (1984) found that cotreatment with fluoranthene and pyrene increased the binding of B[a]P to DNA in mouse skin. These compounds are both cocarcinogens when used with B[a]P in multiple-application cocarcinogenesis assays (Van Duuren and Goldschmidt 1976) and coinitiators when used with B[a]P in initiation-promotion assays (Slaga et al. 1979). The doses of B[a]P used in the DNA binding study by Rice et al. (1984), 15–22 nmol, were much lower than the initiating doses used in the tumor experiments (DiGiovanni and Slaga 1981b; DiGiovanni et al. 1982; Slaga et al. 1979). More recent studies by Smolarek et al. (1987) demonstrated that B[e]P cotreatment increased the binding of B[a]P to DNA only at lower doses of B[a]P, suggesting that the large increases in binding of B[a]P to DNA observed by Rice et al. (1984) may be related to the low doses of B[a]P used in their study. Cotreatment with pyrene, fluoranthene, or catechol with B[a]P resulted in each case in a selective increase in the amount of *anti*-B[a]P diol-epoxide bound to DNA, and catechol cotreatment resulted in a selective increase in *anti*-B[a]P diol-epoxide DNA adduct formation (Rice et al. 1984; Melikian et al. 1986), again using low doses of B[a]P. Stereochemical selectivity of the effects of cocarcinogenic hydrocarbons was also observed in the studies of B[a]P metabolism in hamster embryo cell cultures as noted above (Smolarek and Baird 1984). The recent results of Smolarek et al. (1987) indicate, however, that B[e]P did not induce stereoselective alterations of B[a]P activation in mouse skin treated with initiating doses of B[a]P (i.e., 200 mol per mouse). In addition, B[e]P did not significantly alter the level of B[a]P covalently bound to epidermal DNA. Thus, further work is necessary to determine the mechanism of anticarcinogenesis and cocarcinogenesis by B[e]P, fluoranthene, pyrene, and other weakly or noncarcinogenic hydrocarbons.

V. Miscellaneous Inhibitors of Tumor Initiation

In addition to the modifiers already discussed in this chapter, a wide variety of other, structurally diverse chemicals can modify chemical carcinogenesis, although much less is known about the mechanism(s) of inhibition. This sections examines some of the inhibitors that appear to have effects on the initiation phase of the carcinogenic process, and these compounds are listed in Table 6.

A number of antiinflammatory steroids have modified PAH carcinogenesis in mouse skin. Dexamethasone, a potent corticosteroid, reduced tumor initiation, as well as complete carcinogenesis, by MCA in mouse skin (SLAGA and SCRIBNER 1973; THOMPSON and SLAGA 1976). In addition, skin tumor initiation with DMBA was inhibited by dexamethasone (THOMPSON and SLAGA 1976). Cortisone also inhibited complete carcinogenesis with B[a]P (BOUTWELL 1964), MCA (GHADIALLY and GREEN 1954), and DMBA (BASERGA and SHUBIK 1954). From other studies, the major effect of corticosteroids appears to be inhibition of tumor promotion (see Sect. D). A potential mechanism for their inhibitory effects on both initiation and promotion could be related to their ability to inhibit DNA synthesis (VIAJE and SLAGA 1978).

The effects of the prostaglandin precursor arachidonic acid and various inhibitors of arachidonic acid metabolism on skin tumor initiation by PAH have been investigated (FISCHER et al. 1979, and personal communication, 1988). Mouse epidermis contains significant amounts of the enzymes involved in arachidonic acid metabolism (e.g., cyclooxygenase and lipoxygenase; reviewed in DIGIOVANNI 1989). Flurbiprofen, indomethacin, or 5,8,11,14-eicosatetraynoic acid (ETYA), when given at the same time as B[a]P or B[a]P-7,8-diol, reduced the skin tumor-initiating activity of both compounds (FISCHER et al., personal communication, 1988). Pretreatment with arachidonic acid also reduced the tumor-initiating activity of these two carcinogens (FISCHER et al., personal communication, 1988). With the same experimental protocol, none of the cyclooxygenase inhibitors affected tumor initiation with DMBA. Arachidonic acid pretreatment, however, significantly potentiated DMBA tumor initiation (FISCHER et al. 1979, and personal communication, 1988). It is interesting that antioxidants such as BHA and BHT have been examined for their effects on PAH metabolism and tumor initiation in mouse epidermis. These compounds are known to inhibit peroxidase-dependent metabolism of PAH (ELING et al. 1986; MARNETT et al. 1978; MARNETT 1984) in other tissues. Interestingly, BHA and BHT had little effect on oxidative metabolism or DNA-binding of DMBA when catalyzed by epidermal homogenates in vitro (DIGIOVANNI et al. 1977; SLAGA and BRACKEN 1977). Moreover, BHT did not inhibit but rather enhanced the metabolism of B[a]P catalyzed by epidermal homogenates (BERRY et al. 1977). In addition, BHT had no effect on the epidermal homogenate-mediated binding of B[a]P to calf thymus DNA (BERRY et al. 1977). Thus, the mechanism of anti-initiating activity by inhibitors of enzymes involved in arachidonic metabolism may be complex and involve actions other than inhibition of enzyme activity. In addition, further work is warranted to establish the role of peroxidase-dependent metabolism in skin tumor initiation by PAH. Evidence indicates that enzymes involved in prostaglandin biosynthesis are capable of metabolizing B[a]P-7,8-diol to

Table 6. Miscellaneous inhibitors of chemical carcinogenesis and tumor initiation

Inhibitor	Carcinogen	Species	Tissue	References
Dexamethasone	MCA	Mouse	Skin	Slaga and Scribner (1973); Thompson and Slaga (1976)
	DMBA	Mouse	Skin	Thompson and Slaga (1976)
Cortisone	MCA	Mouse	Skin	Baserga and Shubik (1954)
	DMBA	Mouse	Skin	Ghadially and Green (1954)
	B[a]P	Mouse	Skin	Boutwell (1964)
Indomethacin	B[a]P	Mouse	Skin	Fischer et al. (1988, personal communication)
	B[a]P-7,8-diol	Mouse	Skin	Fischer et al. (1988, personal communication)
	DMBA	Mouse	Skin	Fischer et al. (1979, and personal communication)
Flurbiprofen	B[a]P	Mouse	Skin	Fischer et al. (1988, personal communication)
	DMBA	Mouse	Skin	Fischer et al. (1979, and personal communication)
ETYA	B[a]P-7,8-diol	Mouse	Skin	Fischer et al. (1988, personal communication)
α-Angelicalactone	B[a]P	Mouse	Forestomach	Wattenberg et al. (1979)
Diethyl maleate	MCA	Mouse	Skin	Chuang et al. (1978)
Actinomycin D	DMBA	Mouse	Skin	Bates et al. (1968); Hennings et al. (1968)
Sulfur mustard	DMBA	Mouse	Skin	De Young et al. (1977); Van Duuren and Segal (1976)
Sodium cyanate	DMBA	Rat	Mammary gland	Wattenberg (1980b)
	B[a]P	Mouse	Lung	Wattenberg (1980b)
Polyinosinic-	MCA	Mouse	Skin	Elgjo and Degre (1975)
Polycytidylic acid	DMBA	Mouse	Skin	Gelboin and Levy (1970)
Riboflavin	3′-Me-DAB	Rat	Liver	Lambooy (1976)
Cysteamine HCL	DMBA	Rat	Mammary gland	Marquardt et al. (1974)
D-Glucaro-1,4-lactone	DMBA	Rat	Mammary gland	Walaszek et al. (1984, 1986a)
or related analogues	MNU	Rat	Mammary gland	Walaszek et al. (1986b); Walaszek and Hanausek-Walaszek (1987)
	B[a]P	Rat	Subcutaneous	Walaszek et al. (1988)
	B[a]P	Mouse	Lung	Walaszek et al. (1986c)
	1-Nitropyrene	Mouse	Lung	Walaszek et al. (1988)
	DEN	Rat	Liver	Oredipe et al. (1987)
Norharmon	DBF	Mouse	Skin	Zajdela et al. (1987)
	DBF-3,4-diol	Mouse	Skin	Zajdela et al. (1987)
Trifluralin	B[a]P	Mouse	Lung	Triano et al. (1985)
Dehyroepiandrosterone	DMBA	Mouse	Skin	Pashko et al. (1984)
	DMH	Mouse	Colon	Nyce et al. (1984)
	Urethan	Mouse	Lung	Schwartz and Tannen (1981)

MCA, 3-methylcholanthrene; DMBA, 7,12-dimethylbenz[a]anthracene; B[a]P, benzo[a]pyrene; B[a]P-7,8-diol, (±)7β-8α-dihydroxybenzo[*a*]pyrene; 3′-Me-DAB, 3′-methyl-4-dimethylaminoazobenzene; MNU, methylnitrosourea; DBF, dibenzo[*a*,*e*]fluoranthene; DMH, dimethylhydrazine; ETYA, 5,8,11,14-eicosatetraynoic acid; DEN, diethylnitrosamine.

mutagenic derivatives (MARNETT et al. 1978). In addition, microsomal and solubilized enzyme preparations from rat seminal vesicles catalyzed the conversion of B[a]P-7,8-diol to derivatives that bound covalently to polyguanylic acid (PANTHANANICKAL and MARNETT 1981). In both of these studies, indomethacin was a potent inhibitor. These results have led several investigators to suggest that the enzymes involved in prostaglandin biosynthesis may be important for B[a]P tumor initiation in mouse skin (reviewed in MARNETT 1987).

α-Angelicalactone is a potent inhibitor of B[a]P-induced neoplasia of the mouse forestomach, possibly due to its enzyme-inducing properties (WATTENBERG et al. 1979). Diethyl maleate inhibited MCA-induced skin tumorigenesis in the same species, presumably by altering levels of glutathione (GSH) (CHUANG et al. 1978).

Work in several laboratories has demonstrated that precursors of D-glucaro-1,4-lactone, a natural β-glucuronidase inhibitor, strongly inhibit chemical carcinogenesis in a number of model systems including (i) mammary tumor induction in rats by DMBA (WALASZEK et al. 1984, 1986a b; WALASZEK and HANAUSEK-WALASZEK 1987) and MNU (WALASZEK et al. 1986b; WALASZEK and HANAUSEK-WALASZEK 1987), (ii) sarcoma induction in rats (WALASZEK et al. 1988) and lung adenoma induction in mice by B[a]P (WALASZEK et al. 1986c), and (iii) lung adenoma induction in mice by 1-nitropyrene (WALASZEK et al. 1988). Calcium glucarate also markedly inhibited rat liver carcinogenesis induced by diethylnitrosamine (DEN) (OREDIPE et al. 1987). The mechanism(s) involved in the anticarcinogenic actions of these interesting compounds is not known, although several have been postulated. These compounds appear to inhibit both the initiation (WALASZEK et al. 1984) and promotion (WALASZEK et al. 1986a) stages of chemical carcinogenesis and also inhibit initiation by direct acting carcinogens (WALASZEK et al. 1986b; WALASZEK and HANAUSEK-WALASZEK 1987). Glucarolactones may be capable of altering the metabolism of PAH such as DMBA or B[a]P (WALASZEK et al. 1986a, d). In addition, these compounds may alter the levels of circulating steroids, thus affecting the proliferative capacity of certain target organs, especially the mammary gland (WALASZEK et al. 1986a). Such a mechanism could explain the broad spectrum of anticarcinogenic action possessed by these compounds.

Very little information is available on the mode and mechanism of action of the remaining misecellaneous modifiers listed in Table 6. The reader is referred to the references cited within the table.

D. Inhibition of Tumor Promotion

Many studies on two-stage carcinogenesis in mouse skin as well as specific studies on the mechanisms of skin tumor promotion have been performed, but only recently have studies investigated other tissues (SLAGA et al. 1980a, 1982; SLAGA 1984a). As a consequence, most of the data related to inhibitors of tumor promotion are derived from studies using mouse skin. Up to the present time, relatively few studies have been performed in other two-stage carcinogenesis systems in which an inhibitor was given with a known tumor promoter in that system. In some studies, however, certain inhibitors have been given after carcinogenic treatment (i.e., during the so-called post-initiation phase). In these studies the inhibitor may be affecting events important in tumor promotion and/or progression. The mechanisms by which various agents inhibit tumor promotion are quite diverse owing to the many effects produced by promoting agents such as the widely studied phorbol esters. A summary of some of these effects in mouse skin has already been given in Table 1. The discovery that the phorbol ester receptor is PKC (reviewed in Blumberg 1988) has recently led to the identification of inhibitors of PKC. Nevertheless, many compounds inhibit tumor promotion by phorbol esters through mechanisms other than inhibition of protein kinase C (PKC) activity. In addition, many nonphorbol ester tumor promoters do not directly activate PKC and therefore appear to work initially by a mechanism different from that of the phorbol esters. Little work has been done to determine whether inhibitors of phorbol ester promotion also inhibit promotion by other classes of chemical promoters. Wherever possible these studies will be pointed out.

I. Antiinflammatory Steroids

Belman and Troll (1972) showed that a series of steroidal, antiinflammatory agents inhibited tumor promotion by croton oil in mouse skin in a dose-dependent manner. The relative potency of these steroids correlated with their antiinflammatory activities in mouse skin. These same investigators also showed that dexamethasone inhibited croton oil-induced epidermal hyperplasia. Subsequent studies by Scribner and Slaga (1973) showed that appropriate doses of dexamethasone completely suppressed tumor promotion by TPA in mouse skin for at least 6 months.

The antiinflammatory steroids fluocinolone acetonide (FA), fluocinonide (F), and fluclorolone acetonide (FCA) were later found to be extremely potent inhibitors of tumor promotion in mouse skin (Schwarz et al. 1977). FA, F, and FCA were all potent inhibitors of epidermal DNA synthesis (Schwarz et al. 1977), and the ability of the various steroids to inhibit epidermal DNA synthesis correlated with their antiinflammatory abilities as well as with their abilities to inhibit skin tumor promotion. In addition, when FA, FCA, or F were applied simultaneously with TPA at 10-μg doses, they completely counteracted the TPA-induced increase in nucleated, intrafollicular, epidermal cell layers, inflammation, and edema; 10-μg doses of dexamethasone partially counteracted the increase, and a 10-μg dose of cortisol or tetrahydrocortisol had no effect (Schwarz et al. 1977).

FA has also been examined for its ability to affect specific stages of tumor promotion in mouse skin (Slaga et al. 1982; Slaga 1984b). In this regard, FA was an effective inhibitor of both stage I and stage II of promotion in SENCAR mice (Slaga et al. 1982; Slaga 1984b). In addition, FA has recently been shown to inhibit skin tumor promotion in SENCAR mice by 1,8-dihydroxy-3-methyl-9-anthrone (chrysarobin; DiGiovanni et al. 1988a). Verma et al. (1983) also demonstrated that dexamethasone inhibited skin tumor promotion by 7-bromomethylbenz[*a*]anthracene (7-BRMe-BA). These studies suggest that antiinflammatory steroids may be general inhibitors of tumor promotion by different chemical classes of promoters.

In summary, the available data support the view that the mechanism by which the steroidal, antiinflammatory agents inhibit mouse skin tumor promotion and complete carcinogenesis may be related primarily to their ability to inhibit DNA synthesis. Consequently, this mechanism counteracts the induction of inflammation and hyperplasia by tumor promoters and complete carcinogens. Furthermore, the antiinflammatory ability of a series of steroids was found to correlate with their ability to inhibit mouse skin tumor promotion by TPA and their ability to counteract TPA-induced hyperplasia (F $\geqq$ FA $\geqq$ FCA $\geqq$ dexamethasone > cortisol, and tetrahydrocortisol was inactive). Antiinflammatory steroids also inhibit the activity of phospholipase A2, which may contribute to their antipromoting activity (see Sect. D.IV).

II. Retinoids

The retinoids, as described by Sporn et al. (1976), are a group of molecules composed of vitamin A and its synthetic analogues, which are known to exert profound control on cellular differentiation and growth in epithelial tissues. Interest in retinoids as possible chemopreventive agents in cancer induction started with the observations that retinoid deficiency enhanced the susceptibility of a number of tissues, such as the respiratory tract, bladder, and colon, to chemical carcinogenesis (Netteshein and Williams 1976; Rogers et al. 1973). Investigators also reported that several natural and synthetic retinoids prevented the development of epithelial cancer of the skin (Bollag 1972; Shamberger 1971; Verma and Boutwell 1977), respiratory tract (Saffiotti et al. 1967), mammary gland (Dickens et al. 1979; Grubbs et al. 1977a), urinary bladder (Grubbs et al. 1977b; Sporn et al. 1977), colon (Newberne and Suphakarn 1977), stomach (Chu and Malmgren 1965), vagina, and cervix (Chu and Malmgren 1965) in experimental animals. In most of these experiments, the retinoids were not given to the animals until after completion of administration of the respective carcinogens. Under these conditions, the retinoids did not prevent initiation of carcinogenesis, but rather they modified the post-initiation (i.e., promotional phase) or preneoplastic states during the latent period of cancer development. With the exception of the hamster cheek pouch, the retinoids have an inhibitory effect on epithelial cancer in a number of sites induced by a wide variety of carcinogens (for previous reviews see Slaga and DiGiovanni 1984; Sporn et al. 1976; Sporn 1980). In addition, retinoids counteract the carcinogen-induced hyperplasia and metaplasia of differentiated epithelia in organ-cultured prostate gland (Chopra and Wilkoff 1976; Lasnitzki 1976) and trachea (Crocker and Sanders 1970). Furthermore, retinoids inhibit the oncogenic transformation of fibroblasts by MCA (Merriman and Bertram 1979) and radiation (Harisiadis et al. 1978). In these last two experimental protocols, the retinoids were also effective when given after treatment with the carcinogen. Although retinoids may act by more than one mechanism to modify preneoplastic states, their ability to influence cell differentiation in epithelia is undoubtedly of importance in their suppression of carcinogenesis.

Direct proof for retinoids being potent inhibitors primarily of tumor promotion and not of tumor initiation is derived from studies using two-stage carcinogenesis in mouse skin. A number of retinoids are specific inhibitors of mouse skin tumor promotion. As shown in Table 7, RA is one of the most potent retinoids in counteracting skin tumor promotion by TPA. The effect of the retinoids on skin tumor promotion is considered reversible, because retinoids have to be given with every application of TPA (Weeks et al. 1979). If treatment with RO-10 9359, for example, is terminated after 30 weeks and TPA application continued, the number of tumors begins to increase at 36 weeks; this response is comparable with TPA treatment starting 1 week after initiation (Weeks et al. 1979). Similar reversible effects of retinoids were found by Merriman and Bertram (1979) using C3H/10T/1/2 cells. More recently, it has been reported that inhibition of tumor promotion by 13-*cis*-RA is stable in the absence of further treatments with TPA (Gensler et al. 1986).

The time of treatment with retinoids in relation to the time of treatment with TPA appears to be very important (Verma et al. 1978). The retinoids are most effective at inhibiting skin tumor promotion by TPA when given a few hours

Table 7. Effects of retinoids on tumor promotion in mouse skin

Retinoid	Promoter	Relative potency	References
RA	TPA	Strong	Verma and Boutwell (1977)
13-*cis*-RA	TPA	Strong	Verma et al. (1979)
	Anthralin	Moderate-weak	Dawson et al. (1987); Gensler and Bowden (1984)
RO-10-9359	TPA	Strong	Weeks et al. (1979)
	Croton oil	Strong	Bollag (1972)
Retinal	TPA	Strong	Verma et al. (1979)
5,6-Epoxyretinoic acid	TPA	Strong	Verma et al. (1980c)
5,6-Dihydroretinoic acid	TPA	Strong	Verma et al. (1980c)
Retinol	TPA	Moderate	Verma et al. (1979)
Retinyl acetate	TPA	Moderate	Verma et al. (1977)
Retinyl palmitate	TPA	Moderate	Verma et al. (1977)
13-Trifluoromethyl-TMMP analogue of ethyl retinoate	TPA	Weak	Verma et al. (1979)
RA	7-BRMe-BA	Strong	Verma et al. (1983)
RA	Chrysarobin	Moderate	DiGiovanni et al. (1988a)
RA	Anthralin	Strong	Dawson et al. (1987)
RO-13-7410	Anthralin	Strong	Dawson et al. (1987)
SRI-5896-39	Anthralin	Strong	Dawson et al. (1987)
SRI-5898-53	Anthralin	Moderate	Dawson et al. (1987)

RA, *β*-all *trans*-retinoic acid; TPA, 12-*O*-tetradecanoylphorbol-13-acetate; RO-10-9359, ethyl all *trans*-9-(4-methoxy-2,3,6-trimethylphenyl)-3,7-dimethyl-2,4,6,8-nonatetrenoate; TMMP, trimethylmethoxyphenyl; RO-13-7410, 4-[2-(5,6,7,8-tetrahydro-5,5,8,8-tetramethyl-2-naphthalenyl)-IE-propen-l-yl]benzoic acid; SRI-5896-39, 6[l-(4-carboxyphenyl)-IE-propen-2-yl]-3,4-dihydro-4,4-dimethyl-2*H*-*I*-benzothiopyran; SRI-5898-53, 6-(5,6,7,8-tetrahydro-5,5,8,8-tetramethyl-2-naphthalenyl)-2-naphthalenecarboxylic acid; 7-BRMe-BA, 7-bromomethylbenz[*a*]anthracene.

before, simultaneously, or a few hours after TPA (VERMA et al. 1978). If RA is given 24 h after TPA treatment, it no longer has an inhibitory effect (VERMA et al. 1978). The time frame of treatment with retinoids is also important in counteracting the TPA-induced ODC in mouse epidermis (VERMA et al. 1978). VERMA et al. (1979) reported a strong correlation between the ability of retinoids to inhibit TPA-induced mouse epidermal ODC and their ability to inhibit TPA promotion.

RA has also been examined for its ability to inhibit specific stages of tumor promotion in mouse skin. In this regard, SLAGA and coworkers (1982; SLAGA 1984b) demonstrated that RA was a more effective inhibitor of stage II than of stage I of promotion. More recently, however, VERMA (1987) reported that RA effectively inhibited both stage I and stage II promotion when given at doses similar to those used in studies by SLAGA and coworkers (1982; SLAGA 1984b). In the light of these conflicting results, further work is necessary to establish the promotion-stage specificity of retinoids.

Although the retinoids are potent inhibitors of tumor promotion by TPA and of TPA-induced epidermal ODC, RA does not inhibit epidermal ODC induced by a complete carcinogenic dose of DMBA, nor does it inhibit complete carcinogenesis by DMBA (VERMA et al. 1979). The question remains as to whether retinoids inhibit all classes of skin tumor promoters. Recent studies by DAWSON et al. (1987) and DIGIOVANNI et al. (1988a) indicate that retinoids can inhibit skin tumor promotion by the anthrone class of promoters (e.g., anthralin and chrysarobin). It should be noted that GENSLER and BOWDEN (1984) reported that 13-*cis*-RA, when tested at a single dose, did not inhibit tumor promotion by anthralin in SENCAR mice. Therefore, the type, dose, and time of application of the retinoid relative to promoter treatment may be critical determinants for its inhibition of the effects of different chemical classes of promoters. In addition, RA inhibited skin tumor promotion and ODC induction by 7-BRMe-BA after DMBA initiation (VERMA et al. 1983). FORBES et al. (1979) reported that RA acted as a tumor promoter after UV-induced initiation in mouse skin. In addition, when RA is given at higher doses and is applied more frequently than in the studies in which it inhibited TPA promotion, it can also act as a skin tumor promoter after DMBA initiation (FISCHER et al. 1985; HENNINGS et al. 1982). How RA can act both as an inhibitor of promotion and as a tumor promoter is not known at the present time.

In summary, the results of studies on the inhibition of carcinogenesis by retinoids suggest that they are effective primarily during the promotional stage of carcinogenesis. Although their mechanism of action is not definitely known, evidence suggests that they affect epithelial differentiation and also alter polyamine levels by inhibiting the induction of ODC. Several reports indicate that polyamines are involved in regulating cellular differentiation and growth (BACHRACH 1973; PEGG 1988). Recent evidence indicates that retinoids may bring about many of these effects by interacting with nuclear receptors (GIGUERE et al. 1987; PETKOVICH et al. 1987). These nuclear receptors appear to be *trans*-acting factors which can regulate the expression of specific genes (GIGUERE et al. 1987; PETKOVICH et al. 1987) and this may explain the ability of retinoids to inhibit transcription of ODC mRNA that is induced following treatment with TPA (VERMA 1988).

III. Protease Inhibitors

The use of protease inhibitors as possible chemopreventive agents in cancer induction was started by the observation that TPA and croton oil caused the appearance of a trypsin-like protease in mouse skin (Troll et al. 1970). In order to determine whether this protease was a necessary factor in tumor promotion, protease inhibitors were tested for antipromoting activity in two-stage carcinogenesis in mouse skin. The protease inhibitors tosyllysine chloromethyl ketone (TLCK), TPCK, and tosylalanine methyl ester (TAME) were all potent inhibitors of croton oil and TPA promotion after DMBA initiation (Troll et al. 1970). TPCK was the most effective protease inhibitor that counteracted tumor promotion in these early studies (Troll et al. 1970). The chloromethyl ketone protease inhibitors are also sulfhydryl-reacting agents. Troll et al. (1978) found that TLCK reacts very efficiently and quickly with glutathione, as well as with proteases. It could be argued that the inhibition of tumorigenesis by TLCK and TPCK is due to the sulfhydryl reactions of these compounds; however, because TAME does not react with sulfhydryl groups, the mechanism of action may be truly related to inhibition of protease activity. This view is further supported by the observation that another protease inhibitor, leupeptin, which does not alkylate sulfhydryl groups, is also an inhibitor of croton oil promotion in mouse skin (Hozumi et al. 1972). As shown in Table 8, leupeptin was also found to inhibit, in rats, azoxymethane-induced colon tumors, butylnitrosourethane-induced esophageal tumors, and DMBA-induced mammary tumors (Matsushima et al. 1975). However, leupeptin either had no effect or enhanced tumor growth in the forestomach, liver, and bladder (Matsushima et al. 1975). Yamamura et al. (1978) reported that a synthetic protease inhibitor, *N,N*-dimethylamino-[*p*-(*p*′-guanidinobenzoyloxy)]-benzilcarbonyloxyglycolate, depressed the induction of mammary tumors in rats by DMBA. More recently, consumption by rodents of raw soybeans or of purified fractions from soybeans that contain protease inhibitors has been shown to suppress skin (Troll et al. 1979), breast (Troll et al. 1980), colon (Weed et al. 1985), and liver cancers (Becker 1981). Skin tumors induced in mice by 4-nitroquinoline-*N*-oxide followed by TPA treatment were suppressed when casein in the diet was replaced by raw soybeans with an equivalent protein content. Destruction of the protease inhibitors by heating the raw soybeans abolished their ability to suppress promotion when they were fed to mice (Troll et al. 1979). Moreover, a diet containing raw soybeans suppressed breast cancer caused by ionizing radiation in Sprague-Dawley rats (Troll et al. 1980). Purified fractions from soybeans which contain protease inhibitors suppressed spontaneous liver cancer in the C3H/HeN mice (Becker 1981). A fraction enriched in the Bowman-Birk-inhibitor blocked 1,2-dimethylhydrazine-induced colon cancer in mice (Weed et al. 1985).

The suggestion that protease inhibitors in the diet suppress experimentally induced carcinogenesis was confirmed using the synthetic trypsin inhibitor *N,N*-dimethylamino-[*p*-(*p*′-guanidinobenzoloxy)]-benzilcarbonyloxyglycolate as noted above. [*N,N*-dimethylcarbamoylmethyl-4-(4-guanidinobenzoyloxy)-phenylacetate]-methane sufate, another synthetic protease inhibitor which inhibits both trypsin and chymotrypsin, prevented skin carcinogenesis induced by

Table 8. Effects of protease inhibitors on chemical and radiation carcinogenesis and tumor promotion

Protease inhibitor	Carcinogen and/or promoter	Species	Tissue or Cell type[a]	References
TPCK	TPA	Mouse	Skin	Troll et al. (1970)
TLCK	TPA	Mouse	Skin	Troll et al. (1970)
TAME	TPA	Mouse	Skin	Troll et al. (1970)
Leupeptin	Croton oil	Mouse	Skin	Hozumi et al. (1972)
	Azoxymethane	Rat	Colon	Matsushima et al. (1975)
	Butyl-nitroso-urethane	Rat	Esophagus	Matsushima et al. (1975)
	DMBA	Rat	Mammary gland	Matsushima et al. (1975)
	MCA	Mouse	C3H/10T 1/2	Kuroki and Drevon (1979)
	X-rays	Mouse	C3H/10T 1/2	Kennedy and Little (1978)
	X-rays plus TPA	Mouse	C3H/10T 1/2	Kennedy and Little (1978)
N,N-dimethylamino-[*p*(*p*′-guanidinobenzoyloxy)]-benzilcarbonyloxyglucolate	DMBA	Rat	Mammary gland	Yamamura et al. (1978)
Antipain	MCA	Mouse	C3H/10T 1/2	Kuroki and Drevon (1979)
	X-rays	Mouse	C3H/10T 1/2	Kennedy and Little (1978)
	X-rays plus TPA	Mouse	C3H/10T 1/2	Kennedy and Little (1978)
Chymostatin	MCA	Mouse	C3H/10T 1/2	Kuroki and Drevon (1979)
Elastatinal	MCA	Mouse	C3H/10T 1/2	Kuroki and Drevon (1979)
Pepstatin	MCA	Mouse	C3H/10T 1/2	Kuroki and Drevon (1979)
Raw soybeans	4-NQO plus TPA	Mouse	Skin	Troll et al. (1979)
	Ionizing radiation	Rat	Breast	Troll et al. (1980)
	DMH	Rat	Colon	Weed et al. (1985)
	Spontaneous	Rat	Liver	Becker (1981)
N-N-dimethylcarbamoylmethyl-4-(4-quanidinobenzoyl-oxy)-phenylacetate]methane sulfate	MCA	Mouse	Skin	Ohkoshi and Fujii (1983)
ε-aminocaproic acid	DMH	Mouse	Colon	Corasanti et al. (1982)

DMBA, 7,12-dimethylbenz[*a*]anthracene; TPA, 12-*O*-tetradecanoylphorbol-13-acetate; MCA, 3-methylcholanthrene; 4-NQO, 4-nitroquinoline-*N*-oxide; DMH, dimethylhydrazine; TPCK, tosylphenylalanine chloromethylketone; TLCK, tosyllysine chloromethylketone; TAME, tosylalanine methyl ester.
[a] C3H/10T 1/2 are cells in culture.

3 MCA (Ohkoshi and Fujii 1983). The simplest protease inhibitor used in feeding experiments was ε-aminocaproic acid, a known inhibitor of trypsin and plasminogen activator. It is derived from the amino acid lysine from which the α-amino group has been removed. This inhibitor, when given in the drinking water, effectively suppressed colon cancer induction by 1,2-dimethylhydrazine in mice (Corasanti et al. 1982).

A number of reports have shown that protease inhibitors are capable of suppressing malignant transformation in vitro (Kennedy and Little 1978; Kuroki and Drevon 1979; reviewed in Troll et al. 1987). Kennedy and Little (1978) reported that antipain and leupeptin could inhibit transformation of C3H/10T1/2 cells by X-rays and X-rays plus TPA. Kuroki and Drevon (1979) found that antipain, chymostatin, elastatinal, leupeptin, and pepstatin blocked MCA-induced transformation in C3H/10T1/2 cells when added at a nontoxic dose for the 6-week course of transformation or begun 1 week after carcinogen treatment. These in vitro studies suggested that the protease inhibitors were counteracting the promotional phase of carcinogenesis. Further evidence for this view stems from the observation that protease inhibitors have very little effect on chemical- or radiation-induced mutagenesis in V79 Chinese hamster cells (Kuroki and Drevon 1979). More recently, Kennedy (1985) reported that inhibitors of chymotrypsin are the most effective inhibitors of X-ray-induced transformation in C3H/10T/1/2 cells. In addition evidence was presented suggesting that protease inhibitors may irreversibly revert X-ray-initiated cells to their uninitiated state. The mechanism by which such a reversion takes place remains unknown.

Protease inhibitors have been examined for their effects on specific stages of promotion in mouse skin. In this regard, TPCK was reported to inhibit specifically stage I of promotion in SENCAR mice (Slaga et al. 1982; Slaga 1984b). Interestingly, little is known about the ability of protease inhibitors to modify tumor promotion by other classes of skin tumor promoters. Preliminary studies have indicated that TPCK and TLCK had no inhibitory effect on skin tumor promotion by chrysarobin (DiGiovanni et al. 1987; J. DiGiovanni, unpublished studies).

In summary, the data discussed above, as well as that presented in Table 8, suggest very strongly that protease inhibitors are effective after initiation and during the postinitiation and promotional stages of carcinogenesis, but their possible mechanisms of action remain unknown. The protease inhibitors have very little effect on most of the reported events thought to be important in skin tumor promotion. They do not have any significant effect on TPA-induced epidermal hyperplasia or on TPA-induced ODC activity (Slaga et al. 1980b). However, TPCK does counteract the induction of "dark" basal keratinocytes by TPA, an event that shows a strong correlation with the promoting activities of a number of skin tumor promoters (Slaga et al. 1980b). It has been postulated that protease inhibitors may act through suppression of the formation of active oxygen species. Protease inhibitors are known to prevent the formation of superoxide anion (O_2^-) and hydrogen peroxide (H_2O_2) by polymorphonuclear leukocytes (PMNs) that are activated by TPA (reviewed in Troll et al. 1987). Furthermore, free radials are believed to be involved in tumor promotion (reviewed in Kensler and Jaffe 1986). Recently, several studies have suggested the presence of membrane binding sites (receptors?) which may lead to the internalization of protease inhibitors within cells (Yavelow et al. 1987a, b). These and other studies may ultimately shed light on the exact mechanism of action of protease inhibitors.

IV. Inhibitors of Arachidonic Acid Metabolism

Extensive evidence suggests that prostaglandins or, more correctly, the end products of arachidonic acid metabolism may play an important role in tumor promotion and in carcinogenesis in general. First, tumor promoters and carcinogens, especially in the mouse skin system, induce inflammation and hyperplasia (MILLER 1978; SLAGA et al. 1982). Secondy, skin tumor promoters and carcinogens stimulate phospholipid metabolism (reviewed in SLAGA et al. 1980a, 1982) and prostaglandin biosynthesis (BRESNICK et al. 1979; VERMA et al. 1980a). FISCHER et al. (1980a) reported that, although prostaglandin $F_{2\alpha}$ and E_2 were not skin tumor promoters, they effectively enhanced promotion by TPA. Likewise, LUPULESCU (1978) found that prostaglandin $F_{2\alpha}$ and E_2 enhanced complete skin carcinogenesis in mice by MCA. Third, most tumors have an elevated level of prostaglandins (JAFFE 1974). Finally, a number of prostaglandin synthesis inhibitors are effective in counteracting tumor promotion and carcinogenesis (FISCHER et al. 1979, 1980a; FISCHER and SLAGA 1982; FURSTENBURGER and MARKS 1978; VERMA et al. 1980a). The antiinflammatory steroids are not only very potent inhibitors of promotion but also of phospholipase A_2 activity (BLACKWELL et al. 1978). The latter inhibitory activity could, in theory, decrease the amount of arachidonic acid available for metabolism to the important end products. This decrease may also contribute to the mechanism by which the antiinflammatory steroids counteract the promoter-induced inflammation and hyperplasia. Another phospholipase A_2 inhibitor, dibromoacetophenone, is also a potent inhibitor of skin tumor promotion (FISCHER and SLAGA 1982).

Various prostaglandin synthesis inhibitors that affect tumor promotion and carcinogenesis are listed in Table 9. Compounds such as the antiinflammatory steroids and dibromoacetophenone inhibit very early in the pathway and consequently inhibit the formation of all the important end products of arachidonic metabolism such as hydroxy fatty acids, leukotrienes, thromboxanes, prostaglandins, and prostacyclins (FISCHER et al. 1979; FISCHER and SLAGA 1982). The next most consistent inhibitors of tumor promotion in this general category are phenidone and 5,8,11,14-eicosatetraynoic acid (ETYA); these compounds inhibit both the cyclooxygenase and lipoxygenase pathways. This mode of inhibition also leads to a decrease in all the important end products of arachidonic acid metabolism (FISCHER et al. 1979; FISCHER and SLAGA 1982).

Indomethacin, a potent inhibitor of the cyclooxygenase pathway, has both inhibited and enhanced skin tumor promotion (FISCHER et al. 1980b; FISCHER and SLAGA 1982; FURSTENBURGER and MARKS 1978; VERMA et al. 1980a). Indomethacin is most effective in enhancing or inhibiting promotion when applied 2 h prior to TPA (FISCHER et al. 1979; FISCHER and SLAGA 1982). Enhancement of promotion occurs at doses of 25–100 μg. Doses above 100 μg (e.g., 200 μg) are inhibitory. Flurbiprofen, another cyclooxygenase inhibitor, also has both enhanced and inhibited skin tumor promotion (FISCHER et al. 1980b). The reason for the dual effects of the cyclooxygenase inhibitors is not known, but the inhibitory effect may be related to the toxicity of these compounds. The enhancing effect of the cyclooxygenase inhibitors on promotion may occur because more arachidonic acid is being metabolized by the lipoxygenase pathway while the

Table 9. Effects of Inhibitors of arachidonic acid metabolism on tumor promotion and carcinogenesis

Inhibitor	Pathway inhibited	Carcinogen and/or Tumor promoter	Species	Tissue	References
Antiinflammatory steroid	Phospholipase A_2	TPA	Mouse	Skin	Fischer et al. (1979)
Dibromoacetophenone	Phospholipase A_2	TPA	Mouse	Skin	Fischer and Slaga (1982)
ETYA	Lipoxygenase and cyclooxygenase	TPA	Mouse	Skin	Fischer et al. (1979)
Phenidone	Lipoxygenase and cyclooxygenase	TPA	Mouse	Skin	Fischer et al. (1979); Fischer and Slaga (1982)
Indomethacin	cyclooxygenase	TPA	Mouse	Skin	Fischer et al. (1980b); Furstenberger and Marks (1978); Verma et al. (1980a)
		DMBA	Rat	Mammary gland	McCormick et al. (1985)
RO22-3581	Thromboxane synthetase	TPA	Mouse	Skin	Fischer et al. (1979); Fischer and Slaga (1982)
RO22-3582	Thromboxane synthetase	TPA	Mouse	Skin	Fischer et al. (1979); Fischer and Slaga (1982)
Quercetin	Lipoxygenase	TPA	Mouse	Skin	Kato et al. (1983)
		Teleocidin	Mouse	Skin	Fujiki et al. (1986)
Aspirin	Cyclooxygenase	FANFT	Rat	Urinary bladder	Cohen et al. (1981) Murasaki et al. (1984)
		FANFT-saccharin	Rat	Urinary bladder	Sakata et al. (1986)
NDGA	Lipoxygenase	Teleocidin	Mouse	Skin	Nakadate et al. (1982)
AA861	Lipoxygenase	Teleocidin	Mouse	Skin	Aizu et al. (1986)
3,4,2′,4′-tetrahydrochalcone	Lipoxygenase	Teleocidin	Mouse	Skin	Aizu et al. (1986)

DMBA, 7,12-dimethylbenz[*a*]anthracene; TPA, 12-*O*-tetradecanoylphorbol-13-acetate; ETYA, 5,8,11,14-eicosatetraynoic acid; RO22-3581, imidazolacetophenone; RO22-3582; imidazolphenol; NDGA, nordihydroguaiaretic acid; FANFT, *N*-[4-(5-nitro-2-furyl)-2-thiazolyl]formamide; AA861, 2,3,5-trimethyl-6-(12-hydroxy-5,10-dodecadiynyl)-1,4-benzoquinone.

cyclooxygenase pathway is inhibited (Fischer and Slaga 1982). This hypothesis suggests that the lipoxygenase pathway is important in skin tumor promotion. In support of this hypothesis, a number of lipoxygenase inhibitors are known to be effective inhibitors of TPA-induced ODC and skin tumor promotion including: quercetin (Kato et al. 1983), nordihydroguaiaretic acid (NDGA) (Nakadate et al. 1983), 2,3,5-trimethyl-6-(12-hydroxy-5,10-dodecadiynyl)-1,4-benzoquinone (AA861) (Aizu et al. 1986), and 3,4,2′,4′-tetrahydrochalcone (Aizu et al. 1986).

COHEN et al. (1981) reported that aspirin, another cyclooxygenase inhibitor, inhibited the induction of early hyperplastic lesions of the bladder in rats by *N*-[4-(5-nitro-2-furyl)-2-thiazolyl]formamide (FANFT). However, in this case, COHEN et al. (1981) found that aspirin was a potent inhibitor of the metabolism of FANFT. Subsequently, a more long-term study revealed that, although aspirin inhibited FANFT-induced bladder tumors in rats, the induction of forestomach tumors was significantly enhanced (MURASAKI et al. 1984). Aspirin was shown to inhibit both FANFT initiation and sodium saccharin promotion of bladder tumors in rats (SAKATA et al. 1986). Indomethacin has been shown to protect significantly against rat mammary carcinogenesis induced by DMBA when given during both the early stage and the late stage of the carcinogenic process in this model system (McCORMICK et al. 1985).

Thromboxanes may also be important in skin tumor promotion. The thromboxane synthesis inhibitors imidazolacetophenone and imidazolphenol (RO22-3581 and RO22-3582, respectively) both inhibited skin tumor promotion by TPA (FISCHER et al. 1979; FISCHER and SLAGA 1982).

In summary, the existing data imply that several of the end products of arachidonic acid metabolism play a critical role in tumor promotion through their effects on inflammation and cellular proliferation. It should be noted that the promotion stage specificity, if any, of these compounds has not been adequately investigated. In addition, the ability of these inhibitors to modify tumor promotion in mouse skin by nonphorbol ester tumor promoters [except in the case of the teleocidins which appear to work through the same receptor mechanism as the phorbol esters (BLUMBERG 1988)] remains to be determined.

V. Antioxidants

Although the antioxidants have already been discussed in this chapter in terms of inhibiting complete carcinogenesis and tumor initiation, they are also effective inhibitors of tumor promotion. Table 10 summarizes various antioxidants that have been shown to inhibit skin tumor promotion. BHA and BHT inhibited both TPA and benzoyl peroxide promotion in mouse skin (SLAGA et al. 1983). The major isomer of BHA, 3-t-BHA, was a very effective inhibitor of TPA promotion in mouse skin (SLAGA et al. 1983). In addition, disulfiram and 4-*p*-hydroxyanisole were also effective inhibitors of TPA promotion (SLAGA et al. 1983). KOZUMBO et al. (1983; reviewed in KENSLER and JAFFE 1986) reported that BHA and other antioxidants inhibited TPA-induced epidermal ODC activity over the same dose range as that used for inhibition of tumor promotion.

Although the mechanism by which the antioxidants inhibit tumor promotion in mouse skin is not presently known, they may be scavenging radicals generated directly, in the case of benzoyl peroxide, or indirectly by TPA. The fact that benzoyl peroxide and other free radical-generating compounds, such as lauroyl peroxide and chloroperbenzoic acid, are effective skin tumor promoters suggests that free radicals may be important in tumor promotion (SLAGA et al. 1981). An analogous situation would be the phorbol ester tumor promoters. They can stimulate superoxide anion production in polymorphonuclear leukocytes, and the antipromoters such as RA and dexamethasone and protease inhibitors can

Table 10. Inhibitory effects of antioxidants on carcinogenesis and tumor promotion in the mouse

Antioxidant	Carcinogen and/or tumor promoter	Tissue or cell type	References
BHT	TPA	Skin	Slaga et al. (1983)
	Benzoyl peroxide	Skin	Slaga et al. (1983)
BHA	TPA	Skin	Slaga et al. (1983)
	Benzoyl peroxide	Skin	Slaga et al. (1983)
3-t-BHA	TPA	Skin	Slaga et al. (1983)
2-t-BHA	TPA	Skin	Slaga et al. (1983)
Disulfiram	TPA	Skin	Slaga et al. (1983)
4-p-Hydroxyanisole	TPA	Skin	Slaga et al. (1983)
Cu^{2+}DIPS	TPA	Skin	DiGiovanni et al. (1988b); Egner and Kensler (1985); Kensler et al. (1985)
	Chrysarobin	Skin	DiGiovanni et al. (1988b)
DDTC	TPA	Skin	Perchellet et al. (1987a)
GSH and other GSH elevating agents	TPA	Skin	Perchellet et al. (1985a, 1987b)
Selenium	Croton oil	Skin	Shamberger and Rudolph (1966)
α-Tocopherol	Croton oil	Skin	Shamberger and Rudolph (1966)
	MCA	Fibroblast	Haber and Wissler (1962)
	DMBA	Lung	Sachdev et al. (1980)
Ascorbic acid	Croton oil	Skin	Shamberger (1972)
	MCA	C3H/10T 1/2 cells	Benedict et al. (1980)
	B[a]P	Fibroblast [a]	Kallistratos and Fasske (1980)
	TPA	Skin	Smart et al. (1987)
Ascorbyl palmitate	TPA	Skin	Smart et al. (1987)

BHT, butylated hydroxytoluene; BHA, butylated hydroxyanisole; DMBA, 7,12-dimethylbenz[*a*]anthracene;TPA, 12-*O*-tetradecanoylphorbol-13-acetate; 3-t-BHA, 3-tert-butyl-4-hydroxyanisole; 2-t-BHA, 2-tert-butyl-4-hydroxyanisole; Cu^{2+}DIPS, copper(II) bis(diisopropylsalicylate); DDTC, diethyldithiocarbamate; GSH, glutathione; MCA, 3-methylcholanthrene; B[a]P, benzo[*a*]pyrene.

[a] These studies used rats.

counteract this effect (Goldstein et al. 1979, 1981; reviewed in Troll and Wiesner 1985). Solanki et al. (1981) found that TPA causes a sustained decrease in the basal level of epidermal superoxide dismutase (SOD) and catalase activities. These results indicate that free radicals may increase after TPA because of the low levels of SOD and catalase, which normally detoxify O_2^- and H_2O_2.

Copper (II) bis(diisopropylsalicylate) (Cu^{2+} DIPS), a SOD mimetic, inhibits skin tumor promotion (Kensler et al. 1983) and ODC induction (Egner and Kensler 1985) by TPA. We have found that Cu^{2+} DIPS also moderately inhibited skin tumor promotion by the anthrone derivative chrysarobin (DiGiovanni et al. 1988 b). These results suggest the involvement of O_2^- in skin tumor promotion by both phorbol esters and anthrones.

SHAMBERGER (1970) reported that Se was an effective inhibitor of skin tumor promotion by croton oil. Se is a necessary cofactor for the enzyme GSHP that detoxifies hydrogen peroxide and hydroperoxides within the cell (FREDOVICH 1976). The possibility exists that Se-dependent glutathione peroxidase (GSHP) lowers the level of potentially damaging peroxide radicals that are generated from various carcinogenic and promoting chemicals. However, as stated previously in this chapter, such an action of administered Se has yet to be clearly established. The antiproliferative actions of Se may be involved in its antipromoting effects. SHAMBERGER and RUDOLPH (1966) have also shown that α-tocopheral and ascorbic acid significantly reduced tumor formation induced by DMBA initiation and croton oil promotion. SMART et al. (1987) have more recently demonstrated that topical application of ascorbic acid or ascorbyl palmitate effectively inhibited ODC induction and stimulated epidermal DNA synthesis and tumor promotion by TPA in CD-1 mice. PERCHELLET et al. (1987a) have further shown that combinations of Se and α-tocopherol were highly effective at inhibiting TPA promotion and, in particular, stage II of promotion. Finally, the antioxidant diethyldithiocarbamate (DDTC) has been shown to be a very effective inhibitor of skin tumor promotion by phorbol ester (both stage I and stage II) (PERCHELLET et al. 1987b) although only at very high doses. Certain other agents that raise glutathione levels (including DDTC) appear to be capable of inhibiting TPA promotion, including glutathione itself (PERCHELLET et al. 1985a, b). These agents may be effective inhibitors due to the fact that a wide variety of tumor promoters decrease the ratio of reduced (GSH)/oxidized (GSSG) glutathione in mouse epidermis (PERCHELLET et al. 1986). TPA was also found to stimulate a rapid, transient increase in GSHP followed by a prolonged depression in the activity of this enzyme (PERCHELLET et al. 1986). These changes presumably reflect the induction of a prooxidant state in the epidermal cells by TPA and other tumor promoters.

α-Tocophercol also reduced the number of fibrosarcomas induced by MCA (HABER and WISSLER 1962) and mammary gland adenocarcinomas induced by DMBA (SACHDEV et al. 1980). BENEDICT et al. (1980) reported that ascorbic acid inhibited the transformation of C3H/10T1/2 cells by MCA. The inhibitory effect of ascorbic acid in this study was observed in some cases long after the carcinogen was given. This observation indicates that it was acting during the promotion or progression stage. KALLISTRATOS and FASSKE (1980) reported that high doses of ascorbic acid administered in the drinking water significantly decreased the in vivo induction of sarcomas by B[a]P.

In summary, a number of antioxidants have been found to inhibit tumor promotion in mouse skin. Although their mechanism of action is not definitely known, evidence points to two possibilities: (1) they scavenge various radicals generated directly or indirectly by tumor promoters and/or (2) they increase levels of enzymes that are important in detoxifying cellular radicals.

VI. Polyamine Synthesis Inhibitors

As previously discussed, the induction of ODC activity and elevated polyamines in mouse epidermis by tumor promoters has been proposed as a critical

Table 11. Inhibition of carcinogenesis and tumor promotion by inhibitors of polyamine biosynthesis

Inhibitor	Carcinogen or tumor promoter	Species	Tissue	References
α-DFMO	TPA	Mouse	Skin	Takigawa et al. (1982); Verma et al. (1986); Weeks et al. (1982, 1984)
	Chrysarobin	Mouse	Skin	DiGiovanni et al. (1988a)
	NMU	Rat	Mammary gland	Thompson et al. (1984, 1985, 1986)
	NMU	Rat	Urinary bladder	Homma et al. (1985a)
MGBB	TPA	Mouse	Skin	Hibasami et al. (1988)

TPA, 12-*O*-tetradecanoylphorbol-13-acetate; α-DFMO, α-difluoromethylornithine; NMU, methylnitrosourea; MGBB, methylglyoxal bis(butyl-amidinohydrazone).

phenotypic alteration involved in two-stage carcinogenesis. The retinoids are potent inhibitors of tumor promotion and of tumor promoter-induced ODC and polyamine levels (Verma et al. 1979), but a critical test for the role of polyamines in tumor promotion would be to measure the effect of specific inhibitors of enzymes in the biosynthetic pathway on tumor promotion. Furthermore, retinoids possess multiple actions that may be involved in their inhibitory effects as already noted in the section on retinoids. Weeks et al. (1982) and Takigawa et al. (1982) reported that α-difluoromethylornithine (α-DFMO), a specific and irreversible inhibitor of epidermal ODC and of elevated putrescine levels induced by TPA, was able to inhibit significantly skin tumor promotion by TPA. Verma et al. (1986) more recently demonstrated that small amounts of α-DFMO given in the drinking water could effectively inhibit epidermal ODC induction and skin tumor promotion by TPA. Methylglyoxal bis(butylamidinohydrazone) (MGBB), a reversible inhibitor of both ODC and *S*-adenosylmethionine decarboxylase inhibited TPA-induced ODC, accummulation of putresine and spermidine, and tumor promotion by TPA in mouse skin (Hibasami et al. 1988). Studies to date suggest that polyamine synthesis inhibitors inhibit primarily stage II of promotion in SENCAR mouse skin (reviewed in Weeks et al. 1984) and support the view that elevated ODC activity and polyamine levels are important for later stages of skin tumor promotion (Slaga et al. 1980a). In addition, DiGiovanni and coworkers (1988a) have shown that α-DFMO given as a drinking-water supplement was a very effective inhibitor of chrysarobin promotion of skin tumors in SENCAR mice.

α-DFMO has been examined for its inhibitory effects in a number of other tumorigenesis model systems in addition to the mouse skin system, some of which are summarized in Table 11. In general, all of these studies indicate that α-DFMO inhibits cell proliferation occurring during the post-initiation or promotional stages of chemical carcinogenesis. One study (Homma et al. 1985b) has demonstrated an inhibitory effect of α-DFMO on cell growth in vitro using a rat bladder carcinoma cell line exposed to a tumor-promoting rat urinary fraction.

VII. Miscellaneous Inhibitors of Tumor Promotion

The previous sections of this chapter on inhibitors of tumor promotion have discussed a range of compounds that have been extensively investigated. This section is concerned with inhibitors that have not been as comprehensively studied or are relatively new but that have the potential to represent significant inhibitors of tumor promotion. In general, little is known about their mechanism of action except for the inhibitors of PKC. Table 12 summarizes data on various miscellaneous inhibitors of tumor promotion.

The importance of cyclic nucleotides in tumor promotion has been suggested by several studies. BELMAN and TROLL (1978) reported that cyclic adenosine 5′-monophosphate (cAMP) was an effective inhibitor of skin tumor promotion by TPA. In addition, SLAGA et al. (1980a) found that the phosphodiesterase inhibitor isobutylmethylxanthine (IBMX), which increases the cellular level of

Table 12. Miscellaneous inhibitors of carcinogenesis and tumor promotion

Inhibitor	Carcinogen and/or tumor promoter	Species	Tissue	References
Phosphodiesterase inhibitors (e.g., IBMX)	TPA	Mouse	Skin	SLAGA et al. (1980a)
cAMP		Mouse	Skin	BELMAN and TROLL (1978)
Quercetin	TPA	Mouse	Skin	KATO et al. (1983)
Bryostatin	TPA	Mouse	Skin	HENNINGS et al. (1987)
Palmitoylcarnitine	TPA	Mouse	Skin	NAKADATE et al. (1986)
Glycyrrhetic acid	Teleocidin	Mouse	Skin	NISHINO et al. (1984)
Quinacrine	MNU	Rat	Mammary gland	MCCORMICK (1988)
D-Limonine	DMBA	Rat	Mammary gland	ELEGBEDE et al. (1984); ELSON et al. (1988)
Calcium glucarate	DMBA	Rat	Mammary gland	WALASZEK et al. (1986a)
Tamoxifen	MNU	Rat	Mammary gland	GOTTARDIS and JORDAN (1987)
BCG	Croton oil	Mouse	Skin	SCHINITSKY et al. (1973)
Poly I:C	MCA	Mouse	Skin	ELGJO and DEGRE (1975); GELBOIN and LEVY (1970)
DMSO	TPA	Mouse	Skin	SLAGA et al. (1980a)
Butyrate	TPA	Mouse	Skin	BELMAN and TROLL (1978)
Acetic acid	TPA	Mouse	Skin	SLAGA et al. (1980a)
Low caloric intake	Croton oil	Mouse	Skin	BOUTWELL (1964)
Low-fat diets	DMBA	Rat	Mammary gland	HOPKINS and CARROLL (1979)
	DMH	Rat	Colon	WYNDER et al. (1978)

TPA, 12-*O*-tetradecanoylphorbol-13-acetate; IBMX, isobutylmethylxanthine; cAMP, cyclic-5′-adenosine monophosphate; MNU, Methylnitrosourea; DMBA, 7,12-dimethylbenz[*a*]anthracene; MCA, 3-methylcholanthrene; DMH, dimethylhydrazine; BCG, bacillus Calmette-Guérin vaccine; Poly I:C, polyriboinosinic-polyribocytidylic acid; DMSO, dimethylsulfoxide.

cAMP by inhibiting phosphodiesterase, was an inhibitor of skin tumor promotion by TPA. A number of PKC inhibitors have been discovered. Several of these compounds have been shown to inhibit skin tumor promotion by TPA including: quercetin (Kato et al. 1983), bryostatin 1 (Hennings et al. 1987), and palmitoylcarnitine (Nakadate et al. 1986). However, with the exception of bryostatin 1, it is questionable whether inhibition of PKC is the major mechanism of antipromoting activity for these compounds. Quercetin is an effective inhibitor of lipoxygenase (Kato et al. 1983), and palmitoylcarnitine appears to affect other kinases such as the calmodulin-sensitive, Ca^{2+}-dependent protein kinases (Katoh et al. 1981).

Very little data are available concerning the mechanism of action of most of the remaining inhibitors of promotion listed in Table 12. Glycyrrhetic acid was shown to inhibit skin tumor promotion by teleocidin, possibly through its ability to inhibit promoter-stimulated phospholipid metabolism (Nishino et al. 1984). Quinacrine inhibited the post-initiation phase of MNU-induced mammary carcinogenesis in rats (McCormick 1988). Quinacrine is known to inhibit phospholipase A_2 (Blackwell et al. 1978; Vargaftig and Hai 1972), which may be related to its mechanism of antipromoting action. D-Limonine inhibited both the initiation and postinitiation phases of DMBA-induced rat mammary carcinogenesis (Elegbede et al. 1984; Elson et al. 1988). Dietary calcium glucurate effectively inhibited the post-initiation phase of DMBA-induced rat mammary carcinogenesis (Walaszek et al. 1986a). It was postulated that the lowering of endogenous levels of estradiol and precursors of 17-ketosteroids caused by calcium glucurate may be responsible for its antipromoting action. Tamoxifen, a potent antiestrogen, inhibited the post-initiation phase of MNU-induced mammary carcinogenesis in rats (Gottardis and Jordan 1987). Although tamoxifen has been shown capable of inhibiting PKC activity in vitro (O'Brian et al. 1985), it seems likely that is major effect is due to its antiestrogenic activity. Schinitsky et al. (1973) reported the inhibitory effect of BCG vaccination on skin tumor promotion, and this finding suggests that modulation of the immune system can inhibit tumor promotion. Polyriboinosinic-polyribocytidylic acid (Poly I : C) has an inhibitory effect on carcinogenesis and tumor promotion (Elgjo and Degre 1975; Gelboin and Levy 1970). This affect appears to be mediated by its inhibition of promoter- and carcinogen-induced cell proliferation (Elgjo and Degre 1975). Both butyrate (Belman and Troll 1978) and acetic acid (Slaga et al. 1980a) inhibit skin tumor promotion by TPA, perhaps by their cytotoxic effect on the skin. Dimethyl sulfoxide (DMSO) also inhibits TPA promotion in mouse skin (Slaga et al. 1980a). Because DMSO is a free-radical scavenger, it may inhibit tumor promotion by scavenging important radicals generated during promotion.

The effect of restricted dietary and caloric intake on the development of spontaneous and induced tumors was first reported by Tannenbaum (1940). Initiation-promotion studies in mouse skin suggest that dietary and caloric restriction were effective in the promotional stage of tumor formation (Boutwell 1964). The mechanism by which dietary and caloric restrictions inhibit tumor promotion may be related to the increase in endogenous glucocorticoids that follows such caloric restriction. As previously discussed, anti-inflammatory steroids (glucocorticoids) are potent inhibitors of skin tumor promotion. Several reports have indicated that high-fat diets act as tumor promoters in mammary gland and colon carcinogenesis (HOPKINS and CARROLL 1979; WYNDER et al. 1978). A high-fat diet increases prolactin and estradiol levels that may act as promoters in mammary carcinogenesis (WYNDER et al. 1978). Likewise, a high-fat diet causes bile acids to be produced in excessive amounts, and bile acids in turn can act as promoters in colon carcinogenesis (WYNDER et al. 1978). Low-fat diets, especially those containing polyunsaturated fats, decrease the level of tumors in both the mammary gland and the colon induced by chemical carcinogenesis (HOPKINS and CARROLL 1979; WYNDER et al. 1978).

E. Conclusions

The study of various inhibitors of chemical and radiation carcinogenesis has contributed greatly to our understanding of the important events in carcinogenesis. For inhibition of tumor initiation, the classes of agents discussed in this review include antioxidants, flavones, halogenated hydrocarbons, weakly carcinogenic or noncarcinogenic PAHs, antiinflammatory steroids, prostaglandin synthesis inhibitors, and a number of miscellaneous agents. These chemical agents appear to inhibit the tumor initiation phase of carcinogenesis by either having a direct or indirect effect on (1) the metabolism of the carcinogen, (2) the interaction of the carcinogen with critical macromolecules, or (3) the growth rate of the target cells. The effects of these inhibitors on the metabolism of the carcinogen appear to fall into three major categories: (a) alteration of the metabolism of the carcinogen (decreased activation and/or increased detoxication), (b) scavenging of active molecular species of carcinogens to prevent their reaching critical target sites in the cells, or (c) competitive antagonism. Some of the inhibiting agents appear to inhibit the initiation phase of carcinogenesis by their growth-suppressive effect on the target cells.

Inhibitors of tumor promotion possess diverse mechanisms, although there are some classes of compounds that inhibit both stages of chemical carcinogenesis. The antioxidants are a class of tumor initiation inhibitors that also have an inhibitory effect on the tumor promotion stage of carcinogenesis. They may scavenge important radicals generated during tumor promotion. Retinoids, antiinflammatory steroids, protease inhibitors, prostaglandin synthesis inhibitors, polyamine synthesis inhibitors, cyclic nucleotides, diet restriction, and low-fat diets appear to have a much greater effect on the promotional stage of chemical carcinogenesis than on the initiation stage. Studies on these inhibitors of tumor promotion suggest that cell proliferation, polyamines, prostaglandins, free radicals, primitive stem cells, and a number of other cellular factors are all important in tumor promotion.

In recent years, investigators have examined the potential utility of combinations of anticarcinogenic agents, especially those that presumably work by distinct mechanisms. In several cases, this approach appears to be quite effective at producing chemoprotective regimens against experimental carcinogenesis. In this regard, combinations of Se and vitamin E are more effective than Se alone in the chemoprevention of DMBA-induced mammary carcinogenesis (IP 1985). Vitamin E alone had no effect on DMBA-induced mammary carcinogenesis and therefore appeared to potentiate the actions of Se (Ip 1985). Combinations of vitamin E and Se were also much more effective than either agent alone for the inhibition of skin tumor promotion by TPA (Perchellet et al. 1987a). Although other examples of enhanced effectiveness of combined treatments on chemical carcinogenesis are emerging from the literature, these few examples indicate that such an approach may allow the reduction of dosage necessary for some types of inhibitors that possess toxicity and may ultimately make them more suitable for use in humans. Finally, a recent review of completed and current clinical chemoprevention trials, primarily with retinoids, suggests that some types of human cancer may be responsive to these agents (Bertram et al. 1987). The experimental data summarized in this chapter when considered in light of the preliminary clinical data available (Bertram et al. 1987) imply that the further study of inhibitors of carcinogenesis may ultimately lead to a rational approach to chemoprevention in humans.

Acknowledgements. I wish to thank Ms. Joyce Mayhugh for her excellent secretarial skills and patience in preparing this manuscript. Original, unpublished research was supported by United States Public Health Service grants CA 36979, CA 37111, and CA 38871.

References

Aizu E, Nakadate T, Yamamoto S, Kato R (1986) Inhibition of 12-*O*-tetradecanoylphorbol-13-acetate-mediated epidermal ornithine decarboxylase induction and skin tumor promotion by new lipoxygenase inhibitors lacking protein kinase C inhibitory effects. Carcinogenesis 7:1809–1812

Alworth WL, Slaga TJ (1985) Effects of ellipticine, flavone, and 7,8-benzoflavone upon 7,12-dimethylbenz(a)anthracene, 7,14-dimethylbenz(a,h)anthracene and dibenz(a,h)-anthracene initiated skin tumors in mice. Carcinogenesis 6:487–493

Anderson LM, Seetharam S (1985) Protection against tumorigenesis by 3-methylcholanthrene in mice by β-napthoflavone as a function of inducibility of methylcholanthrene metabolism. Cancer Res 45:6384–6389

Anderson MW, Boroujerdi M, Wilson AGE (1981) Inhibition in vivo of the formation of adducts between metabolites of benzo(a)pyrene and DNA by butylated hydroxyanisole. Cancer Res 41:4309–4315

Arcos JC, Conney AH, Buu-Hoi NP (1961) Induction of microsomal enzyme synthesis by polycyclic aromatic hydrocarbons of different molecular sizes. J Biol Chem 236:1291–1296

Baars AJ, Jansen M, Breimer DD (1978) The influence of phenobarbital, 3-methylcholanthrene, and 2,3,7,8-tetrachlorodibenzo-*p*-dioxin on glutathione-*S*-transferase activity of rat liver cytosol. Biochem Pharmacol 27:2487–2494

Bachrach U (1973) Function of the naturally occurring polyamines. Academic, New York

Bailey GS, Hendricks JD, Shelton DW, Nixon JE, Pawlowski NE (1987) Enhancement of carcinogenesis by the natural anticarcinogen indole-3-carbinol. JNCI 78:931–934

Baird WM, Diamond L (1976) Effect of 7,8-benzoflavone on the formation of benzo(a)-pyrene-DNA-bound products in hamster embryo cells. Chem Biol Interact 13:67–75

Baird WM, Diamond L (1978) Formation of 7,12-dimethylbenz(a)anthracene-DNA adducts in 7,8-benzoflavone-treated hamster embryo cells. Chem Biol Interact 20:181–190

Baird WM, Diamond L (1981) Effects of the cocarcinogen benzo(e)pyrene on metabolism of the carcinogen benzo(a)pyrene in hamster embryo cell cultures. Proc Am Assoc Cancer Res 22:96

Baird WM, Salmon CP, Diamond L (1984) Benzo(e)pyrene-induced alterations in the metabolic activation of benzo(a)pyrene and 7,12-dimethylbenz(a)anthracene by hamster embryo cells. Cancer Res 44:1445–1452

Balmain A, Ramsden M, Bowden GT, Smith J (1984) Activation of the mouse cellular Harvey-*ras* gene in chemically induced benign skin papillomas. Nature 307:658–660

Bates RR, Wortham JS, Counts W, Dingman CW, Gelboin HV (1968) Inhibition by actinomycin D of DNA synthesis and skin tumorigenesis induced by 7,12-dimethylbenz(a)anthracene. Cancer Res 28:27–34

Baserga R, Shubik P (1954) The action of cortisone on transplanted and induced tumors in mice. Cancer Res 14:12–18

Becker FF (1981) Inhibition of spontaneous hepatocarcinogenesis in C3H/HeN mice by Edipro A, an isolated soy protein. Carcinogenesis 2:1213–1214

Belman S, Troll W (1972) The inhibition of croton oil-promoted mouse skin tumorigenesis by steroid hormones. Cancer Res 32:450–454

Belman S, Troll W (1978) Hormones, cyclic nucleotides, and prostaglandins. In: Slaga TJ, Sivak A, Boutwell RK (eds) Carcinogenesis, vol 2. Raven, New York, pp 117–134

Benedict WF, Wheatley WL, Jones PA (1980) Inhibition of chemically induced morphological transformation and reversion of the transformed phenotype by ascorbic acid in C3H/10T1/2 cells. Cancer Res 40:2796–2801

Benson AM, Batzinger RP, Ou S-YL, Beuding E, Cha Y-N, Talalay P (1978) Elevation of hepatic glutathione *S*-transferase activities and protection against mutagenic metabolites of benzo(a)pyrene by dietary antioxidants. Cancer Res 38:4486–4495

Benson AM, Cha Y-N, Bueding E, Heine HS, Talalay P (1979) Elevation of extrahepatic glutathione *S*-transferase and epoxide hydratase activities by 2(3)-tert-butyl-4-hydroxyanisole. Cancer Res 39:2971–2977

Berry DL, Bracken WR, Slaga TJ, Wilson NM, Buty SG, Juchau MR (1977) Benzo(a)pyrene metabolism in mouse epidermis. Analysis by high pressure liquid chromatography and DNA binding. Chem Biol Interact 18:129–142

Berry DL, Slaga TJ, DiGiovanni J, Juchau MR (1979) Studies with chlorinated dibenzo-*p*-dioxins, polybrominated biphenyls, and polychlorinated biphenyls in a two-stage system of mouse skin tumorigenesis: potent anticarcinogenic effects. Ann NY Acad Sci 320:405–414

Bertram B, Tacchi AM, Pool BL, Wiessler M (1982) In vitro formation of methyldiethyldithiocarbamate after the reaction of nitrosoacetoxy-methylmethylamine or methylnitrosourea with disulfiram. Carcinogenesis 3:1361–1366

Bertram B, Frie E, Scherf HR, Schumacher J, Tacchi AM, Wiessler MJ (1985) Influence of a prolonged treatment with disulfiram and D(−)-penicillamine on nitrosodiethylamine induced biological and biochemical effects in rats. J Cancer Res Clin Oncol 109:9–15

Bertram JS, Kolonel LN, Meyskens FL Jr (1987) Rationale and strategies for chemoprevention of cancer in humans. Cancer Res 47:3012–3031

Birt DF, Lawson TA, Julius AD, Runice CE, Salmasi S (1982) Inhibition by dietary selenium on colon cancer induced in the rat by bis(2-oxopropyl)nitrosamine. Cancer Res 42:4455–4459

Blackwell GJ, Flower RJ, Nijkamp FP, Vane JR (1978) Phospholipase A2 activity of guinea pig isolated perfused lungs: stimulation and inhibition by anti-inflammatory steroids. Br J Pharmacol 62:79–80

Blumberg PM (1988) Protein kinase C as the receptor for the phorbol ester tumor promoters. Sixth Rhoads Memorial Award Lecture. Cancer Res 48:1–8

Bollag W (1972) Prophylaxis of chemically induced benign and malignant epithelial tumors by vitamin A acid (retinoic acid). Eur J Cancer 8:689–693

Borchert P, Wattenberg LW (1976) Inhibition of macromolecular binding of benzo-(a)pyrene and inhibition of neoplasia by disulfiram in the mouse forestomach. JNCI 57:173–179

Boutwell RK (1964) Some biological aspects of skin carcinogenesis. Prog Exp Tumor Res 4:207–250

Boutwell RK (1974) The function and mechanism of promoters of carcinogenesis. CRC Crit Rev Toxicol 2:419–443

Bowden GT, Slaga TJ, Shapas BG, Boutwell RK (1974) The role of aryl hydrocarbon hydroxylase in skin tumor-initiation by 7,12-dimethylbenz(a)anthracene and 1,2,5,6-

dibenzanthracene using DNA binding and thymidine-^{3}H incorporation into DNA as criteria. Cancer Res 34:2634–2642
Bresnick E, Meumer P, Lamden M (1979) Epidermal prostaglandins after topical application of a tumor promoter. Cancer Lett 7:121–125
Buening MK, Chang RL, Huang M-T, Fortner JG, Wood AW, Conney AH (1981) Activation and inhibition of benzo(a)pyrene and aflatoxin B1 metabolism in human liver microsomes by naturally occurring flavonoids. Cancer Res 41:67–72
Capel ID, Jenner M, Dorrell HM, Williams DC (1980) The influence of selenium on some hepatic carcinogen metabolizing enzymes of rats. IRCD Med Sci Libr Compend 8:282–283
Cha Y-N, Bueding E (1979) Effect of 2(3)-tert-butyl-4-hydroxyanisole administration on the activities of several hepatic microsomal and cytoplasmic enzymes in mice. Biochem Pharmacol 28:1917–1921
Chang Y, Bjeldanes LF (1987) R-Goitrin- and BHA-induced modulation of aflatoxin B_1 binding to DNA and biliary excretion of thiol conjugates in rats. Carcinogenesis 8:585–590
Chopra DP, Wilkoff LJ (1976) Inhibition and reversal by *β*-retinoic acid of hyperplasia induced in cultured mouse prostate tissue by 3-methylcholanthrene or *N*-methyl-*N'*-nitro-*N*-nitrosoguanidine. JNCI 56:583–589
Christou M, Marcus C, Jefcoate CR (1986) Selective interactions of cytochromes P-450 with the hydroxymethyl derivatives of 7,12-dimethylbenz[*a*]anthracene. Carcinogenesis 7:871–877
Chu WE, Malmgren RA (1965) An inhibitory effect of vitamin A on the induction of tumors of forestomach and cervix in the Syrian hamster by carcinogenic polycyclic hydrocarbons. Cancer Res 25:884–895
Chuang AHL, Mukhtar H, Bresnick E (1978) Effects of diethylmaleate on aryl hydrocarbon hydroxylase and 3-methylcholanthrene induced skin tumorigenesis in rats and mice. JNCI 60:321–325
Chung FL, Wang M, Carmella SG, Hecht SS (1986) Effects of butylated hydroxyanisole on the tumorigenicity and metabolism of *N*-nitrosodimethylamine and *N*-nitrosopyrrolidine in A/J mice. Cancer Res 46:165–168
Clapp NK, Bowles ND, Satterfield LC, Kilman WC (1979) Selective protective effect of butylated hydroxytoluene against 1,2-dimethylhydrazine carcinogenesis in BALB/c mice. JNCI 63:1081–1087
Cohen GM, Bracken WM, Iyer PR, Berry DL, Slaga TJ (1979) Anticarcinogenic effects of 2,3,7,8-tetrachlorodibenzo-*p*-dioxin on benzo(a)pyrene and 7,12-dimethylbenz(a)anthracene tumor-initiation and its relationship to DNA binding. Cancer Res 39:4027–4033
Cohen SM, Zenser RV, Murasaki G, Fukushima S, Mattammal MB, Rapp NS, Davis BB (1981) Aspirin inhibition of *N*-[4-(5-nitro-2-furyl)-2-thiazolyl]-formamide-induced lesions of the urinary bladder correlated with inhibition of metabolism by bladder prostaglandin endoperoxide synthetase. Cancer Res 41:3355–3359
Conney AH (1967) Pharmacological implications of microsomal enzyme induction. Pharmaol Rev 9:317–366
Conney AH (1972) Environmental factors influencing during metabolism. In: La Du BN, Mandel HG, Way EL (eds) Fundamentals of drug metabolism and drug disposition. Williams and Wilkins, Baltimore, pp 253–278
Coombs MM, Bhatt TS, Vose CW (1975) The relationship between metabolism, DNA binding and carcinogenicity of 15,16-dihydro-11-methylcyclopenta(a)phenanthrene-17-one in the presence of a microsomal enzyme inhibitor. Cancer Res 34:305–309
Coombs MW, Kissonerghis A-M, Allen JA (1976) An investigation into the binding of the carcinogen 15,16-dihydro-11-methylcyclopenta(a)phenanthrene-17-one to DNA in vitro. Cancer Res 36:4387–4393
Coombs MM, Bhatt TS, Young S (1979) The carcinogenicity of 15,16-dihydro-11-methylcyclopenta(a)phenanthrene-17-one. Br J Cancer 40:914–921
Coombs MM, Bhatt TS, Vose CW (1981) Microsomal metabolites of the aryl hydrocarbon hydroxylase and tumor inhibitor 7,8-benzoflavone. Carcinogenesis 2:135–140

Coon MJ, Vatsis KP (1978) Biochemical studies on chemical carcinogenesis: role of multiple forms of cytochrome P-450 in the metabolism of benzo(a)pyrene and other foreign compounds. In: Gelboin HV, Ts'o POP (eds) Polycyclic hydrocarbons and cancer: Environment, chemistry and metabolism, vol 1. Academic, New York, pp 336–360

Corasanti JG, Hobika GH, Markus G (1982) Interference with dimethylhydrazine induction of colon tumors in mice by ε-aminocaproic acid. Science 216:1020–1021

Crocker TT, Sanders LL (1970) Influence of vitamin A and 3,7-dimethyl-2,6-octadienal (citral) on the effect of benzo(a)pyrene on hamster trachea in organ culture. Cancer Res 30:1312–1318

Das M, Bickers DR, Mukhtar H (1985) Effect of ellagic acid on hepatic and pulmonary xenobiotic metabolism in mice: studies on the mechanism of its anticarcinogenic action. Carcinogenesis 6:1409–1413

Dashwood RH, Arbogast DN, Fong AT, Hendricks JD, Bailey GS (1988) Mechanisms of anti-carcinogenesis by indol-3-carbinol: detailed in vivo DNA binding dose-response studies after dietary administration with aflatoxin B_1. Carcinogenesis 9:427–432

Dawson MI, Chao WR, Helmes CT (1987) Inhibition by retinoids of anthralin-induced mouse epidermal ornithine decarboxylase activity and anthralin-promoted skin tumor formation. Cancer Res 47:6210–6215

Deitrich RA, Erwin VG (1970) Mechanism of the inhibition of aldehyde dehydrogenase in vivo by disulfiram and diethyldithiocarbamate. Mol Pharmacol 7:301–307

Del Tito BJ Jr, Mukhtar H, Bickers DR (1983) Inhibition of epidermal metabolism and DNA-binding of benzo(a)pyrene by ellagic acid. Biochem Biophys Res Commun 114:388–394

Deutsch J, Leutz JC, Yang SK, Gelboin HV, Chiang YL, Vatsis KP, Coon MJ (1978) Region and stereo selectivity of various forms of purified cytochrome P-450 in the metabolism of benzo(a)pyrene and (±)-*trans*-7,8-dihydroxy-7,8-dihydrobenzo(a)-pyrene as shown by product formation and binding to DNA. Proc Natl Acad Sci USA 76:3123–3127

Deutsch J, Vatsis KP, Coon MJ, Leutz JC, Gelboin HV (1979) Catalytic activity and stereo selectivity of purified forms of rat liver microsomal cytochrome P-450 in the oxygenation of the (−) and (+) enantiomers of *trans*-7,8-dihydroxy-7,8-dihydro-benzo(a)pyrene. Mol Pharmacol 16:1011–1018

DeYoung LM, Mufson RA, Boutwell RK (1977) An apparent inactivation of initiated cells by the potent inhibitor of two-stage mouse skin tumorigenesis, bis-(2-chloroethyl)sulfide. Cancer Res 37:4590–4594

Diamond L, McFall R, Miller J, Gelboin HV (1972) The effects of two isomeric benzoflavones on aryl hydrocarbon hydroxylase and the toxicity and carcinogenicity of polycyclic hydrocarbons. Cancer Res 32:731–736

Dickens MS, Custer RP, Sorof S (1979) Retinoid prevents mammary gland transformation by carcinogenic hydrocarbon in whole-organ culture. Proc Natl Acad Sci USA 76:5891–5895

DiGiovanni J (1978) PhD Thesis, University of Washington, Seattle

DiGiovanni J (1981) Effects of benzo(e)pyrene [B(e)P] and dibenz(a,c)anthracene [DB(a,c)A] on skin tumorigenesis by 7,12-dimethylbenz(a)anthracene and benzo(a)-pyrene. Proc Am Assoc Cancer Res 22:93

DiGiovanni J (1989) Metabolism of polycyclic aromatic hydrocarbons and phorbol esters by mouse skin: relevance to mechanism of action and *trans*-species/strain carcinogenesis: In: Slaga TJ, Klein-Szanto AJP, Boutwell RK, Stevenson D, Spitzer HL, D'Motto B (eds) Progress in clinical and biological research. Skin carcinogenesis: mechanisms and human relevance, vol 298. Liss, New York, pp 167–199

DiGiovanni J, Juchau MR (1980) Biotransformation and bioactivation of 7,12-dimethylbenz(a)anthracene. Drug Metab Rev 11:61–101

DiGiovanni J, Slaga TJ (1981 a) Modification of polycyclic aromatic hydrocarbon carcinogenesis. In: Gelboin HV, Ts'o POP (eds) Polycyclic hydrocarbons and cancer, vol 3. Academic, New York, pp 259–292

DiGiovanni J, Slaga TJ (1981 b) Effects of benzo(e)pyrene (B[e]P) and dibenz(a,c)anthracene (DB[a,c]A) on the skin tumor initiating activity of polycyclic aromatic hydrocarbons. In: Dennis AJ, Cooke WM (eds) Polycyclic aromatic hydrocarbons: chemistry and biological effects. Battelle Press, Columbus, pp 17–29

DiGiovanni J, Slaga TJ, Berry DL, Juchau MR (1977) Metabolism of 7,12-dimethylbenz(a)anthracene in mouse skin homogenates analyzed with high pressure liquid chromatography. Drug Metab Dispos 5:295–301

DiGiovanni J, Slaga TJ, Viaje A, Berry DL, Harvey RG, Juchau MR (1978) Effects of 7,8-benzoflavone on skin tumor-initiating activities of various 7- and 12-substituted derivatives of 7,12-dimethylbenz(a)anthracene in mice. JNCI 61:135–140

DiGiovanni J, Berry DL, Juchau MR, Slaga TJ (1979a) 2,3,7,8-Tetrachlorodibenzo-*p*-dioxin: potent anticarcinogenic activity in CD-1 mice. Biochem Biophys Res Commun 86:577–584

DiGiovanni J, Berry DL, Slaga TJ, Juchau MR (1979b) Studies on the relationships between induction of biotransformation and tumor initiating activity of 7,12-dimethylbenz(a)anthracene in mouse skin. In: Jones PW, Leber P (eds) Polycyclic aromatic hydrocarbons. Ann Arbor Science, Ann Arbor, pp 553–568

DiGiovanni J, Berry DL, Gleason GL, Kishore GS, Slaga TJ (1980a) Time-dependent inhibition by 2,3,7,8-tetrachlorodibenzo-*p*-dioxin of skin tumorigenesis with polycyclic hydrocarbons. Cancer Res 40:1580–1587

DiGiovanni J, Kishore GS, Slaga TJ, Boutwell RK (1980b) 2,3,7,8-Tetrachlorodibenzo-*p*-dioxin (TCDD)-induced alterations in oxidative and non-oxidate biotransformation of PAH in mouse skin: role in anticarcinogenesis by TCDD. In: Bjorseth A, Dennis AJ (eds) Polycyclic aromatic hydrocarbons: chemistry and biological effects. Battelle Press, Columbus, pp 935–954

DiGiovanni J, Slaga TJ, Berry DL, Juchau MR (1980c) Inhibitory effects of environmental chemicals on polycyclic aromatic hydrocarbon carcinogenesis. In: Slaga TJ (ed) Modifiers of chemical carcinogenesis, vol 5. Raven, New York, pp 145–168

DiGiovanni J, Rymer J, Slaga TJ, Boutwell RK (1982) Anticarcinogenic and cocarcinogenic effects of benzo(e)pyrene and dibenz(a,c)anthracene on skin tumor initiation by polycyclic hydrocarbons. Carcinogenesis 3:371–375

DiGiovanni J, Kruszewski FH, Chenicek KJ (1987) Studies on the skin tumor promoting actions of chrysarobin. In: Butterworth B, Slaga TJ (eds) Nongenotoxic mechanisms of carcinogenesis. Cold Spring Harbor Laboratory, New York, pp 25–37 (Banbury Report 25)

DiGiovanni J, Kruszewski FH, Chenicek KJ (1988a) Modulation of chrysarobin skin tumor promotion. Carcinogenesis 9:1445–1450

DiGiovanni J, Kruszewski FH, Coombs MM, Bhatt TS, Pezeshk A (1988b) Structure-activity relationships for epidermal ornithine decarboxylase induction and skin tumor promotion by anthrones. Carcinogenesis 9:1437–1443

Dipple A, Moschel RC, Bigger CAH (1984) Polynuclear aromatic hydrocarbons. In: Searle CE (ed) Chemical carcinogenesis, vol 1. American Chemical Society, Washington DC, pp 41–163 (ACS Monograph 182)

Dixit R, Teel RW, Daniel FB, Stoner GD (1985) Inhibition of benzo(a)pyrene-*trans*-7,8-diol metabolism and DNA binding in mouse lung explants by ellagic acid. Cancer Res 45:2951–2956

Dixit R, Teel RW, Babcock MS, Kim K, Stoner GD (1986) Ellagic acid: an inhibitor of DNA binding of benzo(a)pyrene in cultured mouse and human lung tissues. In: Cooke M, Dennis AJ (eds) Polycyclic aromatic hydrocarbons: chemistry, characterization and carcinogenesis. Batelle Press, Columbus, pp 239–250

Dock L, Cha Y-N, Jernstrom B, Moldeus P (1982) Differential effects of dietary BHA on hepatic enzyme activities and benzo(a)pyrene metabolism in male and female NMRI mice. Carcinogenesis 3:15–19

Dock L, Martinez M, Jernstrom B (1984) Induction of hepatic glutathione *S*-transferase activity by butylated hydroxyanisole and conjugation of benzo(a)pyrene diol-epoxide. Carcinogenesis 5:841–844

DuBois KP, Raymund AB, Hietbrink BE (1961) Inhibitory action of dithiocarbamates on enzymes of animal tissues. Toxicol Appl Pharmacol 3:236–255

Egner PA, Kensler TW (1985) Effects of a biomimetic superoxide dismutase on complete and multistage carcinogenesis in mouse skin. Carcinogenesis 5: 1167–1172

El-Bayoumy K (1985) Effects of organoselenium compounds on induction of mouse forestomach tumors by benzo(a)pyrene. Cancer Res 45:3631–3635

Elegbede JA, Elson CE, Qureshi A, Tanner MA, Gould MN (1984) Inhibition of DMBA-induced mammary cancer by the monoterpene D-limonene. Carcinogenesis 5:661–664

Elgjo K, Degre M (1975) Polyinosinic-polycytidylic acid: inhibition of cell proliferation in carcinogen-treated epidermis and in carcinogen-induced skin tumors in mice. JNCI 54:219–221

Eling T, Curtis J, Battista J, Marnett LJ (1986) Oxidation of (+)-7,8-dihydroxy-7,8-dihydrobenzo(a)pyrene by mouse keratinocytes: evidence for peroxyl radical- and monooxygenase-dependent metabolism. Carcinogenesis 8:1957–1963

Elson CE, Maltzman TH, Boston JL, Tanner MA, Gould MN (1988) Anticarcinogenic activity of D-limonene during the initiation and promotion/progression stages of DMBA-induced rat mammary carcinogenesis. Carcinogenesis 9:331–332

Farber E, Solt D (1978) A new liver model for the study of promotion. In: Slaga TJ, Sivak A, Boutwell RK (eds) Carcinogenesis, vol 2. Mechanisms of tumor promotion and cocarcinogenesis. Raven, New York, pp 443–451

Feuer G, Kellen JA (1974) Inhibition and enhancement of mammary tumorigenesis by 7,12-dimethylbenz(a)anthracene in the female Sprague-Dawley rat. Int J Clin Pharmacol 9:62–69

Fiala ES, Bobotas G, Kulakis C, Wattenberg LW, Weisburger JH (1977) Effects of disulfiram and related compounds on the metabolism in vivo of the colon carcinogen, 1,2-dimethylhydrazine. Biochem Pharmacol 26:1763–1768

Fischer SM, Slaga TJ (1982) Inhibition of mouse skin tumor promotion by several inhibitors of arachidonic acid metabolism. Carcinogenesis 3:1243–1245

Fischer SM, Gleason GL, Bohrman JS, Slaga TJ (1979) Prostaglandin enhancement of skin tumor initiation and promotion. In: Samuelsson B, Ramwell RW, Paoletti R (eds) Advances in prostaglandin and thromboxane research, vol 6. Raven, New York, pp 517–522

Fischer SM, Gleason GL, Bohrman JS, Slaga TJ (1980a) Prostaglandin modulation of phorbol ester skin tumor promotion. Carcinogenesis 1:245–248

Fischer SM, Gleason GL, Mills GD, Slaga TJ (1980b) Indomethacin enhancement of TPA tumor promotion in mice. Cancer Lett 10:343–350

Fischer SM, Klein-Szanto AJP, Adams LM, Slaga TJ (1985) The first stage and complete promoting activity of retinoic acid but not the analog RO-10-9359. Carcinogenesis 6:515–518

Forbes PD, Urbach F, Davies RE (1979) Enhancement of experimental photocarcinogenesis by topical retinoic acid. Cancer Lett 7:85–90

Fowler BA, Hook GER, Lucier GW (1977) Tetrachlorodibenzo-*p*-dioxin induction of renal microsomal enzyme systems: ultrastructural effects on pars recta (S_3) proximal tubule cells of the rat kidney. J Pharmacol Exp Ther 203:712–721

Frank N, Hadjiolov D, Wiessler M (1984) Influence of disulfiram on the metabolism of *N*-nitrosodiethylamine. In: O'Neill IK, Von Borstel RC, Miller CT, Long J, Bartsch H (eds) *N*-Nitroso compounds: occurrence and biological relevance to human cancer. IARC, Lyon, pp 519–523 (Scientific publication no. 57)

Fredovich I (1976) Oxygen radicals, hydrogen peroxide, and oxygen toxicity. In: Pryor WA (ed) Free radicals in biology, vol 1. Academic, New York, pp 239–277

Frei E, Bertram B, Wiessler M (1985) Reduced glutathione inhibits the alkylation by *N*-nitrosodimethylamine of liver DNA in vivo and microsomal fraction in vitro. Chem Biol Interact 55:123–127

Freudenthal R, Jones PW (eds) (1976) Polycyclic aromatic hydrocarbons: chemistry, metabolism and carcinogenesis, vol 1. Raven, New York

Freundt KJ (1978) Variable inhibition of human hepatic drug metabolizing enzymes by disulfiram. Int J Clin Pharmacol 16:323–330

Fujiki H, Horiuchi T, Yamashita K, Hakii H, Suganuma M, Nishino H, Iwashima A, Hirata Y, Sugimura T (1986) Inhibition of tumor promotion by flavonoids. In: Codu

V, Middleton E, Harborne JB (eds) Plant flavonoids in biology and medicine: biochemical, pharmacological, and structure-activity relationships. Liss, New York, pp 429–440

Fukushima S, Totematsu M, Takahashi M (1984) Combined effect of various surfactants on gastric carcinogenesis in rats treated with *N*-methyl-*N'*-nitro-*N*-nitrosoguanidine. Gann 65:371–376

Fukushima S, Sakata T, Tagawa Y, Shibata M-A, Hirose M, Ito N (1987) Different modifying response of butylated hydroxyanisole, butylated hydroxytoluene, and other antioxidants in *N,N*-dibutylnitrosamine esophagus and forestomach carcinogenesis in rats. Cancer Res 47:2113–2116

Furstenburger G, Marks F (1978) Indomethacin inhibition of cell proliferation induced by the phorbol ester TPA is reversed by prostaglandin E_2 in mouse epidermis in vivo. Biochem Biophys Res Commun 84:1103–1111

Furstenberger G, Berry DL, Sorg B, Marks F (1981) Skin tumor promotion by phorbol esters is a two-stage process. Proc Natl Acad Sci USA 78:7722–7726

Gelboin HV, Levy HB (1970) Polyinosinic-polycytidylic acid inhibits chemically induced tumorigenesis in mouse skin. Science 167:205–206

Gelboin HV, Ts'o POP (eds) (1978) Polycyclic hydrocarbons and cancer, vols 1–3. Academic, New York

Gelboin HV, Wiebel FJ, Diamond L (1970) DMBA tumorigenesis and aryl hydrocarbon hydroxylase in mouse skin: inhibition by 7,8-benzoflavone. Science 170:169–171

Gelboin HV, Wiebel FJ, Kinoshita N (1972) Microsomal aryl hydrocarbon hydroxylase: on its role in polycyclic hydrocarbon carcinogenesis and toxicity and the mechanism of enzyme induction. Biochem Soc Symp 34:103–133

Gensler HI, Bowden GT (1984) Influence of 13-*cis*-retinoic acid on mouse skin tumor initiation and promotion. Cancer Lett 22:71–75

Gensler HI, Sim DA, Bowden GT (1986) Influence of the duration of topical 13-*cis*-retinoic acid treatment on inhibition of mouse skin tumor promotion. Cancer Res 46:2767–2770

Ghadially RN, Green HN (1954) The effect of cortisone on chemical carcinogenesis in the mouse skin. Br J Cancer 8:291–295

Giguere V, Ong ES, Segui P, Evans RM (1987) Identification of a receptor for the morphogen retinoic acid. Nature 330:624–629

Goeger DE, Shelton DW, Hendricks JD, Bailey GS (1986) Mechanisms of anticarcinogenesis by indole-3-carbinol: effect on the distribution and metabolism of aflatoxin B_1 in rainbow trout. Carcinogenesis 7:2025–2031

Goldstein JA, Hickman P, Bergman H, McKinney JD, Walker MP (1977) Separation of pure polychlorinated biphenyl isomers into two types of inducers on the basis of induction of cytochrome P-450 or P-448. Chem Biol Interact 17:69–87

Goldstein BD, Witz G, Amoruso M, Stone DS, Troll W (1979) Protease inhibitors antagonize the activation of polymorphonuclear leukocyte oxygen consumption. Biochem Biophys Res Commun 88:854–860

Goldstein BD, Witz G, Amoruso M, Stone DS, Troll W (1981) Stimulation of human polymorphonuclear leukocyte superoxide anion radical by tumor promoters. Cancer Lett 22:257–269

Gottardis MM, Jordan VC (1987) Antitumor actions of keoxifene and tamoxifen in the *N*-nitrosomethylurea-induced rat mammary carcinoma model. Cancer Res 47:4020–4024

Graham S, Dayai H, Swanson M, Mittleman A, Wilkinson G (1978) Diet in the epidemiology of cancer of the colon and rectum. JNCI 61:709–714

Grantham PH, Weisburger JH, Weisburger EK (1973) Effect of the antioxidant butylated hydroxytoluene (BHT) on the metabolism of the carcinogens *N*-2-fluorenylacetamide and *N*-hydroxy-*N*-2-fluorenylacetamide. Food Cosmet Toxicol 11:209–217

Griffin AC, Jacobs MM (1977) Effects of selenium on azo dye hepatocarcinogenesis. Cancer Lett 3:177–181

Grubbs CJ, Moon RC, Sporn MB, Newton DL (1977a) Inhibition of mammary cancer by retinyl methyl ether. Cancer Res 37:599–602

Grubbs CJ, Moon RC, Squire RA, Farrow GM, Stinson SG, Goodman DG, Brown CC, Sporn MB (1977b) 13-*cis*-Retinoic acid inhibition of bladder carcinogenesis induced in rats by *N*-butyl-*N*-(4-hydroxybutyl)nitrosamine. Science 198:743–744

Gurtoo HL, Koser PL, Bansal SK, Fox HW, Sharma SD, Mulhern AI, Pavelic ZP (1985) Inhibition of aflatoxin B1-hepatocarcinogenesis in rats by β-napthoflavone. Carcinogenesis 6:675–678

Haber SL, Wissler RW (1962) Affect of vitamin E on carcinogenicity of methylcholanthrene. Proc Soc Exp Biol Med 111:774–775

Hakadate T, Yamamoto S, Iseki H, Sonoda S, Takemura S, Ura A, Hosoda Y, Kato R (1982) Inhibition of 12-*O*-tetradecanoylphorbol-13-acetate-induced tumor promotion by nordihydroguaiaretic acid, a lipoxygenase inhibitor, and *p*-bromophenacyl bromide, a phospholipase A_2 inhibitor. Gann 73:841–843

Harborne JB (1976) Comparative biochemistry of the flavonoids. Academic, New York

Harisiadis L, Miller RC, Hall EJ, Borek C (1978) A vitamin A analogue inhibits radiation-induced oncogenic transformation. Nature 274:486–487

Harr JR, Exon JH, Whanger PD, Weswig PH (1972) Effect of dietary selenium on *N*-2-fluorenylacetamide (FAA)-induced cancer in vitamin E supplemented, selenium depleted rats. Clin Toxicol 5:187–194

Hecker E (1968) Cocarcinogenic principles from the seed oil of *Croton tiglium* and from other Euphorbiaceae. Cancer Res 28:2338–2349

Hecker E (1971) Isolation and characterization of the cocarcinogenic principles from croton oil. In: Busch H (ed) Methods in cancer research, vol 6. Academic, New York, pp 439–484

Heidelberger C (1975) Chemical carcinogenesis. Ann Rev Biochem 44:79–121

Heidelberger C, Mondal S, Peterson AR (1978) Initiation and promotion in cell cultures. In: Slaga TJ, Sivak A, Boutwell RK (eds) Carcinogenesis, vol 2. Mechanisms of tumor promotion and cocarcinogenesis. Raven, New York, pp 197–202

Hendricks JD, Putnam TP, Bills DD, Sinnhuber RO (1977) Inhibitory effect of a polychlorinated biphenyl (Aroclor 1254) on aflatoxin B_1 carcinogenesis in rainbow trout. JNCI 59:1545–1551

Hennings H, Boutwell RK (1970) Studies on the mechanism of skin tumor promotion. Cancer Res 30:312–320

Hennings H, Smith HC, Colburn NC, Boutwell RK (1968) Inhibition by actinomycin D of DNA and RNA synthesis and of skin carcinogenesis initiated by 7,12-dimethylbenz(a)anthracene and β-propiolactone. Cancer Res 28:543–552

Hennings H, Wenk ML, Donahoe R (1982) Retinoic acid promotion of papilloma formation in mouse skin. Cancer Lett 16:1–5

Hennings H, Blumberg PM, Pettit GR, Herald CL, Shores R, Yuspa SH (1987) Bryostatin 1, an activator of protein kinase C, inhibits tumor promotion by phorbol esters in SENCAR mouse skin. Carcinogenesis 8:1343–1346

Hibasami H, Tsukada T, Maekawa S, Sakurai M, Nakashima K (1988) Inhibition of mouse skin tumor promotion and of promoter-induced epidermal polyamine biosynthesis by methylglyoxal bis(butylamidinohydrazone). Carcinogenesis 9:199–202

Hicks RM, Chowaniec J, Wakefield J St J (1978) Experimental induction of bladder tumors by a two-stage system. In: Slaga TJ, Sivak A, Boutwell RK (eds) Carcinogenesis, vol 2. Mechanisms of tumor promotion and cocarcinogenesis. Raven, New York, pp 475–489

Hirose M, Masuda A, Inoue T, Fukushima S, Ito N (1986) Modification by antioxidants and *p,p'*-diaminodiphenylmethane of 7,12-dimethylbenz(a)anthracene-induced carcinogenesis of the mammary gland and ear duct in CD rats. Carcinogenesis 7:1155–1159

Holder G, Yagi H, Dansette P, Jerina DM, Levin W, Lu AYH, Conney AH (1974) Effects of inducers and epoxide hydrase on the metabolism of benzo(a)pyrene by liver microsomes and a reconstituted system: analysis by high pressure liquid chromatography. Proc Natl Acad Sci USA 71:4356–4360

Homma Y, Ozono S, Numata I, Seidenfeld J, Oyasu R (1985a) Inhibition of carcinogenesis by α-difluoromethylornithine in heterotopically transplanted rat urinary bladders. Cancer Res 45:648–652

Homma Y, Ozono S, Numata I, Seidenfeld J, Oyasu R (1985b) α-Difluoromethylornithine inhibits cell growth stimulated by a tumor-promoting rat urinary fraction. Carcinogenesis 6:159–161

Honjo T, Netter KJ (1969) Inhibition of drug demethylation by disulfiram in vivo and in vitro. Biochem Pharmacol 18:2681–2683

Hopkins GL, Carroll KK (1979) Relationship between amount and type of dietary fat in promotion of mammary carcinogenesis induced by 7,12-dimethylbenz[*a*]anthracene. JNCI 62:1009–1012

Hozumi M, Ogawa M, Sugimura T, Takeuchi T, Umezawa H (1972) Inhibition of tumorigenesis in mouse skin by leupeptin, a protease inhibitor from actinomycetes. Cancer Res 32:1725–1728

Huang M-T, Bhatt RL, Conney AH (1980) Studies on the mechanism of activation of microsomal benzo(a)pyrene hydroxylase activity by 7,8-benzoflavone (BF). Fed Proc 39:2053

Huggins C, Moon RE, Morii S (1962) Extinction of experimental mammary cancer. I. Estradiol-17β and progesterone. Proc Natl Acad Sci USA 48:379–386

Huh TY, McCarter JA (1960) Phenanthrene as an anti-initiating agent. Br J Cancer 14:591–595

Ip C (1981a) Factors influencing the anticarcinogenic efficacy of selenium in dimethylbenz(a)anthracene-induced mammary tumorigenesis in rats. Cancer Res 41:2683–2686

Ip C (1981b) Prophylaxis of mammary neoplasia by selenium supplementation in the initiation and promotion phases of chemical carcinogenesis. Cancer Res 41:4386–4390

Ip C (1985) Selenium inhibition of chemical carcinogenesis. Fed Proc 44:2573–2578

Ip C, Daniel FB (1985) Effects of selenium on 7,12-dimethylbenz(a)anthracene-induced mammary carcinogenesis and DNA adduct formation. Cancer Res 45:61–65

Irving CC, Tice AJ, Murphy WM (1979) Inhibition of *N*-*n*-butyl-*N*-(4-hydroxybutyl) nitrosamine-induced urinary bladder cancer in rats by administration of disulfiram in the diet. Cancer Res 39:3040–3043

Ito N, Nagasaki H, Arai M, Makiura S, Sugihara S, Hirao K (1973) Histopathologic studies on liver tumorigenesis induced in mice by technical polychlorinated biphenyls and its promoting effect on liver tumors induced by benzene hexachloride. JNCI 51:1637–1646

Jacobs MM, Jansson B, Griffin AC (1977) Inhibitory effects of selenium on 1,2-dimethylhydrazine and methyl azoxymethanol acetate induction of colon tumors. Cancer Lett 2:133–138

Jaffe BM (1974) Prostaglandins and cancer: An update. Prostaglandins 6:453–461

Kahl R, Netter KJ (1977) Ethyoxyquin as an inducer and inhibitor of phenobarbital-type cytochrome P-450 in rat liver microsomes. Toxicol Appl Pharmacol 40:473–483

Kallistratos G, Fasske E (1980) Inhibition of benzo(a)pyrene carcinogenesis in rats with vitamin E. J Cancer Res Clin Oncol 98:91–96

Kapitulnik J, Poppers PJ, Buening MK, Fortner JG, Conney AH (1977) Activation of monooxygenases in human liver by 7,8-benzoflavone. Clin Pharmacol Ther 22:475–485

Kapitulnik J, Wislocki PG, Levin W, Yagi H, Jerina DM, Conney AH (1978) Tumorigenicity studies with diol-epoxides of benzo(a)pyrene which indicate that ($\pm$)-*trans*-7β-8α-dihydroxy-9α-10α-epoxy-7,8,9,10-tetrahydrobenzo(a)pyrene is an ultimate carcinogen in newborn mice. Cancer Res 38:354–358

Kato R, Nakadate T, Yamamoto S, Sugimura T (1983) Inhibition of 12-*O*-tetradecanoylphorbol-13-acetate-induced tumor promotion and ornithine decarboxylase activity by quercetin: possible involvement of lipoxygenase inhibition. Carcinogenesis 4:1301–1305

Katoh N, Wrenn RW, Wise BC, Shoji M, Kuo JF (1981) Substrate proteins for calmodulin-sensitive and phospholipid-sensitive Ca^{2+}-dependent protein kinase in heart, and inhibition of their phosphorylation by palmitoylcarnitine. Proc Natl Acad Sci USA 78:4813–4817

Kennedy AR (1985) The conditions for the modification of radiation transformation in vitro by a tumor promoter and protease inhibitors. Carcinogenesis 6:1441–1445

Kennedy AR, Little JB (1978) Protease inhibitors suppress radiation induced malignant transformation in vitro. Nature 276:825–826

Kensler TW, Jaffe BG (1986) Free radicals in tumor promotion. Adv Free Biol Med 2: 347–387

Kensler TW, Bush DM, Kozumbo WJ (1983) Inhibition of tumor promotion by a biomimetic superoxide dismutase. Science 221:75–77
Kensler TW, Egner PA, Trush MA, Bueding E, Groopman JD (1985) Modification of aflatoxin B1-binding to DNA in vivo in rats fed phenolic antioxidants, ethyoxyquin, and a dithiolthione. Carcinogenesis 6:759–763
Kensler TW, Egner PA, Davidson NE, Roebuck BD, Pilul A, Groopman JD (1986) Modulation of aflatoxin metabolism, aflatoxin-N^7-guanine formation and hepatic tumorigenesis in rats fed ethoxyquin: role of induction of glutathione *S*-transferase. Cancer Res 46:3924–3931
Kensler TW, Egner PA, Dolan PM, Groopman JD, Roebuck BD (1987) Mechanism of protection against aflatoxin tumorigenicity in rats fed *S*-(2-pyrazinyl)-4-methyl-1,2-dithiol-3-thione (Oltipraz R) and related 1,2-dithiol-3-thiones and 1,2-dithiol-3 ones. Cancer Res 47:4271–4277
Kimura NT, Kanematsu T, Baba T (1976) Polychlorinated biphenyl(s) as a promoter in experimental hepatocarcinogenesis in rats. Z Krebsforsch 87:257–266
Kinoshita N, Gelboin HV (1972a) Aryl hydrocarbon hydroxylase and polycyclic hydrocarbon tumorigenesis: effect of the enzyme inhibitor 7,8-benzoflavone on tumorigenesis and macromolecular binding. Proc Natl Acad Sci USA 69:824–828
Kinoshita N, Gelboin HV (1972b) The role of aryl hydrocarbon hydroxylase in DMBA skin tumorigenesis: on the mechanism of 7,8-benzoflavone inhibition of tumorigenesis. Cancer Res 32:1329–1339
Kitagawa T, Hino O, Nomura K, Sugano H (1984) Dose-response studies on promoting and anticarcinogenic effects of phenobarbital and DDT in rat hepatocarcinogenesis. Carcinogenesis 5:1653–1656
Kouri RE (1976) Relationship between levels of aryl hydrocarbon hydroxylase activity and susceptibility to 3-methylcholanthrene and benzo(a)pyrene-induced cancers in inbred strains of mice. In: Freudenthal R, Jones PW (eds) Polycyclic aromatic hydrocarbons: chemistry, metabolism and carcinogenesis, vol 1. Raven, New York, pp 139–151
Kouri RE, Rude TH, Joglekar R, Dansette PM, Jerina DM, Atlas SA, Owens IS, Nebert DW (1977) 2,3,7,8-Tetrachlorodibenzo-*p*-dioxin as a cocarcinogen causing 3-methylcholanthrene-initiated subcutaneous tumors in mice genetically "nonresponsive" at Ah locus. Cancer Res 38:2777–2783
Kozumbo WJ, Seed JL, Kensler TW (1983) Inhibition by 2(3)-*tert*-butyl-4-hydroxyanisole and other antioxidants of epidermal ornithine decarboxylase activity induced by 12-*O*-tetradecanoylphorbol-13-acetate. Cancer Res 43:2555–2559
Kuroki T, Drevon C (1979) Inhibition of chemical transformation in C3H/10T1/2 cells by protease inhibitors. Cancer Res 39:2755–2761
Lam LKT, Wattenberg LW (1977) Effects of butylated hydroxyanisole on the metabolism of benzo(a)pyrene by mouse liver microsomes. JNCI 58:413–417
Lam LKT, Pai RP, Wattenberg LW (1979) Synthesis and chemical carcinogen inhibitory activity of 2-tert-butyl-4-hydroxyanisole. J Med Chem 22:569–571
Lam LKT, Fladmoe AV, Hochalter JB, Wattenberg LW (1980) Short time interval effects of butylated hydroxyanisole on the metabolism of benzo(a)pyrene. Cancer Res 40:2824–2828
Lambooy JP (1976) Influence of riboflavin antagonists on azo dye hepatoma induction in the rat. Proc Soc Exp Biol Med 153:532–535
Lane HW, Medina D (1985) Mode of action of selenium inhibition of 7,12-dimethylbenz(a)anthracene-induced mouse mammary tumorigenesis. JNCI 75:675–679
Lasnitzki I (1976) Reversal of methylcholanthrene-induced changes in mouse prostates in vitro by retinoic acid and its analogs. Br J Cancer 34:239–248
LeBoeuf RA, Laishes BA, Hoekstra WG (1985a) Effects of dietary selenium concentration on the development of enzyme-altered liver foci and hepatocellular carcinoma induced by diethylnitrosamine or *N*-acetylaminofluorene in rats. Cancer Res 45:5489–5495
LeBoeuf RA, Laishes BA, Hoekstra WG (1985b) Effects of selenium on cell proliferation in rat liver and mammalian cells as indicated by cytokinetic and biochemical analysis. Cancer Res 45:5496–5504

Lesca P (1981) Influence on the rate of 7,12-dimethylbenz(a)anthracene metabolic activation, in vivo, on its binding to epidermal DNA and skin carcinogenesis. Carcinogenesis 2:199–204

Lesca P (1983) Protective effects of ellagic acid and other plant phenols on benzo(a)pyrene-induced neoplasia in rats. Carcinogenesis 4:1651–1653

Levij IS, Polliack A (1970) Inhibition of chemical carcinogenesis in the hamster cheek pouch by topical chlorpromazine. Nature 228:1096–1097

Lotlikar PD, Clearfield MS, Jhee EC (1984) Effect of butylated hydroxyanisole in vivo and in vitro on hepatic aflatoxin B_1-DNA binding in rats. Cancer Lett 24:241–250

Lu AYH, West SB (1978) Reconstituted mammalian mixed-function oxidases: requirements, specificities and other properties. Pharmacol Ther 2:337–358

Lu AYH, West SB (1980) Multiplicity of mammalian microsomal cytochromes P-450. Pharmacol Rev 31:277–295

Lucier GW, McDaniel OS, Hook GER (1975) Nature of the enhancement of hepatic uridine diphosphate glucuronyltransferase activity by 2,3,7,8-tetra-chlorodibenzo-*p*-dioxin in rats. Biochem Pharmacol 24:325–334

Lupulescu A (1978) Enhancement of carcinogenesis by prostaglandins in male albino Swiss mice. JNCI 61:97–101

Makiura S, Aoe H, Sugihara S, Hirao K, Arai M, Ito N (1974) Inhibitory effect of polychlorinated biphenyl on liver tumorigenesis in rats treated with 3′-methyl-4-dimethylaminoazobenzene, *N*-2-fluorenylacetamide and diethylnitrosamine. JNCI 53:1253–1257

Mandel HG, Manson MM, Judah DJ, Simpson JL, Green JA, Forrester LM, Wolf CR, Neal GE (1987) Metabolic basis for the protective effect of the antioxidant ethoxyquin on aflatoxin B_1 hepatocarcinogenesis in the rat. Cancer Res 47:5218–5223

Manson MM, Green JA, Driver HE (1987) Different modifying response of butylated hydroxyanisole, butylated hydroxytoluene, and other antioxidants in *N,N*-dibutylnitrosamine esophagus and forestomach carcinogenesis of rats. Carcinogenesis 8:723–728

Marnett LJ (1984) Hydroperoxide-dependent oxidations during prostaglandin biosynthesis. In: Pryor WA (ed) Free radicals in biology, vol 6. Academic, Orlando, pp 63–94

Marnett LJ (1987) Peroxy free radicals: potential mediators of initiation and promotion. Carcinogenesis 8:1365–1373

Marnett LJ, Reed GA, Dennison DJ (1978) Prostaglandin synthetase dependent activation of 7,8-dihydro-7,8-dihydroxybenzo(a)pyrene to mutagenic derivatives. Biochem Biophys Res Commun 82:210–216

Marquardt H, Sapozink MD, Zedeck MS (1974) Inhibition by cysteamine-HCL of oncogenesis induced by 7,12-dimethylbenz(a)anthracene without affecting toxicity. Cancer Res 34:3387–3390

Marshall MV, Arnot MS, Jacobs MM, Griffin AC (1979) Selenium effects on the carcinogenicity and metabolism of 2-acetylaminofluorene. Cancer Lett 7:331–338

Matsushima T, Kaziko T, Kawachi T, Hara K, Sugimura T, Takeuchi T, Umezawa H (1975) Effects of protease inhibitors of microbial origin on experimental carcinogenesis. In: Magee PN (ed) Fundamentals in cancer prevention. University of Tokyo Press, Tokyo, pp 57–69

McCord A, Burnett AK, Wolf CR, Morrison V, Craft JA (1988) Role of specific cytochrome P-450 isoenzymes in the regio-selective metabolism of 7,12-dimethylbenz[a]anthracene in microsomes from rats treated with phenobarbital or Sudan III. Carcinogenesis 9:1485–1491

McCormick DL (1988) Anticarcinogenic activity of quinacrine in the rat mammary gland. Carcinogenesis 9:175–178

McCormick DL, Madigan MJ, Moon RC (1985) Modulation of rat mammary carcinogenesis by indomethacin. Cancer Res 45:1803–1808

McLean AEM, Marshall A (1971) Reduced carcinogenic effects of aflatoxin in rats given phenobarbitone. Br J Exp Pathol 52:322–329

Medina D, Lane HW (1983) Stage specificity of selenium-mediated inhibition of mouse mammary tumorigenesis. Biol Trace Elem Res 5:297–306

Medina D, Osborn CJ (1984) Selenium inhibition of DNA synthesis in mouse mammary epithelial cell line YN-4. Cancer Res 44:4361–4365

Meechan RJ, McCafferty DE, Jones RS (1953) 3-Methylcholanthrene as an inhibitor of hepatic cancer induced by 3′-methyl-4-dimethylaminoazobenzene in the diet of the rat: a determination of the time relationships. Cancer Res 13:802–806

Melikian AA, Leszczynshka JM, Hecht SS, Hoffman D (1986) Effects of the cocarcinogen catechol on benzo(a)pyrene metabolism and DNA adduct formation in mouse skin. Carcinogenesis 7:9–15

Merriman RL, Bertram JS (1979) Reversible inhibition by retinoids of 3-methylcholanthrene-induced neoplastic transformation in C3H/10T1/2 clone 8 cells. Cancer Res 39:1661–1666

Miller EC (1978) Some current perspectives on chemical carcinogenesis in humans and experimental animals: presidential address. Cancer Res 38:1479–1496

Miller EC, Miller JA, Brown RR, MacDonald JC (1958) On the protective action of certain polycyclic aromatic hydrocarbons against carcinogenesis by aminoazo dyes and 2-acetylaminofluorene. Cancer Res 18:469–477

Milner JA, Pigott MA, Dipple A (1985) Selective effects of sodium selenite on 7,12-dimethylbenz(a)anthracene-DNA binding in fetal mouse cell cultures. Cancer Res 45: 6347–6354

Miranda CL, Carpenter HM, Cheeke PR, Buhler DR (1981) Effect of ethoxyquin on the toxicity of the pyrrolizidine alkaloid monocrotaline and on hepatic drug metabolism in mice. Chem Biol Interact 37:95–107

Mottram JC (1944) A developing factor in experimental blastogenesis. J Pathol Bacteriol 56:181–187

Mukhtar H, Das M, Del Tito BJ Jr, Bickers DR (1984a) Protection against 3-methylcholanthrene-induced skin tumorigenesis in BALB/c mice by ellagic acid. Biochem Biophys Res Commun 119:751–757

Mukhtar H, Del Tito BJ Jr, Marcelo CL, Das M, Bickers DR (1984b) Ellagic acid: a potent naturally occurring inhibitor of benzo(a)pyrene metabolism and its subsequent glucuronidation, sulfation and covalent binding to DNA in cultured BALB/c mouse keratinocytes. Carcinogenesis 5:1565–1571

Mukhtar H, Das M, Bickers DR (1986) Inhibition of methylcholanthrene-induced skin tumorigenesis in BALB/c mice by chronic oral feeding of trace amounts of ellagic acid in drinking water. Cancer Res 46:2245–2265

Murasaki G, Zenser TV, Davis BB, Cohen SM (1984) Inhibition by aspirin of *N*-[4-(5-nitro-2-furyl)-2-thiazolyl]formamide-induced bladder carcinogenesis and enhancement of forestomach carcinogenesis. Carcinogenesis 5:53–55

Nakadate T, Yamamoto S, Sugimura T (1983) Inhibition of 12-*O*-tetradecanoylphorbol-13-acetate-induced tumor promotion and ornithine decarboxylase activity by quercetin: possible involvement of lipoxygenase inhibition. Carcinogenesis 4:1301–1305

Nakadate T, Yamamoto S, Aizu E, Kato R (1986) Inhibition of 12-*O*-tetradecanoylphorbol-13-acetate-induced tumor promotion and epidermal ornithine decarboxylase activity in mouse skin by palmitoylcarnitine. Cancer Res 46:1589–1593

Nebert DW, Robinson JR, Niwa A, Kimaki K, Poland AP (1975) Genetic expression of aryl hydrocarbon hydroxylase activity in the mouse. J Cell Physiol 85:393–414

Nesnow S, Bergman H, Sovocool W (1980) Metabolism of 7,8-benzoflavone. Proc Am Assoc Cancer Res 21:63

Netteshein P, Williams M (1976) The influence of vitamin A on the susceptibility of the rat lung to 3-methylcholanthrene. Int J Cancer 17:351–357

Newberne P (1976) Environmental modifiers of susceptibility to carcinogenesis. Cancer Detect Prev 1:129–173

Newberne P, Suphakarn V (1977) Preventive role of vitamin A in colon carcinogenesis in rats. Cancer 40:2553–2556

Nishino H, Kitagawa K, Iwashima A (1984) Antitumor-promoting activity of glycyrrhetic acid in mouse skin tumor formation induced by 7,12-dimethylbenz[a]anthracene plus teleocidin. Carcinogenesis 5:1529–1530

Nishizumi M (1976) Enhancement of diethylnitrosamine hepatocarcinogenesis in rats by exposure to polychlorinated biphenyls or phenobarbital. Cancer Lett 2:11–16
Nixon JE, Hendricks JD, Pawlowski NE, Pereira CB, Sinnhuber RO, Bailey GS (1984) Inhibition of aflatoxin B1 carcinogenesis in rainbow trout by flavone and indole compounds. Carcinogenesis 5:615–619
Nyce JW, Magee PN, Hard GC, Schwartz AG (1984) Inhibition of 1,2-dimethylhydrazine-induced colon tumorigenesis in BALB/c mice by dehydroepiandrosterone. Carcinogenesis 5:57–62
O'Brian CA, Liskamp RM, Solomon DH, Weinstein IB (1985) Inhibition of protein kinase C by tamoxifen. Cancer Res 45:2462–2465
Ohkoshi M, Fujii S (1983) Effect of the synthetic protease inhibitor [*N,N*-dimethylcarbamoyl-methyl-4-(4-guanidinobenzoyloxy-phenylacetate]methanesulfate on carcinogenesis by 3-methylcholanthrene in mouse skin. JNCI 71:1053–1057
Okey AB (1972) Dimethylbenzanthracene-induced mammary tumors in rats: inhibition by DDT. Life Sci 11:833–843
Oredipe OA, Barth RF, Hanausek-Walaszek M, Sautins I, Walaszek Z, Webb TE (1987) Effects of calcium glucarate on the promotion of diethylnitrosamine-initiated altered hepatic foci in rats. Proc Am Assoc Cancer Res 28:156
Owens IS (1977) Genetic regulation of UDP-glucuronosyltransferase induction by polycyclic aromatic compounds in mice. J Biol Chem 252:2827–2833
Pamukcu AM, Halciner S, Bryan GT (1977) Inhibition of carcinogenic effect of bracken fern (*Pteridium aquilinum*) by various chemicals. Cancer 40:2450–2454
Panthananickal A, Marnett LJ (1981) Arachadonic acid-dependent metabolism of (±)-7,8-dihydroxy-7,8-dihydrobenzo(a)pyrene to polyguanylic acid-binding derivatives. Chem Biol Interact 33:239–252
Parke DV, Rahim A, Walker R (1974) Reversibility of hepatic changes caused by ethoxyquin. Biochem Pharmacol 23:1871–1876
Pashko LL, Rovito RJ, Williams JR, Sobel EL, Schwartz AG (1984) Dehydroepiandrosterone (DHEA) and 3β-methylandrost-5-en-17-one: inhibitors of 7,2-dimethylbenzo(a)anthracene (DMBA)-initiated and 12-*O*-tetradecanoylphorbol-13-acetate (TPA)-promoted skin papilloma formation in mice. Carcinogenesis 5:463–466
Pegg AE (1988) Polyamine metabolism and its importance in neoplastic growth and as a target for chemotherapy. Cancer Res 48:759–774
Pence BC, Buddingh F, Yang SP (1986) Multiple dietary factors in the enhancement of dimethylhydrazine carcinogenesis: main effect of indole-3-carbinol. JNCI 77:269–276
Peraino C, Fry RJM, Staffeldt E (1971) Reduction and enhancement by phenobarbital of hepatocarcinogenesis induced in the rat by 2-acetylaminofluorene. Cancer Res 31:1506–1512
Peraino C, Fry RJM, Grube DD (1978) Drug-induced enhancement of hepatic tumorigenesis. In: Slaga TJ, Sivak A, Boutwell RK (eds) Carcinogenesis, vol 2. Mechanisms of tumor promotion and cocarcinogenesis. Raven, New York, pp 421–432
Perchellet JP, Perchellet EM, Orten DK, Schneider BA (1985 a) Inhibition of the effects of 12-*O*-tetradecanoylphorbol-13-acetate on mouse epidermal glutathione peroxidase and ornithine decarboxylase activities by glutathione level-raising agents and selenium-containing compounds. Cancer Lett 26:283–293
Perchellet JP, Owen MD, Posey TD, Orten DK, Schneider BA (1985 b) Inhibitory effects of glutathione level-raising agents and D-α-tocopherol on ornithine decarboxylase induction and mouse skin tumor promotion by 12-*O*-tetradecanoylphorbol-13-acetate. Carcinogenesis 6:567–573
Perchellet JP, Perchellet EM, Orten DK, Schneider BA (1986) Decreased ratio of reduced/oxidized glutathione in mouse epidermal cells treated with tumor promoters. Carcinogenesis 7:503–506
Perchellet JP, Abney NL, Thomas RM, Guislain YL, Perchellet EM (1987 a) Effects of combined treatments with selenium, glutathione, and vitamin E on glutathione peroxidase activity, ornithine decarboxylase induction, and complete and multistage carcinogenesis in mouse skin. Cancer Res 47:477–485

Perchellet JP, Abney NL, Thomas RM, Perchellet EM, Maatta EA (1987b) Inhibition of multistage tumor promotion in mouse skin by diethyldithiocarbamate. Cancer Res 47:6302–6309

Petkovich M, Brand NJ, Krust A, Chambon P (1987) A human retinoic acid receptor which belongs to the family of nuclear receptors. Nature 330:444–450

Phillips DH, Sims P (1979) Polycyclic aromatic hydrocarbon metabolites: their reactions with nucleic acids. In: Grover PL (ed) Chemical carcinogens and DNA, vol 2. Boca Raton, pp 29–57

Phillips JC, Lake BG, Gangolli SC, Grasso P, Lloyd AG (1977) Effects of pyrazole and 3-amino-1,2,4-triazole on the metabolism and toxicity of dimethylnitrosamine in the rat. JNCI 58:629–633

Pitot HC, Barsness L, Kitagawa T (1978) Stages in the process of hepatocarcinogenesis in rat liver. In: Slaga TJ, Sivak A, Boutwell RK (eds) Carcinogenesis, vol 2. Mechanisms of tumor promotion and cocarcinogenesis. Raven, New York, pp 433–442

Pitot HC, Goldsworthy T, Campbell HA, Poland AP (1980) Quantitative evaluation of the promotion by 2,3,7,8-tetrachlorodibenzo-*p*-dioxin of hepatocarcinogenesis. Cancer Res 40:3616–3620

Poland A, Glover E (1974) Comparison of 2,3,7,8-tetrachlorodibenzo-*p*-dioxin, a potent inducer of aryl hydrocarbon hydroxylase, with 3-methylcholanthrene. Mol Pharmacol 10:349–359

Poland A, Glover E (1977) Chlorinated biphenyl induction of aryl hydrocarbon hydroxylase activity: a study of the structure-activity relationship. Mol Pharmacol 13:924–938

Poland A, Kende A (1976) 2,3,7,8-Tetrachlorodiobenzo-*p*-dioxin: environmental containment and molecular probe. Fed Proc 35:2404–2411

Preston BD, Van Miller JP, Moore RW, Allen JR (1981) Promoting effects of polychlorinated biphenyls (Arodor 1254) and polychlorinated dibenzofuran-free Aroclor 1254 on diethylnitrosamine-induced tumorigenesis in the rat. JNCI 66:509–515

Rasmussen RE, Wang IY (1974) Dependence of specific metabolism of benzo(a)-pyrene on the inducer of hydroxylase activity. Cancer Res 34:2290–2295

Reddy BS, Weisburger JH, Wynder EL (1978) Colon cancer: bile salts as tumor promoters. In: Slaga TJ, Sivak A, Boutwell RK (eds) Carcinogenesis, vol 2. Mechanisms of tumor promotion and cocarcinogenesis. Raven, New York, pp 453–464

Remmer H (1972) Induction of drug metabolizing enzyme system in the liver. Eur J Clin Pharmacol 5:166–166

Rice JE, Hosted TJ, Lavoie EJ (1984) Fluoranthene and pyrene enhance benzo(a)pyrene-DNA adduct formation in vivo in mouse skin. Cancer Lett 24:327–333

Richardson HL, Stier AR, Borsos-Nachtnebel E (1952) Liver tumor inhibition and adrenal histologic responses in rats to which 3′-methyl-4-dimethylaminoazobenzene and 20-methylcholanthrene were simultaneously administered. Cancer Res 12:356–361

Robinson JR, Considine N, Nebert DW (1974) Genetic expression of aryl hydrocarbon hydroxylase induction. J Biol Chem 249:5851–5859

Rogers AE, Herndon BJ, Newberne PM (1973) Induction by dimethylhydrazine of intestinal carcinoma in normal rats fed high and low levels of vitamin A. Cancer Res 33: 1003–1009

Sachdev GP, Wen G, Martin B, Kishore GS, Fox OF (1980) Effects of dietary fat and alpha-tocopherol on gamma-glutamyltranspeptidase activity of 7,12-dimethylbenz-(alpha)anthracene-induced mammary gland adenocarcinomas. Cancer Biochem Biophys 5:15–23

Saffiotti U, Montesano R, Sellakumar AR, Borg SA (1967) Experimental cancer of the lung. Inhibition by vitamin A of the induction of tracheobronchial metaplasia and squamous cell tumors. Cancer 20:857–869

Sakata T, Hasegawa R, Johansson SL, Zenser TV, Cohen SM (1986) Inhibition by aspirin of *N*-[4-(5-nitro-2-furyl)-2-thiazolyl]formamide initiation and sodium saccharin promotion of urinary bladder carcinogenesis in male F344 rats. Cancer Res 46: 3903–3906

Sato K, Kitahara Y, Yin Z, Waragai F, Nishimura K, Hatayama I, Ebina T, Yamazaki T, Tsuda H, Ito N (1984) Induction by butylated hydroxyanisole of specific molecular forms of glutathione *S*-transferase and UDP-glucuronyl-transferase and inhibition of development of γ-glutamyl transpeptidase-positive foci in rat liver. Carcinogenesis 5: 473–477

Sawicki JT, Dipple A (1983) Effects of butylated hydroxyanisole and butylated hydroxytoluene on 7,12-dimethylbenz(a)anthracene-DNA adduct formation in mouse embryo cell cultures. Cancer Lett 20:165–171

Sayer JM, Yagi H, Wood AW, Conney AH, Jerina DM (1983) Extremely facile reaction between the ultimate carcinogen benzo(a)pyrene-7,8-diol-9,10-epoxide and ellagic acid. J Am Chem Soc 104:5562–5564

Schinitsky MR, Hyman LR, Bazlovec AA, Burkholder PM (1973) Bacillus Calmette-Guerin vaccination and skin tumor promotion with croton oil in mice. Cancer Res 33:659–663

Schmahl D, Kruger FW (1972) Influence of disulfiram (tetraethyl thiuramidsulfide) on the biological actions of *N*-nitrosamines. In: Nakahara W, Takayama S, Sugimura T, Odashima S (eds) Topics in chemical carcinogenesis. University of Tokyo Press, Tokyo, pp 199–210

Schmahl D, Kruger FW, Habs M, Diehl B (1976) Influence of disulfiram on the organotrophy of the carcinogenic effect of dimethylnitrosamine and diethylnitrosamine in rats. Z Krebsforsch 85:271–276

Schurr A, Bo BT, Schooler JC (1978) The effects of disulfiram on rat liver mitochondrial monoamine oxidase. Life Sci 22:1979–1984

Schwartz AG, Tannen RH (1981) Inhibition of 7,12-dimethylbenz(a)anthracene- and urethan-induced lung tumor formation in A15 mice by long-term treatment with dehydroepiandrosterone. Carcinogenesis 2:1335–1337

Schwarz JA, Viaje A, Slaga TJ, Yuspa SH, Hennings H, Lichti U (1977) Fluocinolone acetonide: a potent inhibitor of mouse skin tumor promotion and epidermal DNA synthesis. Chem Biol Interact 17:331–347

Scribner JD, Slaga TJ (1973) Multiple effects of dexamethasone on protein synthesis and hyperplasia caused by a tumor promoter. Cancer Res 33:542–546

Selkirk J (1985) Analogous patterns of benzo(a)pyrene metabolism in human and rodent cells. In: Huberman E, Barr SH (eds) Carcinogenesis, vol 10. The role of chemicals and radiation in the etiology of cancer. Raven, New York, pp 123–133

Selkirk JK, Croy RG, Roller PP, Gelboin HV (1974) High pressure liquid chromatographic analysis of benzo(a)pyrene metabolism and covalent binding and the mechanism of action of 7,8-benzoflavone and 1,2-epoxy-3,3,3-trichloropropane. Cancer Res 34:3474–3480

Shamberger RJ (1970) Relationship of selenium to cancer. I. Inhibitory effect of selenium on carcinogenesis. JNCI 44:931–936

Shamberger RJ (1971) Inhibitory effect of vitamin A on carcinogenesis. JNCI 47:667–673

Shamberger RJ (1972) Increase of peroxidation in carcinogenesis. JNCI 48:1491–1497

Shamberger RJ, Rudolph G (1966) Protection against cocarcinogenesis by antioxidants. Experienta 22:116

Shamberger RJ, Willis CE (1971) Selenium distribution and human cancer mortality. CRC Crit Rev Clin Lab Sci 2:211–221

Shamberger RJ, Tytko SA, Willis CE (1976) Antioxidants and cancer. Part IV. Selenium and age-adjusted human cancer mortality. Arch Environ Health 31:231–235

Shelton DW, Goeger DE, Hendricks JD, Bailey GS (1986) Mechanisms of anticarcinogenesis: the distribution and metabolism of aflatoxin B_1 in rainbow trout fed Aroclor 1254. Carcinogenesis 7:1065–1071

Silinkas KC, Okey AB (1975) Protection by 1,1,1-trichloro-2,2-bis-(*p*-chlorophenyl)-ethane against mammary tumors and leukemia during prolonged feeding of 7,12-dimethylbenz(a)anthracene to female rats. JNCI 55:653

Sims P (1980) The metabolic activation of chemical carcinogens. Br Med Bull 36:11–18

Sladek NE, Mannering GJ (1966) Evidence for a new P-450 hemoprotein in hepatic microsomes from methylcholanthrene treated rats. Biochem Biophys Res Comm 23:668–674

Slaga TJ (ed) (1980) Carcinogenesis, vol 5. Modifiers of chemical carcinogenesis, an approach to the biochemical mechanism and cancer prevention. Raven, New York

Slaga TJ (ed) (1984a) Mechanisms of tumor promotion, vol 1. Tumor promotion in internal organs. CRC, Boca Raton

Slaga TJ (1984b) Multistage skin tumor promotion and specificity of inhibition. In: Slaga TJ (ed) Mechanisms of tumor promotion: tumor promotion and skin carcinogenesis, vol II. CRC, Boca Raton, pp 189–196

Slaga TJ, Boutwell RK (1977) Inhibition of the tumor-initiating ability of the potent carcinogen 7,12-dimethylbenz(a)anthracene by the weak tumor-initiator, 1,2,3,4-dibenzanthracene. Cancer Res 37:128–133

Slaga TJ, Bracken WM (1977) The effects of antioxidants on skin tumor initiation and aryl hydrocarbon hydroxylase. Cancer Res 37:1631–1635

Slaga TJ, DiGiovanni J (1984) Inhibition of chemical carcinogenesis. In: Searle CE (ed) American chemical society monograph 182, vol 2. American Chemical Society, Washington DC, pp 1279–1321

Slaga TJ, Scribner JD (1973) Inhibition of tumor initiation and promotion by antiinflammatory agents. JNCI 51:1723–1725

Slaga TJ, Berry DL, Juchau MR, Thompson S, Buty SG, Viaje A (1976) Effects of benzoflavones and trichloropropene oxide on polynuclear aromatic hydrocarbon metabolism and initiation of skin tumors. In: Freudenthal R, Jones PW (eds) Polynuclear aromatic hydrocarbons: chemistry, metabolism and carcinogenesis, vol 1. Raven, New York, pp 127–137

Slaga TJ, Thompson S, Berry DL, DiGiovanni J, Juchau MR, Viaje A (1977) The effects of benzoflavones on polycyclic hydrocarbon metabolism and skin tumor-initiation. Chem Biol Interact 17:297–312

Slaga TJ, Bracken WM, Viaje A, Berry DL, Fischer SM, Miller DR (1978a) Lack of involvement of 6-hydroxymethylation in benzo(a)pyrene skin tumor-initiation in mice. JNCI 61:451–455

Slaga TJ, Viaje A, Buty SG, Bracken WM (1978b) Dibenz(a,c)anthracene: a potent inhibitor of skin tumor-initiation by 7,12-dimethylbenz(a)anthracene. Res Comm Chem Pathol Pharmacol 19:477–483

Slaga TJ, Jecker L, Bracken WM, Weeks CE (1979) The effects of weak or non-carcinogenic polycyclic hydrocarbons on 7,12-dimethylbenz(a)anthracene and benzo(a)pyrene skin tumor-initiation. Cancer Lett 7:51–59

Slaga TJ, Fischer SM, Weeks CE, Klein-Szanto AJP (1980a). Cellular and biochemical mechanisms of mouse skin tumor promoters. In: Hodgson E, Bend JR, Philpot RM (eds) Reviews in biochemical toxicology, vol 3. Elsevier/North-Holland, New York, pp 231–281

Slaga TJ, Klein-Szanto AJP, Fischer SM, Weeks CE, Nelson K, Major S (1980b) Studies on the mechanism of action of anti-tumor-promoting agents: their specificity in two-stage promotion. Proc Natl Acad Sci USA 77:2251–2254

Slaga TJ, Klein-Szanto AJP, Triplett LL, Yotti LP, Trosko JE (1981) Skin tumor promoting activity of benzoyl peroxide, a widely used free radical generating compound. Science 213:1023–1025

Slaga TJ, Fischer SM, Weeks CE, Nelson K, Mamrack M, Klein-Szanto AJP (1982) Specificity and mechanisms of promoter inhibitors in multistage promotion. In: Hecker E, Kunz W, Fusnig NE, Marks F, Theilman HW (eds) Carcinogenesis – a comprehensive survey, cocarcinogenesis and biological effects of tumor promoters, vol 7. Raven, New York, pp 19–34

Slaga TJ, Solanki V, Logani M (1983) Studies on the mechanism of action of antitumor promoting agents: suggestive evidence for the involvement of free radicals in promotion. In: Nygaard OF, Simic MG (eds) Radioprotectors and anticarcinogens. Academic, New York, pp 471–485

Smart RC, Huang M-T, Chang RL, Sayer JM, Jerina DM, Wood AW, Conney AH (1986) Effect of ellagic acid and 3-*O*-decylellagic acid on the formation on benzo(a)-pyrene-derived DNA adducts in vivo and on the tumorigenicity of 3-methylcholanthrene in mice. Carcinogenesis 7:1669–1675

Smart RC, Huang MT, Han ZT, Kaplan MC, Focella A, Conney AH (1987) Inhibition of 12-*O*-tetradecanoylphorbol-13-acetate induction of ornithine decarboxylase activity, DNA synthesis, and tumor promotion in mouse skin by ascorbic acid and ascorbyl palmitate. Cancer Res 47:6633–6638

Smolarek TA, Baird WM (1984) Benzo(e)pyrene-induced alterations in the binding of benzo(a)pyrene to DNA in hamster embryo cell cultures. Carcinogenesis 5:1065–1069

Smolarek TA, Baird WM (1986) Benzo(e)pyrene-induced alterations in the stereoselectivity of activation of 7,12-dimethylbenz(a)anthracene to DNA-binding metabolites in hamster embryo cell cultures. Cancer Res 46:1170–1175

Smolarek TA, Baird WM, Fisher EP, DiGiovanni J (1987) Benzo(e)pyrene-induced alterations in the binding of benzo(a)pyrene and 7,12-dimethylbenz(a)anthracene to DNA in SENCAR mouse epidermis. Cancer Res 47:3701–3706

Solanki V, Rana RS, Slaga TJ (1981) Diminution of mouse epidermal superoxide dismutase and catalase activities by tumor promoters. Carcinogenesis 2:1141–1146

Solanki V, Yotti L, Logani MK, Slaga TJ (1984) The reduction of tumor initiating activity and cell mediated mutagenicity of dimethylbenz(a)anthracene by a copper coordination compound. Carcinogenesis 5:129–131

Sparnins VL, Venegas PL, Wattenberg LW (1982) Glutathione *S*-transferase activity: enhancement by compounds inhibiting chemical carcinogenesis and by dietary constituents. JNCI 68:493–496

Sparnins VL, Mott AW, Barany G, Wattenberg LW (1986) Effects of allylmethyl trisulfide on gluthatione *S*-transferase activity and BP-induced neoplasia in the mouse. Nutr Cancer 8:211–215

Sparnins VL, Barany G, Wattenberg LW (1988) Effects of organosulfur compounds from garlic and onions on benzo(a)pyrene-induced neoplasia and glutathione *S*-transferase activity in the mouse. Carcinogenesis 9:131–134

Speier JL, Wattenberg LW (1975) Alterations in microsomal metabolism of benzo(a)pyrene in mice fed butylated hydroxyanisole. JNCI 55:469–472

Speier JL, Lam LKY, Wattenberg LW (1978) Effects of administration to mice of butylated hydroxyanisole by oral intubation on benzo(a)pyrene-induced pulmonary adenoma formation and metabolism of benzo(a)pyrene. JNCI 60:605–609

Sporn MB (1980) Retinoids and cancer prevention. Carcinogenesis 5:99–109

Sporn MB, Dunlop NM, Newton DL, Smith JM (1976) Prevention of chemical carcinogenesis by vitamin A and its synthetic analogs (retinoids). Fed Proc 35:1332–1338

Sporn MB, Squire RA, Brown CC, Smith JM, Wenk ML, Springer S (1977) 13-*cis*-Retinoic inhibition of bladder carcinogenesis in the rat. Science 195:487–489

Stripp B, Greene FE, Gillette JR (1969) Disulfiram impairment of drug metabolism by rat liver microsomes. J Pharmacol Exp Ther 170:347–354

Sydor W Jr, Chou MW, Yang SK, Yang CS (1983) Regioselective inhibition of benzo(a)pyrene metabolism by butylated hydroxyanisole. Carcinogenesis 4:131–136

Takigawa M, Verma SK, Simsiman RC, Boutwell RK (1982) Polyamine biosynthesis and skin tumor promotion. Inhibition of 12-*O*-tetradecanoylphorbol-13-acetate-promoted mouse skin tumor formation by the irreversible inhibitor of ornithine decarboxylase α-difluoromethylornithine. Biochem Biophys Res Commun 105:969–976

Tannenbaum A (1940) Relationship of body weight to cancer incidence. Arch Pathol 30:509–517

Teel RW, Dixit R, Stoner GD (1985) The effect of ellagic acid on the uptake, persistence, metabolism and DNA-binding of benzo(a)pyrene in cultured explants of strain A/J mouse lung. Carcinogenesis 6:391–395

Teel RW, Stoner GD, Babcock MS, Dixit R, Kim K (1986) Benzo(a)pyrene metabolism and DNA-binding in cultured explants of human bronchus and in monolayer cultures of human bronchial epithelial cells treated with ellagic acid. Cancer Detect Prev 9: 59–66

Terracini B, Shubik P, Della Porta G (1960) A study of skin carcinogenesis in the mouse with single applications of 9,10-dimethyl-1,2-benzathracene at different dosages. Cancer Res 20:1538–1542

Thakker DR, Yagi H, Akagi H, Koreeda M, Lu AYH, Levin W, Wood AW, Conney AH, Jerina DM (1977) Metabolism of benzo(a)pyrene. VI. Stereoselective metabolism of benzo(a)pyrene and benzo(a)pyrene 7,8-dihydrodiol to diol epoxide. Chem Biol Interact 16:281–300

Thakker DR, Levin W, Buening M, Yagi H, Lehr RE, Wood AW, Conney AH, Jerina DM (1981) Species-specific enhancement by 7,8-benzoflavone of hepatic microsomal metabolism of benzo(a)pyrene 9,10-dihydrodiol to bay-region diol epoxides. Cancer Res 41:1389–1396

Thomas FB, Thomas DM, Shewach DS, Furlong NG (1979) Substrate specificity of aryl hydrocarbon (benzo[a]pyrene) hydroxylase. In: Jones PW, Leber P (eds) Polynuclear aromatic hydrocarbons. Ann Arbor Science, Ann Arbor, pp 669–683

Thompson S, Slaga TJ (1976) The effects of dexamethasone on mouse skin tumor-initiation and aryl hydrocarbon hydroxylase. Eur J Cancer 12:363–370

Thompson HJ, Meeker LD, Becci PJ, Kokoska S (1982) Effect of short-term feeding of sodium selenite on 7,12-dimethylbenz(a)anthracene-induced mammary carcinogenesis in the rat. Cancer Res 42:4954–4958

Thompson HJ, Herbst EJ, Meeker LD, Minocha R, Ronan AM, Fite R (1984) Effect of D,L-α-difluoromethylornithine on murine mammary carcinogenesis. Carcinogenesis 5: 1649–1651

Thompson HJ, Meeker LD, Herbst EJ, Ronan AM, Minocha R (1985) Effect of concentration of D,L-2-difluoromethylornithine on murine mammary carcinogenesis. Cancer Res 45:1170–1173

Thompson HJ, Ronan AM, Ritacco KA, Meeker LD (1986) Effect of tamoxifen and D,L-α-difluoromethylornithine on the growth, ornithine decarboxylase activity and polyamine content of mammary carcinomas induced by 1-methyl-1-nitrosourea. Carcinogenesis 7:837–840

Tottmar O, Marchner H (1976) Disulfiram as a tool in the studies on the metabolism of acetaldehyde in rats. Acta Pharmacol Toxicol 38:366–375

Triano EA, Simpson JB, Kartky M, Lang WR, Triolo AJ (1985) Protective effects of trifluralin on benzo(a)pyrene-induced tumors in A/J mice. Cancer Res 45:601–607

Troll W, Wiesner R (1985) The role of oxygen radicals as a possible mechanism of tumor promotion. Ann Rev Pharmacol Toxicol 25:509–528

Troll W, Klassen A, Janoff A (1970) Tumorigenesis in mouse skin: inhibition by synthetic inhibitors of proteases. Science 169:1211–1213

Troll W, Meyn MS, Rossman TG (1978) Mechanisms of protease action in carcinogenesis. In: Slaga TJ, Sivak S, Boutwell RD (eds) Carcinogenesis, vol 2. Raven, New York, pp 301–312

Troll W, Belman S, Wiesner R, Shellabarger CJ (1979) Protease action in carcinogenesis. In: Holzer H, Tschesche H (eds) Biological function of proteinases. Springer, Berlin Heidelberg New York, pp 165–170

Troll W, Wiesner R, Shellabarger CJ, Holtzman S, Stone JP (1980) Soybean diet lowers breast tumor incidence in irradiated rats. Carcinogenesis 1:469–472

Troll W, Wiesner R, Frenkel K (1987) Anticarcinogenic action of protease inhibitors. Adv Cancer Res 49:265–283

Tsuda H, Sakata T, Masui T, Imaida K, Ito N (1984) Modifying effects of butylated hydroxyanisole, ethoxyquin and acetominophen on induction of neoplastic lesions in rat liver and kidney initiated by *N*-ethyl-*N*-hydroxyethylnitrosamine. Carcinogenesis 5:525–531

Turusov V, Day N, Andrianov L, Jain D (1971) Influence of dose on skin tumors induced in mice by single application of 7,12-dimethylbenz(a)anthracene. JNCI 47:105–111

Ulland BM, Weisburger JH, Yamamoto RS, Weisburger EK (1973) Antioxidants and carcinogenesis: butylated hydroxytoluene, but not diphenyl-*p*-phenylenediamine, inhibits cancer induction by *N*-2-fluorenylacetamide and by *N*-hydroxy-*N*-2-fluorenylacetamide in rats. Food Cosmet Toxicol 11:199–207

Van Duuren BL (1969) Tumor promoting agents in two-stage carcinogenesis. Prog Exp Tumor Res 11:31–68

Van Duuren BL, Goldschmidt BM (1976) Cocarcinogenic and tumor-promoting agents in tobacco carcinogenesis. JNCI 56:1237–1242

Van Duuren BL, Orris L (1965) The tumor enhancing principles of *Croton tiglium* C. Cancer Res 25:1871–1875
Van Duuren BL, Segal A (1976) Inhibition of two-stage carcinogenesis in mouse skin with bis(2-chloroethyl) sulfide. Cancer Res 36:1023–1025
Vargaftig BB, Hai ND (1972) Selective inhibition by mepacrine of the release of rabbit aorta contracting substance evoked by the administration of bradykinin. J Pharm Pharmacol 24:159–161
Verma AK (1987) Inhibition of both stage I and stage II mouse skin tumor promotion by retinoic acid and the dependence of inhibition of tumor promotion on the duration of retinoic acid treatment. Cancer Res 47:5097–5101
Verma AK (1988) Inhibition of tumor promoter 12-*O*-tetradecanoylphorbol-13-acetate-induced synthesis of mouse epidermal ornithine decarboxylase messenger RNA and diacylglycerol-promoted mouse skin tumor formation by retinoic acid. Cancer Res 48:2168–2173
Verma AK, Boutwell RK (1977) Vitamin A acid (retinoic acid), a potent inhibitor of 12-*O*-tetradecanoylphorbol-13-acetate-induced ornithine decarboxylase activity in mouse epidermis. Cancer Res 37:2196–2201
Verma AK, Rice HM, Shapas BG, Boutwell RK (1978) Inhibition of 12-*O*-tetradecanoylphorbol-13-acetate-induced ornithine decarboxylase activity in mouse epidermis by vitamin A analogs (retinoids). Cancer Res 38:793–801
Verma AK, Shapas BG, Rice HM, Boutwell RK (1979) Correlation of the inhibition by retinoids of tumor promoter-induced mouse epidermal ornithine decarboxylase activity and of skin tumor promotion. Cancer Res 39:419–425
Verma AK, Ashendel CL, Boutwell RK (1980a) Inhibition by prostaglandin synthesis inhibitors of the induction of epidermal ornithine decarboxylase activity, the accumulation of prostaglandins, and tumor promotion caused by 12-*O*-tetradecanoylphorbol-13-acetate. Cancer Res 40:308–315
Verma AK, Conrad EA, Boutwell RK (1980b) Induction of mouse epidermal ornithine decarboxylase activity and skin tumors by 7,12-dimethylbenz[a]anthracene: modulation by retinoic acid and 7,8-benzoflavone. Carcinogenesis 1:607–611
Verma AK, Slaga TJ, Wertz PW, Mueller GC, Boutwell RK (1980c) Inhibition of skin tumor promotion by retinoic acid and its metabolite 5,6-epoxyretinoic acid. Cancer Res 40:2367–2371
Verma AK, Garcia CT, Ashendel CL, Boutwell RK (1983) Inhibition of 7-bromomethylbenz[a]anthracene-promoted mouse skin tumor formation by retinoic acid and dexamethasone. Cancer Res 43:3045–3049
Verma AK, Duvick L, Ali M (1986) Modulation of mouse skin tumor promotion by dietary 13-*cis*-retinoic acid and α-difluoromethylornithine. Carcinogenesis 7: 1019–1023
Viaje A, Slaga TJ, Wigler M, Weinstein IB (1978) The effects of antiinflammatory agents on mouse skin tumor promotion, epidermal DNA synthesis, phorbol ester-induced cellular proliferation and production of plasminogen activator. Cancer Res 37:1530–1536
Walaszek Z, Hanausek-Walaszek M (1987) Dietary glucarate inhibits rat mammary tumorigenesis induced by *N*-methyl-*N*-nitrosurea. Proc Am Assoc Cancer Res 28:153
Walaszek Z, Hanausek-Walaszek M, Webb TE (1984) Inhibition of 7,12-dimethylbenz(a)-anthracene-induced rat mammary tumorigenesis by 2,5-di-*O*-acetyl-D-glucaro-1,4:6,3-dilactone, an in vivo *β*-glucuronidase inhibitor. Carcinogenesis 5:767–772
Walaszek Z, Hanausek-Walaszek M, Minton JP, Webb TE (1986a) Dietary glucarate as anti-promoter of 7,12-dimethylbenz(a)anthracene-induced mammary tumorigenesis. Carcinogenesis 7:1463–1466
Walaszek Z, Hanausek-Walaszek M, Webb TE (1986b) Inhibition of *N*-methyl-*N*-nitrosurea induced mammary tumorigenesis in the rat by a beta glucuronidase inhibitor. IRCS Med Sci 14:677–678
Walaszek Z, Hanausek-Walaszek M, Webb TE (1986c) Dietary glucurate-mediated reduction of sensitivity of murine strains to chemical carcinogenesis. Cancer Lett 33:25–32
Walaszek Z, Hanausek-Walaszek M, Webb TE (1986d) A de-glucuronidation inhibitor reduces the induction by benzo(a)pyrene of a 60 kilodalton oncofetal protein and

DNA-binding in vivo. In: Cooke M, Dennis AJ (eds) Polynuclear aromatic hydrocarbons: chemistry, characterization and carcinogenesis. Ninth international symposium. Battelle Press, Columbus, pp 947–959

Walaszek Z, Hanausek-Walaszek M, Webb TE (1988) Repression by sustained-release β-glucuronidase inhibitors of chemical carcinogen-mediated induction of a marker oncofetal protein in rodents. J Toxicol Environ Health 23:15–27

Wargovich MJ (1987) Diallyl sulfide, a flavor component of garlic, inhibits dimethylhydrazine-induced colon cancer. Carcinogenesis 8:487–489

Wattenberg LW (1973) Inhibitions of chemical carcinogen-induced pulmonary neoplasia by butylated hydroxyanisole. J Natl Cancer Inst 50:1541–1544

Wattenberg LW (1971) Inhibition of carcinogenic and toxic effects of diethylnitrosamine and 4-nitroquinoline-*N*-oxide by antioxidants. Fed Proc 31:633

Wattenberg LW (1972) Inhibition of carcinogenic and toxic effects of polycyclic hydrocarbons by phenolic antioxidants and ethoxyquin. JNCI 48:1425–1430

Wattenberg LW (1973) Inhibition of chemical carcinogen-induced pulmonary neoplasia by butylated hydroxyanisole. JNCI 50:1541–1544

Wattenberg LW (1974) Inhibition of carcinogenic and toxic effects of polycyclic hydrocarbons by several sulfur-containing compounds. JNCI 52:1583–1587

Wattenberg LW (1975) Inhibition of dimethylhydrazine-induced neoplasia of the large intestine by disulfiram. JNCI 54:1005–1006

Wattenberg LW (1978) Inhibition of chemical carcinogenesis. JNCI 60:11–18

Wattenberg LW (1980a) Inhibition of chemical carcinogenesis by antioxidants. In: Slaga TJ (ed) Modifiers of chemical carcinogenesis, vol 5. Raven, New York, pp 85–109

Wattenberg LW (1980b) Inhibition of polycyclic aromatic hydrocarbon-induced neoplasia by sodium cyanate. Cancer Res 40:232–234

Wattenberg LW, Bueding E (1986) Inhibitory effects of 5-(2-pyrazinyl)-4-methyl-1,2-dithiol-3-thione (Oltipraz) on carcinogenesis induced by benzo(a)pyrene, diethylnitrosamine and uracil mustard. Carcinogenesis 7:1379–1381

Wattenberg LW, Leong JL (1967) Inhibition of 9,10-dimethylbenzanthracene-induced mammary tumorigenesis. Fed Proc 26:692

Wattenberg LW, Leong JL (1968) Inhibition of the carcinogenic action of 7,12-dimethylbenzo(a)anthracene by beta-napthoflavone. Proc Soc Exp Biol Med 128: 940–943

Wattenberg LW, Leong JL (1970) Inhibition of the carcinogenic action of benzo(a)pyrene by flavones. Cancer Res 30:1922–1925

Wattenberg LW, Loub WD (1978) Inhibition of polycyclic aromatic hydrocarbon-induced neoplasia by naturally occurring indoles. Cancer Res 38:1410–1413

Wattenberg LW, Page MA, Leong JL (1968) Induction of increased benzpyrene hydroxylase activity by flavones and related compounds. Cancer Res 28:934–937

Wattenberg LW, Lam LKT, Fladmore AV, Borchert P (1977) Inhibitors of colon carcinogenesis. Cancer 40:2432–2435

Wattenberg LW, Lam LKT, Fladmoe AV (1979) Inhibition of chemical carcinogen-induced neoplasia by coumarins and α-angelicalactone. Cancer Res 39:1651–1654

Wattenberg LW, Coccia JB, Lam LKT (1980) Inhibitory effect of phenolic compounds on benzo(a)pyrene-induced neoplasia. Cancer Res 40:2820–2823

Weed HG, McGandy RB, Kennedy AR (1985) Protection against dimethylhydrazine-induced adenomatous tumors of the mouse colon by the dietary addition of an extract of soybeans containing the Bowman-Birk protease inhibitor. Carcinogenesis 6:1239–1241

Weeks CE, Slaga TJ, Hennings H, Gleason GL, Bracken WM (1979) Inhibition of phorbol ester-induced tumor promotion in mice by vitamin A analog and antiinflammatory steroid. JNCI 63:401–406

Weeks CE, Herrmann AL, Nelson FR, Slaga TJ (1982) α-Difluoromethylornithine, an irreversible inhibitor of ornithine decarboxylase, inhibits tumor promoter induced polyamine accumulation and carcinogenesis in mouse skin. Proc Natl Acad Sci USA 79:6028–6032

Weeks CE, Slaga TJ, Boutwell RK (1984) The role of polyamines in tumor promotion. In: Slaga TJ (ed) Mechanism of tumor promotion, vol II. CRC Press, Boca Raton, pp 127–139

Weisburger EK, Evarts RP, Wenk MC (1977) Inhibitory effect of butylated hydroxytoluene on intestinal carcinogenesis in rats by azoxymethane. Food Cosmet Toxicol 15:139–141

Wiebel FJ (1980) Activation and inactivation of carcinogens by microsomal monooxygenases: modification by benzoflavones and polycyclic aromatic hydrocarbons. In: Slaga TJ (ed) Modifiers of chemical carcinogenesis, vol 5. Raven, New York, pp 57–84

Wiebel FJ, Leutz JC, Diamond L, Gelboin HV (1971) Aryl hydrocarbon (benzo[a]pyrene) hydroxylase in microsomes from rat tissues: differential inhibition and stimulation by benzoflavones and organic solvents. Arch Biochem Biophys 144:78–86

Wiebel FJ, Gelboin HV, Buu-Hoi NP, Stout MG, Burnham WS (1974) Flavones and polycyclic hydrocarbons as modulators of aryl hydrocarbon[benzo(a)pyrene]hydroxylase. In: Ts'o POP, DiPaolo JA (eds) Chemical carcinogenesis. Dekker, New York, pp 249–270

Wiebel FJ, Selkirk JK, Gelboin HV, Van der Hoeven TA, Coon MJ (1975) Position-specific oxygenation of benzo(a)pyrene by different forms of purified cytochrome P-450 from rabbit liver. Proc Natl Acad Sci USA 72:3917–3920

Willaman JJ (1955) Some biological effects of the flavonoids. J Am Pharm Assoc 44: 404–408

Williams D, Wiebel FJ, Leutz JC, Gelboin H (1971) Effect of polycyclic hydrocarbons in vitro on aryl hydrocarbon (benzo[*a*]pyrene) hydroxylase. Biochem Pharmacol 25: 1431–1432

Williams GM, Tanaka T, Maeura Y (1986) Dose related-inhibition of aflatoxin B1 induced hepatocarcinogenesis by the phenolic antioxidants, butylated hydroxyanisole and butylated hydroxytoluene. Carcinogenesis 7:1043–1050

Wilson NM, Christou M, Turner CR, Wrighton SA, Jefcoate CR (1984) Binding and metabolism of benzo[*a*]pyrene and 7,12-dimethylbenz[*a*]anthracene by seven purified forms of cytochrome P-450. Carcinogenesis 5:1475–1483

Witschi H, Lock S (1978) Butylated hydroxytoluene: a possible promoter of adenoma formation in mouse lung. In: Slaga TJ, Sivak A, Boutwell RK (eds) Carcinogenesis, vol 2. Mechanisms of tumor promotion and cocarcinogenesis. Raven, New York, pp 465–474

Witschi H, Lock S (1979) Enhancement of adenoma formation in mouse lung by butylated hydroxytoluene. Toxicol Appl Pharmacol 50:391–400

Witschi H, Williamson D, Lock S (1977) Enhancement of urethan tumorigenesis in mouse lung by butylated hydroxytoluene. JNCI 58:301–305

Wortzman MS, Besbris HJ, Cohen AM (1980) Effect of dietary selenium on the interaction between 2-acetylaminofluorene and rat liver DNA in vivo. Cancer Res 40: 2670–2676

Wynder EL, Hoffmann D, McCoy DG, Cohen LA, Reddy BS (1978) Tumor promotion and cocarcinogenesis as related to man and his environment. In: Slaga TJ, Boutwell RK (eds) Mechanisms of tumor promotion and cocarcinogenesis, vol 2. Raven, New York, pp 59–77

Yamamoto RS, Weisburger JH, Weisburger EK (1971) Controlling factors in urethan carcinogenesis in mice: effect of enzyme inducers and metabolic inhibitors. Cancer Res 31: 483–486

Yamamura M, Nakamura N, Fukui Y, Takamura C, Yamamoto M, Minato Y, Tamura Y, Fujii S (1978) Inhibition of 7,12-dimethylbenz[*a*]anthracene-induced tumorigenesis in rats by a synthetic protease inhibitor, *N,N*-dimethylamino-[*p*-(*p*'-guanidinobenzoyloxy)]benzilcarbonyloxyglycolate. Gann 69:749–752

Yavelow J, Caggana M, Beck KA (1987a) Proteases occurring in the cell membrane: a possible cell receptor for the Bowman-Birk type of protease inhibitors. Cancer Res 47: 1598–1601

Yavelow J, Scott CB, Mayer TC (1987b) Fluorescent visualization of binding and internalization of the anticarcinogenic Bowman-Birk type protease inhibitors in transformed fibroblasts. Cancer Res 47:1602–1607

Yoshimura H, Ozawa N, Saeki S (1978) Inductive effect of polychlorinated biphenyls mixture and individual isomers on the hepatic microsomal enzymes. Chem Pharm Bull 26: 1215

Yuspa SH, Ben T, Hennings H, Lichti U (1982) Divergent responses of epidermal basal cells exposed to the tumor promoter 12-*O*-tetradecanoylphorbol-13-acetate. Cancer Res 42:2344–2349

Zajdela F, Perin-Roussel O, Saguem S (1987) Marked differences between metagenicity in *Salmonella* and tumor-initiating activities of dibenzo(a,e)fluoranthrene proximate metabolites; initiation inhibiting activity of norharmin. Carcinogenesis 8:461–464

Zemaitis MA, Greene FE (1976) Impairment of hepatic microsomal drug metabolism in the rat during daily disulfiram administration. Biochem Pharmacol 23:1355–1360

CHAPTER 8

Genetic Susceptibility to Chemical Carcinogens

D. G. HARNDEN

A. Introduction

It is very well documented that environmental carcinogens including chemicals, radiation and viruses play a major, perhaps even a dominating, role in the aetiology of the whole range of neoplastic diseases. It is equally clear that factors intrinsic to the patient or animal host also influence both the occurrence of cancer and the biological behaviour of the neoplasm. The principal intrinsic variables are age, hormone balance, immune response and genetic constitution. Moreover, these extrinsic and intrinsic factors interact with each other, sometimes in a quite remarkable way.

When considering the aetiology of a particular cancer it is therefore necessary to consider all the elements which contribute to the emergence of a clinically recognisable cancer. In some instances one of these contributory factors will be overriding. For example, when a group of workers has been exposed to high levels of a powerful carcinogen, the majority, or even all, of those exposed may develop the disease, thus leaving little room for other components in the aetiology. The occurrence of bladder cancer in workers exposed to 2-naphthylamine is such an instance (CASE et al. 1954). However, in most cases the situation is by no means so simple, either because the agent concerned is a weak carcinogen or because the level of exposure is low. Moreover, in many cases exposure is not to one defined chemical entity but to a mixture of many agents which may be present in varying concentrations and exert their effect by a variety of different mechanisms. Then it is vitally important to realise that for human cancer these agents are acting in a population of great variability. Some of that variation will have been acquired during the lifetime of the individual but much of it will be genetic – the genetic diversity of the human population is enormous. One arrives at the conclusion, therefore, that the study of chemical carcinogens must include a consideration of the variation in the response of the individual to exposure to a specific chemical carcinogen. Undoubtedly, there are a variety of different interactions, and, in some cases, the genetic influence will be weak. In others, however, it may well be dominant. The concept that emerges, therefore, is not one of environmentally induced cancers on the one hand and genetic cancers on the other but rather an interaction between these two contributory factors in all cancers. What has to be determined is their relative importance in specific situations.

Cancer in animals is somewhat different. We have so little information on wild animals, as few live long enough to develop cancer, that the discussion must be largely confined to laboratory animals. Most carcinogenesis experiments are carried out on inbred strains of animals for the very good reason that genetically determined variability makes it impossible to interpret experiments carried out on outbred stocks. Almost by definition, therefore, chemical carcinogens interact with the genetic constitution of the animal in the aetiology of specific neoplasms. Indeed many of the currently popular inbred strains were developed because of their unusual susceptibility to a particular chemical carcinogen.

B. Genetic Susceptibility to Cancer in Experimental Animals

I. Inbred Mouse Strains

Historically, inbred mouse strains have been developed either because they have a high spontaneous incidence of cancer or because they are unusually susceptible to the effects of a specific carcinogenic agent. As early as the 1920s crosses between the A strain and the "dilute brown" strain were designed to test the idea that the incidence of cancer is genetically determined. STRONG (1978) records that these experiments led to the conclusion that the occurrence of cancer was inherited as a dominant trait, but it is clear that the situation is more complicated than that, and that many factors influence the occurrence of cancer in these strains of mice; some of these will be genetic and some environmental.

For example, in some mice the high incidence of leukaemia is due to the presence, in that strain, of a leukaemogenic virus. Indeed, the original demonstration of transmission of mouse leukaemia by GROSS in 1951 was made by inoculating cell-free extracts from the AKR strain, which has a high incidence of leukaemia, into newborn mice of the low-leukaemia strain, C3H. In some cases at least, the strain differences are due to the presence within the mouse genome of DNA sequences homologous to the RNA genome of an oncogenic retrovirus. For example, the *AKR-1* locus (which is proviral DNA) has been located on chromosome 7 in the AKR/J strain (ROWE et al. 1972), and the expression of this locus and the subsequent production of virus particles are essential for leukaemogenesis. The provirus, therefore, appears to act as a dominant gene, given that it is in an appropriate genetic background (NAYAR et al. 1980). Other loci may, however, control the expression of the integrated provirus. The presence of the $FV\text{-}1^b$ allele at the *FV-1* locus in BALB/c and other inbred strains causes suppression of the expression of murine leukaemia virus. In AKR-BALB/c crosses, suppression of the expression of the provirus is dominant and depends upon the segregation of the $FV\text{-}1^b$ gene in the offspring (LILLY 1972). Thus, both the presence of the integrated provirus and the genetic background influence the occurrence of leukaemia.

Therefore, when one considers the effect of carcinogenic chemicals, the genetic constitution of the animal must be taken into consideration in evaluating the results. It has already been stressed that, in most experiments, the genetic element has been removed by the use of inbred strains. However, in some cases experimenters have deliberately set out to examine the interaction between the genetic constitution and chemical carcinogens.

II. Interaction Between MHC Type and Chemical Carcinogens

In experimental animal systems there is some evidence that histocompatibility type influences the process of carcinogenesis. Using congenic mouse strains MIYASHITA and MORIKAWA (1987) examined the incidence of pulmonary adenomas after a single subcutaneous injection of either urethan or 4-nitroquinoline-1-oxide. Animals with specific haplotypes (*H-2a* and *H2-t1*) had a significantly higher number of adenoma foci than mice with other haplotypes (*H-2b, H2s, H-2f,* and *H-2t2*). However, there were also significant differences among different strains with the same (*H-2a*) haplotype. The authors concluded that, while the multiplicity of foci is regulated by genes within the *H-2* complex, other non-*H-2* genes also influence the occurrence of adenomas. Similar experiments by GRONBERG et al. (1983) also using congenic mouse strains showed that the incidence of malignancies (predominantly lymphomas and epidermal tumours) following the repeated intragastric administration of 9,10-dimethyl-1,2 benzanthracene (DMBA) was influenced by the *H-2* type. The highest incidence was found in the H-2s strain, while the H-2b strain was the most resistant. The authors offered no mechanistic explanation but noted that there was no correlation with in vitro natural killer (NK) cell activity.

A correlation between histological changes thought to be precancerous and MHC-linked genes was observed after the administration of *N*-2-acetylaminofluorene (AAF) in the diet to closely related strains of rats which differed mainly in the presence or absence of the GRC gene complex (RAO et al. 1984). In animals homozygous for this MHC-linked gene complex, which is concerned with growth and development, a very high proportion had multiple liver foci positive for gamma glutamyltranspeptidase (GGT) and other liver abnormalities, while only 27% of animals without the GRC gene showed small numbers of GGT-positive liver foci with no other abnormalities. Since these lesions are often associated with the subsequent development of cancer, the authors concluded that there is a correlation between susceptibility to cancer induction and these MHC-linked genes.

While the evidence is not extensive, it is reasonable to conclude that in animal systems, as in humans, genes in or near the MHC complex do influence the development of cancer following exposure to chemical carcinogens.

III. Genetic Interactions with Chemical Carcinogens in Animals

Almost all the work in this field has been done with rats and mice, but there are rare studies on other species. A useful model was established in inbred strains of the fish *Oryzias latipes* (medaka); it was shown that the F1 cross between two strains was much more sensitive to the induction of melanoma by *N*-methyl-*N'*-nitro-*N*-nitrosoguanidine (MNNG) than were either of the parental strains (HYODO-TAGUCHI and MATSUDAIRA 1987). There is of course extensive evidence that the spontaneous occurrence of melanoma in *Xiphophorus* species (platy fish) is under genetic control (ANDERS et al. 1984).

In small mammals there are reports of genetic susceptibility to a variety of different neoplasms including lymphoma, pulmonary adenoma, fribrosarcoma, hepatoma and tumours of the bladder, nervous system, skin and mammary gland.

1. Lymphoma

Shisa and Hiai (1985) studied strain differences in the induction of thymic lymphoma in rats following administration of propylnitrosourea. The induced incidence of lymphoma in the six strains studied ranged from 98% in Fischer 344 rats to 10% in rats of the Long Evans strain. Segregation of thymic lymphoma incidence among crosses between these two strains indicated that the increased susceptibility in Fischer rats is due to a dominant gene (termed thymic lymphoma susceptible, *Tls-1*) and that this gene is linked to coat colour markers. The authors noted that another independently assorting gene may also influence susceptibility, but they could find no evidence for expression of murine leukaemia viruses in these susceptible animals.

A slightly different approach was adopted by Ishizaka and Lilly (1987) who studied the induction of thymic lymphoma in mouse strains sensitive or resistant to 3-methylcholanthrene (MCA). In reciprocal bone marrow transfers between mice of these two strains, they were able to show that the sensitivity appeared to reside in the haemopoietic cells rather than in the stromal cells, but in such experiments it is not always easy to determine whether or not stromal cells have engrafted. The principle remains that susceptibility to lymphoma induction by MCA is, at least in part, genetically determined.

2. Hepatoma

The appearance of premalignant changes, benign hepatic foci and hepatocellular carcinoma in mice are all influenced by the genetic constitution of the animals following administration of either promoting agents or complete carcinogens. When C3H/HeJ mice and C57BL/6J mice were treated with *N,N*-diethyl-nitrosamine (DEN) or *N*-ethyl-*N*-nitrosurea (ENU), the mean number of hepatocellular adenomas and carcinomas was approximately 20- to 50-fold higher in the C3H/HeJ mice (Drinkwater and Guster 1986). In breeding experiments, these authors found that the difference in susceptibility to DEN carcinogenesis is controlled by at least two loci. However, one of these loci (termed *Hcs* for hepatocarcinogen sensitivity) is responsible for 85% of the heritability of susceptibility.

Dragani et al. (1987) used C57BL/6J hybrids to study genetic sensitivity to *N*-nitroso-diethylamine (NDEA, alternatively DEN). Hybrids of the C3Hf strain have a high spontaneous incidence of liver tumours, while hybrids of BALB/c mice have a low spontaneous incidence. However, after treatment with the carcinogen followed by the phenobarbital-like promotor, TCPOBOP, the incidence of hepatocellular nodules was the same in both hybrids. The nodules in the C3Hf hybrids were about 10 times the size of those in the BALB/c hybrids after treatment with NDEA alone, and the difference was much greater following the application of the promotor. A few hepatocellular carcinomas were observed, and these were more numerous in the C3Hf-derived hybrids. Once again no evidence of virus involvement was found. A rather different result was obtained by Wolff et al. (1986) who found that, in obese yellow Avy/A (C3HxVY) F1 hybrid mice, the prolonged feeding of sodium phenobarbital in the diet led to an increased prevalence of adenomas but a decreased prevalence of carcinomas as compared with treated agouti (A/a) F1 hybrids. This difference was accompanied by a greater incidence of fatty changes and cytoplasmic vacuolation in the mottled yellow (Avy/A) mice as compared with the agouti (A/a) mice. The authors suggested that these mottled yellow mice may be unusually susceptible to this type of promoting agent.

3. Lung

A series of studies have been carried out on the induction of pulmonary adenomas following urethan treatment in BALB/c-By mice and in other strains by MALKINSON and BEER (1983). Their initial conclusion was that resistance to the induction of pulmonary adenomas by urethane is controlled by a single locus and that resistance is dominant to susceptibility. Crosses between BALB/c-By and the more sensitive A/J and SWR/J strains showed the BALB/c resistance gene to be dominant; however, in crosses between BALB/c and the even more resistant C56BL/6J strain, the progeny resembled the BALB/c strain in their susceptibility to urethan-induced pulmonary adenomas. The authors ruled out variation in NK cell activity as the basis for the genetically determined resistance. In subsequent studies, it was shown that the cell types from which the pulmonary adenomas were derived varied between mouse strains (BEER and MALKINSON 1985) and, further, that a high, in vivo mitotic rate might be correlated with susceptibility to urethan tumorigenesis (THAETE et al. 1986). MALKINSON et al. (1985), in a detailed study corroborated by other centres, later concluded that more than one locus may be involved in determining resistance and sensitivity in this system.

4. Mammary Gland

It has been known for a long time that different strains of rats and mice differed in the inducibility of mammary tumours by 7,12-dimethylbenz[*a*]anthracene (DMBA). In a recent study ISAACS (1986) showed very clearly that in the rat such variation is under genetic control. Outbred rats and a range of inbred rat strains were exposed to a single dose of either DMBA or l-methyl-1-nitrosourea (MNU). NSD, WF and LEW strains, as well as the outbred rats, were found to be highly susceptible to the induction of mammary adenocarcinomas while F344 and AC1 strains were less susceptible. The COP strain was, however, totally resistant to mammary carcinogenesis by either DMBA or MNU, and, moreover, this resistance was found to be controlled by a single autosomal dominant gene. On the other hand GOULD (1986), using a similar system (DMBA-induced tumours in F344 rats, relatively resistant, and W/Fu rats, relatively susceptible), came to the conclusion not only that *susceptibility* is inherited as a dominant trait, but also that it is under the control of more than one autosomal gene. This is not necessarily incompatible with ISAACS' results, since the same strains of rats were not used and it is well established that dominance is a relative phenomenon – any particular allele may be dominant to one allele but recessive to others. Similarly, it is not so hard to imagine one powerful gene with an overriding effect negating the effect of several different loci. However, these differences require further study.

5. Other Tumours

HOSAKA et al. (1985) have shown strain differences in sensitivity to induction of bladder cancer and preneoplastic lesions of the bladder following the oral administration of *N*-(4-(5-nitro-2-furyl)-2-thiazolyl) formamide. Analbuminaemic rats appeared resistant compared with control Sprague-Dawley rats, but no genetic analysis was carried out.

In a study of the induction of neurogenic tumours following neonatal subcutaneous injection of ENU, NAITO et al. (1985) found an excess of trigeminal schwannomas in LE rats (93% in males, 86% in females) as compared with WF rats (24% in males, 20% in females). In a careful series of crossing and backcrossing experiments, this susceptibility was clearly shown to be under genetic control, and the authors concluded that the results may be explained on the complex basis of three independently segregating loci.

OHGAKI et al. (1983) have shown that the induction of gastric carcinoma by MNNG is under genetic control. When MNNG was added to the drinking water of susceptible AC1 rats and resistant Buffalo rats, the incidence of gastric adenocarcinoma in the AC1 strain was 80% in males and 47% in females, while in the Buffalo strain it was only 18% in males and 0% in females.

IV. Transplacental Carcinogenesis and Multigeneration Experiments

In 1966, DRUCKREY and his colleagues demonstrated that tumours could be produced in the offspring of animals treated with ENU. Subsequent studies have shown that the timing is critical and that exposure to a carcinogen early in pregnancy is more likely to lead to teratogenic effects or fetotoxic effects while, in later pregnancy, tumour induction may occur in a variety of species including rat, mouse, hamster and pig. These experiments are comparable to the well-known induction of vaginal adenomatosis and carcinoma of the vagina in the teenage daughters of women who were treated during pregnancy with diethylstilboestrol (HERBST et al. 1976). These observations and the experiments on transplacental carcinogenesis simply show that the carcinogen may cross the placenta and affect cells of the developing fetus in such a way that cancer may be manifest at some time later in life. This, of course, has no genetic implications. However, in an extension of the work, TOMATIS et al. (1975, 1977) showed that when pregnant rats were treated with MNU, the offspring showed an increased incidence of tumours, especially of the kidney and the nervous system; when these (F1) animals were mated to produce an F2 generation, there was an increased incidence of the same tumours in the F2 rats. A similar phenomenon was demonstrated in the 3rd generation. TOMATIS et al. (1981) have also shown an increased incidence of neurogenic tumours in the offspring of untreated female rats and males that had been treated with MNU before mating. In both cases one must suppose that a germ-line mutation had occurred which renders the animals susceptible to the spontaneous occurrence of tissue-specific neoplasms. NOMURA (1983, 1986) has shown that X-ray-induced, germ-line mutations may lead to tumours in mice following post-natal exposure to urethan. When urethan was given to F1 offspring of parents irradiated with 216 R of X-rays, 18% of them developed large clusters of tumour nodules in the lung, whereas only 2.8% of the offspring of the unirradiated control were effected in this way. The incidence of tumour clusters was 2.4 times higher in urethan-treated F1 animals than in untreated F1 progeny of irradiated parents. The author concluded that this is evidence of germ-line mutations causing tumours with an enhancement of penetrance following urethan treatment. These results, perhaps surprising at first sight, are quite in line with the notion that inherited mutations confer susceptibility to specific cancers both in humans and in animals.

C. Genetic Susceptibility to Cancer in Humans

While it is possible to obtain direct evidence of genetically determined susceptibility to chemically induced cancer in experimental animals, most of the evidence for humans is indirect. However, human genetic susceptibility to cancer can be recognised in several different situations, and these will be considered briefly before looking at the somewhat scant evidence of human susceptiblity to chemical carcinogens.

I. Association of Cancer with Normal Traits

The best documented example of the association of cancer with normal traits in humans is the variation in susceptibility to skin cancer according to the degree of skin pigmentation. The incidence of skin cancer is higher in white-skinned people living in sunny climates, e.g. South Africa and Australia, than it is in similar populations in their original northern homeland. This powerful environmental influence is counterbalanced in the black races by their genetically determined skin pigmentation which protects the cellular DNA from the damaging effects of UV light. Further, even amongst white-skinned people the risk of skin cancer varies with the degree of skin pigmentation, the most susceptible being those with red hair and fair skin (IPPEN 1987).

There is some evidence that blood group may affect cancer susceptibility, e.g. individuals with blood group A have a slightly increased risk of developing stomach cancer (MCCONNELL 1966). Similarly, specific histocompatibility types have been associated with an increased risk of specific neoplasms. For example, DAUSSET et al. (1982) reported an association between acute lymphoblastic leukaemia and A2, while Hodgkin's disease is associated with A1,5, B8 and B18. In the case of nasopharyngeal carcinoma, which occurs commonly in south-east Asia, there is again an association with HLA type (SIMONS et al. 1974).

The recent developments in recombinant DNA technology have opened up new ways of recognising genetic heterogeneity in the population. The minisatellite DNA probes developed by JEFFREYS et al. (1985) are multiply polymorphic so that each individual has a unique pattern. But even simple DNA probes may recognise restriction fragment length polymorphisms (RFLPs) when DNA is cut with specific restriction endonucleases. While such polymorphisms are apparently without phenotypic effect, they can be used either to trace closely linked genes in families or to identify subgroups of the population. KRONTIRIS et al. (1985) have described a polymorphic locus close to the c-Ha-*ras* proto-oncogene on human chromosome 11. They suggest that some of the rare alleles are associated with different types of neoplasms, but others (e.g. THEIN et al. 1986) have failed to confirm these findings. HEIGHWAY et al. (1986) have described an unusual distribution of one of the common alleles (A4) in patients with non-small cell carcinoma of the lung, but so far this observation has not been confirmed. The principle is, however, very important in the present context since this would be an example of a genetic polymorphism associated with a cancer which is certainly induced by a chemical carcinogen.

Lastly, there are many instances in which genetic polymorphisms for specific enzymes have been claimed to be associated with cancer proneness. For example, there is a well-characterised polymorphism for the cytochrome P-450 enzyme responsible for the metabolism of the antihypertensive drug debrisoquine. Good metabolisers excrete 10–200 times more of the urinary metabolite than do poor metabolisers (GONZALEZ et al. 1988). It has been suggested that this polymorphism is associated with cancer susceptibility (AYESH et al. 1984), but this is not universally agreed. This point is considered in detail later.

II. Cancer Families

In most of the instances of association with normal traits there is no apparent familial clustering of cancer. There are, however, families in which there is a clear excess of cancer; these fall into two types: (a) families with a classical Mendelian pattern of inheritance of susceptibility to a specific neoplasm or group of neoplasms and (b) those with an unquestionably elevated familial risk but in which no clear pattern of inheritance is found.

1. Mendelian Inheritance

A number of well-defined Mendelian traits carry a greatly increased risk of cancer. Some of these are inherited in an autosomal dominant manner, e.g. familial adenomatous polyposis (FAP or polyposis coli), which is associated with carcinoma of the colon; retinoblastoma, a cancer of the eye in children; multiple endocrine neoplasia type II, which includes both medullary carcinoma of the thyroid and phaeochromocytoma; basal cell naevus syndrome, which predisposes to skin cancers; and neurofibromatosis in which the multiple benign neurofibromas have a low propensity to become malignant. In none of these cases is there any suggestion that there is a specific interaction with a particular chemical carcinogen, though the possibility that this might be so, especially in FAP, must be seriously considered. There is, however, evidence of radiation sensitivity in basal cell naevus syndrome. Autosomal recessive, cancer-prone diseases include xeroderma pigmentosum, in which there is an abnormal sensitivity to UV light and some chemical agents and a susceptibility to develop skin cancer; ataxia-telangiectasia, in which the patients are unusually sensitive to ionising radiation and a limited number of radiomimetic chemical agents and have a high incidence of cancer of several types but particularly leukaemia and lymphoma; the immune deficiency diseases with an elevated risk of leukaemia and lymphoma; the very rare epidermodysplasia verruciformis, in which the patients have an unusual sensitivity to human papillomavirus 5 and the warts tend to become malignant on sun-exposed areas of skin.

There are very few X-linked diseases with an increased cancer risk, including Duncan's disease, an unusual susceptibility to Epstein-Barr virus (EBV) and hence to lymphoma, and the X-linked immunodeficiency diseases such as Wiskott-Aldrich syndrome, in which there is suceptibility to lymphoma.

2. Familial Aggregation

Many families are recorded in which there is a clear excess of cases of either one specific cancer or of several different kinds of cancer. This is particularly true of breast carcinoma and of carcinoma of the colon. While some of these may be due to chance, some follow quite readily definable criteria and so are unlikely to be chance findings. The Li-Fraumeni syndrome shows a pattern of soft tissue sarcomas in children and breast cancers in the mothers and possibly other neoplasms in other family members. It is characteristic of these families that the cancers occur at an earlier age than the corresponding sporadic cancer. Other families have been described in which many members developed cancer of the colon at relatively young ages without any evidence of colonic polyps. This has been termed the "cancer family syndrome" or Lynch syndrome type I, but a more appropriate name is hereditary, non-polyposis, colorectal cancer type a (HNPCCa). Likewise, families showing a similar pattern of disease but in which other cancers occur in addition to colorectal cancer have been termed HNPCCb. Since the occurrence of cancer in these families does not show a precise Mendelian pattern, the relative contributions of hereditary and environmental factors are hard to assess.

III. Evidence in Humans of Genetic Susceptibility to Chemical Carcinogens

It is well established that genetic factors play a part in determining the pattern of cancer in the population. For most cancers the apparent genetic contribution is small, but it has been shown by BODMER (1986) and others that even quite a strong genetic influence may not be apparent as a Mendelian trait or show even familial clustering.

There is no direct evidence at present that genetic variation in response to a chemical carcinogen determines the pattern of cancer incidence in human beings. Indeed, such direct evidence would be very difficult to obtain. It would be necessary to study a family or a population exposed roughly equally to a known carcinogen and to demonstrate that a significantly higher proportion of those carrying a marker gene developed the cancer associated with that particular carcinogen than did those family members or population members who did not carry it. However, examples from other fields suggest that such evidence will eventually be found. There are many instances from the field of pharmacogenetics of genetic variation in the metabolism of specific compounds or in sensitivity to their toxic effects (ROWLANDS et al. 1985). It is much less common to be able to demonstrate genetic control of a susceptibility to an agent which results in a specific pathology at a much later date. Probably the best example is the unusual sensitivity of individuals with low levels of alpha-1 antitrypsin, which results in emphysema following exposure to certain types of dusty environment (WILLIAMS and FAJARDO 1974). The kind of genetic epidemiology which would be necessary to document such genetic variation is now becoming available, and it seems not too optimistic to hope that the use of relevant enzyme polymorphisms or appropriate linked DNA probes will enable this evidence to be forthcoming in the future.

At present there are several lines of circumstantial evidence which support, in a general way, the concept of genetic susceptibility to chemical carcinogens.

1. Second Neoplasms

A patient who is treated for cancer with radiation or with cytotoxic drugs is exposed to carcinogenic doses of these agents and, in some cases, secondary cancers do arise, apparently as a result of such treatment. The question that then arises is "If the first cancer was in some part determined by genetic factors, is there any evidence that the treatment is more likely to give rise to a secondary neoplasm than it would have done in a non-genetically susceptible individual?" The principle can be demonstrated in retinoblastoma in which osteosarcoma may be induced in the orbit following radiotherapy of the eye tumour. All children with retinoblastoma appear to be at risk of radiation-induced osteosarcoma, but those who have the hereditary form of retinoblastoma appear to be especially at risk (STRONG 1977). Evidence for chemical susceptibility is sparse, but HAWKINS (1986) reported a substantial study of second primary tumours among survivors of childhood cancer who had been treated with both anticancer drugs and radiation. He gave suggestive evidence that chemotherapy may be involved in the development of secondary leukaemia. There is also an increased risk of second primary cancers following treatment for retinoblastoma, and this may be due to cyclophosphamide or to radiation or an interaction between these two agents. Similarly, INGRAM et al. (1987) reported a high incidence of second cancers in children treated for non-Hodgkin's lymphoma; while the intensive regime of chemotherapy may have caused the second cancer, radiation therapy may also have contributed.

2. Migrant Studies

Most epidemiological studies on migrants produce evidence of a strong environmental influence on the pattern of occurrence of cancer. The well-known change in the incidence of breast cancer in Japanese women who have migrated to the USA is usually attributed to the adoption of the lifestyle of the new country, and diet is usually held to be responsible, though no specific dietary constituents have been positively identified with the change. There are, however, a few examples of migrant populations retaining their exceptional incidence of a specific cancer in their new location. The high incidence of nasopharyngeal carcinoma (NPC) in Singapore Chinese is a good example. It is generally believed that NPC is the result of a complex interaction between EBV, the genetic make-up of the individual and specific chemical promoting agents (SHANMUGARTNAM and TYE 1970). Another possible example is the high incidence of stomach cancer in certain mid-European stocks in the USA (e.g. STASZEWSKI and HAENSZEL 1965). Unlike most migrant cancer rates, these have remained constant in the new environment. This could be due either to the genetic make-up of the population or to the retention of the old lifestyle or a combination of the two. It is this last possibility that is of interest in the present context.

3. Susceptibility to Specific Carcinogenic Agents

There are several human syndromes in which there is a genetically determined susceptibility to a specific environmental agent. For some of these the evidence is quite clear-cut, for others it is equivocal. In some cases the susceptibility is associated with an increased incidence of neoplasms, but the connection between the genetically determined sensitivity and the elevated cancer incidence is difficult to establish. The best known examples are ataxia-telangiectasia (A-T) and xeroderma pigmentosum (XP).

a) Ataxia-Telangiectasia

A-T is an autosomal recessive disease characterised by a progressive cerebellar degeneration leading to ataxia and differing from other forms of ataxia in the development of telangiectasia on the bulbar conjunctivae, face, ears, hands and feet. Patients with this rare disease also have a variety of other features including immunological deficiency, spontaneous occurrence of chromosome abnormalities, unusual sensitivity to ionising radiation, a defect of DNA repair, and an increased incidence of malignancies particularly, but not solely, of the lymphoid system (BRIDGES and HARNDEN 1982).

The radiosensitivity has been demonstrated at the clinical level and also at the cellular and chromosomal level. Cell survival and chromosomal studies have, however, shown that cells from A-T patients are also more sensitive to a small number of radiomimetic chemicals. For example TAYLOR et al. (1979; TAYLOR 1980) have shown A-T cells to be unusually sensitive to both cell killing and the induction of chromosome aberrations by bleomycin. Lymphoid cells from seven different patients were exposed to bleomycin at G2, and it was shown that both chromatid gaps and chromatid breaks were increased threefold compared with normals and with a control xeroderma patient, showing that this sensitivity to bleomycin appears to be associated with sensitivity to ionising rather than non-ionising radiation. A-T cells are also unquestionably sensitive to neocarzinostatin (SHILOH et al. 1982) and streptonigrin (TAYLOR et al. 1983), and there have been reports, so far unsubstantiated, of sensitivity to methyl methane sulphonate (MMS), actinomycin D, mitomycin C and MNNG (LEHMAN 1982). Attempts to determine further the nature of the basic defect in A-T cells have still not been fully successful, although the demonstration of lack of fidelity in the repair of double-strand breaks by A-T cells is an indication of a possible mechanism (Cox et al. 1986).

Although A-T patients have an unusually high incidence of neoplasms and also have an unusual sensitivity to certain chemical carcinogens as well as ionising radiation, there is no direct link which would enable one to conclude that the gentically determined sensitivity to chemicals is the cause of the excess of the cancers.

b) Xeroderma Pigmentosum

The situation with XP is very similar. XP is a rare autosomal recessive condition which confers unusual sensitivity to exposure to sunlight. Patients homozygous for this gene show progressive skin damage with age, and most will eventually develop multiple skin carcinomas. They also exhibit a marked photophobia. Cleaver (1986) demonstrated that the basic lesion in XP is a defect of the DNA excision repair pathway. The disease is, however, genetically heterogeneous, and there are at least eight different complementation groups, one of which (XP variant) is proficient at excision repair but deficient in post-replication repair. XP cells in culture are unusually sensitive to UV light, and it has also been shown that XP cells are hypermutable following exposure to UV light (Maher and McCormick 1976).

The principal clinical features of the disease can thus be attributed to an unusual sensitivity to UV light, but in experimental systems it has been shown that XP cells are also sensitive to a range of chemical substances, many of them carcinogens such as 4-nitroquinoline-1-oxide (4NQO), which cause bulky lesions in DNA and which are normally dealt with by excision repair (Sasaki 1973). There is, however, no evidence that XP patients are unusually sensitive to specific chemicals, and in one study Kraemer et al. (1984) showed that XP patients did not have an excess of cancers of any internal organs. In view of the widespread suggestion that carcinogens of many kinds are present at low levels in our environment, this is a rather surprising finding. Either these carcinogens are not as important as has been thought, or XP patients have some exceptional mechanism which ensures that the failure to repair chemical damage accurately (observed in vitro) is dealt with satisfactorily in vivo.

c) Other Syndromes

Cells from patient with Fanconi's anaemia are unusually sensitive to the chromosome damaging and cell killing effects of DNA cross-linking agents such as mitomycin C (Sasaki 1975) and diepoxybutane (Auerbach et al. 1981). Patients with this rare autosomal recessive disease are of short stature, develop a profound pancytopaenia and have an elevated risk of developing myelomonocytic leukaemia. Reports of liver tumours in these patients are almost certainly the results of treatment with drugs such as oxymetholone since the liver lesions tend to regress following cessation of treatment (Shannon et al. 1982). It is not clear whether this tumour risk is a genetic susceptibility.

There are reports that the cells from patients with Bloom's syndrome, another rare autosomal recessive disease with a susceptibility to many different kinds of cancer, are unusually sensitive to chemical carcinogens. One of the features of this disease is the spontaneous occurrence of high levels of sister chromatid exchanges (SCEs) in lymphoid cells (RAY and GERMAN 1981). There are some suggestions that Bloom's syndrome cells are sensitive to UV light, but this is not generally agreed (GERMAN and SCHONBERG 1980). However, there is better evidence that lymphocytes from Bloom's syndrome patients are unusually sensitive to the induction of chromosome aberrations and SCEs by ethyl methane sulphonate (KREPINSKY et al. 1979). SHIRAISHI et al. (1985) have studied the effect of 4NQO and MNNG on B-lymphoblastoid cell lines derived from Bloom's syndrome patients. The Bloom's syndrome lines are of two types, those with high SCE levels and those with SCE levels comparable with normal. Following carcinogen treatment of the cells and inoculation into nude mice, tumours were produced only from the Bloom's syndrome lines with high SCEs and not from the normal cell lines or the Bloom's syndrome lines with low SCEs. These observations do suggest that unusual sensitivity to carcinogens play some part in the high susceptibility to cancer in these patients.

IV. Risk in Families of Patients with Chemically Induced Cancer

Levels of genetic risk for some cancers have been identified by studying the incidence of the same cancer in first-degree relatives of the affected individual. Relatives of breast cancer patients, for example, have an elevated risk of breast cancer. However, when one is considering cancer induced by a chemical agent, the situation is complex. Often one is dealing with very small numbers of patients, and the level of exposure to the carcinogen must also be taken into consideration. Occupational cancers have been reported in several members of the same family, but there is really no way, at present, of determining that this is due to anything other than exposure to the same carcinogenic agent. One certainly cannot infer a genetic explanation for such observations.

In the case of lung cancer, several studies have been carried out in order to examine the incidence of cancer in relatives. Many of these failed to take smoking into account, but some studies (e.g. OOI et al. 1986; TOKUHATA and LILIENFELD 1963) do consider smoking habits. Both studies reach the conclusion that there is an excess of cases of lung cancer, both in non-smoking and in smoking relatives, as compared with either spouse controls or neighbourhood controls, the relative risks being about threefold. While this is consistent with there being a genetic risk of cigarette-induced lung cancer, such studies are fraught with difficulties, and there are many sources of bias.

D. Mechanisms of Genetic Susceptibility

It is generally agreed that carcinogenesis is a multistage process. The simplest model for chemical carcinogenesis is the initiation/promotion model of BERENBLUM and SHUBIK (1947) which envisages a change from the normal cell to a potential malignant cell (initiation or induction) followed by a a further stage (promotion) in which another agent, not carcinogenic in its own right, enhances the probability of cancer development. The phenomenon of promotion in chemical carcinogenesis is a specific example of the multistage nature of carcinogenesis: after initiation a series of steps, often termed progression, leads to the development of a clinically recognisable cancer.

Genetic factors can influence the process of carcinogenesis at any of these steps. They may influence initiation in several different ways. The metabolism of the chemical before it reaches the DNA of the target cell will be under genetic control; the damage to the DNA, or rather the repair of that damage, is under genetic control; the genetic environment in which the DNA lesion is induced will also affect the outcome of the initial interaction between a chemical carcinogen and the target cell. Similarly, after the initiation event has occurred to give a potentially neoplastic cell, progression will also be influenced by genetic factors. The immune response to any new antigens on the cell surface, the hormone balance of the individual patient or animal and the immediate local environment will all be subject to genetic variation.

Based on this brief outline genetic susceptibility to chemical carcinogens can be considered in the light of these mechanisms but clear-cut samples are available in only a few instances.

I. Genetic Factors in the Metabolism of the Carcinogen

There is a vast literature on this topic for both human and animal carcinogenesis. Many "carcinogens" are only active after metabolism within the body while others, such as the classical alkylating agents, are direct acting. This is considered in detail elsewhere in this volume. Many interspecies differences in carcinogenic potency are known to be due to differences in the way carcinogenic compounds are handled. Furthermore, some of the tissue specificities which are noted for particular carcinogens are due to the availability of metabolising enzymes or to the particular metabolic pathway in a specific tissue. While these phenomena are of interest and indicate what may be possible, they are not central to the specific question being addressed. What evidence is there that in animals of the same species and in the same tissue there is variability in the metabolism of particular compounds which is genetically based and which, moreover, leads to different levels of interaction between the ultimate carcinogen and the DNA of the target cell?

1. Animal Evidence

While several different groups of carcinogens have been studied, most information is available for the polycyclic aromatic hydrocarbons (PAHs). The first stage of metabolism of PAHs is the mono-oxidation carried out by the cytochrome P-450 system and in particular by the sub-family of this complex system sometimes called aryl hydrocarbon hydroxylase (AHH), which is inducible by MCA and other PAHs. The AHH system is not specific for PAHs nor is it the only cytochrome P-450 sub-family to exhibit activity against PAHs, but it is a valuable and much studied example. Some inbred strains of mice have genetically determined high levels of AHH activity, which correlates with an increased susceptibility to various cancers induced by PAH. For example, KOURI et al. (1980) showed that following intrathecal administration of MCA, 45% of 84 mice with a high, genetically determined AHH level developed lung cancer while only 8% of 25 mice with low AHH levels developed such tumours – a highly significant difference. Control (untreated) mice with high AHH levels did not show an increased incidence of lung tumours.

While the genetic nature of the susceptibility is clear, the mode of inheritance is complex, so that the particular strains of animals must be defined for a specific carcinogenicity test. NEBERT et al. (1974) showed that the causation of fibrosarcomas following subcutaneous administration of MCA varies with the inducibility of AHH activity and that this variation is under genetic control. In crosses between C57BL/6N (high inducibility) and RJ/F (low inducibility) strains the expression of AHH induction by MCA is inherited as an autosomal dominant trait since the F1 hybrid has high levels of inducible activity, indistinguishable from that of the C57BL/6N parent (THORGEIRSSON and NEBERT 1977). However, in other crosses the inducibility of AHH activity by MCA may be codominant or even recessive. It was concluded that AHH inducibility is controlled by at least two different loci, each with multiple alleles. Inheritance of inducibility therefore depends not only upon the alleles present but also upon the genetic background. KOURI et al. (1983) showed clearly that inheritance of susceptibility to dibenz[*a,h*]anthracene carcinogenesis is dependent upon the expression of the gene (*Ahb*) which controls the presence of high-affinity Ah receptors.

While the data indicate that metabolism of the polycyclic hydrocarbon is under genetic control, there is also some evidence that the promotional stage of two-stage chemical carcinogenesis is under genetic control. REINERS et al. (1984) showed that in vitro keratinocytes from SENCAR and C57BL/6 mice were approximately equal in their sensitivity to mutation induction by DMBA. However, in vivo C57BL/6 developed carcinomas sooner and had a greater number of carcinomas per animal than did SENCAR mice following a complete skin carcinogenesis protocol. It was found that SENCAR mice were sensitive to both benzoyl peroxide and 12-*O*-tetradecanoylphorbol-13-acetate (TPA) while C57BL/6 mice were sensitive to benzoyl peroxide and refractory to TPA.

2. Human Evidence of Genetic Control of Aryl Hydrocarbon Hydroxylase

Many attempts have been made to demonstrate that in humans, as in animals, inducibility of AHH is under genetic control and associated with a susceptibility to neoplasia. In 1973 KELLERMAN et al. published a study which appeared to show that the inducibility of AHH varied in the human population, that this distribution was trimodal – compatible with inheritance as a simple Mendelian trait – and that those who had moderate or high inducibility were unusually liable to develop carcinoma of the lung as compared with individuals with low inducibility. Since then a number of studies have failed to confirm this observation (e.g. WARD et al. 1978; MCLEMMORE et al. 1977). Others, however, have found inducibility of AHH to be correlated with the incidence of cancer. For example, in a very careful case-control study KOURI et al. (1982) showed a positive correlation between AHH inducibility and primary carcinoma of the lung. The assays used in these studies are difficult, particularly at the low levels being studied, and while it is now generally agreed that AHH inducibility does vary between individuals, there is no clear evidence that it is under genetic control in humans. Similarly, there is some evidence, not yet conclusive, that inducibility correlates with cancer induction.

3. Genetic Variation in the Cytochrome P-450 System in Humans

There is ample evidence from pharmacogenetic studies that there is a wide-ranging genetic control over many drugs that have been studied in humans. While the AHH studies have been controversial, good progress has been made in related fields. For example, the metabolism of specific drugs which are modified by the cytochrome P-450 system has been intensively studied.

Antipyrine (AP), for example, has been used as an index of the activity of the liver, cytochrome P-450-dependent, drug oxidising system. AP is a salicylate-like drug which is metabolised almost completely in the liver, and several studies have been carried out to determine whether the half-life of AP correlates with other measures of AHH activity. In one study a good correlation was found, only in some cases, between AP hydroxylation and benzo[*a*]pyrene hydroxylation by homogenates of human liver (KAPITULNIK et al. 1977). Further studies have shown, however, that AP is metabolised by several different pathways and, moreover, that many environmental influences and physiological changes influence its half-life. It is unlikely, therefore, to be useful for detailed genetic studies. Another drug system likely to be of use in such studies is the hydroxylation of the antihypertensive drug debrisoquine. The 4-hydroxy metabolite of debrisoquine can be easily measured in urine after a single dose of the drug and can be expressed as a ratio of the unchanged drug to the 4-hydroxy metabolite (IDLE et al. 1979). There is a population polymorphism, which is bimodal, with about 90% of the population being extensive metabolisers, i.e. a low metabolic ratio, and 10% who are poor metabolisers, i.e. a high metabolic ratio. Extensive metabolisers excrete 10–200 times more 4-hydroxydebrisoquine than do poor metabolisers.

Debrisoquine is metabolised by a cytochrome P-450 sub-family which also handles a number of other commonly prescribed drugs, though this sub-family is not the same as the AHH complex. The enzyme is controlled in most cases by a single pair of allelic genes with the poor metaboliser gene being recessive; there is some evidence, however, that in a subgroup of families the enzyme is under the control of a different pair of genes (PRICE-EVANS et al. 1980). Case-control studies have shown that poor metabolisers are less common among lung cancer patients than among controls (approx. 2% and 9%), suggesting that poor metabolisers may be less likely to develop lung cancer (HETZEL et al. 1980; AYESH et al. 1984).

At the molecular level it has recently been recognised that poor metabolisers have negligible amounts of a cytochrome P-450 enzyme, P-450dbl. This gene has now been cloned (GONZALEZ et al. 1988), and three variant messenger RNAs have been isolated, suggesting a splicing defect of the pre-mRNA.

This poses a difficulty in interpreting the epidemiological studies since the dbl enzyme is not known to metabolise any of the well-established environmental carcinogens, but it may, of course, metabolise as-yet-unrecognised carcinogens (e.g. in the diet). A second problem, from the public health point of view, is that since approximately 90% of the population are in the high risk category identified by debrisoquine metabolism, it would not be a suitable marker if the ratios were reversed and only 10% or so were in the high-risk group. The importance of the debrisoquine polymorphism is that it clearly identifies a genetic variation in the population which affects the metabolism of commonly used drugs. The possibility that such polymorphisms exist for the metabolism of carcinogens is therefore very real.

4. Agents Other Than Polycyclic Aromatic Hydrocarbons

A polymorphism is present in the human population for the acetylation of isoniazid (INH) and other drugs. This polymorphism has been exploited by CARTWRIGHT and his colleagues (1982) to study the phenotypes of bladder cancer patients and controls. The population can be divided into fast acetylator and slow acetylator phenotypes, which appear to be under the control of a single autosomal recessive gene with slow acetylation being recessive. INH and acetyl-INH have both been shown to be carcinogenic in mice, but there is no evidence of human carcinogenicity.

II. DNA Repair

Variations in the likelihood of a carcinogenic metabolite reaching the DNA of the cell is one source of genetic variation in the initiation process. Once the active carcinogen has reached the target, there is no evidence to suggest that the number of primary lesions actually caused in the DNA is subject to any genetic variation. However, since most of these lesions will be repaired, at least at low doses, it is variation in the repair processes which is of interest, and these are subject to genetic influences.

Surprisingly little work has been done with small mammals, but many mouse, hamster and rat cell lines have been derived, often from a common parental cell strain, which differ in their sensitivity to radiation or to mutagenic chemicals (e.g. SCOTT et al. 1974). The basis of this differential sensitivity has been the subject of much study, and a variety of different retention, of the drug. In other cases there is differential uptake, or a differential retention, of the drug. In other cases the metabolism of the cell line becomes altered so that the drug is inactivated, e.g. the amplification of the dihydrofolate reductase gene in cell lines unusually resistant to the toxic effects of methotrexate. In some instances, however, it appears that cell lines do differ in their repair functions. Of particular interest in this respect are cell lines differing in their capacity to repair damage by alkylating agents. These agents are known to cause promutagenic lesions, which are almost certainly also procarcinogenic lesions. Recent studies by BRENNAND and MARGISON (1986a, b) have shown that when a bacterial gene which codes for an alkyl transferase is transfected into Chinese hamster cells which are normally deficient in alkyl transferase, the cells not only express the new repair gene but are more resistant to the toxic and mutagenic effects of these agents. Ockey et al. (1986) using the same transfected Chinese hamster cells showed that the presence of the *E. coli* O^6-alkylguanine alkyltransferase renders the cells less susceptible to the induction of SCEs and chromosome damage by methylating agents. From these studies it is concluded that the biological effects of these alkylating agents can be attributed to the presence of either O^6-alkylguanine or O^4-alkylthymine in the DNA. While this does not prove that, in nature, genetic variation can alter susceptibility to chemical carcinogenesis, it does demonstrate the principle that cells differing in a single DNA repair gene have quite different capacities to deal with promutagenic and procarcinogenic DNA lesions.

In the case of human cells the evidence that genetic variation in DNA repair capacity is causally related to differences in cancer susceptibility is firmer but still not conclusive. In XP the nature of the DNA repair defect has been very thoroughly studied and associated quite clearly with the mutagenicity response to both UV irradiation and chemical carcinogens (Maher and McCormick 1976). However, it is much harder to link the DNA repair defect directly to the increased cancer risk. It seems highly probably, however, that such a link does exist.

Similarly in A-T the existence of a DNA repair defect is now generally agreed (Lehmann 1982), but the precise nature of the defect is not known. Furthermore, there is no evidence that A-T cells are hypermutable following exposure to either X-rays or radiometric chemicals (Arlett and Harcourt 1983). There is, however, evidence that the clonal chromosome aberrations observed in A-T patients, and which are presumed to be the end result of inaccurate repair, are associated with neoplastic change (Kaiser-McCaw and Hecht 1982). More recently, it has been shown that, in both A-T cells and T-lymphoid neoplasms, the specific break points are within the T-cell receptor genes and an unknown locus proximal to the IgH gene on chromosome 14 (Kennaugh et al. 1986, Baer et al. 1987).

Once again, the link between a genetically determined defect in DNA repair and the induction of neoplasia is not firmly established, but the circumstantial evidence is very strong.

III. Genetic Effects on the Environment of the Chemically Transformed Cell

While the immediate environment of the transformed cell and systemic influences will undoubtedly affect the probability with which such a cell will progress to become a clinically recognisable cancer, and further genetic factors will influence such progression, it is hard to point to instances specific for chemical carcinogenesis. There are many examples of chemical carcinogenesis being influenced by hormone balance, and, inasmuch as hormone levels vary both with sex and genetic constitution, this does illustrate the principle that genetic effects on chemical carcinogenesis need not be confined to the initiation stage.

Similarly, the effect of specific promotors may vary with the strain of animal as shown, for example, by the study on strain differences in response to TPA and benzoyl peroxide by Reiners et al. (1984) already mentioned.

E. Consequences of Genetic Variation in Chemical Carcinogenesis

The conclusion from both human and animal work is that, following exposure to a chemical carcinogen, some individuals in the population may be more likely to develop cancer than others. Since chemical carcinogenesis in humans encompasses many occupational, drug-induced, smoking-related, alcohol- and diet-related cancers as well as cancers caused by casual exposure of the population to other carcinogenic agents, this could be important in a variety of different ways. At present, we do not have any hard data from human studies to give an estimate of the magnitude of the effect.

For the extreme cases like A-T and XP, exposure to the chemicals to which these subjects are sensitive, e.g. during therapy, should obviously be minimised. Since these are rare, this should be quite feasible. It is already known that XP patients protected from UV light do not develop skin cancers (Lynch et al. 1977), so the principle of prevention in this situation is well established. Further, in these extreme cases, antenatal diagnosis is possible (Ramsey et al. 1974), and so selective termination of pregnancy is one of the genetic options (Harnden 1987).

For cancer patients, especially children and young people who have a long life expectation, the possibility that they may have some special cancer susceptibility following exposure to mutagenic agents provides one more reason why mutagenic and carcinogenic treatment regimes hould be superceded whenever and wherever possible by less harmful treatment modalities. Such a recommendation can be made without there being a special marker for susceptibility. However, in considering occupational cancer in which the nature of the hazard is usually known in, at least, a general way (e.g exposure to mineral oils or asbestos), it is, at present, not possible to go beyond the general safety recommendation that *all* workers should be protected from the hazard. But current safety procedures are based not on elimination of risk but of reducing risk to levels believed to be acceptable. If some people are especially at risk, the acceptable level for those individuals may be considerably lower. Unless there is a marker for risk, however, nothing can be done, and at the moment there is no marker sufficiently well-defined that action could be taken. If such markers emerge (and it is likely that they will), one has to contemplate screening the entire workforce for genetic susceptibility. Screening for genetic variation at the point of entry is already a well-established practice for some occupations (red/green colour blindness in engine drivers, height in guardsmen, psyachological aptitude in airline pilots). If an association between a testable marker and a particular occupational cancer were to become well-established, screening should present no practical or ethical problems. Screening an existing workforce, however, would be a different matter since many "at risk" jobs are likely to be dirty and therefore highly paid. Delicate handling by management and unions as well as occupational health officers will be necessary.

For smoking-related cancers, any association with, say, a restriction fragment length polymorphism, would pose many interesting questions (e.g. if Heighway et al.'s 1986 work is confirmed). Those in the susceptible group could reasonably be advised to avoid smoking, but would this give an incentive to those in the less susceptible group to continue or even to begin smoking? The question of individual freedom would also be raised in some quarters. The problems are very similar for any susceptibility which is found in the general population as distinct from those within a specific risk group. The problem is, however, not a new one. We are already facing up to a very similar situation in the case of sun-induced cancers which are strongly dependent in (genetically determined) skin pigmentation. Various strategies have been adopted. Many individuals at risk choose to restrict exposure. The Australian Government has recently introduced an extremely vigorous campaign to encourage fair-skinned citizens to protect themselves and particularly children from direct sun. As a back-up to primary prevention, clinics for early detection of skin lesions have been set up. Even in the UK, where sun exposure is not so extreme and where vacation can be 2 weeks' sun worship in southern climes, such public campaigns ("Be a mole watcher") have been met with general public acceptability and some success in locating early cases of melanoma.

In this case, of course, not only is the hazard rather obvious but no special test is required to identify the genetic variation. Were such a test necessary to identify a cancer-associated genetic marker, would it be acceptable to the public? Would it be a sound economic proposition? We may shortly be obliged to find answers to these questions.

References

Anders F, Schartle M, Barnekow A, Anders A (1984) Xiphophorus as an in vivo model for studies on normal and defective control of oncogenes. Adv Cancer Res 42:191–275

Arlett CF, Harcourt SA (1983) Variation in response to mutagens amongst normal and repair defective human cells. In: CW Lawrence (ed) Induced mutagenesis. Plenum, New York, pp 249–270

Auerbach AD, Adler B, Chaganti RSK (1981) Prenatal and postnatal diagnosis and carrier detection of Fanconi anaemia by a cytogenetic method. Pediatrics 67:128–135

Ayesh R, Idle JR, Ritchie JG, Crothers MJ, Hetzel MR (1984) Metabolic oxidation phenotypes as markers for susceptibility to lung cancer. Nature 312:169–171

Baer R, Heppell A, Taylor AMR, Rabbitts PH, Boullier B, Rabbits TH (1987) The breakpoint of an inversion in a T cell leukaemia, sequences downstream of the immunoglobulin heavy chain locus implicated in tumorigenesis. Proc. Natl Acad Sci USA 84:9069–9073

Beer DG, Malkinson AM (1985) Genetic influence in type 2 or Clara cell origin of pulmonary adenomas in urethan-treated mice. JNCI 75:963–969

Berenblum I, Shubik P (1947) The role of croton oil applications associated with a single painting of a carcinogen in tumour induction of the mouse's skin. Br J Cancer 1:379–382

Bodmer WF (1986) Inherited susceptibility to cancer. In: Franks LM, Tiech N (eds) Introduction to cellular and molecular biology of cancer. Oxford University Press, Oxford, pp 93–110

Brennand J, Margison GP (1986 a) Expression in mammalian cells of a truncated *E. coli* gene coding for ^6O-alkylguanine alkyltransferase reduces the toxic effects of alkylating agents. Carcinogenesis 7:2081–2084

Brennand J, Margison GP (1986 b) Reduction of the toxicity and mutagenicity of alkylating agents in mammalian cells harboring the *E. coli* alkyltransferase gene. Proc Natl Acad Sci 83:6292–6296

Bridges BA, Harnden DG (eds) (1982) Ataxia-telangiectasia: a cellular and molecular link between cancer, neuropathology, and immune deficiency. Wiley, Chichester

Cartwright RA, Glashan RW, Rogers HJ (1982) *N*-Acetyltransferase phenotypes in bladder carcinogenesis: a pharmacogenetic epidemiological approach to bladder cancer. Lancet ii:842–844

Case RAM, Hosker ME, McDonald DB, Pearson JT (1954) Tumours of urinary bladder workmen engaged in manufacture and use of certain dye-stuff intermediators in British chemical industry: role of aniline benzidine, alpha-naphthylamine and beta naphthylamine. Br J Ind Med 11:75

Cleaver JE (1968) Defective repair replication of DNA in xeroderma pigmentosum. Nature 218:652

Cox R, Debenham P, Masson WK, Webb MBT (1986) Ataxia telangiectasia: a human mutation giving high frequency misrepair of DNA double strand scissions. Mol Biol Med 3:229–244

Dausset J, Columbani J, Hors J (1982) Major histocompatibility complex and cancer with special reference to human familial tumours (Hodgkin's disease and other malignancies). Cancer Surveys 1:119–149

Dragani TA, Manenti G, Della-Porta G (1987) Genetic susceptibility to murine hepatocarcinogenesis is associated with high growth rate of NDEA initiated hepatocytes. J Cancer Res Clin Oncol 113:223–229

Drinkwater NR, Guster JJ (1986) Genetic control of hepatocarcinogenesis in C57BL/6J and C3H/HeJ inbred mice. Carcinogenesis 7:1701–1707

Druckrey H, Ivankovic S, Preussmann R (1966) Teratogenic and carcinogenic effects in the offspring after single injection of ethylnitrosourea to pregnant rats. Nature 210:1378–1379

German J, Schonberg S (1980) Bloom's syndrome IX. Reviews of cytological and biochemical aspects. In: Gelboin HV et al. (eds) Genetic and environmental factors in experimental and human cancer. Jpn Sci Soc Press, Tokyo, pp 175–186

Gonzales FJ, Skoda RC, Kimura S, Umeno M, Zanger UM, Nebert DW, Gelboin HV, Hardwick JP, Meyer UA (1988) Characterisation of the common genetic defect in humans deficient in debrisoquine metabolism. Nature 331:442–446

Gould MN (1986) Inheritance and site of expression of genes controlling susceptiblity in mammary cancer in an inbred rat model. Cancer Res 46:1199–1202

Gronberg A, Cochran AJ, Karre K, Klein G, Klein GO, Kiessling R (1983) Incidence and type of tumours induced by DMBA carcinogenesis in H-2 congenic strains on a B10 background. Int J Cancer 32:247–252

Gross L (1951) "Spontaneous" leukaemia developing in C3H mice following inoculation in infancy with AK leukaemic extracts of AK embryos. Proc Soc Exp Biol 76:27–32

Harnden DG (1987) Genetic predisposition for cancer risks in man. In: Bannasch P (ed) Cancer risks: strategies for elimination. Springer, Berlin Heidelberg New York, pp 3–13

Hawkins MM (1986) Second primary tumours among survivors of childhood cancer treated with anti cancer drugs. IARC Sci Publ 78:231–252

Heighway J, Thatcher N, Cerney T, Hasleton PS (1986) Genetic predisposition to human lung cancer. Br J Cancer 53:453–457

Herbst AL, Scully RE, Robboy SJ (1976) Prenatal diethylstilboestrol exposure and human genital tract abnormalities. Natl Cancer Inst Monogr 51:25–35

Hetzel MR, Law MR, Keal EE, Sloan TP, Idle JR, Smith RL (1980) Is there a genetic component in bronchial carcinoma in smokers? Thorax 35:709

Hosaka Y, Muramatsu M, Matsushima T, Niijama T, Nagase S (1985) Low susceptibility of analbuminaemic rats to induction of bladder cancer by *N*-(4-(5-nitro-2-furyl)-2-thiazolyl)formamide. Jpn J Cancer Res 76:577–582

Hyodo-Taguchi V, Matsudaira H (1987) Higher susceptibility to *N*-methyl-*N*-nitro-*N*-nitrosoguanidine-induced tumorigenesis in an interstrain hybrid of the fish *Oryzias latipes* (medaka). Jpn J Cancer Res 78:487–493

Idle JR, Mahgoub A, Angelo MM, Dring LG, Rawlins MD (1979) The metabolism of (^{14}C)-debrisoquine in man. Br J Clin Pharmacol 7:257–266

Ingram L, Mott MG, Mann JR, Raafat F, Darbyshire PJ, Morris-Jones PH (1987) Second malignancies in children treated for non-Hodgkin's lymphoma and T-cell leukaemia with the UKCCCSG regimens. Br J Cancer 55:463–466
Ippen H (1987) Cancer risk from ultraviolet radiation. In: Bannasch P (ed) Cancer risks: strategies for elimination. Springer, Berlin Heidelberg New York, pp 137–142
Isaacs JT (1986) Genetic control of resistance of chemically induced mammary adenocarcinogenesis in the rat. Cancer Res 46:3958–3963
Ishizaka ST, Lilly F (1987) Genetically determined resistance to 3-methylcholanthrene induced lymphoma is expressed at the level of bone marrow derived cells. J Exp Med 166:565:570
Jeffreys A, Brookfield JFY, Semenoff R (1985) Positive identification of an immigration test case using human DNA fingerprints. Nature 317:818–819
Kaiser-McCaw B, Hecht F (1982) Ataxia-telangiectasia: chromosomes and cancer. In: Bridges BA, Harnden DG (eds) A cellular and molecular link between cancer, neuropathology and immune deficiency. Wiley, Chichester, pp 243–257
Kapitulnik J, Poppers PJ, Conney AH (1977) Comparative metabolism of benzo(a)pyrene and drugs in human liver. Clin Pharmacol Ther 21:166–176
Kellerman G, Shaw CR, Luten-Kellerman M (1973) Aryl hydrocarbon hydroxylase inducibility and bronchogenic carcinoma. N Engl J Med 289:934–937
Kennaugh A, Butterworth S, Hollis R, Baer P, Rabbits TH, Taylor AMR (1986) The chromosome breakpoint at 14q32 in an ataxia telangiectasia t(14;14) T cell clone is different from the 14q32 breakpoint in Burkitt's and an inversion T cell lymphoma. Hum Genet 73:254–259
Kouri RE, Billups LH, Rude TH, Whitmere DE, Sass B, Henry CJ (1980) Correlation of inducibility of aryl hydrocarbon hydroxylase with susceptibility to 3-methylcholanthrene induced lung cancers. Cancer Lett 9:277–284
Kouri RE, McKinney CE, Slomiany DR, Snodgrass NP, Wray NP, McLemmore TL (1982) Positive correlation between high aryl hydrocarbon hydroxylase activity and primary lung cancer as analysed in cryopreserved lymphocytes. Cancer Res 42:5040–5037
Kouri RE, Connolly GM, Nebert DW, Lubet RA (1983) Association between susceptibility to anthracene induced fibrosarcoma formation and the Ah locus. Int J Cancer 32:765–768
Kraemer KH, Lee MM, Scotto J (1984) DNA repair protects against cutaneous and internal neoplasm: evidence from xeroderm pigmentosum. Carcinogenesis 5:511–514
Krepinsky AB, Heddle JA, German J (1979) Sensitivity of Bloom's syndrome lymphocytes to ethyl methanesulphonate. Hum Genet 50:151–156
Krontiris TG, Di Martino NA, Colb M, Parkinson DR (1985) Unique allelic restriction fragments of the human Ha-*ras* locus and tumour DNAs of cancer patients. Nature 313:369–374
Lehmann AR (1982) The cellular and molecular responses of ataxia-telangiectasia cells to DNA damage: In: Bridges BA, Harnden DG (eds) Ataxia telangiectasia: a cellular and molecular link between cancer, neuropathology and immune deficiency. Wiley, Chichester, pp 83–101
Lilly F (1972) Mouse leukaemia: a model of a multiple gene disease. JNCI 49:927–934
Lynch HT, Frichot BC, Lynch JF (1977) Cancer control in xeroderma pigmentosum. Arch Dermatol 113:193–195
Maher VM, McCormick JJ (1976) Effect of DNA repair on the cytotoxicity and mutagenicity of UV irradiation and of chemical carcinogens in normal and xeroderma pigmentosum cells. In: Yuhars JM, Tennant RJ, Regan ID (eds) Biology of radiation carcinogenesis. Raven, New York, pp 129–145
Malkinson AM, Beer DS (1983) Major effect on susceptibility to urethan induced pulmonary adenoma by a single gene in BALB/cBY mice. JNCI 70:931–936
Malkinson AM, Nesbitt MN, Skamene E (1985) Susceptibility to urethan induced pulmonary adenomas between A/J and C57BL/6J mice: use of AXB and BXA recombinant inbred line indicating a three locus genetic model. JNCI 75:971–974
McConnell RB (1966) The genetics of gastro-intestinal disorders. Oxford University Press, London

McLemmore TL, Martin RR, Busbee DL (1977) Aryl hydrocarbon hydroxylase activity in pulmonary macrophages and lymphocytes from lung cancer and non-cancer patients. Cancer Res 37:1175–1181
Miyashita N, Morikawa K (1987) H-2-controlled genetic susceptibility to pulmonary adenomas induced by urethane and 4-nitroquinoline-1-oxide in A/Wy congenic strains. Jpn J Cancer Res 78:494–498
Naito M, Ito A, Aoyama A (1985) Genetics of susceptibility of rats to trigeminal schwannomas induced by neonatal administration of *N*-ethyl-*N*-nitrosourea. JNCI 74:241–245
Nayar KT, O'Neill B, Kouri RE (1980) Inheritance of murine endogenous RNA viruses. In: RE Kouri (ed) Genetic differences in chemical carcinogenesis. CRC, Boca Raton, pp 93–128
Nebert DW, Benedict WF, Kouri RE (1974) In: Ts'o POP, DiPaolo JA (eds) Chemical carcinogenesis. Dekker, New York, pp 271–288
Nomura T (1983) X-ray induced germ-line mutation leading to tumours: its manifestation in mice given urethan post-natally. Mutat Res 121:59–65
Nomura T (1986) Further studies on X-ray and chemically induced germ-line alterations causing tumours and malformations in mice. Proc Clin Biol Res 209B:13–20
Ockey CG, White GRM, Brennand J, Margison GP (1986) Chinese hamster cells harbouring the *E. coli* ^6O-alkylguanine alkyltransferase gene are less susceptible to SCE induction and chromosome damage by methylating agents. Carcinogenesis 7:2077–2080
Ohgaki H, Kawach T, Matsukura N, Morino K, Miyamoto M, Sugimura T (1983) Genetic control of susceptibility of rats to gastric carcinoma. Cancer Res 43:3663–3667
Ooi WL, Elston RC, Chen VW, Bailey-Wilson JE, Rothschild H (1986) Increased familial risk for lung cancer. JNCI 76:212–217
Price-Evans DA, Mahgoub A, Sloan TP, Idle JR, Smith RL (1980) A family and population study of the genetic polymorphism of debrisoquine oxidation in a white British population. J Med Genet 17:102–105
Ramsey CA, Coltart TM, Blunt S, Pawsey SA (1974) Prenatal diagnosis of xeroderma pigmentosum, report of the first successful case. Lancet ii:1109
Rao KN, Shinozuka H, Kunz HW, Gill TJ (1984) Enhanced susceptibility to a chemical carcinogen in rats carrying MHC-linked genes influencing development (GRC). Int J Cancer 34:113–120
Ray JH, German J (1981) The chromosome changes in Bloom's syndrome, ataxia-telangiectasia and Fanconi's anaemia. In: Arrighi FE, Rao PN, Stubblefield E (eds) Genes, chromosomes and neoplasia. Raven, New York, pp 357–378
Reiners JJ, Nesnow S, Slaga TJ (1984) Murine susceptibility to two-stage skin carcinogens is influenced by the agent used for promotion. Carcinogenesis 5:301–307
Rowe WP, Hartley JW, Bremner T (1972) Genetic mapping of a murine leukaemia virus inducing locus of AKR mice. Science 178:860–862
Rowlands M, Sheiner LB, Steimer JL (1985) Variability in drug therapy description estimation and control. Raven, New York
Sasaki MS (1973) DNA repair capacity and susceptibility to chromosome breakage in xeroderma pigmentation cells. Mutat Res 20:291–293
Sasaki MS (1975) Is Fanconi's anaemia defective in a process essential to the repair of DNA crosslinks. Nature 257:501–503
Scott D, Fox M, Fox BW (1974) The relationship between chromosomal aberrations survival and DNA repair in tumour cell lines of differential sensitivity to X-rays and sulphur mustard. Mutat Res 22:207–221
Shanmugartnam K, Tye CY (1970) A study of nasopharyngeal cancer among Singapore Chinese with special reference to migrant status and specific community. J Chronic Dis 23:433
Shannon RS, Mann JR, Harper E, Harnden DG, Morten JEN, Herbert A (1982) Wilms tumour and aniridia: clinical and cytogenetic features. Arch Dis Child 57:685–690
Shiloh Y, Tabor E, Becker Y (1982) The response of ataxia telangiectasia homozygous and heterozygous skin fibroblasts to neocarzinostatin. Carcinogenesis 3:815–820

Shirashi Y, Yosida TH, Sandberg AA (1985) Malignant transformation of Bloom syndrome B-lymphoblastoid cell lines by carcinogens. Proc Natl Acad Sci USA 82:5102–5106

Shisa H, Hiai H (1985) Genetically determined susceptibility of Fischer 344 rats to propylnitrosourea induced thymic lymphomas. Cancer Res 45:1483–1487

Simons MJ, Wee GB, Day NE, Morris JP, Shanmugaratnam K, De The GB (1974) Immunogenetic aspects of nasopharyngeal carcinoma. Differences in HL-A antigen profiles between patients and comparison groups. Int J Cancer 13:122

Staszewski J, Haenszel WJ (1965) Cancer mortality among the Polish born in the United States. JNCI 35:271

Strong LC (1977) Theories of pathogenesis; mutation and cancer. In: Mulvihill JJ, Miller RW, Fraumeni JF (eds) Genetics of human cancer. Raven, New York, pp 401–415

Strong LC (1978) Inbred mice in science. In: Morse HC (ed) Origins of inbred mice. Academic, New York, pp 45–67

Taylor AMR (1980) Evidence for a DNA repair defect following gamma irradiation of cells from patients with ataxia telangiectasia. In: Proceedings Conference of structural pathology in DNA and the biology of aging. Dt Forschungsgemeinschaft, Boppard Boldt, pp 252–262

Taylor AMR, Rosney CM, Campbell JB (1979) Unusual sensitivity of ataxia telangiectasia cells to bleomycin. Cancer Res 39:1046–1050

Taylor AMR, Flude E, Garner CR, Campbell J, Edwards MJ (1983) The effect of the DNA strand cleaving anti-tumour agent streptomigrin in ataxia telangiectasia cells. Cancer Res 43:2700–2703

Thaete LG, Beer DG, Malkinson AM (1986) Genetic variation in the proliferation of murine pulmonary type II cells; basal rates and alterations following urethan treatment. Cancer Res 46:5335–5338

Thein SL, Oscier DG, Flint J, Wainscoat JS (1986) Ha-*ras* hypervariable alleles in myelodysplasia. Nature 321:84–85

Thorgeirsson SS, Nebert DW (1977) The Ah locus and the metabolism of chemical carcinogens and other foreign compounds. Adv Cancer Res 25:149–194

Tokuhata GK, Lilienfeld AM (1963) Familial aggregation of lung cancer among hospital patients. Public Health Rep 78:277–283

Tomatis L, Hilfrich J, Turusov V (1975) The occurrence of tumours in F_1, F_2, and F_3 descendants of BD rats exposed to *N*-nitrosomethylurea during pregnancy. Int J Cancer 15:385:390

Tomatis L, Ponomarkov V, Turusov V (1977) Effects of ethylnitrosourea administration during pregnancy on three subsequent generations of BDVI rats. Int J Cancer 19:240–248

Tomatis L, Cabral JRP, Likhachev AJ, Ponomarkov V (1981) Increased cancer incidence in the progeny of male rats exposed to ethylnitrosurea before mating. Int J Cancer 28:475–478

Ward E, Paigen B, Steenland K (1978) Aryl hydrocarbon hydroxylase in persons with lung or laryngeal cancer. Int J Cancer 22:384–389

Williams WD, Fajardo LF (1974) Alpha-1-antitrypsin deficiency. A hereditary enigma. Am J Clin Pathol 61:311

Wolff GL, Morrissey RL, Chen JJ (1986) Amplified response to phenobarbital promotion of hepatotumorigenesis in obese yellow Avy/A (C3H XVY) F-1 hybrid mice. Carcinogenesis 7:1895–1898

CHAPTER 9

Interactive Effects Between Viruses and Chemical Carcinogens

A. HAUGEN and C. C. HARRIS

A. Introduction

The concept of multistage carcinogenesis was largely formulated by investigators observing the interactive effects of viruses and chemicals. ROUS and his coworkers (ROUS and KIDD 1938; ROUS and FRIEDEWALD 1944) found that virally induced skin papillomas that had regressed could be made to reappear by chemical irritants and concluded that the induced growth of these latent or dormant tumor cells occurred by a different mechanism (tumor promotion) than tumor induction (tumor initiation). BERENBLUM and SHUBIK (1947) more precisely divided carcinogenesis into initiation and promotion stages, and today the development of cancer is considered to be a multistep process involving several genetic and epigenetic steps (BOUTWELL 1974; CAIRNS 1975). Viruses are able to act at various stages of carcinogenesis, and interactive effects between environmental chemical carcinogens and viruses may be of substantial importance in human carcinogenesis (Fig. 1). Viral infections are widespread in humans, and

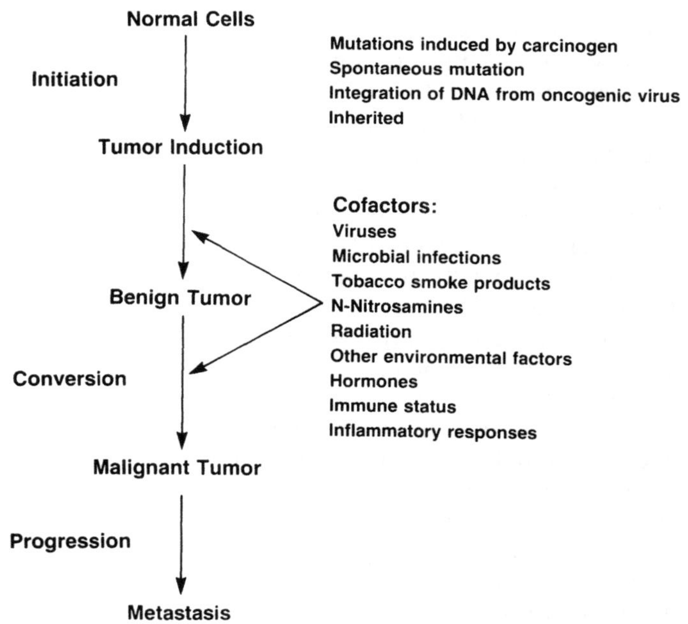

Fig. 1. Model of multistep carcinogenesis

Table 1. Examples of putative virus-chemical carcinogen interactive effects in human cancer

Cancer	Virus	Environmental factors
Cervix	HPV	Tobacco smoke products
Oral cavity	HPV	Tobacco smoke products
Larynx	HPV	Tobacco smoke products X-Irradiation
Lung	Unknown	Tobacco smoke products
Skin	HPV	Sunlight
Nasopharynx	EBV	*N*-Nitrosamines
Liver	HBV	Aflatoxin B_1, alcohol

HPV, human papillomavirus; EBV, Epstein-Barr virus; HBV, hepatitis B virus.

viruses are factors in several types of human cancer, e.g., Burkitt's lymphoma, nasopharyngeal cancer, liver cancer, T-cell leukemia, skin cancer, and cervical cancer (for review, see PHILLIPS 1983). The purpose of our review is to summarize the current status of laboratory and epidemiological studies designed to investigate interactive effects of viruses and chemicals in carcinogenesis (Table 1).

B. Cervical Cancer

Epidemiological studies have revealed that wide differences in the geographic distribution of cervical cancer exist between different countries, within one country, and between different groups in a geographic area (PETO 1986). The highest incidences of cervical cancer are found in South America, Hong Kong, East Germany, Romania, and the Maori population in New Zealand. During the past 20 years the incidence of cervical cancer has been decreasing in Europe (JOOSSENS and GEBOERS 1984). Neoplasia of the uterine cervix occurs most commonly in women who have had early intercourse, sexual intercourse with multiple partners, and many pregnancies (ROTKIN 1973). These observations suggest that an infectious agent that is sexually transmitted could be the risk factor in the development of this cancer. The overwhelming majority of cervical carcinomas arise within the area of the transformation zone in the cervix, which also undergoes formation at the time of menarche and remodeling after pregnancy (COPPELSON and REID 1968).

Cervical cancer is a disease in which viruses are thought to play a major pathogenic role. Several potentially oncogenic viruses have been implicated. The major suspected candidates are herpes simplex virus type 2 (HSV-2), human papillomavirus (HPV), and, to a certain degree, human cytomegalovirus (CMV). Numerous epidemiological and serological studies have observed an association between HSV-2 infection and cervical carcinoma (KESSLER 1974; THIRY et al. 1977; NAHMIAS et al. 1974), and HSV RNA and proteins have been detected within premalignant and malignant cervical tissue (ROYSTON and AURELIAN

1970; McDOUGALL et al. 1980; MAITLAND et al. 1981). However, HSV DNA has only rarely been found in biopsies from cervical neoplasia (FRENKEL et al. 1972; CASSAI et al. 1981; ZUR HAUSEN et al. 1981). Failure to demonstrate HSV DNA in these specimens led GALLOWAY and McDOUGALL (1983) to propose a "hit and run" mechanism for HSV oncogenesis that does not require persistence of viral DNA in the carcinoma cells.

ZUR HAUSEN (1982) has argued that the failure to detect HSV-2 DNA in cervical neoplasia indicates that the virus may not be the major etiological agent for cancer at this site. Increasing evidence has associated HPV with cervical neoplasia, dysplasia, and carcinoma-in-situ. HPV is known to consist of a heterogenic group of viruses that commonly infects the uterine cervix; over 50 types of HPV are currently known (GISSMANN 1984; SMITH and CAMPO 1985; PFISTER 1984; ZUR HAUSEN, personal communication). HPV DNA sequences have been detected in more than 80% of precancerous lesions of the cervix (DURST et al. 1983; GISSMANN 1984; GISSMANN et al. 1984; BOSHART et al. 1984). Reports suggest that certain HPV types (HPV-6, 11, 16, and 18) are closely linked to intraepithelial neoplasia (CIN), carcinoma-in-situ (CIS), and cervical cancer (GISSMANN et al. 1982a, b, 1984; SYRJANEN et al. 1985). Furthermore, specific HPV strains have been associated with certain histological changes; for example, the genomes of HPV-6 and 11 have been found primarily in condylomata, whereas HPV-16 and 18 are associated primarily with invasive carcinoma and high-grade cancers (GISSMANN et al. 1983; DURST et al. 1983).

The progression and regression of the various cervical lesions may depend on the oncogenic potential of the HPV types and possible interactive effects with chemical carcinogens in cigarette smoke such as *N*-nitrosamines (WINKELSTEIN et al. 1984; HARRINGTON et al. 1973). Recent data on increased risk of cervical cancer in cigarette smokers are of interest. A convincing and growing body of evidence shows that heavy smoking significantly increases the risk of cervical cancers, independent of sexual habits and other known confounding factors. TREVATHAN et al. (1983) conducted a case-control study indicating a relative risk of 12.7 for carcinoma-in-situ and 10.2 for severe dysplasia in smokers. A Swedish cohort study conducted by CEDERLOFF et al. (1976) among 27732 women showed that smokers have a 7.2-fold higher risk of cervical cancer compared with nonsmokers. However, the excess risk of cervical cancer among smokers is smaller in several case-control studies (THOMAS 1973; STELLMAN et al. 1980; MARSHALL et al. 1983), and there is no clear-cut, dose-exposure relationship with the amount of smoking (STELLMAN et al. 1980). Recently, nicotine and cotinine were detected in the cervical mucus of smokers, indicating that constituents in tobacco smoke reach the cervix and thus could have a genotoxic effect in the cervical epithelium (SASSON et al. 1985). Exposure of the male at the workplace to chemicals, oils, soot, and dirt might play a role in the etiology of cervical cancer, but at the moment there is no convincing evidence for such an association (WAKEFIELD et al. 1973; WILLIAMS and HORN 1977; ROBINSON 1982; ZAKELJ et al. 1984).

Another interesting hypothesis concerns possible synergistic actions between HSV and HPV as postulated by ZUR HAUSEN (1982), in which HSV acts as a tumor initiator. The induction of chromosomal aberrations, DNA repair, and mutagenicity by HSV infection supports this hypothesis (HAMPAR and ELLISON

1963; NISHIYAMA and RAPP 1981; SCHLEHOFER and ZUR HAUSEN 1982). Such changes within the DNA resolve the conflict between the serological studies favoring previous HSV infections and the failure to detect HSV-2 DNA in the tumors. Recent data show that the amplification of c-*myc* and c-Ha-*ras* oncogenes and overexpression of the c-*myc* gene characterize the most advanced tumors, indicating a possible interaction between viral genes and oncogenes activated by mutations, gene rearrangement, or amplification (RIOU et al. 1984).

Unfortunately, a suitable animal model for investigating the possible interactive effects of these viruses and chemicals has not been developed. Induction of cervical carcinomas by topical application of chemical carcinogens in rodents has been reported, however.

C. Skin Cancer

Certain epidermal squamous carcinomas might be attributable to HPV and exposure to various cocarcinogenic environmental factors, and there is evidence suggesting that papillomavirus can show oncogenic potential in human skin (ORTH et al. 1979; IKENBERG et al. 1983; SMITH and CAMPO 1985). Cofactor effects between specific papillomaviruses, sunlight, and various chemicals appear to be a precondition for neoplastic transformation in certain skin neoplasia.

Epidermodysplasia verruciformis (EV) is a rare, autosomal, and recessive disease in humans characterized by flat warts, macular lesions, and early development of skin carcinomas in about 30% of EV patients (JABLONSKA et al. 1972; LUTZNER 1978). Recent studies have implied that multiple factors are involved such as hereditary predisposition, the type of HPV infection, possible changes in cell-mediated immunology, and sun exposure (LUTZNER 1978). With the finding of HPV-5, 8, and 3 related copies in warts and carcinomas from EV patients, it seems likely that HPV may play a role in skin carcinogenesis (GREEN et al. 1982; OSTROW et al. 1982; KREMSDORF et al. 1983; GASSENMEIER et al. 1984; KREMSDORF et al. 1984).

Rabbits, cattle, horses, dogs, elk, and deer are known to be infected by papillomavirus, and spontaneously occurring papillomavirus-associated skin cancers have been reported to occur in these animals (LANCASTER and OLSON 1982; MORENO-LOPEZ et al. 1986). Progression from virally induced papilloma to carcinoma was first demonstrated with Shope's rabbit papilloma. The Shope papillomavirus induces papilloma in its natural host, the cottontail rabbit, and also in domestic rabbits (ROUS and BEARD 1934; SYVERTON 1952). KREIDER et al. (1979) reported neoplastic transformation by Shope papillomavirus of normal rabbit skin grafted in athymic mice. Recently, they described an in vivo system for studying the interaction of HPV with human skin (KREIDER et al. 1986). In experimentally produced papillomas and carcinomas in the domestic rabbit, there seems to be a cocarcinogenic interaction between the Shope papillomavirus and 3-methylcholanthrene (ROUS and KIDD 1938; ROUS and FRIEDEWALD 1944). Bovine ocular squamous cell carcinomas occur either on the eyelids or on the epithelial surface lining the conjunctival sac, and sunlight and papillomavirus may be cocarcinogenic in animals that suffer from ocular carcinoma syndrome (FORD et al. 1982; VANSELOW et al. 1982). Ovine facial and aural carcinomas have a similar complex etiology involving unpigmented skin, exposure to sunlight, and infection with a papillomavirus. In the rodent *Mastomys natalensis*, virus-producing tumors begin to appear when the animals are about 1 year old, and malignant conversion is enhanced by chronic treatment with 12-*O*-tetradecanoylphorbol-13-acetate (TPA) (AMTMANN et al. 1984). They found extra-chromosomal papilloma DNA in various tissues and that transcription of the viral genomes was required for tumor formation. In the two-stage mouse skin carcinogenesis model, epidermal cells are not completely transformed when an activated *ras* gene, in the form of the Balb murine sarcomavirus, is applied to mouse skin; however, subsequent treatment with TPA induces benign papillomas or invasive carcinomas (BROWN et al. 1986).

D. Alimentary and Respiratory Tract Cancer

The airways leading to the lung are in continuous contact with the external environment. The mucosal cells lining the oral cavity, the larynx, and the respiratory tract come into contact with numerous chemical, physical, and infectious agents, many of which may be involved in the pathogenesis of cancer.

Multiple squamous cell papillomas and carcinomas may occur in the oral cavity. Although the warts are etiologically heterogenous, HPV (strains 1, 2, 6, 11, 12, and 16) accounts for some of the warts (EISENBERG et al. 1985; LONING et al. 1985; NAGHASHFAR et al. 1985; SYRJANEN et al. 1986). SYRJANEN et al. (1983) have demonstrated productive HPV infection close to the carcinomas of the oral cavity. HPV-16 and 2 have been found in three of seven tongue carcinomas examined (DE VILLIERS et al. 1985). HPV-6, 11, and 16 appear to be associated with laryngeal papilloma (GISSMANN et al. 1982a; MOUNTS et al. 1982; STEINBERG et al. 1983; ABRAMSON et al. 1985), and 20% of adult laryngeal papillomas undergo malignant transformation. Treatment of laryngeal warts with X-irradiation leads to a large number of carcinomas after 5–40 years (MAJOROS et al. 1963; PEREZ et al. 1966; FOUTS et al. 1969; PROFFITT et al. 1970). Smoking and alcohol are thought to contribute to malignant transformation in the oral cavity and the upper digestive tract (PINBORG 1980).

The Epstein-Barr virus (EBV), which is a herpesvirus, is linked by both serioepidemiological and biochemical data to nasopharyngeal carcinoma. In north Africa and east Africa it accounts for approximately 7% of all cancers in males and, in some areas of southeast Africa, nasopharyngeal carcinoma accounts for about 20% of cancers. However, the highest rates of incidence occur in the southern parts of China, Hong Kong, and Singapore. Other factors such as *N*-nitrosamines in food, e.g., "soft" fish (YU et al. 1981), appear to play a role in the development of nasopharyngeal carcinoma.

Squamous cell papillomas are uncommon in the respiratory tract. There are histological similarities between squamous cell papillomas in the bronchial epithelium and condyloma lesions in the uterine cervix, and HPV might be involved (RUBEL and REYNOLDS 1979; SYRJANEN 1980). HPV DNA has been found in an anaplastic carcinoma of the lung (HPV-16) and in bronchial squamous cell carcinoma using a mixed probe for HPV-6, 11, 16, 18, and 30 (STREMLAU et al. 1985).

Chronic respiratory infections can enhance the neoplastic response of the lungs of laboratory animals to carcinogens. The effect of influenza infections in the respiratory system of mice has been investigated (KOTIN and WISELEY 1963). Squamous carcinomas were induced in C57 black mice after infection with influenza virus and exposure to aerosols of gasoline, but influenza virus alone did not induce tumors. SCHREIBER et al. (1972) also reported that chronic respiratory infection can enhance the neoplastic response caused by chemical carcinogens in rat lung. In conclusion, animal experiments indicate that viral respiratory infections might act as a cofactor in the development of respiratory-tract neoplasia.

Condylomatous changes in the esophageal epithelium and the presence of HPV antigens are evidence that HPV infection of the esophagus may occur (SYRJANEN 1982; WINKLER et al. 1985). This hypothesis is supported by the bovine model studied in areas of Scotland and northern England, in which squamous cell carcinoma of the bovine alimentary tract is caused by the combined effects of a specific papillomavirus (BPV-4) and a carcinogen present in bracken fern (JARRETT et al. 1978).

E. Liver Cancer

The incidence of primary hepatocellular carcinoma (PHC) is highest in Africa and Asia (WATERHOUSE et al. 1982). Attention has been focused on hepatitis B virus (HBV) because (a) prospective epidemiological studies have shown a high incidence of PHC among carriers in these HBV-endemic areas (BEASLEY et al. 1981; SAKUMA et al. 1982; LU et al. 1983); (b) clinical observation reveals that most of the patients with PHC are carriers of hepatitis B surface antigen (HBsAg) and have chronic active hepatitis (BLUMBERG and LONDON 1981; SAKUMA et al. 1982; LU et al. 1983; HINO et al. 1984b), and (c) recently, integrated HBV sequences have been found in the hepatocyte genome in patients with chronic hepatitis, hepatocellular carcinoma, and in HBV carriers (CHAKRABORTY et al. 1980; SHAFRITZ et al. 1981; BRECHOT et al. 1982; HINO et al. 1984a, b).

PHC is less common and occurs at an older age in urban populations than in rural populations. HBV infection at an early age in rural areas is considered to be the explanation for these observations. However, recent evidence suggests that other environmental factors are important. For example, a case control study of PHC patients of whom all were born and reared in rural areas but half had then moved to an urban setting was conducted by KEW et al. (1983). They concluded that HBV status, which was similar in both groups, was not responsible for the differences in incidence and age at onset of PHC. Other risk factors include exposure to aflatoxin B_1 (AFB_1) (IARC 1976; WOGAN 1976; VAN RENSBURG et al.

1985), consumption of alcoholic beverages (LIEBER et al. 1979; MARTINI 1980; OHNISHI et al. 1982) or contaminated water (SU 1979), low selenium levels (LI et al. 1985; YU et al. 1985), occupation (STEMHAGEN et al. 1983), tobacco smoking (TRICHOPOULOS et al. 1980; LAM et al. 1982; YU et al. 1983; KEW et al. 1985; AUSTIN et al. 1986), and androgen therapy (WESTABY et al. 1983). In the People's Republic of China, the correlation between PHC incidence and estimated dietary mycotoxin intake is statistically higher than the correlation between PHC incidence and the geographic distribution of HBV infection (WANG et al. 1983). Whereas the incidence of HBV infection does not vary among people living in the low- and high-altitude areas of Kenya, both the incidence of food contamination by AFB_1 and the frequency of PHC are higher in the low altitude areas. Lack of correlation between PHC incidence and HBV infection was also found in Eskimo populations in whom a high prevalence of HBV carriers but a low incidence of PHC was reported (MELBYE et al. 1984).

Consumption of alcoholic beverages has long been considered to be a major etiological factor in the pathogenesis of PHC in Western countries (TUYNS and OBRADVIC 1975; KELLER 1978; MARTINI 1980). A recent case control study has also suggested that the consumption of alcoholic beverages may be an important factor for PHC in the Philippines (BULATAO-JAYME et al. 1982). Individuals consuming daily more than 21 g of alcohol and more than an estimated 4 µg of AFB_1 in contaminated food, primarily cassava and corn, had a 35-fold increased relative risk of PHC. The effects of alcohol intake and dietary mycotoxin intake were synergistic. The results suggest that the role of alcoholic beverages, both free of and contaminated with mycotoxins, should be further examined in other Asian countries and in Africa.

The evidence, especially that obtained through long-term prospective follow-up of carriers vs non-carriers in areas of prevalence, strongly supports HBV as a major etiological agent for PHC. Accordingly, prevention programs involving universal immunization of newborns in high-risk areas have been initiated for the control of chronic hepatitis and hepatocellular carcinoma (SUN et al. 1986, 1987). Greater efforts to reduce consumption of carcinogen-contaminated food and drinking water are also being made.

Liver carcinogenesis, including the identification of putative precursor lesions of PHC, has been extensively studied in laboratory animals (see for review, FARBER and CAMERON 1980; PITOT et al. 1982; PERAINO et al. 1983). Many carcinogens from different chemical classes, e.g., *N*-nitrosamines and aromatic amines, cause PHC in several animal species. One of the most remarkable features of the putative preneoplastic cells found in the carcinogen-induced liver nodules is their resistance to different types of cytotoxic agents. Regardless of the model system of liver carcinogenesis used to produce the nodules, a set of common phenotypic changes, perhaps an adaptive response analogous in principle to that found in the heat-shock response of *Drosophila* (ASHBURNER and BONNER 1979), occurs in the resistant nodular cells, including increased levels of glutathione, epoxide hydrolase, and xenobiotic conjugating enzymes and decreased amounts of cytochrome P450. These phenotypic changes most likely alter the metabolic balance between activation and deactivation of procarcinogens and other xenobiotics and make the cell more resistant to injury caused

Table 2. Examples of possible selective advantages of preneoplastic and neoplastic cells

A. Defect in control of differentiation, e.g., resistance to induction of terminal differentiation by endogenous or exogenous factors
B. Defect in control of growth
1. Autocrine production of growth factors
2. Increased sensitivity to growth factors produced by other cells
3. Decreased sensitivity to inhibitors of growth
C. Differential response to cytotoxic agents
1. Inhibition of viral cytopathological response
2. Resistance to damage by electrophiles
3. Resistance to oxidative stress
D. Other
1. Escape from intercellular control mediated by cell-to-cell communication of hormones
2. Increased capacity to repair DNA damage

by electrophilic reactants and other cytotoxic metabolites (Table 2). AFB_1 is one of the most potent hepatocarcinogens known in experimental animals (IARC 1976; WOGAN 1976) and also produces cancers in other tissues, including bile duct, pancreas, and bone in monkeys (SIEBER et al. 1979). Whereas viral hepatitis does not cause PHC in marmoset monkeys, AFB_1, either with or without virus, is carcinogenic, and the combination of AFB_1 and virus causes a more severe hepatitis and cirrhosis than either agent alone (LIN et al. 1974). In rats, a shift in cancer incidence from PHC to intestinal carcinomas occurs in vitamin A deficiency (POIRIER et al. 1977; TAKASHI et al. 1982; GHOSHAL and FARBER 1983). Because of variations in organ specificity in different species, chemical hepatocarcinogens that are active in experimental animals may also contribute to the etiology of other types of cancer in man in geographic areas contaminated by these compounds.

In the areas showing a high incidence of human PHC, among the domestic animals only the duck (*Anas domesticus*) had a high incidence of liver tumours (CHANG 1973). Duck hepatoma deserves special attention owing to these facts: (a) its prevalence correlates with the increasing incidence of human PHC in the different regions of Qidong (CHANG 1973); (b) ducks are highly susceptible to the carcinogenic effects of food contaminated by mycotoxins, including AFB_1 (QIAN et al. 1976); and (c) HBV-like virus has been found in the ducks (WANG et al. 1980; ZHOU et al. 1980; SUMMERS and MASON 1982; YOKOSUKA et al. 1985). Similar viral particles have been observed in hepatocytes and hepatoma cells from ducks with viremia, and the involvement of food contaminated by aflatoxins is suggested by an analysis of the data reported by QIAN et al. (1976). When ducks in an area where flocks are commonly infected with the HBV-like virus were fed aflatoxin-contaminated corn, 18% (25 of 140 animals) developed hepatomas, whereas no hepatomas were found in the 40 ducks fed uncontaminated corn. These experiments need to be repeated in virus-infected vs virus-free flocks.

In addition to the duck, HBV-like viruses have also been detected in the eastern woodchuck (*Marmota monas*) (KODAMA et al. 1985), the tree squirrel (*Sciurus carolinensis pensylvanicus*) (FEITELSON et al. 1986), and the ground squirrel (*Spermophils beecvheyi*) (SUMMERS and MASON 1982). Chronic infection by HBV-like virus appears to be causally related to the subsequent development of liver cancer in the woodchuck (SNYDER et al. 1982; GERIN 1984; MARION et al. 1986).

In summary, liver cancer is one of the most prevalent forms of cancer in the world, and HBV is considered to be a major etiological factor. Evidence from epidemiological studies has also indicated that environmental contaminants such as mycotoxins may, either in combination with HBV or independently, be important etiological factors in the pathogenesis of PHC. Laboratory data also suggest an interplay between viral and chemical factors in the multifactorial etiology of PHC (Fig. 2; see HARRIS and SUN 1987 for a review). AFB_1, the chemical carcinogen most frequently implicated in the etiology of hepatocellular carcinoma, is a procarcinogen that must be activated by mixed-function oxidases to an electrophilic metabolite before it can exert its carcinogenic effects. Interindividual differences (>10-fold) in the metabolic activation of AFB_1 have been observed. These differences may play a part in an individual's susceptibility to AFB_1 carcinogenesis. Chemical carcinogens and integrated HBV may function via the activation of cellular protooncogenes, e.g., N-*ras*, and the inactivation of tumor suppressor genes.

F. In Vitro Studies

It is well established that viruses and carcinogens can interact synergistically in inducing specific cancers in vivo. In cell culture, virus-chemical carcinogen inter-

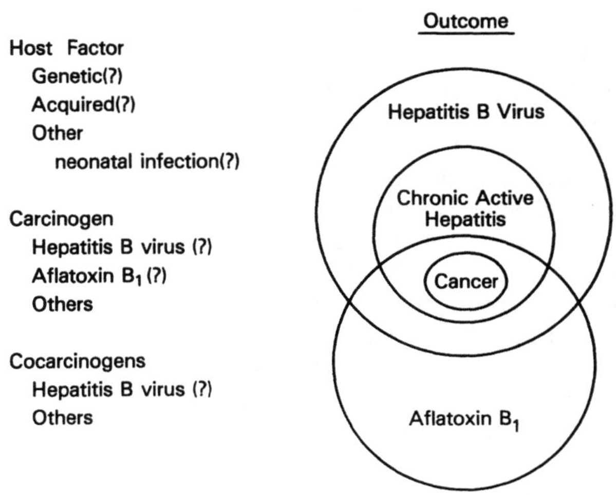

Fig. 2. Hypothetical diagram of the cocarginogenicity of hepatitis B virus and carcinogens in human liver carcinogenesis

active effects have been studied mainly in rodent fibroblasts. Numerous studies in vitro have demonstrated synergistic effects between chemical carcinogens and either RNA tumor virus or DNA tumor virus (see review by FISHER and WEINSTEIN 1980). Cells pretreated with benzo[*a*]pyrene, 4-nitroquinoline-*N*-oxide, 3-methylcholanthrene, X-rays, or thymidine analogues increased the transformation frequency by SV40, whereas pretreatment with 7,12-dimethylbenz[*a*]anthracene (DMBA) had no effect on SV40 transformation (TODARO and GREEN 1964; COGGIN 1969; DIAMOND et al. 1974; HIRAI et al. 1974; MILO et al. 1978). An enhancement of transformation by several polycyclic aromatic hydrocarbons (PAH) has been demonstrated in cells subsequently infected with adenovirus SA7, mutant adenovirus type 5, or HSV-2 (CASTO et al. 1973, 1976, 1979; FISHER et al. 1978; JOHNSON 1982). In this system enhancement occurs regardless of whether 3T3 Swiss mouse cells were exposed to PAH either before or after HSV-2 infection. Mouse mammary epithelial cells were neoplastically transformed after combined treatment with a mouse mammary tumor virus followed by DMBA (HOWARD et al. 1983). Pretreatment of rat embryo fibroblasts with an alkylating agent (methylmethane sulfonate) prior to infection with wild-type adenovirus 5 resulted in the earlier appearance of transformed foci and an enhanced frequency of transformation (FISHER et al. 1978).

The ability to analyze the molecular basis of viral-chemical carcinogen interactive effects has been aided by the development of well-characterized cell culture systems which respond to the combination of virus and carcinogens, tumor promoters, or hormones/growth factors. Furthermore, transfection of cells has made it possible to identify viral and cellular DNA sequences that transform cells in culture. It is important to make a distinction between immortalization of cells in culture and neoplastic transformation. The complete transformation of primary cells in culture to tumor cells has been shown to involve several genetic steps (LAND et al. 1983; NEWBOLD and OVERELL 1983; RULEY 1983). The first step can be induced by treatment with viruses and carcinogens or by oncogenes (polyoma large T, adenovirus E1A). Cells immortalized by this procedure can be fully transformed by transfection with oncogenes such as those belonging to the *ras* family, adenovirus E1B, and polyoma middle T (RASSOULZADEGAN et al. 1982, 1983; VAN DEN ELSEN et al. 1983; LOGAN et al. 1984; CUZIN et al. 1985; LAND et al. 1983; RULEY 1983).

Since most neoplasms in humans are carcinomas, in vitro systems for studying the role of virus-chemical carcinogen interactive effects in human epithelial cells are particularly important. However, human cells are more resistant to tumorigenesis than rodent cells (DIPAOLO 1983; KAKUNAGA et al. 1983; HARRIS 1987). Except for a few reports, the synergistic mechanism of transformation by a virus and a chemical carcinogen remains to be elucidated in human systems. RHIM et al. (1986) recently reported that MNNG or 4NQO can cause neoplastic transformation of epidermal keratinocytes previously exposed to adeno-12 SV40 hybrid virus; treatment with chemical carcinogens alone did not immortalize the cells. Gamma rays transformed normal human fibroblasts into immortalized cells (NAMBA et al. 1978), which were fully transformed to tumorigenic cells with Harvey murine sarcomavirus (Ha-MSV) (NAMBA et al. 1986). Ha-MSV alone did not transform the cells into immortal or tumorigenic cells. Transformation to anchorage independence was observed when primary human embryonic kidney cells were transfected with a combination of human papovavirus BK and the Ha-*ras* oncogene (PATER and PATER 1986). Normal human epithelial cells can be transformed by SV40 into continuous cell lines. Recently, CHRISTIAN et al. (1987) showed that subsequent treatment with 3-methylcholanthrene resulted in neoplastic transformation of these cells. Treatment of EBV-immortalized human B cells by *N*-acetoxy-2-acetylaminofluorene fully transformed the cells (KESSLER et al. 1987). In this system MNNG treatment did not lead to cells capable of tumor formation in vivo. These studies indicate that cells appear to require first an immortalization step to be susceptible to further progression towards malignancy and that more than one gene is involved. However, it is difficult to assess whether the acquisition to immortality in vitro corresponds in vivo to a significant step in human carcinogenesis.

G. Mechanisms of Transformation

During the 1970s infection by viruses was proposed as the single main etiological factor in human carcinogenesis (DULBECCO 1973; GALLO et al. 1977). The current view implicates both viruses and chemical carcinogens as important in carcinogenesis, and they may act synergistically. As reviewed in this chapter, today we know that viruses are associated with genital cancer, nasopharyngeal cancer, liver cell carcinoma, and certain skin cancers. However, demonstration of viruses, virus-specific antigens, and viral nucleic acids in cancer tissue or cancer patients is not a direct etiological proof of oncogenesis. On the other hand, viruses might be involved in causing cancer even though the viral genome is not integrated. While viruses have demonstrated oncogenic properties under in vitro conditions (HSV-1 and 2, CMV, and EBV), virus alone probably only rarely causes cancer. Additional factors, including chemicals, exposure to sun, X-rays, or another virus are frequently required to transform the cells neoplastically. This is exemplified by studies suggesting that HSV could act as an initiator and HPV as a promoter in cervical carcinogenesis, the evidence that cigarette smoking is an important risk factor for cervical cancer, and by many of the other studies mentioned above.

Metaplastic and dysplastic lesions are fertile sites for the development of neoplasia where the cancer might be triggered by virus infection. For example, the transformation zone (squamous columnar junction) in the cervix or metaplastic lesions in the large airways (AUERBACH et al. 1961; JACKSON and MULDOON 1973) may be targets for viral action. Following infection, viruses may integrate into the cellular genome causing protooncogene activation or inactivation of tumor suppressor genes. Aberrant control of viral genome expression might be a factor in the neoplastic process (ZUR HAUSEN 1986). Furthermore, many chemical carcinogens and viruses induce chromosome breaks. These breaks could provide an increased opportunity for either virus DNA integration or possible synergistic chromosomal damage.

Epigenetic mechanisms may play an important role in tumor promotion. Viruses have been shown to increase the rate of DNA, RNA, and protein synthesis, causing hyperplasia that could increase the number of target cells and the sensitivity of the cells to the chemical carcinogens, or the cells might become resistant to growth control mechanisms. For example, products of viral genes, e.g., SV40 large T antigen, or platelet-derived growth factor from *sis* may enhance cellular growth in an autocrine manner, and certain products from activated protooncogenes such as *myc* may interfere with pathways of cellular differentiation. This dysregulation of the pathways of cellular growth and differentiation could lead to selective clonal expansion of preneoplastic cells initiated by exposure to genotoxic chemicals and/or viruses. Differential viral cytotoxicity of normal vs preneoplastic cells is another, but largely unexplored, possible mechanism of tumor promotion.

H. Conclusions

Substantial evidence from both laboratory and epidemiological studies supports the hypothesis that viruses and chemicals have interactive effects in both the initiation and promotion stages of carcinogenesis. However, the importance of these interactive effects at certain cancer sites may be underestimated by epidemiological investigations due to the frequency of viral infections such as those in the respiratory tract. Chemicals and viruses can both directly cause DNA damage, leading to activation of protooncogenes and inactivation of putative tumor suppressor genes. These chemical and viral agents can also act by nongenotoxic mechanisms, by deregulating growth and differentiation pathways and by allowing selective clonal expansion of preneoplastic cells.

Acknowledgement. This paper was written during the tenure of an American Cancer Society-Eleanor Roosevelt International Cancer Fellowship awarded to A. H. by the International Union Against Cancer.

References

Abramson AL, Brandsma J, Steinberg B, Winkler B (1985) Verrucous carcinoma of the larynx. Arch Otolaryngol 111:709–715

Amtmann E, Volm M, Wayss K (1984) Tumour induction in the rodent *mastomys natalensis* by activation of endogenous papilloma virus genomes. Nature 308:291–292

Ashburner M, Bonner JJ (1979) The induction of gene activity in *Drosophila* by heat shock. Cell 17:241–254

Auerbach O, Sout A, Hammond E, et al. (1961) Changes in bronchial epithelium in relation to lung cancer. N Engl J Med 265:253–259

Austin H, Deizell E, Grufferman S, Levine R, Morrison AS, Stolley P, Cole P (1986) A case-control study of hepatocellular carcinoma and the hepatitis B virus, cigarette smoking and alcohol consumption. Cancer Res 46:962–966

Beasley RP, Hwang LY, Lin CC, Chien CS (1981) Hepatocellular carcinoma and hepatitis B virus. A prospective study of 22,707 men in Taiwan. Lancet 2:1129–1133

Berenblum I, Shubik P (1947) A new quantitative approach to the study of the stages of chemical carcinogenesis in the mouse skin. Br J Cancer 1:383–391

Blumberg BS, London WT (1981) Hepatitis B virus and the prevention of primary hepatocellular carcinoma. N Engl J Med 304:782–784

Boshart M, Gissman L, Ikenberg H, Kleinheinz A, Scheurlen W, zur Hausen H (1984) A new type of papillomavirus DNA, its presence in genital cancer biopsies and in cell lines derived from cervical cancer. EMBO J 3:1151–1157

Boutwell RK (1974) The function and mechanism of promoters of carcinogenesis. CRC Crit Rev Toxicol 2:419–431

Brechot C, Nalpas B, Courouce AM, Duhamel C, Callard P, Carnot F, Tiollas P, Berthelot P (1982) Evidence that hepatitis B virus has a role in liver-cell carcinoma in alcoholic liver disease. N Engl J Med 306:1384–1387

Brinton LA, Fraumeni JF (1986) Epidemiology of uterine cervical cancer. J Chron Dis 39: 1051–1065

Brinton LA, Schairer C, Haenszel W, Stolley P, Lehman HF, Levine R, Savitz DA (1986) Cigarette smoking and invasive cervical cancer. JAMA 255:3265–3269

Brown K, Quintanilla M, Ramsden M, Kerr IB, Young S, Balmain A (1986) V-ras genes from Harvey and balb murine sarcoma viruses can act as initiators of two-stage mouse skin carcinogenesis. Cell 46:447–456

Bulatao-Jayme J, Almero EM, Castro CA, Jardeleza TH, Salamat LA (1982) A case-control dietary study of primary liver cancer risk from aflatoxin exposure. Int J Epidemiol 11:112–119

Cairns J (1975) Mutation selection and the natural history of cancer. Nature 255:197–200

Cassai E, Rotola A, Meneguzzi G, et al. (1981) Herpes simplex virus in human cancer. I. Relationship between human cervical tumours and herpes simplex type. II. Eur J Cancer 17:685–693

Casto BC, Pieczynski WJ, DiPaolo JA (1973) Enhancement of adenovirus transformation by pretreatment of hamster cells with carcinogenic polycyclic hydrocarbons. Cancer Res 33:819–824

Casto BC, Pieczynski WJ, Janosko N, DiPaolo JA (1976) Significance of treatment interval and DNA repair in the enhancement of viral transformation by chemical carcinogens and mutagens. Chem Biol Interact 13:105–125

Casto BC, Miyagi M, Meyers J, DiPaolo JA (1979) Increased integration of viral genome following chemical and viral treatment of hamster embryo cells. Chem Biol Interact 25: 255–269

Cederloff R, Friberg I, Hrubec Z, et al. (1976) The relationship of smoking and some social covariables to mortality and cancer morbidity: a ten-year follow-up in a probability sample of 55000 Swedish subjects age 18–69. Department of Environmental Hygiene, Karolinska Institute, Stockholm

Chakraborty PR, Ruiz-Opano N, Shouval D, Shafritz DA (1980) Identification of integrated hepatitis B virus DNA and expression of viral RNA in an HBsAg producing human hepatocellular carcinoma cell line. Nature 286:531–533

Chang L (1973) Investigation of liver cancer in ducks of Qidong County. Qidong Liver Cancer Research Report

Christian BJ, Wu SQ, Ignjatonic MM, Meisner LF, Rezinkoff CA (1987) Cytogenic changes accompanying stepwise neoplastic transformation in vitro of normal human uroepithelial cells. Proc Am Assoc Cancer Res 28:121

Coggin JH (1969) Enhanced virus transformation of hamster embryo cells in vitro. J Virol 3:458–462

Coppleson M, Reid B (1968) The etiology of squamous carcinoma of the cervix. Obstet Gynecol 32:432–436

Cuzin F, Meneguzzi G, Binetruy B, Cerni C, Connan G, Gusoni M, de Lapeyriere O (1985) Stepwise tumoral progression in rodent fibroblasts transformed with bovine papilloma virus type 7 (BPV7) DNA. In: Howley PM, Biohet TR (eds) Papilloma viruses: molecular and clinical aspects. Liss, New York, pp 473–486

de Villiers E-M, Weidauer H, Otto H, zur Hausen H (1985) Papillomavirus DNA in tongue carcinomas. Int J Cancer 36:575–578

Diamond L, Knorr R, Shimizu Y (1974) Enhancement of Simian virus 40-induced transformation of Chinese hamster embryo cells by 4-nitroquinoline 1-oxide. Cancer Res 34:2599–2604

DiPaolo JA (1983) Relative difficulties in transforming human and animal cells in vitro. JNCI 70:3–7

Docherty JJ, Ludovici PP, Schloss GT (1972) Inhibitory effect of carcinogenic aromatic hydrocarbons on transformation of 3T3 cells by SV40. Proc Soc Exp Biol Med 140:969–973

Dulbecco R (1973) Viruses and the mechanisms of cancer. In: Doll R, Vodopja I (eds) Host-environmental interactions in the etiology of cancer in man. IARC, Lyon, pp 357–366
Durst M, Gissmann L, Ikenberg H, zur Hausen H (1983) A papillomavirus DNA from a cervical carcinoma and its prevalence in cancer biopsy samples from different geographic regions. Proc Natl Acad Sci USA 80:3812–3815
Eisenberg E, Rosenberg B, Knitchkoff DJ, Conn F (1985) Verrucous carcinoma: a possible viral pathogenesis. Oral Surg 59:52–57
Farber E, Cameron R (1980) The sequential analysis of cancer development. Adv Cancer Res 31:125–226
Feitelson MA, Millman I, Blumberg BS (1986) Tree squirrel hepatitis B virus: Antigenic and structural characterization. Proc Natl Acad Sci USA 83:2994–2997
Frenkel N, Roizman B, Cassai E, Nahmias A (1972) A DNA fragment of herpes simplex 2 and its transcription in human cervical cancer tissue. Proc Natl Acad Sci USA 69: 3784–3789
Fisher PB, Weinstein IB (1980) Chemical viral interactions and multiple aspects of cell transformation. In: Montesano R, Bartsch H, Tomatis L (eds) Molecular and cellular aspects of carcinogen screening tests. International Agency for Research on Cancer, Lyons, pp 113–131
Fisher PB, Weinstein IB, Eisenberg D, Ginsberg HS (1978) Interactions between adenovirus, a tumor promoter, and chemical carcinogens in transformation of rat embryo cell cultures. Proc Natl Acad Sci USA 75:2311–2314
Ford JN, Jennings PA, Spradbrow PB, Francis J (1982) Evidence for papillomaviruses in ocular lesions of cattle. Res Vet Sci 32:257–258
Fouts EA, Greenlaw RH, Rush BF, et al. (1969) Vernicous squamous cell carcinoma of the oral cavity. Cancer 23:152–160
Gallo RC, Saxinger WC, Gallangher RE, et al. (1977) Some ideas on the origin of leukemia in man and recent evidence for the presence of type-C viral related information. In: Hiatt H, Watson J, Winsten J (eds) Origin of human cancer. J Cold Spring Harbor Lab Prev, New York, pp 1253–1285
Galloway DA, McDougall JK (1983) The oncogenic potential of herpes simplex viruses: evidence for a "hit-and-run" mechanism. Nature 302:21–24
Gassenmaier A, Lammel M, Pfister H (1984) Molecular cloning and characterization of the DNAs of human papilloviruses 19, 20 and 25 from a patient with epidermodysplasia verruciformis. J Virol 52:1019–1023
Gerin JL (1984) The woodchuck (*Marmota monax*): an animal model of hepatitis. In: Chisari FV (ed) Advances in hepatitis research. Masson, New York, pp 40–48
Ghoshal AK, Farber E (1983) The induction of resistant hepatocytes during initiation of liver carcinogenesis with chemicals in rats fed a choline-deficient, low methionine diet. Carcinogenesis 4:801–804
Gissmann L (1984) Papillomaviruses and their association with cancer in animals and man. Cancer Surv 3:161–181
Gissmann L, Diehl V, Schultz-Coulon, zur Hausen H (1982a) Molecular cloning and characterization of human papilloma virus DNA derived from a laryngeal papilloma. J Virol 44:393–400
Gissmann L, de Villiers E-M, zur Hausen H (1982b) Analysis of human genital warts (condylomata acuminata) and other genital tumors in human papillomavirus type 6 DNA. Int J Cancer 29:143–146
Gissmann L, Wolnik L, Ikenberg H, et al. (1983) Human papillomavirus types 6 and 11 DNA sequences in genital and laryngeal papillomas and in some cervical cancer biopsies. Proc Natl Acad Sci USA 80:560–563
Gissmann L, Boshart M, Durst M, Ikenberg H, Wagner D, zur Hausen H (1984) Presence of human papillomavirus in genital tumors. J Invest Dermatol 83:26–28
Green M, Brackmann KH, Sanders PR, Loewentein PM, Freel JH, Eisinger M, Switlyk SA (1982) Isolation of a human papillomavirus from a patient with epidermodysplasia vernicilovirus: presence of related viral DNA genomes in human urogenital tumors. Proc Natl Acad Sci USA 79:4437–4441

Hampar B, Ellison SA (1963) Cellular alterations in the MCH line of Chinese hamster cells following infection with herpes simplex virus. Nature 192:145–147
Harrington JS, Nunn JR, Irwig L (1973) Dimethylnitrosamine in the human vaginal vault. Nature 241:49–50
Harris CC (1987) Human tissues and cells in carcinogenesis research. Cancer Res 47:1–10
Harris CC, Sun TT (1987) Interactive effects of chemical carcinogens and hepatitis B virus in the pathogenesis of human hepatocellular carcinoma. Cancer Surv 5:765–780
Hino O, Kitagawa T, Koike K, Kobayishi M, Hara M, Mori W, Nakashima T, Hattori N, Sugano H (1984a) Detection of hepatitis B virus DNA in hepatocellular carcinomas in Japan. Hepatology 4:90–95
Hino O, Kitagawa T, Sugano H (1984b) Relationship between serum and histochemical markers for hepatitis B virus and rate of viral integration in hepatocellular carcinomas in Japan. Int J Cancer 35:5–10
Hirai K, Defendi V, Diamond L (1974) Enhancement of simian virus 40 transformation and integration by 4-nitroquinoline 1-oxide. Cancer Res 34:3497–3500
Howard DK, Schlom J, Fisher PB (1983) Chemical carcinogen-mouse mammary tumor virus interactions in cell transplantation. In Vitro 19:58–66
IARC (1976) Evaluation of carcinogenic risk of chemicals to man. International Agency for Research on Cancer, Lyon (IARC monograph series, vol 10)
Ikenberg H, Gissmann L, Gross G, Grussendorf-Conen E-I, zur Hausen H (1983) Human papillomavirus type 16-related DNA in genital Bowen's disease and in bowenoid papulosis. Int J Cancer 32:563–565
Jablonska S, Dabrowski J, Jakubowicz K (1972) Epidermodysplasia verruciformis as a model in studies on the role of papovaviruses in oncogenesis. Cancer Res 32:583–589
Jablonska S, Orth G, Lutzner MA (1982) Immunopathology of papillomavirus induced tumors in different tissues. Springer Semin Immunopathol 5:33–62
Jackson GG, Muldoon RL (1973) Viruses causing common respiratory infection in man. IV. Reoviruses and adenoviruses. J Infect Dis 128:811–866
Jarrett WFH, McNeil PE, Grimshaw WTR, et al. (1978) High incidence area of cattle cancer with a possible interaction between an environmental carcinogen and a papilloma virus. Nature 274:215–217
Johnson FB (1982) Chemical interaction with herpes simplex type 2 virus: enhancement of transformation by selected chemical carcinogenesis and procarcinogens. Carcinogenesis 3:1235–1240
Joossens JV, Geboers J (1984) Epidemiology and mortality trends of sex-related cancers. In: Wolff JP, Scott JS (eds) Hormones and sexual factors in human cancer aetiology. Elsevier, Amsterdam, pp 3–18
Kakunaga T, Crow JD, Hamada H, Hirakawa T, Leavitt J (1983) Mechanisms of neoplastic transformation in human cells. In: Harris CC, Autrup H (eds) Human carcinogenesis. Academic, New York, pp 371–400
Keller AZ (1978) Liver cirrhosis, tobacco, alcohol and cancer among blacks. J Am Med Assoc 70:575–589
Kessler DJ, Heilman CA, Cossman J, Maguire RT, Torgeirsson SS (1987) Transformation of Epstein-Barr virus immortalized human B-cells by chemical carcinogens. Cancer Res 47:527–531
Kessler II (1974) Perspectives in the epidemiology of cervical cancer with special reference to the herpes virus hypothesis. Cancer Res 34:1091–1110
Kew MC, Rossouw E, Hodkinson J, Paterson A, Dusheiko GM, Whitcutt JM (1983) Hepatitis B virus status of Southern African blacks with hepatocellular carcinoma: comparison between rural and urban patients. Hepatology 3:65–68
Kew MC, Dibisceglie AM, Paterson AC (1985) Smoking as a risk factor in hepatocellular carcinoma. Cancer 56:2315–2317
Kodama K, Ogasawar N, Yoshikawa H, Murakami S (1985) Nucleotide sequence of a cloned woodchuck hepatitis virus genome: evolutional relationship between hepadenoviruses. J Virol 56:978–987
Kotin P, Wiseley DV (1963) Production of lung cancer in mice by inhalation exposure to influenza virus and aerosols of hydrocarbons. Prog Exp Tumor Res 3:186–215

Kreider J, Bartlett GL, Sharkey FE (1979) Primary neoplastic transformation in vivo of xenogeneic skin grafts in nude mice. Cancer Res 39:272–276

Kreider J, Howett MK, Lill NL, Bartlett GL, Zaino RJ, Sedlacek TV, Montel R (1986) In vivo transformation of human skin with human papillomavirus type II from condylomata acuminata. J Virol 59:369–376

Kremsdorf D, Jablonska S, Favre M, Orth G (1983) Human papillomaviruses associated with epidermodysplasia verruciformis. II. Molecular cloning and biochemical characterization of human papillomavirus 3a, 8, 10, and 12 genomes. J Virol 48:340–351

Kremsdorf D, Favre M, Jablonska S, Obalek S, Rueda LA, Lutzner HA, Blanchet-Bardon C, van Voorst Vader PC, Orth G (1984) Molecular cloning and characterization of the genomes and nine newly recognized human papillomavirus types associated with epidermodysplasia verruciformis. J Virol 52:1013–1018

Lam KC, Yu MC, Leung JWC, Henderson BE (1982) Hepatitis B virus and cigarette smoking: risk factors for hepatocellular carcinoma in Hong Kong. Cancer Res 42:5246–5248

Lancaster WD, Olson C (1982) Animal papillomaviruses. Microbiol Rev 46:191–207

Land H, Parada LF, Weinberg RA (1983) Tumorigenic conversion of primary embryo fibroblasts requires at least two cooperating oncogenes. Nature 304:596–602

Li W, Xie J, Yu S, Zhu Y, Gong X, Hou C, Wu B, Cao L (1985) Correlation between geographical distribution of hepatocellular carcinoma and selenium level. Qidong Liver Cancer Research, Qidong Liver Cancer Institute, Jiangsu, pp 43–46

Lieber CS, Seitz HK, Garro AJ, Worner TM (1979) Alcohol-related diseases and carcinogenesis. Cancer Res 39:2863–2886

Lin JJ, Liu C, Svoboda DJ (1974) Long-term effects of aflatoxin B_1 and viral hepatitis in marmoset liver. Lab Invest 30:167–178

Logan J, Pilder S, Shenk T (1984) Cancer Cells 2:527–532. Cold Spring Harbor Lab

Loning T, Ikenberg H, Becker J, Gissmann L, Hoepler I, zur Hausen H (1985) Analysis of oral papillomas, leukoplasias, and invasive carcinomas for human papillomavirus type related DNA. J Invest Dermatol 84:417–420

Lu J, Li W, Qin C, Nee C, Huang F (1983) Matched prospective study on chronic carriers of HBsAg and primary hepatocellular carcinoma. Chin Oncol 5:406–409

Lutzner MA (1978) Epidermodysplasia verruciformis. An autosomal recessive disease characterized by viral warts and skin cancer. Bull Cancer (Paris) 65:169–182

Maitland NJ, Kinross JH, Busuttil A, Ludgate SM, Smart GE, Jones KW (1981) The detection of DNA tumor virus-specific RNA sequences in abnormal human cervical biopsies by in situ hybridization. J Gen Virol 55:123–137

Majoros M, Devine KD, Parkhill EM (1963) Malignant transformation of benign laryngeal papillomas in children after radiation therapy. Surg Clin North Am 43:1049–1061

Marion PL, van Davelaus MJ, Knight SS, Salazar FH, Garcia G, Popper H, Robinson WS (1986) Hepatocellular carcinoma in ground squirrels persistently infected with ground squirrel hepatitis virus. Proc Natl Acad Sci USA 83:4543–4546

Marshall JR, Graham S, Byers T, Swanson M, Brasure J (1983) Diet and smoking in the epidemiology of cancer of the cervix. JNCI 70:847–851

Martini GA (1980) The role of alcohol in the etiology of cancer of the liver. In: Nieburgs HE (ed) Prevention and detection of cancer. Part II. Detection, vol 2. Cancer Detection in specific sites. Dekker, New York, pp 2163–2175

McDougall JK, Galloway DA, Fenoglio CM (1980) Cervical carcinoma: detection of herpes simplex virus RNA in cells undergoing neoplastic change. Int J Cancer 25:1–8

Melbye M, Skinhoj P, Nielsen NH, Vestergaad BF, Ebbesen P, Hansen PH, Biggar RJ (1984) Virus-associated cancers in Greenland: frequent hepatitis B virus infection but low primary hepatocellular carcinoma incidence. JNCI 73:1267–1272

Milo GE, Blakeslee JR, Hart R, Yohu DS (1978) Chemical carcinogen alteration of SV40 virus induced transformation of normal human cell populations in vitro. Chem Biol Interact 22:185–197

Moreno-Lopez J, Morner T, Pettersson U (1986) Papillomavirus DNA associated with pulmonary fibromatosis in European elks. J Virol 57:1173–1176

Mounts P, Shah KV, Kashima H (1982) Viral etiology of juvenile and adult-onset squamous papilloma of the larynx. Proc Natl Acad Sci USA 79:5424–5429
Naghashfar Z, Sawada E, Kutcher MJ, Swancar J, Gupta J, Daniel R, Kashima H, Woodruff JD, Shah K (1985) Identification of genital tract papillomaviruses HPV-6 and HPV-16 in warts of the oral cavity. J Med Virol 17:313–324
Nahmias H, Naib ZM, Josey W (1974) Epidemiological studies relating genital herpes to cervical cancer. Cancer Res 34:1111–1117
Namba M, Nishitani K, Kimoto T (1978) Carcinogenesis in tissue culture. 29. Neoplastic transformation of a normal human diploid cell strain, WI-38, with Co-60 gamma rays. Jpn J Exp Med 48:303–311
Namba M, Nishitani K, Fukushima F, Kimoto T, Nose K (1986) Multistep process of neoplastic transformation of normal human fibroblasts by ^{60}Co gamma rays and Harvey sarcoma viruses. Int J Cancer 37:419–423
Newbold RF, Overell RW (1983) Fibroblast immortality is a prerequisite for transformation by EJ c-Ha-*ras* oncogene. Nature 304:648–651
Nishiyama Y, Rapp F (1981) Repair replication of viral and cellular DNA in herpes simplex virus type 2-infected human embryonic and xeroderma pigmentosum cells. Virology 110:466–475
Ohnishi K, Lida S, Furama S, Goto N, Nomura F, Takashi M, Mishima A, Kono K, Kimura K, Musha H, Kotota K, Okuda K (1982) The effect of chronic habitual alcohol intake on the development of liver cirrhosis and hepatocellular carcinoma: Relation to hepatitis B surface antigen carriage. Cancer Res 49:672–677
Orth G, Jablonska S, Jarzabek-Chorzelska M, Obalek S, Rzesa G, Favre M, Croissant O (1979) Characteristics of the lesions and risk of malignant conversion as related to the type of human papillomavirus involved in epidermodysplasia verruciformis. Cancer Res 39:1074–1082
Ostrow RS, Bender M, Niimura M, Seki T, Kawashima M, Pass F, Faras AJ (1982) Human papilloma virus DNA in cutaneous primary and metastasized squamous cell carcinoma from patients with epidermodysplasia verruciformis. Proc Natl Acad Sci USA 79:1634–1638
Pater A, Pater MM (1986) Transformation of primary human embryonic kidney cells to anchorage independence by a combination of BK virus DNA and the Ha-*ras* oncogene. J Virol 58:680–683
Peraino C, Richards WL, Stevens FJ (1983) Multistage hepatocarcinogenesis. In: Slaga TJ (ed) Mechanisms of tumor promotion, vol I. CRC Press, Boca Raton, pp 1–54
Perez CA, Kraus FT, Evans JC, et al. (1966) Anaplastic transformation in verrucous carcinoma of the oral cavity after radiation therapy. Radiology 86:108–115
Peto R (1986) Introduction: geographic pattern and trends. In: Peto R, zur Hausen H (eds) Viral etiology of cervical cancer. Cold Spring Harbor Laboratory, New York,pp 3–15 (Banbury Report 21)
Pfister H (1984) Biology and biochemistry of papillomaviruses. Rev Physiol Biochem Pharmacol 99:111–181
Phillips LA (1983) Viruses associated with human cancer. Dekker, New York
Pinborg J (1980) Oral cancer and precancer. Wright, Bristol
Pitot HC, Goldsworthy T, Moram S (1982) The natural history of carcinogenesis: implications of experimental carcinogenesis in the genesis of human cancer. In: Harris CC, Cerutti PA (eds) Mechanisms of chemical carcinogenesis. Liss, New York, pp 141–154
Poirier LA, Grantham PH, Rogers AE (1977) The effects of a marginally lipotrope-deficient diet on the hepatic levels of *S*-adenosylmethionine and in the urinary metabolites of 2-acetylaminofluorene in rats. Cancer Res 37:744–748
Proffitt SD, Spooner TR, Korek JC (1970) Origin of undifferentiated neoplasm from verrucous epidermal carcinoma of oral cavity following irradiation. Cancer 26:389–393
Qian GS, Xu GS, Liu YF (1976) Cancer induction trials in animals by molded corns and insecticides. Qidong Liver Cancer Res 68–76
Rassoulzadegan M, Cowie A, Carr A, Glaichenhaus N, Kamen R, Cuzin F (1982) The roles of individual polyoma virus early proteins in oncogenic transformations. Nature 300:713–718

Rassoulzadegan M, Naghashfar Z, Couri A, Carr A, Grisoni M, Kamen R, Cuzin F (1983) Expression of the large T protein of polyoma virus promotes the establishment in culture of "normal" rodent fibroblast cell lines. Proc Natl Acad Sci USA 80:4354–4358

Rhim JS, Fujita J, Arnstein O, Aaronson SA (1986) Neoplastic conversion of human keratinocytes by adenovirus 12-SV40 virus and chemical carcinogens. Science 227:1250–1252

Riou G, Barrois M, Tordjman I, Dutronquay V, Orth G (1984) Presend de genomes de papillomavirus et amplification des oncogenes c-myc et c-Ha-ras dans der cancers envahissants du col de l'uterus. CR Acad Sci Paris [ser 3] 299:575–580

Robinson J (1982) Cancer of the cervix: occupational risks of husbands and wives and possible preventive strategies. In: Jordan JA, Sharp F, Singer A (eds) Pre-clinical neoplasia at the cervix. Royal College of Obstetricians and Gynaecologists, London, p 11

Rotkin ID (1973) A comparison review of key epidemiological studies in cervical cancer related to current searches for transmissible agents. Cancer Res 33:1353–1367

Rous P, Beard JW (1934) The progression to carcinoma of virus-induced rabbit papillomas (Shope). J Exp Med 62:523–542

Rous P, Friedewald WF (1944) The effect of chemical carcinogens on virus induced rabbit carcinomas. J Exp Med 79:511–537

Rous P, Kidd JG (1938) The carcinogenic effect of a papilloma virus on the tarred skin of rabbits. I. Description of the phenomenon. J Exp Med 67:399–422

Royston I, Aurelian L (1970) Immunofluorescent detection of herpes virus antigens in exfoliated cells from human cervical carcinoma. Proc Natl Acad Sci USA 67:204–212

Rubel LR, Reynolds RE (1979) Cytologic description of squamous cell papilloma of the respiratory tract. Acta Cytol (Baltimore) 23:227–230

Ruley HE (1983) Adenovirus early region 1A enables viral and cellular transforming genes to transform primary cells in culture. Nature 304:602–606

Sakuma K, Takahara T, Okuda K, Tsuda F, Mayumi M (1982) Prognosis of hepatitis B virus surface antigen carriers in relation to routine liver function tests: a prospective study. Gastroenterology 83:114–117

Sasson IM, Haley NJ, Hoffmann D, Wynder EL, Hellberg DH, Nilsson S (1985) Cigarette smoking and neoplasia of the uterine cervix: smoke constituents in cervical mucus. N Engl J Med 312:315–316

Schlehofer JR, zur Hausen H (1982) Induction of mutations within the host cell genome by partially inactivated herpes simplex virus type I. Virology 122:471–475

Schreiber H, Nettesheim P, Lijinsky W, Richter CB, Walburg HE (1972) Induction of lung cancer in germfree, specific-pathogen-free, and infected rats by *N*-nitrosoheptamethyleneimine: enhancement by respiratory infection. JNCI 49:1107–1111

Shafritz DA, Shouval D, Sherman HI, Hadziyannis SJ, Kew MC (1981) Integration of hepatitis B virus DNA into the genome of liver cells in chronic liver disease and hepatocellular carcinoma. N Engl J Med 305:1067–1073

Sieber SM, Correa P, Dalgard DW, Adamson RH (1979) Induction of osteogenic sarcomas and tumors of the hepatobiliary system in nonhuman primates with aflatoxin B_1. Cancer Res 39:4545–4554

Smith KT, Campo MS (1985) The biology of papillomaviruses and their role in oncogenesis. Anticancer Res 5:31–48

Snyder RL, Tyler G, Summers J (1982) Chronic hepatitis and hepatocellular carcinoma associated with woodchuck hepatitis virus. Am J Pathol 107:422–425

Steinberg BM, Topp WC, Schneider PS, et al. (1983) Laryngeal papillomavirus infection during clinical remission. N Engl J Med 308:1261–1264

Stellman ST, Austin H, Wynder EL (1980) Cervix cancer and cigarette smoking: a case control study. Am J Epidem 111:383–388

Stemhagen A, Slade J, Altman R, Bill J (1983) Occupational risk factors and liver cancer. A retrospective case-control study of primary liver cancer in New Jersey. Am J Epidemiol 117:443–454

Stremlau A, Gissman L, Ikenberg H, Stark M, Bannasch P, zur Hausen H (1985) Human papilloma virus type 16 related DNA in an anaplastic carcinoma of the lung. Cancer 55:1737–1740
Su D (1979) Drinking water and liver cell cancer. An epidemiologic approach to the etiology of this disease in China. Chin Med J 92:748–756
Summers J, Mason WS (1982) Properties of the hepatitis B liver viruses related to taxonomic classification. Hepatology 2:61S–55S
Sun T, Chu Y, Hsia C, Wei P, Wu S (1986) Strategies and current trends of etiologic prevention of liver cancer. In: Harris CC (ed) Biochemical and molecular epidemiology of human cancer. Liss, New York, pp 283–293
Sun T, Chu Y, Ni Z, Lu J, Huang F, Ni ZP, Pei X, Yu L, Lu G (1987) Cancer: perspective to control. J Cell Physiol (Suppl 4)
Syrjanen KJ (1980) Bronchial squamous cell carcinomas associated with epithelial changes identical to condylomatous lesions of the uterine cervix. Lung 158:131–142
Syrjanen KJ (1982) Histological changes identical to those of condylomatous lesions found in esophageal squamous cell carcinomas. Arch Geschwulstforsch 52:283–292
Syrjanen KJ, Pyrhoren S, Syrjanen SM, Lamberg M (1983) Immunohistochemical demonstration of human papillomavirus (HPV) antigens in oral squamous cell lesions. Br J Oral Surg 21:147–153
Syrjanen KJ, de Villiers E-M, Vayrynen M, Mantyjarvi R, Parkkinen S, Saarikoski S, Castre O (1985) Double infection of uterine cervix by human papillomavirus (HPV) 16 and HPV 18 progressing into an invasive cancer in less than three years of prospective follow-up. Lancet i:510–511
Syrjanen SM, Syrjanen KJ, Lamberg MA (1986) Detection of human papillomavirus DNA in oral mucosal lesions using in situ DNA-hybridization applied on paraffin sections. Oral Surg 62:660–667
Syverton JT (1952) The pathogenesis of the rabbit papilloma in carcinoma sequence. Ann NY Acad Sci 54:1126–1137
Takashi S, Lombardi B, Shinozuka H (1982) Progression of carcinogen induced foci of a-glutamyltranspeptidase-positive hepatocytes to hepatomas in rats fed a choline deficient diet. Int J Cancer 29:445–450
Thiry L, Sprecher-Goldberger S, Hannecart-Pokorni E, Gould I, Bosseus M (1977) Specific non-immunoglobulin G antibodies and cell mediated responses to herpes simplex virus antigens in women with cervical carcinoma. Cancer Res 37:1301–1306
Thomas DB (1973) An epidemiologic study of carcinoma in situ and squamous dysplasia of the uterine cervix. Am J Epidemiol 98:10–28
Todaro GJ, Green H (1964) Enhancement by thymidine analogs of susceptibility of cells to transformation by SV40. Virology 24:393–400
Trevathan E, Layde P, Webster LA, et al. (1983) Cigarette smoking and dysplasia and carcinoma in situ of the uterine cervix. J Am Med Assoc 250:499–502
Trichopoulos D, McMahon B, Sparros L, Merikas G (1980) Smoking and hepatitis B-negative primary hepatocellular carcinoma. JNCI 65:111–114
Tuyns AJ, Obradovic M (1975) Brief communication with unexpected high incidence of primary liver cancer in Geneva, Switzerland. JNCI 54:61–64
Van den Elsen PJ, Houweling A, van der Eb A (1983) Expression of region E1b of human adenovirus in the absence of region E1a is not sufficient for complete transformation. Virology 128:377–390
van Rensburg SJ, Cook-Mozaffari P, van Schalkwyk DJ, van Der Watt JJ, Vincent TJ, Purchase IF (1985) Hepatocellular carcinoma and dietary aflatoxin in Mozambique and Transkei. Br J Cancer 51:713–726
Vanselow BA, Spradbrow PB, Jackson ARB (1982) Papillomaviruses, papillomas and squamous cell carcinomas in sheep. Vet Rec 110:561–562
Wakefield J, Yule R, Smith A, Adelstein AM (1973) Relation of abnormal cytological smears and carcinoma at cervix uteri to husband's occupation. Br Med J 2:142–143
Wang N, Sun Z, Pan Q, Zhu Y, Xia Q (1980) Liver cancer, liver disease background, and virus-like particles in serum among ducks from high incidence area of human hepatocellular carcinoma. Chin J Oncol 2:174–176

Wang Y, Yeh P, Li W, Liu Y (1983) Correlation between geographical distribution of liver cancer and aflatoxin B_1 climate conditions. Scientia Sinica, series B, pp 431–437

Waterhouse J, Muir C, Powell J (1982) Cancer incidence in five continents, vol IV. World Health Organization, Lyon (IARC publications no 42)

Westaby D, Portmann B, Williams R (1983) Androgen related primary hepatic tumors in non-Fanconi patients. Cancer 51:1947–1962

Williams RR, Horn JW (1977) Association of cancer sites with tobacco and alcohol consumption and socio-economic status of patients: interview study from the third national cancer survey. JNCI 58:525–547

Winkelstein W, Shillitoe EJ, Brand R, Johnson KK (1984) Further comments on cancer at the uterine cervix, smoking, and herpes virus infection. Am J Epidemiol 119:1–8

Winkler B, Capo V, Reumann W, Ma A, LaPorta R, Reilly S, Green PMR, Richart RM, Crum CP (1985) Human papillomavirus infection of the esophagus. Cancer 55: 149–155

Wogan GN (1976) Aflatoxins and their relationship to hepatocellular carcinoma. In: Okuda K, Peters RL (eds) Hepatocellular carcinoma. Wiley, New York, pp 25–42

Yokosuka O, Omata M, Zhou Y-Z, Imazeki F, Okuda K (1985) Duck hepatitis B virus DNA in liver and serum of Chinese ducks: integration of viral DNA in a hepatocellular carcinoma. Proc Natl Acad Sci USA 82:5180–5184

Yu MC, Ho JHC, Ross RK, Henderson BE (1981) Nasopharyngeal carcinomas in Chinese-salted fish or inhaled smoke? Prev Med 10:15–24

Yu MC, Mack T, Hanisch R, Peters RL, Henderson BE, Pike MC (1983) Hepatitis, alcohol consumption, cigarette smoking and hepatocellular carcinoma in Los Angeles. Cancer Res 43:6077–6079

Yu S, Chu Y, Gong X, Hou C, Li W, Gong H, Xie J (1985) Regional variation of cancer mortality incidence and its relation to selenium levels in China. Biological Trace Element Research 7:21–29

Zakelj MP, Fraser P, Inship H (1984) Cervical cancer and husband's occupation. Lancet I:510

Zhou I, Kou P, Shao L (1980) A virus possibly related to hepatitis and hepatoma in ducks. Shanghai Med 3:1–11

zur Hausen H (1982) Human genital cancer: synergism between two virus infections or synergism between a virus infection and initiating events? Lancet ii:1370–1372

zur Hausen H (1986) Intracellular surveillance of persisting viral infections. Lancet 2: 489–491

zur Hausen H, Schulte-Holthausen H, Klein G, et al. (1970) EBV DNA in biopsies of Burkitt tumours and anaplastic carcinoma of the nasopharynx. Nature 228:1056–1058

zur Hausen H, de Villiers EM, Gissmann L (1981) Papillomavirus infections and human genital cancer. Gynecol Oncol 12:124–128

Part III. Oncogenes in Tumour Development

CHAPTER 10

Human Oncogenes

A. HALL

A. Introduction

Oncogenes were first characterised as distinct genes in acutely transforming RNA tumour viruses. A combination of genetic and molecular techniques confirmed that a single gene carried by the virus, the viral oncogene, was solely responsible for the malignant growth of cells in the infected host (for review, see BISHOP and VARMUS 1984). There is no evidence to suggest that acutely transforming retroviruses play any role in the development of human malignancies, and a broader definition of an oncogene has evolved to encompass cellular sequences that are affected by genetic changes and thereby contribute to the initiation or maintenance of the malignant phenotype. The clearest example of this occurs when rearrangement, amplification or mutation of a gene is present in a tumour cell. In this case the gene is considered to be a proto-oncogene, the implication being that the genetic changes observed represent the conversion of this normal gene into a dominantly acting oncogene. In such cases, stronger justification for the term oncogene can be obtained by reintroducing the cloned gene into non-transformed cells and demonstrating a biological effect.

The identification of genes aberrantly expressed in tumour cells has also been taken as circumstantial evidence that they may be oncogenes. This is much more problematical, since it is often impossible to define the exact progenitor of a tumour cell and therefore the level of gene expression prior to malignant conversion. In addition, oncogenes by their nature are expected to change the expression of many other genes. Some of these genes may even contribute to the observed phenotype, but unless they have been shown to have biological activity in their own right, they are not generally considered as oncogenes.

There is growing evidence for the involvement of recessive oncogenes in human tumours. In this case it is the absence of a gene product essential for the normal development of a cell type which is thought to lead to abnormalities that may be manifested as a malignancy. To date recessive oncogenes have mainly been implicated in tumours associated with an hereditary predisposition, but this need not always be the case.

Finally, it is worth considering the kinds of biological changes that oncogenes might be expected to elicit. The generally held view is that human cancer develops as a consequence of a series of genetic changes, i.e. alterations to proto-oncogenes. Whether the steps often attributed to tumour development, such as immortality, lack of growth control, escape from immunological surveillance, metastasis, etc. involve discrete families of oncogenes or whether there is overlap between the phenotypes is not clear at the moment. What is known is that almost all oncogenes so far characterised have profound effects on cellular growth control and act in a dominant fashion. It is too early to say to what extent this bias reflects the limitations of the assays currently available for detecting oncogenes.

The first part of this chapter will deal with the various ways in which 50 or so proto-oncogenes have so far been identified, mainly through analysing genetic events leading to malignancy in animal systems. Although they are present in the human genome, by no means all of these proto-oncogenes have been implicated in human malignant disease. The second section will review the evidence for the involvement of particular genes in human cancer.

B. Identification of Oncogenes

I. Viral Oncogenes and Proto-Oncogenes

Since the identification of the first viral oncogene, v-*src*, carried by a chicken RNA tumour virus, over twenty different retroviruses each carrying a different viral oncogene have been characterised (see Table 1) (for reviews, see WEISS et al. 1984; HALL 1986). The origin of the viral oncogenes has been the topic of much discussion, but it is now widely accepted that v-*src* was derived from a cellular gene c-*src* after transduction of the cellular sequences by a chicken retrovirus (VARMUS 1984). The v-*src* and c-*src* genes differ not only in the level of expression from their viral and cellular promoters but also in their protein products which have some amino acid differences (27 out of 533) (TAKEYA and HANAFUSA 1983); hence, c-*src* is regarded as a proto-oncogene.

It is now clear that each of the 24 viral oncogenes shown in Table 1 was derived from a cellular proto-oncogene (BISHOP 1983; VARMUS 1984). Furthermore, these proto-oncogenes are highly conserved amongst all vertebrate species, leading to much speculation that they may be involved in human cancer (BISHOP 1983). Table 1 shows the chromosomal location of the human proto-oncogenes corresponding to the known viral oncogenes and the biochemical properties and cellular locations of the protein products.

Three acutely transforming retroviruses have been found that contain two viral genes, each derived from a distinct cellular locus (SAULE et al. 1987). In the case of the MC29 chicken virus, both genes, v-*myc* and v-*mil*, carried by the virus have been shown to be capable of acting independently as oncogenes. Of the two viral genes carried by the avian erythroblastosis virus, v-*erb*A and v-*erb*B, only the latter can act alone. *erb*A appears to block the differentiation of the erythroblasts, after they have been transformed by *erb*B, to produce a particular, aggressive type of leukaemia. Although *erb*A cannot alone transform cells, it is regarded as an oncogene (and its progenitor c-*erb*A as a proto-oncogene). It is now known that c-*erb*A is the thyroid hormone receptor gene (SAP et al. 1986), and this is, so far, the only example of this kind of receptor classified as a proto-oncogene. A third virus, E26, contains the v-*myb* and v-*ets* genes. Although v-*myb* can function independently as an oncogene, the situation with v-*ets* is not yet clear.

Table 1. Viral oncogenes identified from acutely transforming retroviruses. The corresponding human proto-oncogene and its chromosomal location (with reference) are shown along with the known biochemical activities and locations of the oncogene products

Viral oncogene	Chromosomal location of corresponding human proto-oncogene	Activities of protein product	Cell location
v-*src*	c-*src*; 20q12–q13 (LEBAU et al. 1984)	tyr kinase	PM
v-*fps* (fes)	c-*fps*; 15q25–q26 (HARPER et al. 1983)	tyr kinase	PM
v-*yes*	c-*yes*1; 18q21.3 (YOSHIDA et al. 1985)	tyr kinase	PM
v-*abl*	c-*abl*; 9q34 (GROFFEN et al. 1986)	tyr kinase	PM
v-*fgr*	c-*fgr*; 1p36.1–36.2 (TRONICK et al. 1985)	tyr kinase	PM
v-*mos*	c-*mos*; 8q22 (PRAKASH et al. 1982)	ser/thr kinase	C
v-*raf* (mil)	c-*raf*1; 3p25 (BONNER et al. 1984)	ser/thr kinase	C
v-*rel*	c-*rel*; 2p13cen (WALRO et al. 1987)	ser/thr kinase	C
v-*fms*	c-*fms*;5q34 (LEBEAU et al. 1986)	Receptor-like	PM
v-*erb*B	c-*erb*B1; 7p13–q11.2 (FRANCKE et al. 1986)	Receptor-like	PM
v-*ros*	c-*ros*; 6q22 (SATOH et al. 1987)	Receptor-like	PM
v-*kit*	c-*kit*; 4q11–q21 (YARDEN et al. 1987)	Receptor-like	PM
v-*myc*	c-*myc*; 8q24 (DALLA FAVERA et al. 1982)	Nuclear	N
v-*myb*	c-*myb*; 6q22–q24 (HARPER et al. 1983)	Nuclear	N
v-*ski*	c-*ski*; 1q22–qter (BALAZS et al. 1984)	Nuclear	N
v-*fos*	c-*fos*; 14q21–q31	Nuclear	N
v-*jun*	c-*jun* (BOMANN et al. 1987)	Nuclear transcription factor	N
v-*erb*A	c-*erb*A1; 17q12 (SHEER et al. 1985)	Nuclear receptor	N
v-Ha-*ras*	c-Ha-*ras*1; 11p15 (DE MARTINVILLE et al. 1983)	GTP binding	PM
v-Ki-*ras*	c-Ki-*ras*2, 12p12–pter (O'BRIEN et al. 1983)	GTP binding	PM
v-*sis*	c-*sis*; 22q12–q13 (BARTRAM et al. 1983)	Growth factor	S
v-*ets*	c-*ets*1; 11q23–q24 (SACCHI et al. 1986)	–	–
	c-*ets*2; 21q22.3 (SACCHI et al. 1986)	–	–
v-*akt*	c-*akt*1 (STAAL 1987)	–	–
v-*sea*	c-*sea*	–	–

(–), unknown; PM, plasma membrane; C, cytoplasm; N, nucleus; S, secreted.

II. Viral Integration Sites

A group of chronically transforming RNA tumour viruses also produces tumours in animals but with a much longer latency period than the acute viruses. These viruses do not carry an oncogene but have been found to integrate within the host genome at specific sites. It is thought that subsequent independent genetic events are then required to produce full malignancy.

The integration of the avian leukosis virus in B-cell lymphomas was the first well-characterised example of how chronically transforming RNA tumour viruses could activate a proto-oncogene (HAYWARD et al. 1981). In this case, activation of the c-*myc* locus was found to occur after integration of the virus, the viral LTR providing new transcription control signals for c-*myc*, resulting in over-expression. This is an important observation since it shows that proto-oncogenes can be activated in situ to provide one of the steps in tumour progression. Activation of the c-*erb*B gene by avian leukosis virus has also been observed in erythroblastosis (FUNG et al. 1983). In this case, however, integration occurs within the coding sequence of the gene and leads to a truncated protein being produced.

Table 2. Proto-oncogenes detected by viral *cis* activation through analysis of integration sites of chronically transforming retroviruses. Only well-characterised cases and where the genes had not previously been detected as viral oncogenes are shown

Proto-oncogene (virus)	Chromosomal location of human gene		Protein product	Cell location
int-1 (MMTV)	12q12–q13	(TURC-CAREL et al. 1987)	Growth factor	S
int-2 (MMTV)	11q13	(SEARLE et al. 1988)	Growth factor	S
pim-1 (MuLV)	6p21	(NAGARAJAN et al. 1986)	ser/thr kinase	PM
lck (MuLV)	1p32–p35	(MARTH et al. 1986)	Tyrosine kinase	PM
spi-1 (SFFV)	–	–	–	–
evi-1 (ecotropic provirus)	–	–	–	–

MMTV, mouse mammary tumour virus; MULV, murine leukaemia virus; SFFV, spleen focus forming virus; S, secreted; PM, plasma membrane.

Table 2 shows loci which have been characterised using this type of analysis but which were previously unknown. Integrated copies of the mouse mammary tumour virus, MMTV, have been found at two loci in mouse breast tumours (NUSSE et al. 1984; DICKSON et al. 1984). The loci *int*-1 and *int*-2 are regarded as proto-oncogenes even though they have not been detected independently as viral oncogenes. Subsequent analysis of *int*-1 and *int*-2 has shown that the genes are important in early embryonic development and are usually silent in adult tissues, but that, in MMTV-induced tumors, the viral integration event activates transcription of the genes. More recently, it has been shown that *int*-1, when carried on a retrovirus vector, can partially transform a mammary epithelial cell line (BROWN et al. 1986) providing stronger evidence that *int*-1, at least, is a true oncogene.

Analysis of murine leukaemia virus-induced T-cell lymphoma has identified a locus *pim*, which encodes a ser/thr protein kinase and which is transcriptionally activated after viral insertion (SELTEN et al. 1985; MEEKER et al. 1987). Another protein kinase-encoding gene, *lck* is activated in murine T-cell lymphomas by viral insertional activation (VORONOVA and SEFTON 1986), though this protein has been shown to be a tyrosine kinase. More recently, two new putative proto-oncogenes have been cloned from virally induced murine myeloid tumours (*evi*-1) and erythroleukaemias (*spi*-1) (MUCENSKI et al. 1988; MOREAU-GACHELIN et al. 1988), though nothing is yet known about their protein products.

III. Transfection Assay

The great interest in the genetic basis of human cancer led to the search for biological assays capable of detecting oncogenes directly in human-tumour DNA samples. In 1979, two groups reported the successful use of an established mouse fibroblast cell line, NIH 3T3, as a recipient for DNA transfer using tumour DNA (SHIH et al. 1979; COOPER et al. 1980). A little later, DNA from a human source, the bladder carcinoma cell line EJ, was used successfully to transform NIH 3T3 cells (KRONTIRIS and COOPER 1981; SHIH et al. 1981). This was observed as foci of transformed cells in a background confluent monolayer of untransformed NIH 3T3 cells. Over the past 6 years several hundred human tumour DNAs from cell lines and fresh biopsies have been assayed in this way. Well over 95% of the dominantly transforming oncogenes detected with this biological assay have been found to be activated versions of one of the three members of the *ras* gene family, c-Ha-*ras*1, c-Ki-*ras* or N-*ras* (SANTOS et al. 1982; McCOY et al. 1983). Ha-*ras* and Ki-*ras* were first characterised as cellular homologues of two viral *ras* genes (CHANG et al. 1982); N-*ras* was discovered by the transfection assay and is the third member of this closely related family (SHIMIZU et al. 1983; HALL et al. 1983).

There has been one report of a high incidence of activation of a new (unnamed) gene in thyroid carcinomas (FUSCO et al. 1987), but no further analysis has yet been reported. A few isolated examples of other novel genes have been detected using the transfection focus assay (see Table 3), but these have so far been sporadic, *dbl* (SRIVA-STAVA et al. 1986; EVA et al. 1987), *lca* (OCHIYA et al. 1986), *trk* (MARTIN-ZANCA et al. 1986), *met* (COOPER et al. 1984; DEAN et al. 1985), *hst* (SAKAMOTO et al. 1986), and *ret* (TAKAHASHI et al. 1985). In the case of *dbl, hst,* and *ret* at least, the oncogene was generated from its proto-oncogene during the transfection procedure and was not present in the original tumour. One instance in which a non-*ras* gene is frequently activated in a particular tumour is the *neu* oncogene in chemically induced rat neuroblastomas (SCHECHTER et al. 1984, 1985). So far, however, activated c-*erb*B2 (the human equivalent of rat c-*neu*) has not been detected by transfection.

Variations of the transfection assay have been attempted. One particularly fruitful approach has been to co-transfect tumour DNA with an antibiotic resistance marker gene, usually neomycin resistance (neo^R). Neo-resistant colonies are then selected in vitro, pooled and injected into nude mice where tumour formation can be observed. This has revealed at least one new oncogene, *mas* (YOUNG et al. 1986), as well as activations of the *ras* genes which went undetected in the focus assay (BOS et al. 1985). It appears that this nude mouse assay is more sensitive in detecting oncogenes than the focus-forming assay. Another

Table 3. Proto-oncogenes detected by NIH 3T3 DNA transfection assay. Only cases in which novel genes have been detected are shown. The activated *neu* oncogene was detected in rat neuroblastoma, all other genes were from human tumour cell lines or biopsies

Proto-oncogene	Chromosomal location of human gene		Protein product	Cell location
N-*ras*	1pcen–p21	(DAVIS et al. 1983)	GTP binding	PM
trk	1q31–q41		Tyrosine kinase	C
hst	–		Growth factor	S
met	7q21–q31	(DEAN et al. 1985)	Receptor-like	PM
dbl	–	–	–	C
ret	–	–	–	–
mas	6	(RABIN et al. 1987)	Receptor-like	PM
lca	2	(OCHIYA et al. 1986)	–	–
neu (c-*erb*B2)	17q21	(COUSSENS et al. 1985)	Receptor-like	PM

PM, plasma membrane; C, cytoplasm; S, secreted.

variation of the transfection assay reported recently makes use of a defined culture medium to select for genes affecting growth (ZHAN et al. 1987).

NIH 3T3 is an established cell line and as such would not be expected to respond to genes responsible for immortalisation. Non-established cell lines, e.g. primary rat embryo fibroblasts and Syrian hamster dermal fibroblasts, have been used in attempts to identify immortalisation genes. This has been totally unsuccessful using tumour DNA, though it has been employed to show that some of the already cloned oncogenes have an immortalisation capability. In particular, two groups reported co-transfection of primary fibroblasts with combinations of oncogenes and showed two-step, full transformation to malignancy (LAND et al. 1983; RULEY 1983). However, the biochemical basis of immortalisation is still very unclear, and all oncogenes shown to be capable of immortalising cells (e.g. *myc*, p53) also have effects on growth control.

IV. Translocational Breakpoints

Many tumours, especially leukaemias, have well-characterised translocations. It is tempting to assume that the breakpoints represent the site of an oncogene. The two best-characterised examples of breakpoint analysis are in Burkitt's lymphoma, where a translocation event brings the c-*myc* gene (on chromosome 8) and one of the three immunoglobin loci together (for review, see KLEIN 1983), and in chronic myeloid leukemia (CML), where a translocation event moves the c-*abl* gene from chromosome 9 to a locus called *bcr* on chromosome 22 (GROFFEN et al. 1984). The involvement of these two proto-oncogenes was identified using cloned probes, but novel potential oncogenes have been recognised by cloning sites of breakpoints (see Table 4A). Although it need not always be the case (NB c-*abl*/bcr), the general principle has been that one part of the translocation is likely to involve a gene which is capable of undergoing rearrangement during normal development (CROCE 1987). So far, this is restricted to the immunoglobulin heavy chain (chromosome 14q32), the two immunoglobulin light chains (chromosomes 2p11 and 22q11) and the T-cell receptor genes (α, chromosome 14q11-q12; β, chromosome 7q32-q36; γ, chromosome 7p15). By using probes derived from one of these genes (depending on which chromosome is involved), clones can be obtained from libraries which encompass the breakpoint. In this way three putative proto-oncogenes have been identified, *bcl-1, bcl-2* (TSUJIMOTO et al. 1985), and

Table 4. Proto-oncogenes detected by cloning. A. Breakpoint regions of well-characterised translocations in human leukaemia. B. Amplified sequences. C. Others

Proto-oncogene	Chromosomal location of human gene		Protein product
A *bcl*-1	11q13	(TSUJIMOTO et al. 1985)	–
bcl-2	18q21	(TSUJIMOTO et al. 1985)	–
tcl-1	14q32.2	(CROCE et al. 1985)	–
B *gli*	12q13–q14.3	(KINZLER et al. 1987)	–
C p53	17p13	(ISOBE et al. 1986)	Nucleus

tcl-1 (CROCE et al. 1985). No independent biological evidence is yet available to suggest what, if any, role they might play in the transformation process.

V. Other Routes

1. Structural Homology

Several of the proto-oncogenes identified as described earlier have turned out to be members of a gene family (Table 5). Naturally, the other members of the family are also prime candidates for oncogenes. For example, using probes derived from the c-*myc* gene, at least three other related genes have been identified (N-*myc*, L-*myc*, and R-*myc*). Subsequent cloning and sequence analysis have revealed that these four genes are closely related and constitute a gene family (DEPINHO et al. 1987). Furthermore, gene transfer experiments using cotransfection into primary fibroblasts have shown that all four *myc* genes have transformation activity (SCHWAB et al. 1985; DEPINHO et al. 1987).

The three *ras* genes belong to a very closely conserved family. However, using *ras* probes, at least five more distantly related genes have been identified, three *rho* genes (MADAULE and AXEL 1985), R-*ras* (LOWE et al. 1987) and *ral* (CHARDIN and TAVITIAN 1986), with significant amino acid homology to *ras*, though there is yet no evidence available to say whether they can function as oncogenes.

Although all the protein kinases belong to a super family in the sense of having homologies within the kinase domain, *src* appears to belong to a more closely related family of which at least six others are known (COOPER 1987). Three of these, *src*, *fgr*, and *yes*, have been found to be viral oncogenes (Table 1), and *lck* is activated by viral insertion (Table 2). The other members (*hck*, *lyn*, and *syn*/*slk*) should therefore be regarded as candidate proto-oncogenes (Table 5).

Table 5. Proto-oncogene gene families: the prototype is shown along with related genes detected by DNA or protein homology

Prototype	Other members of family	Chromosomal location of human gene	
c-*myc*	N-*myc*	2p23–pter	(SCHWAB et al. 1984)
	L-*myc*	1p32	(NAU et al. 1985)
	R-*myc*	–	(DEPINHO et al. 1987)
c-*ras* (Ha, Ki, N)	*rho* ($\times$ 3)	–	(MADAULE and AXEL 1985)
	ral	–	(CHARDIN and TAVITIAN 1986)
	R-*ras*	–	(LOWE et al. 1987)
c-*src*/*fgr*/*yes*1	*lck*	1p32–35	(MARTH et al. 1986)
	hck	–	
	lyn	–	
	fyn (*syn*/*slk*)	6q21	(YOSHIDA et al. 1986)
c-*ets*1/2	*erg*	–	(REDDY et al. 1987)
c-*raf*1	*pks*	Xp11.4	(MARK et al. 1986)
c-*abl*	*arg*	1q24–25	(KRUH et al. 1986)
c-*ros*1	*flt* (*frt*)	13q12	(SATOH et al. 1987)

2. Functional Homology

Since the genes encoding at least six growth-factor receptors, namely c-*erb*B1 (ligand EGF), c-*fms* (ligand CSF1), c-*ros*, c-*erb*B2, c-*kit*, and c-*met* (ligands unknown), can function as oncogenes (DOWNWARD et al. 1984; SHERR et al. 1985; BIRCHMEIER et al. 1986; SCHECHTER et al. 1985; YARDEN et al. 1987; DEAN et al. 1985), it is tempting to speculate that genes encoding other growth-factor receptors such as those for PDGF, insulin or bombesin may also be proto-oncogenes. The genes for the PDGF and insulin receptors have been cloned, and, in the case of the insulin receptor, in vitro mutagenesis can lead to the generation of a gene which, when introduced into NIH 3T3 cells, results in morphological transformation (WANG et al. 1987).

Similarly, genes for growth factors themselves can act as oncogenes, e.g. v-*sis* which codes for a PDGF molecule (WATERFIELD et al. 1983). Abnormal production of a growth factor by a cell also expressing a receptor for that growth factor would be expected to lead to autocrine stimulation of growth, and there is much evidence that human tumours produce growth factors. As yet, however, the role they play in maintaining the transformed phenotype is unclear.

3. Cloning Amplified Sequences

Many human tumours are associated with amplified sequences. These have been observed cytologically as double minute (DM) chromosomes or as homogenous staining regions (HSRs) of chromosome. More recently, using denaturation-renaturation gel techniques, it has proved possible to clone regions of amplified sequences. In one reported case a gene, *gli*, was cloned from a human glioma cell line containing a DM (KINZLER et al. 1987) (Table 4B). It is not at all clear whether *gli* can be regarded as a proto-oncogene, and although it is expressed at high levels in this particular glioma cell line, amplification or over-expression has not been detected in other lines.

4. Transformation by DNA Tumour Viruses

DNA tumour viruses such as SV40, polyoma and papilloma contain genes which act as oncogenes. However, attempts to isolate cellular sequences with homology to these have been unsuccessful. The analysis of polyoma and SV40 transforming proteins has, however, led to the identification of a nuclear protein p53 which associates with viral oncoproteins in transformed cells (LANE and CRAWFORD 1979). The gene for p53 has been cloned (Table 4C) and has the capacity to immortalise cells and cooperate with *ras* in the co-transfection assay (PARADA et al. 1984).

5. Recessive Oncogenes

Several cancers are associated with hereditary predispositions. Retinoblastoma and Wilms' tumour, for example, are associated with specific chromosomal deletions, and it is assumed that one copy of a recessive oncogene has been deleted (the hereditary component) and that initiation of the tumour requires inactivation of the second allele (the somatic component). The identification and cloning of the recessive oncogene in retinoblastoma has been reported and has involved cloning large stretches of DNA that have been deleted in the syndrome and then attempting to define the critical gene (FRIEND et al. 1986). This will be dealt with in depth in a later chapter and will not be discussed further here. Although the analysis of recessive oncogenes is still in its infancy, rapid progress has been made in the chromosomal localisation of the genetic defect in several cancer-prone syndromes. The cloning of recessive oncogenes will undoubtedly be a very exciting and rapidly evolving area in the near future.

VI. Conclusions

It might be argued that almost any protein involved in controlling cell proliferation could function as an oncogene if its regulation or level of expression were disturbed. It is too early to tell whether this is the case or not, but certainly any newly cloned gene which might be involved in cell proliferation is tested for its biological effect by transfection into NIH 3T3 cells.

It will become obvious from the next section, however, that there is an enormous gap between the ever-expanding list of oncogenes as defined in animal systems or experimentally in tissue culture and the still relatively small list of oncogenes for which there is any evidence for a role in human malignancy.

C. Activation of Proto-Oncogenes in Human Malignancies

I. Viruses

The human T-cell lymphotropic viruses I and II are the only retroviruses known to be directly responsible for a human malignant disease. They have been shown to be the aetiological agents in rare forms of adult T-cell leukaemia (POIESZ et al. 1980), but their detailed mechanism of action is unknown. They do not carry a conventional viral oncogene nor do they integrate at specific chromosomal loci. It is believed that a protein product of the virus (*tat*) activates cellular gene expression in *trans*. In agreement with this, it has been shown that transgenic mice expressing the *tat* gene under the control of the HTLV-1 LTR show a very high incidence of mesenchymal tumours (NERENBERG et al. 1987). The *tat* gene can therefore be regarded as a viral oncogene which can act in *trans* on cellular genes. No cellular counterpart of *tat* has been found.

A variety of DNA tumour viruses are thought to play some role in human cancer (see MARSHALL and RIGBY 1984), in particular, papillomaviruses in genital malignancies, hepatitis B in liver carcinomas, and EBV in Burkitt's lymphoma. These viruses do not contain oncogenes which have sequence homology to cellular genes, but they do contain transforming genes, some of which have been shown to be capable of transforming NIH 3T3 cells. The oncogenes carried by these viruses will not be discussed further here.

II. Growth Factors and Their Receptors

Four viral and two cellular oncogenes derived from growth-factor receptor genes are known. The formation of these oncogenes from their normal cellular counterparts has been found to occur by (1) removal of the external ligand binding domain only (e.g. v-*ros*, v-*kit* or viral integration at the c-*erb*B1 locus in chickens), (2) removal of the external domain and the carboxy-terminal region (v-*erb*B), (3) point mutation in the transmembrane domain (activation of c-*neu* in rat neuroblastomas), and (4) multiple amino acid changes especially at phosphorylation sites (v-*fms*). So far, no examples of any of these kinds of genetic alteration have been found in a receptor gene in a human malignancy. What has been found, however, is ectopic or high-level expression of specific receptors in certain types of cancer.

1. EGF Receptor (c-*erb*B1)

Amplification of the c-*erb*B1 gene has been detected in several tumour types (Table 6). Around 40% of brain tumours of glial origin have been reported to contain higher than normal levels of EGF receptor, and analysis of the DNA showed gene amplification (between 10 and 50-fold) in these tissue samples (LIBERMANN et al. 1985; WONG et al. 1987). In most cases the level of gene amplification is over eightfold, and, in several cases, rearrangement of the amplified gene was also observed (WONG et al. 1987).

A much greater proportion of squamous cell carcinomas (>90%) have increased levels of EGF receptor protein compared with normal skin (HENDLER

Table 6. Oncogene amplification in human malignancies. Percentages are given only when a substantial number of tumours have been examined

Gene	Tumour type	Percentage amplification	References
c-*erb*B1	Brain (glial)	40	LIBERMANN et al. (1985); WONG et al. (1987)
	Squamous cell carcinoma	10	OZANNE et al. (1986)
	Lung squamous cell carcinoma	20	BERGER et al. (1987)
c-*erb*B2	Breast adenocarcinoma	25	SLAMON et al. (1987); VENTER et al. (1987)
	Stomach adenocarcinoma	2/9	TOYOSHIMA et al. (1986)
	Kidney adenocarcinoma	1/4	TOYOSHIMA et al. (1986)
c-*myc*	Small cell lung carcinoma	20	NAU et al. (1986)
	Breast adenocarcinoma	10	VARLEY et al. (1987)
	Gastric adenocarcinoma	3/16	SHIBUYA et al. (1985)
N-*myc*	Late stage (III + IV) neuroblastoma	50	SEEGER et al. (1985)
	Small cell lung carcinoma	20	NAU et al. (1986)
	Retinoblastoma	20	LEE et al. (1984)
L-*myc*	Small cell lung carcinoma	15	NAU et al. (1985)
ras	Colon/lung	1	PULCIANI et al. (1985)
c-*akt*	Gastric adenocarcinoma	1/5	STAAL (1987)

and OZANNE 1984; OZANNE et al. 1986). In many cases this amounts to around 10^6 receptors/cell. However, unlike the brain glial tumours, this does not correlate with gene amplification, which is found in only around 10% of squamous carcinoma biopsies (YAMAMOTO et al. 1986). Interestingly, 10 of 12 cell lines tested by YAMAMOTO et al. had an amplified EGF receptor gene, suggesting that squamous carcinomas with this amplification event are better suited to growth in vitro.

The role of EGF receptor amplification, if any, is still not clear. Little is known about the levels of the receptor protein in subpopulations of normal cells, and it is possible that at least some of the carcinomas were derived from normal cells already expressing high levels of the protein. On the other hand, no normal cell populations have yet been found that express the very high level of receptor found in some of these carcinomas. Recently, it has been reported that high level expression in NIH 3T3 cells of the normal EGF receptor leads to cell transformation and malignancy (VELU et al. 1987) but only when the cells are cultured in the presence of EGF. In addition, antibodies to the EGF receptor have been shown to block the growth of some squamous carcinomas grown in nude mice (RODECK et al. 1987). These experimental results would certainly be consistent with the idea that over-expression of the EGF receptor, coupled to stimulation by EGF or TGFα through an autocrine or paracrine route, is an important part of the maintenance of the tumour phenotype.

2. c-*erb*B2 Receptor

An activated c-*neu* proto-oncogene was first identified in an experimentally derived rat neuroblastoma cell line using the NIH 3T3 DNA transfection assay (SCHECHTER et al. 1984). The normal and activated genes have been cloned, and the only difference between them is a single amino acid substitution (val→glu) in the trans-membrane domain (BARGMANN et al. 1986). No activated c-*erb*B2 genes have been detected in human tumours using the transfection assay, but since it is known that the gene is 120 kb long (the rat c-*neu* is only 40 kb), it is unlikely that this would survive the manipulations of the DNA transfection procedure. There have been no reports of more detailed sequence analyses of c-*erb*B2 in human tumours aimed at looking for point mutations analogous to those in activated c-*neu*.

There is, however, evidence for gene amplification of c-*erb*B2, especially in human breast adenocarcinomas. Around 25% of these malignancies have an amplified gene (3–50 copies) with concomitant over-expression of the receptor protein (SLAMON et al. 1987; VENTER et al. 1987). More interestingly, there is some indication that amplification of c-*erb*B2 in breast adenocarcinoma correlates with a poor prognosis (SLAMON et al. 1987; VARLEY et al. 1987). c-*erb*B2 amplification has been detected in other kinds of adenocarcinomas (e.g. kidney and stomach, see Table 6) but not in squamous cell carcinomas (TOYOSHIMA et al. 1986). This is in contrast to c-*erb*B1 amplification which is found relatively frequently in squamous but only very rarely in adenocarcinomas. It has been shown that overexpression of c-*erb*B2 in NIH 3T3 cells can lead to full malignancy, which is consistent with the idea that amplification of this gene in carcinomas may be important (HUDZIAK et al. 1987). However, it is still not clear whether transformation by over-expression of the nornaml c-*erb*B2 receptor is dependent in stimulation by the ligand; indeed the ligand for this receptor is not known.

3. Other Receptors

Using the DNA transfection assay, a novel receptor-like oncogene, *mas*, has been identified in the breast carcinoma cell line MCF-7 (YOUNG et al. 1986) (Table 7). Unlike all the other receptors so far mentioned, which have a single membrane-spanning domain, *mas* has seven potential membrane-spanning domains resembling rhodopsin and the β-adrenergic receptor. The same transfection procedure revealed two other oncogenes in the MCF-7 cell line, an amplified N-*ras* gene and a rearranged c-*ros* gene. In the case of the two receptor-encoding genes, *mas* and *ros*, it appears that these genes were activated during the transfection procedure and were not altered in the original tumour cell line. Interestingly, both these genes are on chromosome 6 and are located in regions often associated with tumour-specific rearrangements (RABIN et al. 1987).

A chemically induced tumorigenic cell line derived from a human osteosarcoma cell was found to have an activated oncogene using the DNA transfection assay (COOPER et al. 1984). This gene, *met*, has been found to encode a putative growth-factor receptor (DEAN et al. 1985) (the ligand is not known), but no changes in *met* have been detected in human malignancies. It has been shown, however, that spontaneous transformation of NIH 3T3 cells in culture is often associated with amplification of this gene (COOPER et al. 1986).

Table 7. Oncogenes detected using human tumour DNA and the NIH 3T3 transfection assay

Gene	Tumour type	Percentage tumours	Mechanism	References
ras	Most types	20–30	Point mutation	BARBACID (1986)
*onc*D	Colon	1 case	Tropomyosin/tyrosine kinase fusion	MARTIN-ZANCA et al. (1986)
lca	Liver cancer	2/10	Unknown	OCHIYA et al. (1986)
unnamed	Thyroid papillary carcinoma	2/5	Unknown	FUSCO et al. (1987)
raf	Stomach cancer	Several cases	Rearrangement at 5′ and 3′ ends	NAKATSU et al. (1986)
	Laryngeal cancer	1 case	Enhanced rearrangement during transfection	KASID et al. (1987)
met	In vitro-transformed osteosarcoma	1 case	Rearrangement	COOPER et al. (1984)
dbl	Diffuse B-cell lymphoma	Several cases	Rearranged during transfection	SRIVASTAVA et al. (1986)
hst	Stomach/Kaposi	Several cases	Rearranged during transfection	TAIRA et al. (1987)
mas	MCF-7 line	1 case	Rearranged during transfection	YOUNG et al. (1986)
ros	MCF-7 line	1 case	Rearranged during transfection	YOUNG et al. (1986)

4. Growth Factors

One viral oncogene, v-*sis*, is known to code for a PDGF-like molecule. This oncogene transforms cells of mesenchymal origin by an autocrine stimulation of growth. Many examples are known of human tumours abnormally expressing growth factors, e.g. PDGF from sarcomas (EVA et al. 1982), bombesin-like peptides from small cell lung tumours (SAUSVILLE et al. 1986) and TGFα in a variety of carcinomas (DERYNCK et al. 1987). However, there is no evidence for any genetic alterations at these loci in tumour cells, and the biological significance of these observations is still not clear.

DNA from several types of human tumour, including a Kaposi's sarcoma and stomach tumours, have yielded an oncogene in the NIH 3T3 transfection assay, called *hst* (for *h*uman *st*omach) (NAKATSU et al. 1986). This gene was found to encode a novel growth factor with around 40% homology to acidic or basic fibroblast-derived growth factors (BOVI et al. 1987). [Interestingly, *int*-2 seems to belong to the same FGF-like family, although expression of *int*-2 in human adult tissues (normal or transformed) has not been detected.] Subsequent analysis of *hst* genes in the original stomach tumours revealed no alterations, and it is now clear that the gene was activated during the transfection assay. The same appears to be true for the Kaposi's sarcomas (BOVI et al. 1987). Thus far, no examples of rearranged or altered growth factor genes have been detected in human tumours.

III. Protein Kinases

The largest group of viral oncogene products are the tyrosine kinases. Some of these are growth-factor receptors as discussed previously, but others are intracellular proteins (see Table 1) attached to the plasma membrane. In addition, three viral oncogene products encode ser/thr protein kinases, and these are located in the soluble cytoplasmic fraction of the cell. The way in which both these types of protein kinases and their normal counterparts affect cell signalling pathways is unclear at the moment (see HALL 1989 for a review).

Activation of protein kinases has been found to occur by deletion or mutation of regions which cover regulatory phosphorylation sites in the protein. This appears to be an important mechanism for generating a protein with a deregulated kinase activity. Although the intracellular tyrosine and ser/thr kinases are commonly found with viral oncogenes, there are very few examples in which activation of this type of protein is specifically implicated in a human malignancy. There is, however, one notable exception, c-*abl*.

1. c-*abl*

The rearrangement of the c-*abl* gene in chronic granulocytic leukaemia (CGL) is the only case in a human malignancy in which a known protein kinase is genetically altered. In around 95% of CGL patients, the disease is associated with the Philadelphia chromosome (Ph^+) (9: 22 translocation). Cloning of breakpoints in the Ph chromosome has revealed that the c-*abl* gene is translocated from chromosome 9 to 22 at a region of 22 called the breakpoint cluster region (*bcr*) (GROFFEN et al. 1984) (Table 8). The result of this rearrangement is that the c-*abl*

Table 8. Oncogene rearrangement in human malignancies

Gene	Tumour type	Locus rearranged to	Percentage tumours	References
c-*myc*	Burkitt lymphoma	Igs	100	DALLA FAVERA et al. (1982)
	Acute T-cell leukaemia	Tα	10–20	ERIKSON et al. (1986)
c-*abl*	CGL (Ph^+)	*bcr*	90	GROFFEN et al. (1984)
	ALL (Ph^+)	*bcr*	10	CLARK et al. (1987)
*bcl*1	B-Cell lymphomas	Igs	80	TSUJIMOTO et al. (1985)
*bcl*2	Follicular lymphomas	Igs	80	TSUJIMOTO et al. (1985)
*tcl*1	T-Cell chronic lymphocytic leukaemia	Tα	–	CROCE et al. 1985)

Igs, immunoglobulin loci; Tα, T-cell receptor α chain.

gene loses its first coding exon, and the rest of the gene is fused to a gene termed *bcr* or *phl*. A hybrid RNA and a chimaeric protein are produced, and, in contrast to the normal p145 *abl* protein, the *bcr*/*abl* p210 protein has a high in vitro kinase activity. It is still not clear whether the *bcr*-derived amino-terminus of the novel tumour-specific protein has a functional role, or whether it is deregulation of the kinase, by removal of the amino-terminus of *abl*, that is important. More recent analysis of Ph^+ acute lymphocytic leukaemia (ALL) also shows a *bcr-abl* fusion, but rearrangement occurs at a different site within the *bcr* gene to generate a p190 *bcr*/*abl* chimaeric protein (CLARK et al. 1987; HERMANS et al. 1987). The substrates for the *abl* kinase are unknown, which makes it difficult to predict the biochemical consequences of this deregulated kinase in the cell.

2. Others

Several oncogenes detected by DNA transfection of human tumour DNA have kinase activity. The *onc*D gene has so far been identified in only a single colon carcinoma. This gene is the product of two rearranged genes with a tropomysin-derived amino-terminus fused to a tyrosine kinase carboxy-terminus (MARTIN-ZANCA et al. 1986), and this rearrangement was present in the original tumour cell line and is not an artefact of the transfection procedure. The tyrosine kinase domain is derived from a tyrosine kinase gene not previously characterised. The *dbl* oncogene, originally detected in a single case of diffuse B-cell lymphoma, has been sequenced and is predicted to produce a 66K protein (SRIVASTAVA et al. 1986). The protein is cytoplasmic and phosphorylated on serine, though it is not yet clear whether it is itself a kinase. It has recently been shown that the *dbl* gene was rearranged during the transfection procedure (EVA et al. 1987).

The c-*raf* oncogene encodes a soluble thr/ser protein kinase activity. Activated c-*raf* genes have been detected, using DNA transfection, in DNA from a few patients with stomach cancer (NAKATSU et al. 1986). It appears that the gene is rearranged at both the 5′ and 3′ ends, but the biochemical consequences are notknown. More recently, an altered *raf* locus has been detected using DNA isolated from a radiation-resistant laryngeal cancer cell line (KASID et al. 1987). Activation was shown to have occurred through rearrangement during the transfection procedure. Interestingly, this happened with a very high frequency, much higher than is found with normal DNA, leading the authors to speculate that there might be some alteration of the locus in the tumour cells which makes the gene more susceptible to rearrangement during transfection. Another report found an activated c-*raf* gene in non-cancerous fibroblasts from a patient with the Li-Fraumeni cancer-prone syndrome (CHANG et al. 1987).

There has been one report of high pp60*src* kinase activity in 21 of 21 colon carcinoma cell lines and 15 of 15 colon tumour biopsies (BOLEN et al. 1987). No alteration in the gene organisation was detected, though there was some evidence that the increased kinase activity could not be totally accounted for by increased levels of protein, suggesting an alteration in the protein. More work will be required to clarify whether pp60*src* is important in colon cancer.

IV. Nuclear Proteins

The genes v-*myc* and v-*fos* are the best-studied examples of viral oncogenes encoding nuclear products. Although the biochemical function of the proteins is unclear, it is known that the expression of the cellular proto-oncogenes c-*myc* and c-*fos* is tightly coupled to the growth state of the cells. c-*fos*, for example, is normally induced by growth factor stimulation of cells after about 30 min and then disappears after 2 h (KRUIJER et al. 1984). c-*myc* is expressed in quiescent cells but at much lower levels than in growing cells. In this case expression increases about 60 min after addition of growth factors to quiescent cells and stays high throughout the cell cycle (MULLER et al. 1984). Both v-*fos*- and v-*myc*-transformed cells express continuously high levels of the proteins. Although no alterations at the c-*fos* locus are known in human malignancies, many examples of altered c-*myc* have been reported.

1. c-*myc*

Rearrangement at the c-*myc* locus is found in 100% of Burkitt's lymphomas: 90% involve an 8:14 translocation, and c-*myc*, which is normally on chromosome 8, is translocated to the immunoglobulin heavy chain locus on 14q32 (DALLA FAVERA et al. 1982). In the other 10% of cases (containing 8:2 or 8:22 translocation) the two immunoglobulin light chains are transferred from 2 or 22 to 8 (see CROCE et al. 1986 for review). There is still much controversy over the consequences of this translocational event, but it is generally thought that it leads to deregulated expression of the c-*myc* gene in B cells. Rearrangements of c-*myc* in T-cell leukaemia are less well characterised, but 10%–20% of acute T-cell leukaemias have a translocation of the c-*myc* gene to the T-cell receptor α chain gene (ERIKSON et al. 1986) (Table 8).

Rearrangements involving chromosome 14q32 are also observed in non-Burkitt's B-cell neoplasms. In the most common human B-cell lymphoma, follicular lymphoma, the breakpoint position has been cloned and a locus, *bcl*-2, identified (TSUJIMOTO et al. 1985), which is rearranged (at the 3′ end) in around 80% of follicular lymphomas. The sequence of the *bcl*-2 gene is now known (TSUJIMOTO and CROCE 1986), and open reading frames for two protein products can be predicted. The predicted structure of either protein would not be affected by the translocation event. *bcl*-2 does not appear to encode a membrane or a secreted protein, but whether or not its product(s) is nuclear like c-*myc* remains to be seen. Another locus *bcl*-1 has been cloned from a translocation present in diffuse B-cell lymphoma and in chronic lymphocytic leukaemia (TSUJIMOTO et al. 1985).

2. *myc* Family

Amplification of the c-*myc* gene has been detected in several tumour types (YOKOTA et al. 1986; NAU et al. 1986). In particular, around 20% of breast carcinoma has an amplified (two- to tenfold) c-*myc* gene, and this seems to correlate with a poor short-term prognosis (VARLEY et al. 1987). Amplification of a gene hybridising to a c-*myc* probe was detected in human neuroblastomas (SCHWAB et al. 1983). In fact it is a related gene, N-*myc*, which is amplified in around 50% of late, stage III and IV, neuroblastomas. The presence of amplified N-*myc* correlates with rapid progression of the disease (SEEGER et al. 1985). N-*myc* amplification has also been found in two other tumours with neural or neuroendocrine properties; a subset of small cell lung carcinomas (NAU et al. 1986) and in retinoblastomas (LEE et al. 1984). In the latter case, although only 3 of 11 samples had an amplified gene, 7 of 11 evidenced over-expression of mRNA. In the case of Wilms' tumours, although gene amplification is not observed, all the tumours show high level mRNA expression (ALT et al. 1986).

Using *myc* probes, two other members of this family have been detected, L-*myc* and R-*myc* (NAU et al. 1985; ALT et al. 1986; DEPINHO et al. 1987). The four genes have similar structural organisation, although N-, L- and R-*myc* have a much more restricted expression compared with c-*myc* (ALT et al. 1986; DEPINHO et al. 1987). Amplification of c-, N- and L-*myc* genes has been detected in small cell lung carcinoma, and of 31 such lines tested, 9 had amplified c-*myc*, 7 amplified N-*myc* and 5 amplified L-*myc* (NAU et al. 1986). No cases were seen of amplification of two *myc* genes in the same cell line.

V. Guanine Nucleotide Binding Proteins

1. *ras*

The most common genetic change known in human malignancy is an alteration in one of the three human *ras* genes, c-Ha-*ras*1, c-Ki-*ras*2 or N-*ras* (see BARBACID 1986 for review). Altered *ras* genes were first detected in human tumours using the NIH 3T3 transfection assay (KRONTIRIS and COOPER 1981; SHIH et al. 1981), and since then many hundreds of tumours have been screened in this way. Around 20% of most types of human malignancies have been found to contain an altered *ras* gene, though one notable exception to this is breast carcinomas, which only very rarely have been found to contain altered *ras* genes. Cloning and sequencing of the activated genes have revealed that, in almost all cases, the mechanism of activation is a point mutation altering the codon at either position 12, 13, or 61 in the protein (TABIN et al. 1983; BOS et al. 1985). In around 1% of tumours (primarily colon and lung), an amplified *ras* gene has been detected (PULCIANI et al. 1985).

Sensitive biochemical approaches to look for point mutations have revealed more detailed information on the frequency of *ras* activation. Thus BOS et al, using a polymerase chain reaction to amplify colon tumour DNA followed by oligonucleotide probe analysis, found that around 40% of colon carcinomas contained an altered *ras* gene (mainly Ki−) (BOS et al. 1987). A similar analysis of lung carcinomas showed that around 20% had an altered *ras* gene (mainly Ki−) (RODENHUIS et al. 1987). However, closer inspection of the tumour types showed that all the activations were in lung adenocarcinomas. No *ras* mutations were detected in large or small cell carcinomas of the lung. DNA from patients with acute myeloid leukemia analysed by the same procedure showed around 25% involvement of *ras* (usually N−) (FARR et al. 1988).

The common occurrence of somatic *ras* mutations in human tumours makes it very probable that the mutational event is an important step in progression to malignancy, although it is still not clear whether it is an early or late event. In vivo models strongly suggest that *ras* mutations occur early in carcinogen-induced rodent tumours (see BARBACID 1986 for review), and some recent work with human material would appear to be consistent with this idea. Thus BOS et al. (1987) reported that they were able to obtain premalignant adenomas from patients with colon cancer. In five of six patients both the colon cancer and a premalignant adenoma contained the same *ras* mutation. In another study using cells obtained from patients with the pre-leukaemic disease, myelodysplastic syndrome (MDS), *ras* mutations were found at around 30% frequency (HIRAI et al. 1987).

Cloned mutant *ras* genes or purified mutant *ras* protein, when reintroduced into established rodent fibroblasts, produce a fully malignant phenotype, and the biochemical basis for this transformation is being intensively studied (see HALL 1989 for review). It is likely that the *ras* proteins p21*ras* are regulatory G-proteins involved in controlling a critical aspect of cell proliferation. Mutations in the protein lead to a breakdown in this regulation, resulting in uncontrolled cell growth. The proliferative signalling systems that might be affected by p21*ras* will be discussed in a following chapter on signal transduction.

2. Others

Although G proteins have a crucial role in regulating many cellular processes and, in particular, signal transduction events at the plasma membrane, only the *ras* genes have been shown to act as oncogenes. There has been one recent report that Gs (the G-protein regulating stimulation of adenylate cyclase activity) is altered in growth hormone-secreting pituitary adenomas (VALLER et al. 1987), and it appears to be constitutively activated. It is not yet clear whether the gene for the α (GTP-binding) subunit of Gs has been altered in these tumours.

D. Conclusions

Well over 50 proto-oncogenes have been described in this chapter, and the characterisation of these genes has had a major impact on our understanding of cellular growth control. Furthermore, many of these genes, when introduced experimentally into cells in tissue culture, have been shown to have some biological activity which is relevant to the transformed phenotype. In some respects, therefore, it is disappointing that so few genetic changes have been identified in human malignancies. An alteration in the *ras* oncogene is clearly the most important genetic defect that is known in human cancer and is found in well over 20% of most kinds of human tumours. Understandably then, this is a major focus for chemotherapeutic anticancer research. Rearrangements of c-*myc*, c-*abl*, and *bcl*-2 in specific types of leukaemia are very frequent (>90%), and in those particular kinds of diseases, these genes are the focus of much attention, both diagnostically and from the point of view of therapy.

There are some encouraging signs that amplification of specific genes might be important in human cancer (e.g. N-*myc* in neuroblastoma, c-*erb*B2 in breast adenocarcinomas, and c-*erb*B1 in squamous carcinomas), but it is too early to tell in most cases what the biological consequences of this are likely to be for the tumour.

It is open to speculation as to why so few of the known proto-oncogenes have been altered in human cancers. It is possible that some of these genes have been altered but in subtle ways which have so far gone undetected. For example, point mutations would be very difficult to detect unless a biological assay were available to select for them in the same way as *ras*. It may be that many oncogene products function in a tissue-specific way and would have no effect on NIH 3T3 cells. An alternative explanation is that the types of oncogenes involved in human cancer are of a different nature to those involved in animal cancer, and, in particular, perhaps there is a greater involvement of recessive oncogenes.

Whatever the reason, it is still a strong belief that human cancer occurs after an accumulation of several genetic alterations. A number of these defects have been defined in the past 5 years, but there is still a long way to go.

References

Alt FW, Depinho R, Zimmerman K, Tesfaye A, Yarcopoulos G, Nisen P (1986) The human *myc* gene family. Cold Spring Harbor Symp Quant Biol 51:931–941

Balazs I, Grzeschik KH, Stavnezer E (1984) Assignment of the human homologue of a chicken oncogene to chromosome 1. Cytogenet Cell Genet 37:410–411

Barbacid M (1986) *ras* Genes. Ann Rev Biochem 56:779–827

Bargmann CJ, Hung MC, Weinberg RA (1986) Multiple independent activations of the *neu* oncogene by a point mutation altering the transmembrane domain of p185. Cell 45:649–657

Bartram C, de Kleen A, Hagemeyer A, Van Agthoven T, Van Dessel AG, Bootsma D, Grosveld G, Ferguson Smith MA, Davies T, Stone M, Heisterkamp N, Stephenson JR, Groffen J (1983) Translocation of the human c-*abl* oncogene correlates with the presence of a Philadelphia chromosome in chronic myelocytic leukemia. Nature 306:277–287

Berger MS, Gullick WS, Greenfield C, Evans S, Addis BJ, Waterfield MD (1987) Epidermal growth factor receptors in lung tumours. J Pathol 152:297–307

Birchmeier C, Young D, Wigler M (1986) Characterisation of two new human oncogenes. Cold Spring Harbor Symp Quant Biol 51:993–1000

Bishop JM (1983) Cellular oncogenes and retroviruses. Ann Rev Biochem 52:301–354

Bishop JM, Varmus HE (1984) Functions and origins of retroviral transforming genes. In: Weiss R, Teich N, Varmus H, Coffin J (eds) Molecular biology of tumor viruses, RNA tumor viruses. Cold Spring Harbor Press, New York

Bohmann D, Bos TJ, Admon A, Nishimura T, Vogt PK, Tjian R (1987) Human proto-oncogenic c-*jun* encodes a DNA binding protein with structural and functional properties of transcription factor AP-1. Science 238:1386–1392

Bolen JB, Veillette A, Schwartz AM, DeSean V, Rosen N (1987) Activation of p60c-*src* protein kinase activity in human colon carcinoma. Proc Natl Acad Sci USA 84:2251–2255

Bonner T, O'Brien S, Nash WG, Rapp UR, Morton CC, Leder P (1984) The human homologs of the raf (mil) oncogene are located in chromosomes 3 and 4. Science 223:71–74

Bos JL, Toksoz D, Marshall CJ, de Vries MV, Veeneman GH, Van der Eb AT, van Bloom JH, Janssen JWG, Steenvoorden ACM (1985) Amino acid substitutions at codon 13 of the N-*ras* oncogene in human acute myeloid leukaemias. Nature 315:726–730

Bos JL, Fearon ER, Hamilton SR, de Vries MV, van Bloom JM, van der Eb AJ, Vogelstein B (1987) Prevalence of *ras* gene mutations in human colorectal cancer. Nature 327:293–297

Bovi PD, Curatola AM, Kern FG, Greco A, Iltmann M, Basilico C (1987) An oncogene isolated by transfection of Kaposi's sarcoma DNA encodes a growth factor that is a member of the FGF family. Cell 50:729–737

Brown AMC, Wildin RS, Prendergast TJ, Varmus HE (1986) A retrovirus vector expressing the putative mammary oncogene *int*-1 caused partial transformation of a mammary epithelial cell line. Cell 46:1001–1009

Chang EM, Gonda MA, Ellis RW, Scolnick EM, Lowy DR (1982) Human genome contains four genes homologous to transforming genes of Harvey and Kirsten murine sarcoma viruses. Proc Natl Acad Sci USA 79:4848–4852

Chang EM, Pirollo KF, Zou ZQ, Cheung HY, Lawler EL, Garner R, White E, Bernstein WB, Fraunesi JW, Blattner WA (1987) Oncogenes in radio-resistant non-cancerous skin fibroblasts from a cancer-prone family. Science 237:1036–1038

Chardin P, Tavitian A (1986) The *ral* gene: a new *ras* related gene isolated by the use of a synthetic probe. EMBO J 5:2203–2208

Clark SS, McLaughlin J, Crist WM, Champlin R, Witte ON (1987) Unique forms of the *abl* tyrosine kinase distinguish Ph′-positive CML from Ph′-positive ALL. Science 235:85–88

Cooper CS, Park M, Blair D, Tainsky MA, Huebner K, Croce C, Van de Woude GF (1984) Molecular cloning of a new transforming gene from a chemically transformed human cell line. Nature 311:29–33

Cooper CS, Tempest PR, Beckman MP, Heldin CH, Brookes P (1986) Amplification and overexpression of the *met* gene in spontaneously transformed NIH-3T3 mouse fibroblasts. EMBO J 5:2623–2628

Cooper GM, Okenquist S, Silverman L (1980) Transforming activity of DNA of chemically transformed and normal cells. Nature 284:419–421

Cooper JA (1987) The *src* family of protein kinases. In: Kemp B, Alewood PF (eds) Peptides and protein phosphorylation. CRC Press, Boca Raton

Coussens L, Yangfeng TL, Liao YC, Chen E, McGrath J, Seeburg PM, Libermann TA, Schlessinger J, Franke U, Levinson A, Ullrich A (1985) Tyrosine kinase receptor with extensive homology to EGF receptor shows chromosomal location with *neu* oncogene. Science 230:1132–1139

Croce CM (1987) Role of chromosome translocations in human neoplasia. Cell 49:155–156

Croce CM, Isobe M, Palumbo A, Puck J, Ming J, Tweardy D, Erikson J, Davis M, Roverra G (1985) Gene for α-chain of human T cell-receptor: location on chromosome 14 region involved in T cell neoplasms. Science 227:1044–1047

Croce CM, Erikson J, Haluska FG, Finger LR, Showe LC, Tsujimoto T (1986) Molecular genetics of human B and T-cell neoplasms. Cold Spring Harbor Symp Quant Biol 51:891–898

Dalla Favera R, Bregni M, Erikson J, Patterson DE, Gallo RC (1982) Assignment of the human c-*myc* oncogenes to the region of chromosome 8 which is translocated in Burkitt's lymphoma cells. Proc Natl Acad Sci USA 79:7824–7827

Davis M, Malcolm S, Hall A, Marshall CJ (1983) Localisation of the human N-*ras* oncogene to chromosome 1cen p21 by in situ hybridisation. EMBO J 2:2281–2283

Dean TY, Park M, Beau MM, Robins TS, Diaz MO, Rowley JD, Blair DE, Van de Woude GF (1985) The human *met* oncogene is related to the tyrosine kinase oncogenes. Nature 318:385–388

de Martinville B, Giacalone J, Shih C, Weinberg RA, Francke U (1983) Oncogene from human EJ bladder carcinoma is located on the short arm of chromosome 11. Science 219:498–501

DePinho R, Mitzock L, Hatton K, Ferrier P, Zimmerman K, Legony E, Tesfaye A, Collum R, Yancopolous E, Nisen P, Kriz R, Alt F (1987) *myc* family of cellular oncogenes. J Cell Biochem 33:257–266

Derynck R, Goeddel DV, Ullrich A, Gutterman JU, Williams RD, Brigman TS, Berger WH (1987) Synthesis of messenger RNAs from transforming growth factors α and β and the epidermal growth factor receptor by human tumors. Cancer Res 47: 707–712

Dickson C, Smith R, Brookes S, Peters G (1984) Tumorigenesis by mouse mammary tumor virus: proviral activation of a cellular gene in the common integration region *int*-2. Cell 37:529–536

Downward J, Yarden Y, Mayes E, Scrace G, Totty N, Stockwell P, Ullrich A, Schlessinger J, Waterfield MD (1984) Close similarity of epidermal growth factor receptor and c-*erb*B oncogene protein sequences. Nature 307:512–527

Erikson J, Finger L, Sun L, Ar-Rushdi-A, Nishikura K, Minowada J, Finan J, Emanuel B, Nowell PC, Croce C (1986) Deregulation of the αt locus of the T-cell receptor in T-cell leukemia. Science 232:884–886

Eva A, Robbins KG, Andersen PR, Srinivasan A, Tronick S, Reddy PE, Ellneore NW, Gate AT, Lautenberger JA, Papas TS, Westin EH, Wong-Staal F, Gallo RC, Aaronson SA (1982) Cellular genes analogous to retroviral *onc* genes are transcribed in human tumours. Nature 295:116–119

Eva A, Vecchio G, Diamond M, Tronick SR, Dina R, Cooper GM, Aaronson SA (1987) Independently activated *dbl* oncogenes exhibit similar yet distinct structural alterations. Oncogene 1:355–360

Farr CJ, Saiki RK, Erlich HA, McCormick F, Marshall CJ (1988) Analysis of *ras* gene mutations in acute myeloid leukemias using the polymerase chain reaction and oligonucleotide probes. Proc Natl Acad Sci USA 85:1629–1633

Francke V, Yang-Feng TL, Brissender JE, Ullrich A (1986) Chromosomal mapping of genes involved in growth control. Cold Spring Harbor Symp Quant Biol 51:855–860

Friend SH, Bernards R, Rogelj S, Weinberg RA, Rapaport JM, Albert DM, Dryja TP (1986) A human DNA segment with properties of the gene that predisposes to retinoblastoma and osteosarcoma. Nature 323:643–646

Fung YKT, Lewis WG, Kung HJ, Crittenden LB (1983) Activation of the cellular oncogene c-*erb*B by LTR insertion: molecular basis for induction of erythroblastosis by avian leukosis virus. Cell 33:357–368

Fusco A, Grieco M, Santaro M, Berlingieri MT, Pilotti S, Pierotti MA, Della Porta G, Vecchio G (1987) A new oncogene in human thyroid papillary carcinoma and their lymph node metastases. Nature 328:170–172

Groffen J, Stephenson JR, Heisterkamp N, de Klein A, Bartram CE (1984) Philadelphia chromosomal breakpoints are clustered within a limited region, *bcr*, on chromosome 22. Cell 36:93–99

Groffen J, Heisterkamp N, Starm K (1986) Oncogene activation by chromosomal translocation in chronic myelocytic leukaemia. Cold Spring Harbor Symp Quant Biol 51: 911–921

Hall A (1986) Oncogenes. In: Rigby PWJ (ed) Genetic engineering, vol 5. Academic, London, pp 61–177
Hall A (1989) Oncogene products involved in signal transduction. In: Naccache PH (ed) G proteins and calcium mobilisation. CRC Press, Boca Raton
Hall A, Marshall CJ, Spurr N, Weiss RA (1983) Identification of transforming genes in two human sarcoma cell lines as a new member of the *ras* gene family located on chromosome 1. Nature 303:396–400
Harper MG, Franchini G, Love J, Simon MI, Gallo RL, Wong-Staal F (1983) Chromosomal sublocalisation of human c-*myb* and c-*fes* cellular oncogenes. Nature 304:169–171
Hayward WS, Neel BG, Astrin SM (1981) Activation of a cellular *onc* gene by promoter insertion in ALV-induced lymphoid leukosis. Nature 290:475–479
Hendler FJ, Ozanne BW (1984) Human squamous cell lung cancers express increased epidermal growth factor receptors. J Clin Invest 74:647–651
Hermans A, Heisterkamp N, von-Linden M, van Baal S, Meijer D, van der Plas D, Wiedemann L, Groffen J, Bootsma D, Grosveld G (1987) Unique fusion of *bcr* and c-*abl* genes in Philadelphia chromosome positive acute lymphoblastic leukemia. Cell 51:33–40
Hirai H, Kobayashi Y, Mano H, Hagiwara K, Maru Y, Omine M, Mizoguchi H, Nishida J, Takaku F (1987) A point mutation at codon 13 of the N-*ras* oncogene in myelodysplastic syndrome. Nature 327:430–434
Hudziak RM, Schlesinger J, Ullrich A (1987) Increased expression of the putative growth factor receptor $p185^{Herz}$ causes transformation and tumorigenesis of NIH 3T3 cells. Proc Natl Acad Sci USA 84:7159–7163
Isobe M, Emanuel BS, Givol D, Oren M, Croce CM (1986) Localisation of gene for human p53 to band 17p13. Nature 320:84–85
Kasid U, Pfeifer A, Weichselbaum RR, Dritschelo A, Mark GE (1987) The *raf* oncogene is associated with radiation-resistant human laryngeal cancer. Science 237:1039–1041
Kinzler KW, Bigner SH, Bigner DD, Trent JM, Law ML, O'Brien SJ, Wong AJ, Vogelstein B (1987) Identification of an amplified, highly expressed gene in a human genome. Science 236:70–73
Klein G (1983) Specific chromosomal translocations and the genes of B cell derived tumors in mice and men. Cell 32:311–315
Krontiris TE, Cooper GM (1981) Transforming activity of human tumor DNAs. Proc Natl Acad Sci USA 78:1181–1184
Kruh GD, King CR, Kraus MM, Popescu NC, Amsbaugh SC, McBride ND, Aaronson SA (1986) A novel human gene closely related to the *abl* proto-oncogene. Science 234: 1545–1548
Kruijer W, Cooper JA, Hunter T, Verma IM (1984) PDGF induces rapid but transient expression of the c-*fos* gene and protein. Nature 312:711–716
Land H, Parada LF, Weinberg RA (1983) Tumorigenic conversion of primary embryo fibroblasts requires at least two cooperating oncogenes. Nature 304:596–601
Lane DP, Crawford LV (1979) T antigen is bound to a host protein in SV40-transformed cells. Nature 278:261–263
LeBeau MM, Westbrook CA, Diaz MO, Rowley JD (1984) In situ hybridization studies of c-*src*: evidence for two distinct loci on human chromosomes number 1 and number 20. Nature 312:70–77
LeBeau HM, Pettenati MJ, Lemons RS, Diaz MO, Westbrook CA, Lawson RA, Sherr CJ, Rowley JD (1986) Assignment of GM-CSF, CSF-1, and FMS genes to human chromosome 5 provides evidence for linkage of a family of genes regulating hematopoiesis and for their involvement in the deletion (5q) in myeloid disorders. Cold Spring Harbor Symp Quant Biol 51:899–909
Lee W, Murphree A, Benedict W (1984) Expression and amplification of the N-*myc* gene in primary retinoblastoma. Nature 309:458–460
Libermann TA, Nusbaum HR, Razon N, Kris R, Lax I, Soreq H, Whittle N, Waterfield MD, Ullrich A, Schlessinger J (1985) Amplification, enhanced expression and possible rearrangement of EGF receptor gene in primary human brain tumours of glial origin. Nature 313:144–147

Lowe DG, Capon DT, Delvart E, Sakaguchi AY, Naylor SL, Goeddel DV (1987) Structure of the human and murine R-*ras* genes, novel genes closely related to *ras* proto-oncogenes. Cell 48:137–146

Madaule P, Axel R (1985) A novel *ras* related gene family. Cell 51:31–40

Mark GE, Seeley TW, Shows TB, Mountz SD (1986) *pks*, a *raf*-related sequence in humans. Proc Natl Acad Sci USA 83:6312–6316

Marshall CJ, Rigby PWJ (1984) Viral and cellular oncogenes. Cancer Surv 3:183–214

Marth JD, Disteche C, Pravtcheva D, Ruddle F, Krebs EG, Perlmutter RM (1986) Localisation of a lymphocyte-specific protein trypsine kinase gene (*lck*) at a site of frequent chromosomal abnormalities in human lymphomas. Proc Natl Acad Sci USA 83:7400–7404

Martin-Zanca D, Hughes SM, Barbacid M (1986) A human oncogene formed by fusion of truncated tropomyosin and protein tyrosine kinase sequences. Nature 319:743–748

McCoy MS, Toole JJ, Cunningham JM, Chang EM, Lowy DR, Weinberg RA (1983) Characterisation of a human colon/lung carcinoma oncogene. Nature 302:79–81

Meeker TC, Nagarajan L, Rushdi A, Rovera G, Huebner K, Croce C (1987) Characterization of the human PlM-1 gene – a putative protooncogene coding for tissue specific member of the protein kinase family. Oncogene Res 1:87–101

Moreau-Gachelin F, Tavitian A, Tambourin P (1988) *Spi*-1 is a putative oncogene in virally induced murine erythroleukemias. Nature 331:277–280

Mucenski ML, Taylor BA, Ihle JN, Hartley JW, Morse HC, Jenkins NA, Copeland NG (1988) Identification of a common ecotropic viral integration site *evi*-1 in the DNA of AKXD murine myeloid tumours. Mol Cell Biol 8:301–308

Muller R, Bravo R, Burckhardt J, Curran T (1984) Induction of c-*fos* gene and protein by growth factors precedes activation of c-*myc*. Nature 312:716–720

Nagarajan L, Louie E, Tsuikimoto Y, Rushdi A, Huebner K, Croce CM (1986) Localisation of the human *pim* oncogene (PIM) to a region of chromosome 6 involved in translocations in acute leukemias. Proc Natl Acad Sci USA 83:2556–2560

Nakatsu Y, Nomoto S, Oh-uchida M, Shimizu K, Sekijuchi M (1986) Structure of the activated c-*raf*-1 gene from human stomach cancer. Cold Spring Harbor Symp Quant Biol 51:1001–1007

Nau MN, Brookes BJ, Battey J, Sansville E, Gazdar AF, Kirsch IR, McBride OW, Bertress V, Hollis GF, Minna JD (1985) L-*myc* and new *myc* related gene amplified and expressed in human small cell lung cancer. Nature 318:69–73

Nau MN, Brooks BJ, Carney DN, Gazdar AF, Battey JF, Sansville EA, Minna JD (1986) Human small cell lung cancers show amplification and expression of the N-*myc* gene. Proc Natl Acad Sci USA 83:1092–1096

Nerenberg M, Hinrichs SH, Reynolds RK, Khowy G, Jay G (1987) The *tat* gene of human T-lymphotropic virus type 1 induces mesenchymal tumors in transgenic mice. Science 237:1324–1329

Nusse R, van Ooyen A, Cox D, Fung YK, Varmus HE (1984) Mode of proviral activation of a putative mammary oncogene (*int*-1) on mouse chromosome 15. Nature 307:131–136

O'Brien SJ, Nash WG, Goodwin JL, Lowy DR, Chang EM (1983) Dispersion of the *ras* family of transforming genes to four different chromosomes in man. Nature 302:839–842

Ochiya T, Fujiyama A, Fukushige A, Hatada I, Matsubara K (1986) Molecular cloning of an oncogene from a human hepatocellular carcinoma. Proc Natl Acad Sci USA 83:4993–4997

Ozanne B, Richards CS, Hendler F, Burns D, Gusterson B (1986) Overexpression of the EGF receptor is a hallmark of squamous cell carcinomas. J Pathol 149:9–14

Parada LF, Land H, Weinberg RA, Wolf D, Rotter V (1984) Cooperation between gene encoding p53 tumour antigen and *ras* in cellular transformation. Nature 312:649–651

Poiesz BJ, Ruscetti FW, Gazda AF, Bunn PA, Minna JD, Gallo RC (1980) Detection and isolation of type C retrovirus particles from fresh and cultured lymphocytes of a patient with cutaneous T-cell lymphoma. Proc Natl Acad Sci USA 77:7415–7419

Prakash K, McBride OW, Swan DC, Tronick SR, Aaronson SA (1982) Molecular cloning and chromosomal mapping of a human locus related to the transforming gene of Moloney murine sarcoma virus. Proc Natl Acad Sci USA 79:5210–5214

Pulciani S, Santos E, Long EK, Sorrentino V, Barbacid M (1985) *ras* gene amplification and malignant transformation. Mol Cell Biol 5:2836–2841

Rabin M, Birnbaum D, Young D, Birchmeier C, Wigler M (1987) Human *ros* 1 and *mas* 1 oncogenes located in regions of chromosomes 6 associated with tumour specific rearrangements. Oncogene Res 1:169–183

Reddy ESP, Rao VN, Papas TS (1987) The *erg* gene: a human gene related to the *ets* oncogene. Proc Natl Acad Sci USA 84:6131–6135

Rodeck U, Herlyn M, Herlym D, Motthoff C, Atkinson B, Vavello M, Steplewski Z, Koprowski H (1987) Tumour growth modulation by a monoclonal antibody to the epidermal growth factor receptor: immunologically mediated and effector cell-independent effects. Cancer Res 47:3692–3696

Rodenhuis S, Van de Wettering ML, Moot WJ, Evers SG, Van Zandwijk N, Bos JL (1987) Mutational activation of the K-*ras* oncogene. New Engl J Med 317:929–935

Ruley EH (1983) Adenovirus early region 1A enables viral and cellular transforming genes to transform primary cells in culture. Nature 304:602–606

Sacchi N, Watson DK, Guertsvan Kessel H, Hagemeifer A, Kersey J, Drabkin HO, Patterson D, Papas TS (1986) Hu-*ets*-1 and the *ets*-2 genes are transposed in acute leukemias with (4,11) and (8 : 21) translocations. Science 280:379–381

Sakamoto H, Mori M, Taira M, Yoshida T, Matsukawa S, Shimizu K, Sekiguchi M, Terada M, Sugimura T (1986) Transforming gene from human stomach cancers and a non-cancerous portion of stomach mucosa. Proc Natl Acad Sci USA 83:3997–4001

Santos E, Tronick SR, Aaronson SA, Pulciani S, Barbacid M (1982) T24 human bladder carcinoma oncogene is an activated form of the normal human homologue of BALB- and Harvey-MSV transforming genes. Nature 298:343–347

Sap J, Munoz A, Damon K, Goldberg Y, Ghysdael J, Leutz A, Beug H, Vennstrom B (1986) The c-*erb*A protein is a high affinity receptor for thyroid hormone. Nature 324:635–640

Satoh H, Yoshida MC, Sasake M (1987) Regional localization of the human c-*ros*-1 on 6q22 and *flt* on 13q12. Jpn J Cancer Res 78:772–775

Saule S, Dozier C, Denhe ZF, Martin P, Steheln D (1987) Retroviruses with two oncogenes. Int J Rad Appl Instrum 14:441–444

Sausville E, Labacq-Verheyden AM, Spindel ER, Cuttita F, Gazdar AF, Battey JF (1986) Expression of the gastrin-releasing peptide gene in human small cell lung cancer. J Biol Chem 261:2451–2457

Schechter AL, Stern DF, Vaidyanathan L, Decker SJ, Drebin JA, Greene MI, Weinberg RA (1984) The *neu* oncogene: an *erb*B related gene encoding a 185,000 Mr tumour antigen. Nature 312:513–516

Schechter AL, Hung MC, Vaidyanathan L, Weinberg RA, Yanh-Feng TL, Franke V, Ullrich A, Coussens L (1985) The *neu* gene: an *erb*B homologous gene distinct from and unlinked to the gene encoding the EGF receptor. Science 229:976–978

Schwab M, Alitalo K, Klemonauer KH, Varmus HE, Bishop JM, Gilbert F, Brodeur G, Goldstein M, Trent J (1983) Amplified DNA with limited homology to *myc* cellular oncogene showed by human neuroblastoma cell lines and a neuroblastoma tumour marker. Nature 305:245–247

Schwab M, Varmus HE, Bishop JM, Grzeschikk H, Naylor SL, Sakaguchi AY, Brodeur G, Trent J (1984) Chromosome localization in normal human cells and neuroblastomas of a gene related to c-*myc*. Nature 308:288–291

Schwab M, Varmus HE, Bishop JM (1985) Human N-*myc* gene contributes to neoplastic transformation of mammalian cells in culture. Nature 316:160–162

Searle AG, Peters J, Lyon MF, Evans EP, Edwards JH, Buckle VJ (1988) Chromosome maps of man and mouse. Genomics 1:3–18

Seeger R, Brodeur G, Sather M, Dutton A, Siegel S, Wong K, Hammond O (1985) Association of multiple copies of N-*myc* with the rapid progression of neuroblastomas. N Engl J Med 313:1111–1116

Selten G, Cuypers HT, Berns A (1985) Proviral activation of the putative oncogene *pim*-1 in MuLV induced T-cell lymphomas. EMBO J 4:1793–1798
Sheer D, Sheppard DM, LeBeau MM, Rowley JD, Roman C, Solomon E (1985) Localization of the oncogene c-*erb* A1 immediately proximal to the acute promyelocytic leukemia breakpoint on chromosome 17. Ann Hum Genet 49:167–171
Sherr CJ, Rettenmier CW, Sacca R, Roussel MF, Look AT, Stanley ER (1985) The c-*fms* protooncogene product is related to the receptor for the mononuclear phagocyte growth factor CSF-1. Cell 41:665–676
Shibuya M, Yokota J, Ueyama Y (1985) Amplification and expression of a cellular oncogene (c-*myc*) in human gastric adenocarcinoma cells. Mol Cell Biol 5:414–418
Shih C, Shilo BZ, Goldfarb MP, Dannenberg A, Weinberg RA (1979) Passage of phenotypes of chemically transformed cells via transfection of DNA and chromatin. Proc Natl Acad Sci USA 76:5714–5718
Shih C, Padhy LC, Murray M, Weinberg RA (1981) Transforming genes of carcinomas and neuroblastomas introduced into mouse fibroblasts. Nature 290:261–264
Shimizu K, Goldfarb M, Suard Y, Perucho M, Li Y, Wigler M (1983) Three human transforming genes are related to the viral *ras* oncogenes. Proc Natl Acad Sci USA 80:2112–2116
Slamon DJ, Clark GM, Wong SG, Levin WL (1987) Human breast cancer: correlations of relapse and survival with amplification of the HER-2/neu oncogene. Science 235:177–182
Srivastava SK, Wheelcock RP, Aaronson SA, Eva A (1986) Identification of the protein encoded by the human diffuse B-cell lymphoma (dbl) oncogene. Proc Natl Acad Sci USA 83:8868–8872
Staal SP (1987) Molecular cloning of the *akt* oncogene and its human homologues AKT1 and AKT2. Amplification of AKT1 in a primary human gastric adenocarcinoma. Proc Natl Acad Sci USA 84:5034–5037
Tabin CJ, Bradley SM, Bargmann CI, Weinberg RA, Papageorge AG, Scolnick EM, Dhar R, Lowy DR, Chang EH (1982) Mechanism of activation of a human oncogene. Nature 300:143–149
Takahashi M, Ritz J, Cooper GM (1985) Activation of a novel human transforming gene, *ret*, by DNA rearrangement. Cell 42:581–588
Takeya T, Hanafusa H (1983) Structure and sequence of the cellular gene homologous to RSV *src* and the mechanism for generating the transforming virus. Cell 32: 881–890
Toyoshima K, Sember K, Akiyama T, Ikawa A, Yamamoto T (1986) The c-*erb*B gene encodes a receptor-like protein with tyrosine kinase activity. Cold Spring Harbor Symp Quant Biol 51:977–982
Tronick SR, Popescu NC, Cheah MS, Swan DC, Amsbaugh SC, Lengel CR, Dipaulo JA, Robbins KC (1985) Isolation and chromosomal localisation of the human *fgr* protooncogene. Proc Natl Acad Sci USA 82:6595–6599
Tsujimoto Y, Croce CM (1986) Analysis of the structure, transcripts, and protein products of *bcl*-2, the gene involved in human follicular lymphoma. Proc Natl Acad Sci USA 83:5214–5218
Tsujimoto Y, Jaffe E, Cossman J, Gorham CM, Nowell PC, Croce CM (1985) Clustering of breakpoints on chromosome 11 with the t (11;14) chromosome translocation. Nature 315:340–343
Turc-Carel C, Pietrzak E, Kakati S, Kenniburgh AJ, Sandberg AA (1987) The human int-1 gene is located at chromosome region 12q12-12q13 and is not rearranged in mixoid liposarcoma with t(12;16) (q13;p11)I. Oncogene Res 1:397–406
Vallar L, Spada A, Giannattasio G (1987) Altered Gs and adenylate cyclase activity in human GH-secreting pituitary adenomas. Nature 330:566–568
Varley JM, Swallow JC, Brammer WJ, Whittaker JI, Walker RA (1987) Alterations to either c-*erb*-2 (neu) or c-*myc* proto-oncogenes in breast carcinomas correlate with poor short-term prognosis. Oncogene 1:423–430
Varmus HE (1984) The molecular genetics of cellular oncogenes. Ann Rev Genet 18:553–612

Velu T, Begeunot L, Vass WC, Willingham MC, Merlino GT, Pastan I, Lowy DR (1987) Epidermal growth factor-dependent transformation by a human EGF receptor proto-oncogene. Science 238:1408–1410

Venter DJ, Kumar S, Tyzi NL, Gullick WJM (1987) Overexpression of the c-*erb*B-2 oncoprotein in human breast carcinomas. Lancet ii:69–71

Voronova AF, Sefton BM (1986) Expression of a new tyrosine kinase is stimulated by retrovirus promoter insertion. Nature 319:682–685

Walro DS, Herzog NK, Zhang J, Lim MY, Bose HR (1987) The transforming protein of avian reticuloendotheliosis virus is a soluble cytoplasmic protein which is associated with a protein kinase activity. Virology 160:433–444

Wang LH, Lin B, Jong SMJ, Dixon D, Ellis L, Roth RA, Rutter WJ (1987) Activation of transforming potential of the human insulin receptor gene. Proc Natl Acad Sci USA 84:5725–5729

Waterfield MD, Scrace GT, Whittle N, Stroobant P, Johnson A, Wasteson A, Westermark B, Heldin CH, Huang JS, Deuel TF (1983) Platelet derived growth factor is structurally related to the putative transforming protein p28*sis* of simian sarcoma virus. Nature 304:35–39

Weiss RA, Teich N, Varmus H, Coffin J (1984) RNA tumor viruses, 2nd edn. Cold Spring Harbor Laboratory, New York

Wong AJ, Bigner SH, Bigner DD, Kinzker KW, Hamilton SR, Vogelstein B (1987) Increased expression of the epidermal growth factor receptor gene in malignant gliomas is invariably associated with gene amplification. Proc Natl Acad Sci USA 84: 6899–6903

Yamamoto T, Kamata N, Kaircuo H, Shimizu S, Kuroki T, Toyoshima K, Rikimaru K, Nomura N, Ishizaki R, Paston I, Gamon S, Shimizu N (1986) High incidence of amplification of the EGF receptor gene in human squamous carcinoma cell lines. Cancer Res 46:414–416

Yarden Y, Kuang WJ, Yang-feng YT, Coussens L, Munemitsu S, Dull TJ, Chen E, Schlessinger J, Francke U, Ullrich A (1987) Human protooncogene c-*kit*: a new cell surface receptor tyrosine kinase for an unidentified ligand. EMBO J 6:3341–3351

Yokota J, Yokota YT, Battifora H, LeFevre C, Clue MJ (1986) Alterations of *myc, myb,* and *ras* protooncogenes in cancer are frequent and show clinical correlation. Science 231:261–265

Yoshida MC, Sasaki M, Nuse K, Semba K, Nishizawa M, Yamamoto T, Toyoshima K (1985) Regional mapping of the human protooncogene c-*yes*-1 to chromosome 18 at band q21.3. Jpn J Cancer Res 76:559–562

Yoshida MC, Satoh H, Sasaki M, Semba K, Yamamoto T, Toyoshima K (1986) Regional location of a novel *yes*-related proto-oncogene *syn* on human chromosome 6 at band q21. Jpn J Cancer Res 77:1059–1061

Young D, Waitches G, Birchmeier C, Fasano O, Wigler M (1986) Isolation and characterization of a new cellular oncogene encoding a protein with multiple potential transmembrane domains. Cell 45:711–719

Zhan X, Culpepper A, Reddy M, Loveless J, Goldfarb M (1987) Human oncogenes detected by a defined medium culture assay. Oncogene 1:369–376

CHAPTER 11

Recessive Oncogenes and Anti-Oncogenes

J. K. COWELL

A. Introduction

The idea that inappropriate functions of specific genes can cause cancer has gained a lot of support from the study of acutely transforming retroviruses which carry oncogenes. Initially these oncogenes were isolated from avian retroviruses. Cellular homologues of these genes, termed proto-oncogenes, have been identified in eukaryotes. The demonstration that the introduction of activated oncogenes can confer a tumorigenic phenotype in certain cells has given rise to the concept that oncogenes act in a dominant manner. This appears to be true in chicken cells, for example, in which introduction of the *src* oncogene induces cancerous changes in host cells. This dominant transforming activity is not, however, a feature of all classes of genes that are involved in tumour development. Indeed, there is now a growing body of evidence to suggest that other classes of genes known as anti-oncogenes or suppressor genes may also function in the progressive series of events which leads to tumorigenesis.

B. Malignancy in Cell Hybrids

I. Dominant vs Recessive

The dominant/recessive relationship of malignant and normal phenotypes has been studied extensively using somatic cell hybrids. This hypothesis in its simplest form predicts that if hybrids produced between transformed and normal cells can produce progressively growing tumours, then the malignant phenotype is genetically dominant. Suppression of malignancy in such hybrids would suggest that the malignant phenotype is recessive. The chromosomes of two very different cell types can be combined in a single cell by the process of cell fusion (see COWELL 1986), which can cross species barriers. Early experiments involved fusions between mouse and human cells, simplifying chromosome analysis since the origin of each can easily be determined.

At first there were controversies, with some authors claiming that the malignant phenotype was dominant (AVILES et al. 1977; KUCHERLAPATI and SHIN 1979; CROCE et al. 1975) and others, that it was recessive (STANBRIDGE 1976). Although initially in vitro assays were used to assess the malignant phenotype, tumorigenicity was eventually defined as the ability to produce tumours in syngeneic or in the "nude" mouse strains, a system that was not itself without critics (see FRANKS 1982). From the vast amount of data presented in the 1970s, the picture which most often emerged was that the malignant phenotype was, in fact, recessive but that, from suppressed hybrid cells, variants arose which were tumorigenic (STANBRIDGE 1976; KLINGER 1980; SAGER 1985). In human/rodent cell hybrids, human chromosomes are selectively eliminated, and the reappearance of the malignant phenotype was associated with the loss of chromosomes. It appeared, therefore, that certain chromosomes might carry genes which could suppress the malignant phenotype in the hybrid cells (STANBRIDGE et al. 1981; KLINGER and SHOWS 1983).

The problems encountered through chromosome instability in interspecific cell hybrids can be overcome, more or less, by creating intraspecific cell hybrids in which chromosome loss is usually minimal. In these cases suppression was almost always observed unless extensive chromosome loss had occurred. Karyotyping, however, in intraspecific cell hybrids was a problem, since it was not always possible to determine which chromosomes were derived from the malignant and which from the normal parental cells. In some cases the chromosomes from one of the parents were marked with cytogenetic "tags", such as heterochromatic variations, and in one case (EVANS et al. 1982) it was demonstrated that loss of suppression was due to loss of chromosomes from the normal parent.

In some cases it was reported that loss of specific chromosomes led to the removal of regulatory control in suppressed hybrids, allowing re-expression of the tumorigenic phenotype. In a karyotypic study of rare tumorigenic segregants from suppressed hybrids, STANBRIDGE et al. (1981) reported a statistical correlation between the loss of human chromosomes 11 and 14 from HeLa × normal fibroblast hybrids and re-expression of the malignant potential. Chromosome 11 was also implicated in the suppression of malignancy by KLINGER and colleagues, who performed similar experiments (KLINGER and SHOWS 1983), although they reported that specific combinations of chromosomes were more likely responsible for the suppression rather than any single chromosome. Evidence to suggest that a single copy of chromosome 11 could suppress the malignant phenotype was presented by WEISSMAN et al. (1987) and will be discussed later with the relevance of these experiments to the analysis of Wilms' tumour.

To determine whether the presence of dominantly transforming oncogenes is required for the maintenance of the malignant phenotype, several groups created somatic cell hybrids between oncogene-transformed and normal cells. GEISER et al. (1986) fused normal fibroblasts with EJ bladder carcinoma cells which contain an activated H-*ras* gene. The EJ/fibroblast hybrid was non-tumorigenic even though the H-*ras* gene was still being expressed in tumorigenic segregants at normal levels. Similar observations were made by NODA et al. (1983) who showed that, even though the K-*ras* gene continued to be expressed, suppression resulted when K-*ras*-transformed mouse cells were fused with non-tumorigenic mouse cells. CRAIG and SAGER (1985) transformed non-tumorigenic hamster cells with the EJ activated H-*ras* gene, and again non-tumorigenic hamster cells could suppress tumorigenicity.

II. Role of Differentiation

During the course of many of these experiments it became clear that the parental origin of the cells used in the fusions could be an important factor. The most commonly used malignant parent cell lines, such as HeLa, are highly evolved tumorigenic cell lines, and the normal cells are usually fibroblasts or lymphocytes. In many cases the histogenic origin of the cells used was ignored. In an attempt to investigate the role of the differentiation status of the normal cells over the transformed phenotype, we established hybrids between a newly derived bladder epithelial cell line (SUMMERHAYES and FRANKS 1979; COWELL 1982) and normal cells from histologically different tissues. The transformed cell line had a near normal karyotype (COWELL 1980) and expressed many of the different characteristics of its histological origin, as well as producing tumours in syngeneic hosts. When transformed cells were fused with undifferentiated mouse embryo fibroblasts which were non-tumorigenic, many of the individual hydrid clones showed an enhanced tumorigenic potential (COWELL and FRANKS 1984). The tumour histology was mixed but predominantly undifferentiated. Chromosome loss was minimal, but in all cases a single copy of chromosome 4 was missing. It appeared therefore that the undifferentiated phenotype was dominant. Either the loss of a single copy of chromosome 4 allowed re-expression of the malignant phenotype in these hybrids or the phenotype of the rapidly growing embryonic cells was acting synergistically with the transformed epithelial cell. When the same bladder cell line was fused with normal mouse bladder epithelial cells from the same mouse strain, the tumorigenic potential was suppressed (COWELL and FRANKS 1984). In many cases chromosome analysis was complicated by the large numbers of chromosomes present, but, even in the presence of considerable chromosome loss, the differentiated phenotype was dominant. We concluded that gene expression in normal bladder cells could suppress the genes which were promoting malignancy, whereas undifferentiated fibroblasts could not. It appears that the differentiated phenotype overruled the transformed phenotype. When the transformed bladder cells were fused with epithelial cells from other tissues such as gut and salivary gland, the same result was observed (Cowell, unpublished observations). Similar observations were also made by STANBRIDGE and colleagues (WEISSMAN and STANBRIDGE 1983). They noted that fusion of tumour cells from the same origin (epithelium) resulted in stable expression of the malignant phenotype, and this was also true when carcinoma cells were fused with lymphoma cells. When the same carcinoma cells were fused with sarcoma or melanoma cells, suppression was sometimes observed, although chromosome instability in these hybrids resulted in the emergence of tumorigenic variants. In the nude mouse, the non-tumorigenic variants apparently respond to differentiation-inducing signals and differentiate rapidly, whereas the tumorigenic segregants produced large, undifferentiated carcinomas.

It is clear from the preceeding discussion that the analysis of malignancy by cell fusion is a complex system. Much of the variation probably depends on uncontrolled variables such as the type of parental cell used, chromosome instability and the system used to test tumorigenicity. What has emerged, however, is that the ability of malignant cells to produce tumours can be suppressed by normal cells, and that this ability is due to as yet unidentified genes either on a single chromosome or on limited groups of chromosomes. The fact that suppression occurs in the presence of active, dominantly transforming oncogenes suggests that these genes are not sufficient, or required, for transformation but more likely contribute to progression in tumour development. Finally, it appears that the selective silencing of genes during cell commitment and terminal differentiation can overrule the genetic signals committing them to a transformed phenotype. This suggestion has also come from the observation that some tumours, particularly neuroblastomas and possibly retinoblastomas (Gallie et al. 1982), can spontaneously regress. Histopathological examination showed that the regressed tumour cells had a differentiated phenotype, thereby eliminating their proliferative potential. The suggestion that tumorigenesis is determined by recessive genes has led to the suggestion that the normal function of certain genes can prevent the development of specific tumours associated with their loss, and, as such, they should be classified as anti-oncogenes, as described by Knudson (1985, 1986). Viral oncogenesis appears to play an important role in cell proliferation and, as judged by their expression, act in a wide range of tissues. Anti-oncogenes, on the other hand, are more tissue specific and may be linked to differentiation. The anti-oncogenic action of certain genes has been highlighted recently by the analysis of two paediatric cancers, retinoblastoma and Wilms' tumour, in which again the role of differentiation will be shown to be an important factor in their aetiology.

C. Paediatric Cancer Predisposition Syndromes

I. Retinoblastoma

Retinoblastoma (Rb) is an intraocular eye tumour which predominantly affects children under the age of 5 years. Whilst within the confines of the eye, it is the most curable of all childrens' cancers; once extraocular spread has occurred, it is lethal, there never having been a successful treatment of metastatic retinoblastoma. Like many cancers there are both sporadic and hereditary cases of Rb which occur with roughly equal frequencies (Jay et al. 1988). The sporadic form of the disease is characterized by unifocal, unilateral tumours with a mean age of onset of 18 months. The hereditary form of the tumour, however, is almost always bilateral and multifocal with a mean of 4–5 tumours per eye and a mean age of onset of 10 months. Knudson (1971) noted that the number of tumours which developed in hereditary cases closely fitted a Poisson distribution, and since not all retinal cells produce tumours, at least one second event was required for tumour development; a Poisson distribution suggests that this second event is random. Based on these observations Knudson suggested that only two events were necessary for tumour development, both in the hereditary and non-hereditary forms. In the hereditary one the first event is carried in the germ line and the second is a random somatic mutation whereas, in sporadic tumours, both events occur in somatic cells. Both mutations were deemed to occur in the same, homologous gene; thus tumour cells become homozygous for the mutation.

The successful treatment of Rb throughout most of this century meant that it soon became obvious that affected individuals had affected children and that the inheritance followed an autosomal dominant pattern. For individuals carrying the Rb mutation their children are at 50% risk of developing the tumour. In rare cases, however, the penetrance of the gene is incomplete, and unaffected individuals can have affected parents and children. It is clear, therefore, that "hereditary cancer" is a misnomer since it is only a predisposition to tumour development that is inherited.

The frequency of Rb is approximately 1:20000 births each year. The possibility that two mutations could occur sporadically in more than one retinal precursor cell is remote, and therefore bilateral and multifocal tumours are considered to be hereditary. Unilateral, unifocal tumours are, for the most part, sporadic, although unilaterally affected individuals occasionally have affected children who are more usually bilaterally affected. Estimates that 15% of unilateral cases are hereditary (VOGEL 1979) probably reflect a bias in reporting; in our study of over 1300 cases (JAY et al. 1988) only 4% of hereditary cases were unilaterally affected. In this same study, 47% were bilaterally affected, giving a hereditary fraction of 0.51.

The location of the gene for hereditary Rb has long been inferred (see LELE et al. 1963) to be on one of the D-group chromosomes because of the presence of a constitutional chromosome deletion in a small proportion of Rb patients (COWELL et al. 1986a). With the advent of chromosome banding techniques, the deletion was shown to be on chromosome 13. Over the past 20 years the number of reported deletions has exceeded 100 cases, and although the extent of the deletion varies from patient to patient, in all cases all, or part, of chromosome band 13q14 is missing. In keeping with Knudson's predisposition hypothesis, the constitutional deletions only affect one of the chromosome 13 homologues, and a second event is still required for tumour development. The exact location of the Rb gene, as determined by the minimum region of overlap of the deletions, is still controversial, with some authors favouring a more proximal location (SPARKES et al. 1984; WARD et al. 1984) and others a more distal location (COWELL et al. 1987b; DUNCAN et al. 1987). The precise localisation depends on the interpretation of sub-band chromosome deletions, the identification of which has been made possible by the development of high resolution chromosome banding protocols. Deletion carriers can also be identified using phenotypic criteria and gene dosage studies. When the chromosome deletion is large, there are invariably other characteristic congenital abnormalities, notably mental retardation and dysmorphic features. Based on these phenotypes alone, therefore, it is often possible to predict the presence of a 13q- deletion, although, on rare occasions, we have observed patients with mental retardation who were not deletion carriers. Deletions can also be identified through quantitation of the esterase-D enzyme (ESD). The ESD gene was localised to chromosome region 13q14 by SPARKES et al. (1980), who showed that deletion carriers had only 50% of the enzyme levels seen in normal controls. Even cells with the smallest deletions had reduced enzyme levels, suggesting that the ESD and Rb genes are very close, at least within half of band 13q14. Quantitation of ESD levels has now become a routine method for the detection of deletion carriers, and in a series of 500 patients we have detected 18 deletions so far, giving a frequency of 3%–4% (COWELL et al. 1989). In one of these patients the enzyme assay prevailed when conventional chromosome analysis failed (COWELL et al. 1987b). It was clear that the Rb and ESD genes were in the same subregion of 13q14. A report by SPARKES et al. (1984) suggested that they were in the proximal 13q14 region since, in a patient with a 13q14.1-31 deletion, red blood cell ESD levels were reportedly normal. In situ hybridisation using the ESD gene (DUNCAN et al. 1987) to chromosomes from a lymphoblastoid cell line derived from this patient (EL), however, could not demonstrate an ESD gene on the deletion chromosome. Our own analysis of the EL cell line shows that the cells only have 50% ESD activity. We have reported on a similar patient (COWELL et al. 1987b) with normal ESD levels and a 13q14-31 deletion, but in this instance the breakpoint was in the distal part of 13q14. This supports our previous localisation of the Rb gene in the distal part of 13q14 (COWELL et al. 1987b). We have recently confirmed this localisation by isolating the 13q14-31 deletion in a somatic cell hybrid and have shown that the ESD gene is present and that the Rb gene is not.

The close physical proximity of the Rb and ESD genes raised the possibility that natural variations in the ESD gene, characterised by different electrophoretic mobilities, could be used in linkage studies to determine whether the hereditary, non-deletion form of the disease was also located in region 13q14. The ESD enzyme shows two electrophoretic variants, type 1 and type 2 (HOPKINSON et al. 1973). The generally low frequency of the rarer type 2 allele, however, severely hindered this linkage analysis. Despite this, SPARKES and colleagues (1983) were able to identify three families in which the inheritance of the Rb predisposition was correlated with particular ESD alleles and obtained a lod score of 3.5, confirming linkage. Thus, the hereditary form of Rb was also localised to 13q14. Several other authors have added valuable linkage data since this first report, and the lod score at present stands at 13.8 with no reported recombinations between these two loci (COWELL et al. 1987c). The demonstration that the same locus was involved in the development of sporadic as well as hereditary cases of Rb soon followed and is discussed later.

II. Wilms' Tumour

Wilms' tumour (WT) is a paediatric kidney tumour affecting 1:10000 births each year and accounts for 6% of all children's tumours. As with Rb, the survival rate is high if the tumour remains within the kidney capsule, but the prognosis is poorer if the capsule is invaded. In a recent survey by the United Kingdom Children's Cancer Study Group (UKCCSG) only 4% of cases were bilateral and less than 1% were familial (Pritchard, personal communication). Although often considered a hereditary tumour (KNUDSON and STRONG 1972), the genetic component does not appear to be as obvious as in Rb, which may simply be due to the low penetrance of the WT gene. In the UK the mean age of onset of the bilateral cases is 30 months, compared with 42 months for unilateral cases, which supports a two-hit theory (KNUDSON 1971). Special attention has been paid to WT since the observation that it was frequently associated with a congenital absence of irises, i.e. aniridia. In a study by MILLER (1964) it was noted that 1 of 75 aniridia patients developed WT – the so-called aniridia-Wilms' tumour association (AWTA). When this sub-group of patients was investigated further, it was shown that many of them also had either mental handicap and/or abnormal genitalia (RICCARDI et al. 1978). This syndrome has been referred to as the AGR triad (for aniridia, gonadal dysplasia, retardation) and confers a 50% risk to the individual for the development of WT (NARAHARA et al. 1984). Chromosome analysis in these patients has shown in the vast majority that they have a constitutional deletion on the short arm of chromosome 11. Although the extent of the deletion varies from patient to patient, all or part of region 11p13 is missing. In some only a small part of 11p13 was missing, and, in these patients, mental handicap and gonadal dysplasia were absent. These observations place the aniridia and WT genes in the same subregion of 11p13. The order of these genes was determined by the analysis of a deletion patient with WT, mental handicap and genital dysplasia but not aniridia (TURLEAU et al. 1984b). The deletion was of 11p11-p13 with the breakpoint in the distal part of band p13. Thus, the aniridia locus is placed distal to the WT, locus and in the distal part of p13, probably 11p13.3-13.6. Other genes which have been placed in 11p13 are catalase, which is in the distal part of the band (JUNIEN et al. 1980), and the beta-subunit of follicle-stimulating hormone (GLASER et al. 1986). The probable gene order for region 11p13 is therefore; cen-cat-WT-An-FSH-Tel.

In the absence of any large pedigrees, it is difficult to assess the penetrance of the WT gene, but a clue to a reduced penetrance can be obtained from a study of the deletion patients. NARAHARA et al. (1984) presented a review of 11p13 deletion patients of whom only 52% have developed the tumour, suggesting that the second hit is a rarer event than in Rb. Alternatively, more than one gene may be important in the genesis of WT. WT is not the only malignancy associated with deletion of region 11p13 and the AGR triad. ANDERSEN et al. (1978) reported one AGR patient, a girl, who developed bilateral gonadoblastoma at 21 months. Particularly interesting is that both the kidneys and the gonads are of mesodermal origin and derive from embryologically adjacent sites involving the mesonephros. These observations imply that genes in 11p13 may act pleiotrophically and affect the development of a number of different tissues.

The presence of chromosome deletions has indicated the location of genes which are important in the predisposition to Rb and WT, but the majority of tumours are apparently sporadic, and the question arises whether the events leading to tumour development result from genetic events at the same loci. Two types of analysis have been used to investigate this question: the first utilises traditional analyses of chromosomes in tumour cells, and the second employs more refined techniques of molecular biology using DNA probes to look at specific chromosomal loci.

III. Chromosome Analysis of Tumour Cells

1. Retinoblastoma

It is often very difficult to obtain good spreads and elongated chromosomes in order to generate good banding directly from chromosome analysis of tumour cells. Several groups have used short-term cultures to try to increase the number of mitoses, and in some cases the use of feeder layers of mouse cells has allowed the establishment of permanent cell lines. Early reports suggested that chromosome abnormalities, such as deletions, inversions and translocations, with breakpoints in 13q14, were the most consistent abnormality in Rb tumour cells (Balaban-Malenbaum et al. 1981). Later studies on larger numbers of tumours, however, showed that whilst 13q14 abnormalities were the most common, these were only present in the minority of tumours. No tumour has been reported in which deletion of 13q14 is the only abnormality, and often there are many chromosome abnormalities, making it difficult to assign causality to any of them. In an extensive analysis Squire et al. (1985) showed rearrangement of chromosome 13 in only 3 of 27 tumours. In contrast, aneuploidy for chromosome arms 6p and 1q was observed in 15 of 27 and 21 of 27, respectively. The involvement of iso-6p has also been reported previously (Kusnetsova et al. 1982). It should be noted that the quality of G-banding in tumours is often poor, and the possibility of subtle changes in 13q14 cannot always be excluded. Analysis of tumour/normal pairs with polymorphic sequences is likely to provide a better indication of the reorganisation of the genetic material on chromosome 13 (see below).

2. Wilms' Tumour

Chromosome analysis of WTs has shown the same overall pattern as seen in Rb, in which, although abnormalities involving the short arm of chromosome 11 were the most frequently observed, the vast majority of tumours have normal copies of chromosome 11. Kaneko et al. (1981) reported a tumour in which the only abnormality was del(11p13), although the quality of the banding was not good enough to exclude other subtle rearrangements. In two other tumours both copies of chromosome 11 were normal, but abnormalities involving chromosomes 1 and 16 were observed (Kaneko et al. 1983). Kondo et al. (1984) could only demonstrate 11p abnormalities in 1 of 9 WTs, whereas Douglass et al. (1986) showed 6 of 14 tumours carried abnormalities involving 11p. In this study, simple deletions as well as deletions produced as a result of complex chromosome rearrangements were seen. Slater and colleagues (Slater et al. 1985; Slater and De Kraker 1982) showed aberration in 3 of 11 WTs studied by her and in a later review (Slater 1986) reported that chromosome 11p abnormalities were the most common abnormality, being present in 13 of 38 tumours. The next most commonly involved region was 1p. Chromosome 16 was also frequently involved in rearrangements. Even though 11p is frequently involved in rearrangements, it is often within a background of other chromosome abnormalities such that it is not possible to establish the relative importance of these karyotypic changes. In our own group, one cell line has been established which, when injected into the renal capsule of nude mice, produces tumours. However, despite high resolution chromosome analysis, no chromosome abnormality has been observed. In our own study of WTs, we have so far seen only minor chromosome changes in normally diploid cells and, in only 4 of 20 cases, did these involve chromosome 11.

The possibility that there may be two different genes involved in the development of WT was suggested from observation of the somatic overgrowth condition (BECKWITH 1963; WIEDEMANN 1964) known as Beckwith-Wiedemann syndrome (BWS). In rare cases, patients with BWS have constitutional chromosome abnormalities involving the distal tip of the short arm of chromosome 11 (WAZIRI et al. 1983; TURLEAU et al. 1984a). One of the features of this condition, which has a complex phenotype, is that there is often an associated predisposition to the development of rare paediatric tumours, most usually WT but also hepatoblastoma, rhabdomyosarcomas, adrenal adenocarcinomas and non-Burkitt's lymphoma (WIEDEMANN 1983). In some patients combinations of these tumours have been reported, suggesting a common aetiological event arising as a result of a mutation at the same locus. HAAS et al. (1986) reported a BWS patient who developed nephroblastoma and carried a constitutional chromosome deletion of region 11p11.1-p13. These studies suggest a common pathogenic mechanism for these clinically associated tumour types.

It seems, therefore, that there is evidence that chromosome regions 13q14 and 11p13 might be important in tumorigenesis in sporadic forms of the tumour, but amid a background of random changes, it is not possible to determine cause and effect. A more refined analysis has been presented by several groups demonstrating that, even in the absence of obvious chromosome changes, there has been genetic reorganisation, specifically of chromosomes 11 and 13, in tumour cells from WT and Rb, respectively.

The role of chromosome 11 in Wilms' tumorigenesis was tested by WEISSMAN et al. (1987) using microcell-mediated chromosome transfer techniques which can introduce single chromosomes into selected recipient cells. A normal chromosome 11, in the form of an X:11 translocation to assist selection in HAT medium (COWELL 1986), was introduced into a WT cell line. Those cells which took up the X:11 chromosome lost the ability to form tumours in nude mice. The introduction of an X:13 translocation into the same WT cell line had no effect on tumorigenicity. Interestingly, the in vitro characteristics of cellular transformation were unaffected, suggesting that they are under separate control. The fact that the X:13 chromosome did not affect tumorigenicity demonstrates that genes on chromosome 11 are responsible for the suppression.

IV. Homozygosity Studies

Knudson's hypothesis predicts that the Rb and WT genes require recessive mutations. For the tumours to develop they must become homozygous for the mutation, which can be any disruption of the gene affecting function including deletion, rearrangement or mutation. So far, large homozygous deletions do not appear to be the means of generating loss of function of these genes. More refined techniques using DNA probes may, however, detect small homozygous deletions in some tumours (see below). For example, we have looked at the presence of the catalase gene in the tumour from one of our del(11p13) patients and showed that the gene was present in the tumour. However, a homozygous deletion may have involved a smaller region of 11p13 and not including the catalase locus. An earlier report by BENEDICT et al. (1983) that tumour cells from a 13q− patient showed no ESD activity suggested that there had been a homozygous deletion. Analysis of this tumour using the ESD gene probe showed however that, in fact, the ESD gene was present (Lee et al. 1987b). It is possible that homozygous deletions are lethal for the cell since other, adjacent genes are important for cell survival. It has been suggested that, in non-deletion hereditary cases, the first hit is either a sub-microscopic deletion or a mutation within the gene and occurs at a relatively high frequency. Hemizygosity could be created as a result of the loss of the chromosome containing the normal alleles. Chromosome analysis and dosage studies using 11p and 13q DNA probes have shown that this is not the case. Using polymorphic DNA probes from 13q14, Cavenee et al. (1983) showed that individuals who were constitutionally heterozygous at particular chromosome 13 loci became homozygous for the same loci in the Rb tumours. When loci on other chromosomes were tested, the generation of homozygosity seemed to be restricted to 13q. Godbout et al. (1983) also demonstrated the generation of homozygosity in tumours using the electrophoretic variants of the ESD protein. Other reports of homozygosity in Rb cells soon followed (Dryja et al. 1984; Benedict et al. 1987). The same aberrations were also seen in WT (Orkin et al. 1984; Fearon et al. 1984; Koufos et al. 1984) when observations were made; the allele loss was not due to chromosome loss, and homozygosity was restricted to 11p. Cavenee et al. (1983) proposed two mechanisms for the generation of homozygosity, nondisjunction and mitotic recombination. Mitotic recombination has been demonstrated in both Rb (Cavenee et al. 1985) and WT (Raizis et al. 1985), and non-disjunction has been inferred in many other tumours. Furthermore, Cavenee et al. (1986) were able to show that, when non-disjunction did occur, it was the chromosome which was inherited from the affected parent which was retained in the tumour.

The association between cancer predisposition and BWS and WT was discussed earlier, and several groups have demonstrated the development of homozygosity in the rare, paediatric tumours associated with BWS. Koufos et al. (1986) analysed three hepatoblastomas and showed that homozygosity for 11p15 markers was generated in two, whilst in the third heterozygosity was retained. There were similar findings in two rhabdomyosarcomas. Markers from other chromosomes were the same in tumour and normal tissues, showing that loss of alleles is restricted to chromosome 11. Recently, Scrable et al. (1987) have shown that the development of homozygosity in rhabdomyosarcomas is restricted to the distal half of the short arm of chromosome 11, and mitotic recombination mapping suggests that the predisposition locus is in the 11p15.5-pter region. Similar experiments by Henry et al. (1987) suggest that this same region is involved in the development of some WT, possibly implying a second locus for the WT predisposition locus on the short arm of chromosome 11.

Survivors of the inherited form of Rb are at significant risk for development of second tumours, the most common being osteosarcoma (OS). This does not appear to be the case in the non-hereditary form of the tumour. Radiation treatment of Rb undoubtedly contributes to the high incidence of OS, many tumours arising within the irradiated field, but more than one-third of OS in Rb patients occur in unirradiated bones. The high incidence of OS in Rb patients suggests that the same genetic locus could be responsible for the predisposition to both tumours. HANSEN et al. (1985) investigated OS tumours from both Rb and sporadic patients by chromosome 13-specific DNA probes. They demonstrated homozygosity for the 13pter-q21 region in one patient who had previously had bilateral Rb and also in a sporadic OS patient. Many loci on other chromosomes were also analysed but shown to be unaffected, demonstrating the specificity of the genetic reorganisation. The presumptive Rb gene was also investigated in OS, and in two cases reorganisation of the gene was found. Similar results were also reported by DRYJA et al. (1986a).

In conclusion, therefore, the two-hit hypothesis has been supported by homozygosity studies, and sporadic tumours arise probably as a result of genetic reorganisation in the same regions implicated in the hereditary forms of Rb and WT. These observations have led several groups towards the goal of isolating the genes conferring the predisposition to both tumours.

D. Isolation of Cancer Genes

The histopathology of Rb and WT demonstrates essentially poorly differentiated cells. In Rb there is some evidence for pseudorosette formation, but the tumour cells are essentially immature retinoblasts. WT is a catchall for children's kidney tumours, and as a group they have variable histology. However, nephron and glomerulus development is absent, and the tumours are often of undifferentiated blastemic cells, stroma and epithelium. In both tumours it would appear that differentiation signals normally operating in development are disturbed in some way, and the embryonic cells continue to divide uncontrollably. A clue to the identity of signals which might be operating in differentiation was provided by GROBSTEIN (1953), who studied the in vitro differentiation of the kidney. Normally, the ureteric bud invades the renal blastema early in development, which provides the signal to differentiate. Differentiation occurs at the advancing tip of the invading ureteric tissue, which eventually forms the collecting ducts in the kidney. Grobstein introduced a membrane between the blastemal cells and the ureteric bud tissue and showed that local differentiation occurred at the point of contact with the membrane. This observation argues for a diffusible substance from the ureteric tissue which induces differentiation in the blastemic cells. The only other tissue which could replace the role of the ureteric bud was spinal cord. It is attractive to imagine that WT precursor cells fail to respond to this signal and continue to divide uncontrollably.

The identification of specific chromosome regions which contain genes responsible for an inherited predisposition to the development of a specific tumour raised the possibility of cloning these genes. Any gene isolated from these regions would be a candidate and could then be critically evaluated for a role in tumorigenicity. The series of experiments which set out to isolate these genes can be likened to a fishing expedition. Since the nature of these genes is completely unknown, random DNA sequences needed to be isolated from the chromosome in question and mapped to the critical region. The statistics governing the fishing expedition turn out in favour of the fish. With 3×10^9 base pairs (bp) in the haploid genome, each chromosome contains approximately 7.2×10^7 bp and the

13q14 region, which comprises 4% of the chromosome, contains 2.8×10^6 bp = 2800 Kb. Since cosmid vectors only carry 40 Kb, the task is considerable. However, although the odds clearly favour the fish, luck has favoured the fisherman, for the Rb gene at least.

The search for 13q14 probes required DNA libraries enriched for chromosome 13 sequences. Two groups used somatic cell hybrids which contained a small number of human chromosomes including 13 (Cavenee et al. 1984; Dryja et al. 1984) to generate lambda libraries from which recombinant phage containing human DNA could be identified by virtue of the characteristic repetitive sequences they contain. Unique sequences from some of these clones were then isolated and regionally localised on chromosome 13 using either somatic cell hybrids with overlapping deletions or by in situ hybridisation. Lalande et al. (1984) constructed a human chromosome 13-specific library from flow-sorted chromosomes, from which unique DNA sequences could be isolated directly and again mapped using panels of somatic cell hybrids. Similar experiments were undertaken by Scheffer et al. (1986). It is the chromosome 13 probes generated from these groups which have been used in the majority of analyses of Rb.

I. Retinoblastoma

It was one of the probes isolated by Lalande et al. (1984) which eventually led to the isolation of a candidate for the Rb gene. Clone H3-8 was shown to map to a submicroscopic deletion in region 13q14 in a Rb patient. Using this probe, adjacent DNA sequences were generated using walking strategies (Dryja et al. 1986b; Friend et al. 1986). One of these sequences was shown to be highly conserved between species, suggesting that it was localised in a coding sequence. Using this DNA sequence Friend et al. (1986) identified an mRNA which was 4.7 Kb long. A cDNA clone was subsequently isolated from fetal retinal cells transformed with adenovirus 12. This cDNA was apparently not present in either fetal retina or Rb cells but was present in a wide variety of other tumour types. Similar experiments using H3-8 as the starting sequence led to the isolation of the same gene by other groups (Lee et al. 1987a; Fung et al. 1987). In one study (Lee et al. 1987a), using cultured cells from Rbs, all tumours showed either abnormal length transcripts or absence of transcripts. Fung et al. (1987) also demonstrated abnormal/absence of transcripts in all tumours, although only 40% of tumours analysed showed abnormalities in the genomic sequence as determined by Southern blot analysis. The demonstration of an apparently normal DNA sequence in one tumour in which the mRNA was missing possibly suggests a promotor mutation (Lee et al. 1987a). Deletions of the genomic sequence were sometimes homozygous and sometimes hemizygous for the whole gene (Fung et al. 1987), but in some tumours partial, internal deletions were seen, which usually involved the 7.5 Kb *Hin*dIII fragment and sometimes the adjacent 9.8 Kb, 6.2 Kb, and 2.1 Kb fragments. In one instance in which the 9.8 Kb fragment was missing, it was replaced by a novel 8.0 Kb fragment, suggesting a small deletion within this region of the gene.

Lee et al. (1987a) sequenced 4523 nucleotides from their cDNA which contained 12 exons and covered at least 200 Kb of genomic DNA. A comparison with known gene sequences in data banks failed to show any homology with any known gene. From the DNA sequence, LEE et al. (1987a) observed a long, open reading frame encoding a hypothetical protein with features suggestive of a DNA-binding protein. Following cloning in an appropriate vector, a fusion protein was made, from which antibodies were prepared (LEE et al. 1987c). This antibody detected a protein which was present in those cells with the mRNA transcript and absent in those tumours which had no transcript. Characterisation of the protein showed that it was a phosphoprotein confined to the nucleus and bound to DNA. These observations support the theory that the Rb gene may function by regulating other genes in the cell.

II. Prenatal Diagnosis

The isolation of a candidate gene raised the possibility of using it for prenatal diagnosis since the potential for recombination is eliminated. However, despite the large size of the cDNA and the related genomic sequence, no RFLPs have been identified to date with the cDNA, despite screening over 50 enzymes (Dryja, personal communication). However, using the cDNA to screen lambda libraries, almost the entire genomic sequence has been isolated (WIGGS et al. 1988). From within the introns unique sequences, which recognise RFLPs, have been isolated. One probe in particular, p68RS2, identifies at least 8 alleles based on a polymorphic, tandemly repeating unit, 50–55 bp long, a so-called variable-number tandem repeat (VNTR). It has been estimated that this sequence is informative in over 70% of Rb families. In a limited study of 18 families, no recombination was found in 17 of them, giving a lod score of 10.8. In the 18th family, however, there were two obligate crossovers in a multigeneration family. A similar family has recently been identified in Holland (Buys and Scheffer, personal communication), although there is a possibility that the diagnosis of Rb in these cases is questionable. In the absence of a reliable and generally applicable test for prenatal screening, all sibs and children of Rb patients are screened regularly throughout the first few years of life (COWELL et al. 1988). These examinations take the form of ophthalmoscopy under anaesthetic. Early diagnosis means that early conservative treatment is possible. More advanced tumours require more drastic treatment and sometimes removal of the eye. In the UK this dedicated service has undoubtedly led to greater than 90% survival rate of patients with Rb. A reliable prenatal testing service would allow attention to be focussed on those children who need counselling since only half of the children will have inherited the Rb predisposition gene.

III. Wilms' Tumour

Molecular cloning strategies have also been designed to try to identify the WT predisposition gene and mostly involve random-mapping strategies of chromosome-11-specific DNA sequences isolated from a variety of chromosome-specific DNA libraries and different somatic cell hybrids with different overlapping chromosome 11 deletions. One important source of chromosome 11 sequences has been from the flow-sorted chromosome library constructed by the Lawrence Livermore laboratory. Created in the charon 21 vector, these libraries contain inserts which are, on average, 4 Kb in size and designed specifically to isolate unique sequences directly by screening the library with repetitive human DNA and selecting those clones which do not hybridise. Recently, three groups have reported the isolation of clones using this strategy (LEWIS and YEGER 1987; DAVIS et al. 1987; BRUNS et al. 1987). It is clear from the accumulated data that a large number of clones have been screened to obtain only a few clones which are in the 11p13 region. A similar low incidence of 11p13 clones was found by COWELL et al. (1987a) screening a cosmid library made from a somatic cell hybrid with only human chromosome 11. JUNIEN and colleagues (HUERRE-JEANPIERRE et al. 1987) used a library which was more enriched for 11p DNA sequences by flow sorting an 11:22 translocation that contained only the short arm of chromosome 11 and a small piece of chromosome 22. From the 35 clones isolated, 11 were shown positively to be from 11p and two were in the catalase-FSH interval.

PORTEUS and colleagues have used another approach to make libraries enriched for 11p sequences by chromosome-mediated gene transfer. The EJ bladder cell line contains an activated H-*ras* gene which can transform mouse 3T3 cells following DNA transfection experiments (PORTEOUS et al. 1986). Chromosomes from the EJ bladder cell line were prepared and introduced into a suitable host mouse cell line and transformants selected by virtue of their ability to form foci of transformed cells. All transformants contained the H-*ras* gene and various other pieces of DNA from the short arm of chromosome 11. One clone in particular contained both the catalase and FSH genes and, presumably, the intervening sequences. A lambda library was prepared from this transfectant and individual DNA clones isolated (PORTEOUS et al. 1987). Mapping was performed using a panel of hybrids which could subregionally localise DNA sequences to the WT locus. The first thing that became obvious was that the majority of clones were from the distal tip of chromosome 11, but 17 clones have been isolated from the 11p13 region, and several of these are considered to be close to the WT locus (PORTEUS et al. 1987).

Chromosome analysis of T-cell leukaemia (ALL) demonstrated a fairly consistent translocation between chromosomes 11 and 14 with breakpoints in 11p13 and 14q11 (WILLIAMS et al. 1984). Although the breakpoint on chromosome 14 is invariably at 14q11, being the site of the alpha chain of the T-cell receptor (LEWIS et al. 1985; ERIKSON et al. 1985), region 11p15 may be involved in some translocations instead of 11p13 (Rabbitts, personal communication). In one case reported by LEWIS et al. (1985), the 11:14 translocation has been isolated in somatic cell hybrids, and the breakpoints are between the catalase and FSH genes. Chromosome walking experiments by Rabbitts and colleagues (personal communication) have isolated DNA sequences from the breakpoint junctions from 11p15 and 11p13. The sequences in 11p15 are from an as yet unidentified gene sequence, but the sequences in 11p13 have not yet been characterised. In the light of the involvement of these regions in a predisposition to BWS and WT, respectively, these sequences must be strong candidates for the tumour predisposition loci.

IV. N-*myc* Expression

If the assumption that oncogenes contribute to the progression of tumorigenicity is correct, then it might be expected that specific patterns of oncogene expression could be detected in recessive paediatric cancers. Analysis of chromosome rearrangements has failed to implicate the involvement of any oncogene in their genesis, although dosage studies have shown elevated levels of the N-*myc* oncogene in a variety of tumours. In neuroblastomas, for example, elevated expression is seen, but only in advanced-stage tumours, again suggesting a role in progression. There have also been reports of N-*myc* amplification in Rb cells (LEE et al. 1985). SQUIRE et al. (1986) were able to examine different tumour foci from the same patient but did not find that N-*myc* amplification was a consistent feature. They were also able to show N-*myc* expression in normal adult brain and retina and the same levels of expression in normal fetal tissues as in the tumours. They concluded that N-*myc* expression in tumours probably reflects the origin of the cells rather than indicating a role in oncogenic transformation. This same conclusion was reached for the elevated expression of IGF II in WT (SCOTT et al. 1985). This explanation does not seem so easy to apply for N-*myc* expression in WT. NIESEN et al. (1986) showed that 29 of 30 WT as well as other embryonic tumours showed amplification of N-*myc* expression. Unlike neuroblastomas, the pattern of expression in WT does not seem to be stage-specific or related to a particular histology.

E. Conclusions

Rb is the prototype for the study of hereditary cancer because of the early pinpointing of the predisposition locus in the genome through linkage analysis and deletion mapping. There are at least 50 known hereditary cancers (McCUSICK 1983) and, undoubtedly, even more that are unrecognised. The most extensively studied are those, such as Rb, with a clearly definable phenotype and pattern of inheritance. In others, although the inheritance is clear, the phenotype can be variable, e.g. neurofibromatosis (RICCARDI and EICHNER 1986) and multiple endocrine neoplasia (MEN) (SCHIMKE 1984). In both of these forms of hereditary cancer it has not been possible to demonstrate constitutional chromosome deletions in any patient, but for neurofibromatosis, for example, the location of the predisposition gene has been assigned to chromosome 17 using classical linkage studies (BARKER et al. 1987). Similarly, the gene for MEN 2 has been localised to chromosome 10 (MATHEW et al. 1987; SIMPSON et al. 1987), and the polyposis coli gene, responsible for one hereditary form of colon cancer, has been placed on chromosome 5 (BODMER et al. 1987). It is likely that the location of all hereditary forms of cancer will soon be found, even when the phenotype and penetrance are variable, as in breast cancer.

The demonstration of the development of homozygosity for particular DNA "markers" is proving most effective for identifying that part of the genome which may carry recessive genes related to the development of particular cancers. In addition to Rb and WT, this approach has been used recently to confirm the localisation of the polyposis coli predisposition gene on chromosome 5

(SOLOMON et al. 1987) and implicate the short arm of chromosome 3 in the development of small lung cell carcinoma (MOOIBROEK et al. 1987) and chromosome 22 in meningiomas (SEIZINGER et al. 1987). This approach may be more informative than conventional linkage analysis since homozygosity can be generated over relatively long stretches of DNA, whereas linkage analysis requires that the probe is fairly close to the tumour locus and that large family pedigrees are available. One potential pitfall in homozygosity analysis is that several chromosomes may be involved; thus, ideally, all chromosome arms should be analysed since it has been shown, in ductal breast tumours (LUNDBERG et al. 1987) and melanomas (DRACOPOLI et al. 1985, 1987), for example, that the generation of homozygosity was seen for most of the chromosomes analysed. Finally, the importance of chromosome analysis should not be overlooked whilst the molecular analysis of tumours is in its ascendancy. Small changes such as the deletion on the short arm of chromosome 3 in small cell lung carcinoma (WHANG-PENG et al. 1982) and the loss of chromosome 22 in meningiomas (ZANKL and ZANG 1980) provided important clues to the location of cancer genes before molecular genetic analysis was performed. The localisation of the tumour predisposition genes described in this article has been achieved in a relatively short period, over the past 2 years, and it is likely that the isolation and characterisation of these genes will soon follow. The time is rapidly approaching when it should be possible to obtain a clearer picture of the defects in these genes that result in tumour development.

References

Andersen S, Geertingen P, Larsen HW, Mikkelson M, Parbing A, Vestermark S, Warburg M (1978) Aniridia, cateract and gonadoblastoma in a mentally retarded girl with deletion of chromosome 11. Ophthalmologica 176:171–177

Aviles D, Jami J, Rousset JP, Ritz E (1977) Tumour × host cell hybrids in the mouse: chromosomes from the normal parent maintained in malignant hybrid tumours. JNCI 58:1391–1397

Balaban-Malenbaum G, Gilbert F, Nichols WW, Hill R, Shields J, Meadows AT (1981) A deleted chromosome number 13 in human retinoblastoma cells: relevance to tumorigenesis. Cancer Genet Cytogenet 3:243–250

Barker D, Wright E, Nguyen K, Cannon L, Fain P, Goldgar D, Bishop DT, Carey J, Baty B, Kivlin J, Willard H, Waye JS, Greig G, Leinwand L, Nakamura Y, O'Connell P, Leppert M, Lalouel JM, White R, Scolnick M (1987) Gene for von Recklinghausen neurofibromatosis is in the pericentromeric region of chromosome 17. Science 236:1100–1102

Beckwith (1963) Extreme cytomegaly of the adrenal fetal cortex, omphalocele hyperplasia of kidneys and pancreas, and Leydig-cell hyperplasia: another syndrome? Western Soc Pediat Res, Los Angeles, Nov 11th

Benedict WF, Murphree AL, Banerjee A, Spina CA, Sparkes MC, Sparkes RS (1983) Patient with 13 chromosome deletion: evidence that the retinoblastoma gene is a recessive cancer gene. Science 219:973–975

Benedict WF, Srivatsan ES, Mark C, Banerjee A, Sparkes RS, Murphree AL (1987) Complete or partial homozygosity of chromosome 13 in primary retinoblastoma. Cancer Res 47:4189–4191

Bodmer WF, Bailey CJ, Bodmer J, Bussey HJR, Ellis A, Gorman P, Lucibello FC, Murday VA, Rider SH, Scambler P, Sheer D, Solomon E, Spurr NK (1987) Localisation of the gene for familial adenomatous polyposis on chromosome 5. Nature 328:614–616

Bruns GAP, Barnes SD, Gessler M, Brennick JB, Weiner MJ (1987) DNA probes for chromosome 11 and the WAGR deletion. Human Gene Mapping 9:240

Cavenee W, Leach R, Mohandas T, Pearson P, White R (1984) Isolation and regional localisation of DNA segments revealing polymorphic loci for human chromosome 13. Am J Hum Genet 36:10–24

Cavenee WK, Dryja TP, Phillips RA, Benedict WF, Godbout R, Gallie BL, Murphree AL, Strong LC, White R (1983) Expression of recessive alleles by chromosomal mechanisms in retinoblastoma. Nature 305:779–784

Cavenee WK, Hansen MF, Nordenskjold M, Kock E, Maumenee I, Squire JA, Phillips RA, Gallie BL (1985) Genetic origin of mutations predisposing to retinoblastoma. Science 228:501–503

Cavenee WK, Murphree AL, Schull MM, Benedict WF, Sparkes RS, Kock E, Nordenskjold M (1986) Prediction of familial predisposition to retinoblastoma. N Engl J Med 314: 1201–1207

Cowell JK (1980) Consistent chromosome abnormalities associated with mouse bladder epithelial cell lines transformed in vitro. JNCI 65:955–961

Cowell JK (1982) Appearance of double minute chromosomes in somatic cell hybridisation experiments involving the HAT selection system. Cell Biol Int Rep 6:393–399

Cowell JK (1986) Manipulation of somatic cell hybrids for the analysis of the human genome. In: Rooney DE, Czepulkowski BH (eds) Humyn cytogenetics, a practical approach. IRL Press, Oxford, pp 202–218

Cowell JK (1988) Molecular approaches towards the isolation of pediatric cancer predisposition genes. In: Kemshead JT (ed) Monoclonal antibodies and nuclear probes in childhood solid tumours. CRC Press, Boca Raton

Cowell JK, Franks LM (1984) The ability of normal mouse cells to reduce the malignant potential of transformed mouse bladder epithelial cells depends on their somatic origin. Int J Cancer 33:657–667

Cowell J, Rutland P, Jay M, Hungerford J (1986 a) Deletions of the esterase-D locus from a survey of 200 retinoblastoma patients. Hum Genet 72:164–167

Cowell JK, Thompson E, Rutland P (1986 b) The need to screen all retinoblastoma patients for esterase-D activity; detection of submicroscopic deletions. Arch Dis Child 62:8–11

Cowell JK, Wadey RB, Pritchard J, Little PFR (1987 a) Isolation and regional localisation of unique sequences from a chromosome 11 specific cosmid library. Human Gene Mapping 9:385

Cowell JK, Hungerford J, Rutland P, Jay M (1987 b) A chromosomal breakpoint which separates the esterase-D and retinoblastoma predisposition loci in a patient with del (13) (q14-q31). Cancer Genet Cytogenet 27:27–31

Cowell JK, Jay M, Rutland P, Hungerford J (1987 c) An assessment of the usefulness of the esterase-D protein polymorphism in the antenatal prediction of retinoblastoma in the United Kingdom. Br J Cancer 55:661–664

Cowell JK, Hungerford J, Jay M, Rutland P (1987 d) Retinoblastoma-clinical and genetic aspects: a review. J Roy Soc Med 81:220–223

Cowell JK, Hungerford J, Rutland P, Jay M (1989) Genetic and cytogenetic analysis of patients showing reduced esterase-D levels and mental retardation from a survey of 500 individuals with retinoblastoma. Ophthal Ped Genet 10:117–127

Craig RW, Sager R (1985) Suppression of tumorigenicity in hybrids of normal and oncogene-transformed CHEF cells. Proc Natl Acad Sci USA 82: 2062–2066

Croce CM, Aden D, Koprowski H (1975) Tumorigenicity of mouse-man diploid hybrids in nude mice. Science 190:1200–1202

Davis LM, Nowak NJ, Shows TB (1987) Seven new single copy DNA segments map to 11p13. Human gene Mapping 9:69

Douglass EC, Look AT, Webber B, Parham D, Wilimas JA, Green AA, Roberson PK (1986) Hyperdiploidy and chromosomal rearrangements define the anaplastic variant of Wilms' tumour. J Clin Oncol 4:975–981

Dracopoli NC, Houghton AN, Old LJ (1985) Loss of polymorphic restriction fragments in malignant melanoma: implications for tumour heterogeneity. Proc Natl Acad Sci USA 82:1470–1474

Dracopoli NC, Alhadeff B, Houghton AN, Old LJ (1987) Loss of heterozygosity at autosomal and X-linked loci during tumour progression in a patient with melanoma. Cancer Res 47:3995–4000

Dryja TP, Cavenee WK, White R, Rapaport JM, Peterson R, Albert DM, Bruns GA (1984) Homozygosity of chromosome 13 in retinoblastoma. N Engl J Med 310: 550–553

Dryja TP, Rapaport JM, Epstein J, Goorin AM, Weichselbaum R, Koufos A, Cavenee WK (1986a) Chromosome 13 homozygosity in osteosarcoma without retinoblastoma. Am J Hum Genet 38:59–66

Dryja TP, Rapaport JM, Joyce JM, Petersen RA (1986b) Molecular detection of deletions involving band q14 of chromosome 13 in retinoblastomas. Proc Natl Acad Sci USA 83:7391–7394

Duncan AMV, Morgan C, Gallie BL, Phillips RA, Squire J (1987) Re-evaluation of the sublocalisation of esterase-D and its relation to the retinoblastoma locus by in situ hybridisation. Cytogenet Cell Genet 44:153–157

Erikson J, Williams DL, Finan J, Nowell PC, Croce CM (1985) Locus of the a-chain of the T-cell receptor is split by chromosome translocation in T-cell leukemias. Science 229: 784–786

Evans EP, Burtenshaw MD, Brown BB, Hennion R, Harris H (1982) The analysis of malignancy by cell fusion. IX. Reexamination and clarification of the cytogenetic problem. J Cell Sci 56:113–130

Fearon ER, Vogelstein B, Feinberg AP (1984) Somatic deletion and duplication of genes on chromosome 11 in Wilms' tumours. Nature 309:176–178

Franks LM (1982) The use of cell hybrids to study phenotypic expression in ageing and tumour cells. In: Celis JE (ed) Gene expression in normal and transformed cells. Plenum, New York

Friend SH, Bernards R, Rogelj S, Weinberg RA, Rapaport JM, Albert DM, Dryja TP (1986) A human DNA segment with properties of the gene that predisposes to retinoblastoma and osteosarcoma. Nature 323:643–646

Fung YT, Murphree AL, T'Ang A, Qian J, Hinrichs SH, Benedict WF (1987) Structural evidence for the authenticity of the human retinoblastoma gene. Science 236:1657–1661

Gallie BL, Ellsworth RM, Abramson DH, Phillips RA (1982) Retinoblastoma: spontaneous regression of retinoblastoma or benign manifestation of the mutation? Br J Cancer 45:513–521

Geiser AG, Der CJ, Marshall CJ, Stanbridge EJ (1986) Suppression of tumorigenicity with continued expression of the c-Ha-*ras* oncogene in EJ bladder carcinoma-human fibroblast hybrid cells. Proc Natl Acad Sci USA 83:5209–5213

Glaser T, Lewis WH, Bruns GAP, Watkins PC, Rogler CE, Shows TB, Powers VE, Willard HF, Goguen JM, Simola KOJ, Housman DE (1986) b-Subunit of follicle-stimulating hormone is deleted in patients with aniridia and Wilms' tumour, allowing a further definition of the WAGR locus. Nature 321:882–887

Godbout R, Dryja TP, Squire J, Gallie BL, Phillips RA (1983) Somatic inactivation of genes on chromosome 13 is a common event in retinoblastoma. Nature 304:451–453

Grobstein C (1953) Morphogenetic interaction between embryonic mouse tissue separated by a membrane filter. Nature 172:869–871

Haas OA, Zoubek A, Grumayer ER, Gadner H (1986) Constitutional interstitial deletion of 11p11 and pericentric inversion of chromosome 9 in a patient with Wiedemann-Beckwith syndrome and hepatoblastoma. Cancer Genet Cytogenet 23:95–104

Hansen MF, Koufos A, Gallie BL, Phillips RA, Fodstad O, Brogger A, Gedde-Dahl T, Cavenee WK (1985) Osteosarcoma and retinoblastoma: a shared chromosomal mechanism revealing recessive predisposition. Proc Natl Acad Sci USA 82:6216–6220

Henry I, Grandjouan S, Azoulay M, Huerre-Jeanpierre C, Couillin P, Junien C (1987) Mitotic recombination with loss of heterozygosity distal to FSHB in a WAGR associated nephroblastoma. Human Gene Mapping 9:628

Hopkinson DA, Mestriner MA, Cortner J, Harris H (1973) Esterase-D: a new human polymorphism. Ann Hum Genet 37:119–137

Huerre-Jeanpierre C, Azoylay M, Henry I, Serre JL, Lavedan C, Coullin P, Antignac C, Salles AM, Lewis WH, Glaser T, Bernheim A, Junien C (1987) Malformative syndromes and predisposition to tumour: isolation and localisation of 11p markers. Human Gene Mapping 9:489

Jay M, Cowell JK, Hungerford J (1988) Register of retinoblastoma: preliminary results. Eye 2:102–105

Junien C, Turleau C, De Grouchy J, Said R, Rethore MO, Tenconi R, Dufier JL (1980) Regional assignment of catalase(CAT) gene to band 11p13. Association with the aniridia-Wilms' tumour-gonadoblastoma (WAGR) complex. Ann Genet 28:165–168

Kaneko Y, Euges MC, Rowley JD (1981) Interstitial deletion of short arm of chromosome 11 limited to Wilms' tumour cells in a patient without aniridia. Cancer Res 41: 4577–4578

Kaneko Y, Kondo K, Rowley JD, Moohr JW, Maurer HS (1983) Further chromosome studies on Wilms' tumour cells of patients without aniridia. Cancer Genet Cytogenet 10:191–197

Klinger H, Shows T (1983) Suppression of tumorigenicity in somatic cell hybrids. II. Human chromosomes implicated as suppressors of tumorigenicity in hybrids with Chinese hamster ovary cells. JNCI 71:559–569

Klinger HP (1980) Suppression of malignancy in somatic cell hybrids. I. Suppression and reexpression of tumorigenicity in diploid human × D98AH2 hybrids and independent segregation of tumorigenicity from other cell phenotypes. Cytogenet Cell Genet 27: 254–266

Knudson AG (1971) Mutation and cancer: statistical study of retinoblastoma. Proc Natl Acad Sci USA 68:820–823

Knudson AG (1985) Hereditary cancer, oncogenes, and antioncogenes. Cancer Res 45: 1437–1443

Knudson AG (1986) Genetics of human cancer. Ann Rev Genet 20:231–251

Knudson AG, Strong LC (1972) Mutation and cancer: a model for Wilms' tumour of the kidney. JNCI 40:313–324

Kondo K, Chilcote RR, Maurer HS, Rowley J (1984) Chromosome abnormalities in tumour cells from patients with sporadic Wilms' tumour. Cancer Res 44:5376–5381

Koufos A, Hansen MF, Lampkin BC, Workman ML, Copeland NG, Jenkins NA, Cavenee WK (1984) Loss of alleles at loci on human chromosome 11 during genesis of Wilms' tumour. Nature 309:170–172

Koufos A, Hansen MF, Copeland NG, Jenkins MA, Lampkin BC, Cavenee WK (1986) Loss of heterozygosity in 3 embryonal tumours suggests a common pathogenetic mechanism. Nature 316:330–334

Kucherlapati R, Shin S (1979) Genetic control of tumorigenicity in interspecific mammalian cell hybrids. Cell 16:639–648

Kusnetsova LE, Prigogina EL, Pogosianz HE, Belkina BM (1982) Similar chromosome abnormalities in several retinoblastomas. Hum Genet 61:201–204

Lalande M, Dryja TP, Schreck RR, Shipley J, Flint A, Latt SA (1984) Isolation of human chromosome 13-specific DNA sequences cloned from flow sorted chromosomes and potentially linked to the retinoblastoma locus. Cancer Genet Cytogenet 13:283–295

Lee WH, Murphree AL, Benedict WF (1985) Expression and amplification of the N-*myc* gene in primary retinoblastoma. Nature 309:458–460

Lee WH, Bookstein R, Hong F, Young L, Shew JY, Lee EP (1987a) Human retinoblastoma susceptibility gene: cloning, identification and sequence. Science 235: 1394–1399

Lee WH, Bookstein R, Wheatley W, Benedict WF, Lee EP (1987b) A null allele of esterase-D is a marker for genetic events in retinoblastoma formation. Hum Genet 76:33–36

Lee WH, Shew JY, Hong FD, Sery TW, Donoso LA, Young LJ, Bookstein R, Lee YP (1987c) The retinoblastoma susceptibility gene encodes a nuclear phosphoprotein associated with DNA binding activity. Nature 329:642–645

Lele KP, Penrose LS, Stallard HB (1963) Chromosome deletion in a case of retinoblastoma. Ann Hum Genet 27:171–174

Lewis WH, Yeger H (1987) Characterisation of the aniridia-Wilms' tumour association region from chromosome 11p. Human Gene Mapping 9:72

Lewis WH, Goguen JM, Powers VE, Willard HF, Michalopoulos EE (1985) Gene order on the short arm of human chromosome 11: regional assignment of the LDH A gene distal to catalase in two translocations. Hum Genet 71:249–253

Lundberg C, Skoog L, Cavenee WK, Nordenskjold M (1987) Loss of heterozygosity in human ductal breast tumours indicates a recessive mutation on chromosome 13. Proc Natl Acad Sci USA 84:2372–2376

Mathew CGP, Chin KS, Easton EF, Thorpe K, Carter C, Liou GI, Fong S-L, Bridges CDB, Haak H, Nieuwenhuijzen Kruseman AC, Schifter S, Hansen HH, Telenius H, Telenuis-Berg M, Ponder BAJ (1987) A linked genetic marker for multiple endocrine neoplasia type 2A on chromosome 10. Nature 328:527–528

McCusick VA (1983) Mendelian inheritance in man, 6th edn. Johns Hopkins University Press, Baltimore

Miller RW (1964) Association of Wilms' tumour with aniridia, hemihypertrophy and other congenital malformations. N Engl J Med 270:922

Mooibroek H, Osinga J, Postmus PE, Carritt B, Buys CHMC (1987) Loss of heterozygosity for a chromosome 3 sequence presumably at 3p21 in small cell lung cancer. Cancer Genet Cytogenet 27:361–365

Narahara K, Kikkawa K, Kimira S, Kimoto H, Ogata M, Kasai M, Matsuoka K (1984) Regional mapping of catalase and Wilms' tumour-aniridia, genitourinary abnormalities, and mental retardation triad loci to the chromosome segment 11p1305-p1306. Hum Genet 66:181–185

Niesen PD, Zimmermann KA, Cotter SV, Gilbert F, Alt FW (1986) Enhanced expression of the N-*myc* gene in Wilms' tumour. Cancer Res 46:6217–6222

Noda M, Selinger Z, Scolnick EM, Bassin RH (1983) Flat revertants isolated from Kirsten sarcoma virus-transformed cells are resistant to the action of specific oncogenes. Proc Natl Acad Sci USA 80:5602–5606

Orkin SH, Goldman DS, Sallan SE (1984) Development of homozygosity for chromosome 11p markers in Wilms' tumour. Nature 309:172–174

Porteous DJ, Morten JEN, Cranston G, Fletcher JM, Mitchell A, van Heyningen V, Fantes JA, Boyd PA, Hastie ND (1986) Molecular and physical arrangements of human DNA in HRAS1-selected, chromosome-mediated transfectants. Mol Cell Biol 6:2223–2232

Porteous DJ, Bickmore W, Hirst M, Boyd P, Little P, Gosden J, Hastie N (1987) Cotransfer of defined genes and cloning of anonymous DNA markers following HRAS-1 selected chromosome mediated gene transfer. Human Gene Mapping 9:51

Raizis AM, Becroft DM, Shaw RL, Reeve AE (1985) A mitotic recombination in Wilms' tumour occurs between the parathyroid hormone locus and 11p13. Hum Genet 70: 344–346

Riccardi VM, Eichner JE (1986) Neurofibromatosis: phenotype, natural history and pathogenesis. Johns Hopkins University Press, Baltimore

Riccardi VM, Sujansky E, Smith AC, Francke U (1978) Chromosome imbalance in the aniridia-Wilms' tumour association: 11p interstitial deletion. Pediatrics 61:604–610

Sager R (1985) Genetic suppression of tumour formation. Adv Cancer Res 44:43–68

Scheffer H, Van Der Lelie D, Aanstoot GH, Good N, Nienhaus AJ, Van Der Hout AH, Pearson PL, Buys CHCM (1986) A straightforward approach to isolate DNA sequences with potential linkage to the retinoblastoma gene. Hum Genet 74: 249–255

Schimke RN (1984) Genetic aspects of multiple endocrine neoplasia. Ann Rev Med 35: 25–31

Scott J, Cowell JK, Robertson ME, Priestley LM, Wadey R, Hopkins B, Pritchard J, Bell GI, Rall LB, Graham CF, Knott TJ (1985) Insulin-like growth factor-II gene expression in Wilms' tumour and embryonic tissues. Nature 317:260–262

Scrable HJ, Witte DP, Lampkin BC, Cavenee WK (1987) Chromosomal localisation of the human rhabdomyosarcoma locus by mitotic recombination mapping. Nature 329: 645–647

Seizinger BR, De La Monte S, Atkins L, Gusella JF, Martuza RL (1987) Molecular genetic approach to human meningioma: loss of genes on chromosome 22. Proc Natl Acad Sci USA 84:5419–5423
Simpson NE, Kidd KK, Goodfellow PJ, McDermid H, Myers S, Kidd JR, Jackson CE, Duncan AMV, Farrer LA, Brasch K, Castiglione C, Genel M, Gertner J, Greenberg CR, Gusella JF, Holden JJA, White BN (1987) Assignment of multiple endocrine neoplasia type 2A to chromosome 10 by linkage. Nature 328:528–530
Slater RM (1986) The cytogenetics of Wilms' tumour. Cancer Genet Cytogenet 19:37–41
Slater RM, De Kraker J (1982) Chromosome number 11 and Wilms' tumour. Cancer Genet Cytogenet 5:237–245
Slater RM, De Kraker J, Voute PA, Delemare JFM (1985) A cytogenetic study of Wilms' tumour. Cancer Genet Cytogenet 14:95–109
Solomon E, Voss R, Hall V, Bodmer WF, Jass JR, Jeffreys AJ, Lucibello FC, Patel I, Rider SH (1987) Chromosome 5 allele loss in human colorectal carcinomas. Nature 328:616–619
Sparkes RS, Sparkes MC, Wilson MG, Towner JW, Benedict W, Murphree AL, Yunis JJ (1980) Regional assignment of genes for human esterase D and retinoblastoma to chromosome band 13q14. Science 208:1042–1044
Sparkes RS, Murphree AL, Lingua RW, Sparkes MC, Field LL, Funderbuck S, Benedict WF (1983) Gene for hereditary retinoblastoma assigned to human chromosome 13 by linkage to esterase-D. Science 217:971–973
Sparkes RS, Sparkes MC, Kalina RE, Pagon RA, Salk DJ, Disteche CM (1984) Separation of the retinoblastoma and esterase D loci in a patient with sporadic retinoblastoma and del (13)(q14.1q22.3). Hum Genet 68:258–259
Squire J, Gallie BL, Phillips RA (1985) A detailed analysis of chromosomal changes in heritable and non-heritable retinoblastoma. Hum Genet 70:291–301
Squire J, Goddard AD, Canton M, Becker A, Phillips RA, Gallie BL (1986) Tumour induction by the retinoblastoma mutation is independent of N-*myc* expression. Nature 322:555–557
Stanbridge EJ (1976) Suppression of malignancy in human cells. Nature 260:17–20
Stanbridge EJ, Flandermeyer RR, Daniels DW, Nelson-Rees WA (1981) Specific chromosome loss associated with the expression of tumorigenicity in human cell hybrids. Somatic Cell Genet 7:699–712
Summerhays IC, Franks LM (1979) Effect of donor age on neoplastic transformation of adult mouse bladder epithelium in vitro. JNCI 62:1017–1023
Turleau C, De Grouchy J, Chavin-Colin F, Martelli H, Voyer M, Charlas R (1984a) Trisomy 11p15 and Beckwith-Wiedemann syndrome. A report of two cases. Hum Genet 67:219–221
Turleau C, DeGrouchy J, Nihoul-Fekete C, Dufier JL, Chavin-Colin F, Junien C (1984b) Del 11p13/nephroblastoma without aniridia. Hum Genet 67:455–456
Vogel W (1979) The genetics of retinoblastoma. Hum Genet 52:1–54
Ward P, Packman S, Loughman W, Sparkes M, Sparkes R, McMahon A, Gregory T, Ablin A (1984) Location of the retinoblastoma susceptibility gene(s) and the human esterase D locus. J Med Genet 21:92–95
Waziri M, Patil SR, Hanson JW, Bartley JA (1983) Abnormality of chromosome 11 in patients with features of Beckwith-Wiedemann syndrome. J Pediatr 102:873–876
Weissman BE, Stanbridge EJ (1983) Complementation of the tumorigenic phenotype in human cell hybrids. JNCI 70:667–672
Weissman BE, Saxon PJ, Pasquale SR, Jones GR, Geiser AG, Stanbrodge EJ (1987) Introduction of a normal human chromosome 11 into a Wilms' tumour cell line controls its tumorigenic expression. Science 236:175–180
Whang-Peng J, Bunn JPA, Kao-Shan CS, Lee EC, Carney DN, Gazdar A, Minna JD (1982) A nonrandom chromosomal abnormality, del 3p(14-23), in human small cell lung cancer (SCLC). Cancer Genet Cytogenet 6:119–134
Wiedemann HR (1964) Complexe malformatif familial avec hernie ombilicle et macroglossie; un syndrome nouveau? J Genet Hum 13:223–232

Wiedemann HR (1983) Tumours and hemihypertrophy associated with Wiedemann-Beckwith syndrome. Eur J Pediatr 141:129
Wiggs J, Nordenskjeld M, Yandell D, Rapaport J, Grondin V, Janson M, Werelius B, Petersen R, Craft A, Riedel K, Lieberfarb R, Walton D, Wilson W, Dryja TP (1988) Prediction of the risk of hereditary retinoblastoma using DNA polymorphisms within the retinoblastoma gene. N Engl J Med 318:151–157
Williams DL, Look T, Melvin SL, Roberson PK, Dahl G, Flake T, Stass S (1984) New chromosomal translocations correlate with specific immunophenotypes of childhood acute lymphoblastic leukaemia. Cell 36:101–109
Zankl H, Zang KD (1980) Correlations between clinical and cytogenetical data in 180 meningiomas. Cancer Genet Cytogenet 1:351–356

CHAPTER 12

The Role of Oncogene Activation in Chemical Carcinogenesis

C. S. COOPER

A. Introduction

Over the past 8 years there has been a revolution in our perception of the ways in which exposure to chemical carcinogens causes cell transformation and tumour induction. The recent exciting advances in this field are rooted in earlier observations that identified genetic material as the probable target during carcinogenesis. Boverii, who in 1914 noted frequent karyotypic abnormalities in tumour cells, is usually credited as the first person to recognise the potential importance of genetic changes in tumour development (reviewed in STRONG 1949). More recently, the discovery for particular groups of chemicals of correlations between carcinogenicity and mutagenicity (MCCANN et al. 1975; MCCANN and AMES 1976) and between carcinogenicity and the extent of covalent binding of the cancer-causing chemicals to DNA (BROOKES and LAWLEY 1964; FREI et al. 1978) provided important indirect support for the idea that DNA is the critical target during chemical carcinogenesis. This field of endeavour blossomed with the discovery that DNA from lines of chemically transformed cells could be used to transform a line of immortal NIH 3T3 mouse fibroblasts in DNA transfection experiments (SHIH et al. 1979; COOPER et al. 1980), an observation that provided direct support for the concept that cell transformation may involve the generation of dominantly acting transforming genes (oncogenes) by mutation of particular normal cellular genes (proto-oncogenes). The fruits yielded by the DNA transfection procedure and by other techniques that are now used for detecting activated cellular genes form the subject matter of this chapter.

The new technologies of DNA transfection and molecular biology have helped to answer the following key questions: what are the identities of the genes activated during chemical carcinogenesis? what genetic changes (point mutations, rearrangements, etc.) are responsible for gene activation? at what stages in carcinogenesis does gene activation occur? and what is the precise mechanism of gene activation following chemical exposure? At least a dozen model systems of tumour induction and cell transformation have now been examined, and it is becoming apparent that each model provides a different combination of answers to these four questions. For this reason I have elected to discuss each model system separately. However, before I start this survey of the roles of oncogene activations in chemical carcinogenesis, it is appropriate to discuss briefly the techniques now available for detecting and characterising activated cellular genes.

B. Detection of Activated Genes

I. DNA Transfection

Several methods are now used routinely to identify activated cellular genes. The DNA transfection-transformation assay is the most widely employed one. In this procedure cellular DNA is introduced into NIH 3T3 mouse fibroblasts as a coprecipitate with calcium phosphate. Transformation of the recipient cells, which is a consequence of the stable uptake of activated transforming genes, is then monitored either by examining cell cultures for the appearance of foci of morphologically transformed cells (the focus assay) (SHIH et al. 1979; COOPER et al. 1980) or by examining the ability of populations of transfected cells to induce tumours following subcutaneous injection into immunosupressed mice (the nude mouse assay) (BLAIR et al. 1982). The transforming genes transferred in these experiments are usually identified by hybridising Southern blots of DNA from transformed NIH 3T3 cells to the appropriate oncogene probes. Activated versions of many different cellular genes can be detected using this method, but the genes most frequently found are the members of the *ras* gene family (H-*ras*, N-*ras*, and K-*ras*), which are almost invariably activated by point mutations (BARBACID 1987). In early studies on *ras* gene activation, the identity of the genetic changes responsible for activation were determined by comparing the sequence of the molecularly cloned, activated *ras* gene with that of the normal unactivated homologue. Confirmation that activation was indeed the result of the point mutations detected in these studies was obtained by examining the transforming properties of contructs made by replacing a small region of the normal gene with the corresponding mutated region of the activated gene (REDDY et al. 1982; TABIN et al. 1982).

II. Oligonucleotide Probes

More recently, this laborious method of analysing the point mutations responsible for *ras* gene activation has been replaced by the use of oligonucleotide probes. In this procedure Southern blots of restriction enzyme-digested cellular DNA are hybridised to the appropriate oligonucleotide probe (usually 20-mers) under conditions that only allow hybridisation when there is complete homology between the probe and cellular DNA sequences; even a single mismatch prevents hybridisation (Bos et al. 1984). Hybridisation of DNA from transformed NIH 3T3 cells or from the original tumour to a series of oligonucleotide probes representing all of the possible point mutations at the *ras* codon under examination (usually codons 12, 13 or 61) allows the activating mutation to be identified. The oligonucleotide hybridisation technique has now been modified to include a DNA polymerase chain reaction (pcr) step in which DNA polymerase in combination with the appropriate oligonucleotide primers is used in vitro to amplify selectively, up to 100000-fold, the region of the *ras* gene under investigation (Bos et al. 1987). This modification results in a dramatic increase in the sensitivity of the technique, allowing large numbers of tumour DNA samples to be rapidly screened for mutations with oligonucleotide probes following transfer of small samples of the "amplified DNA" to hybridisation membranes using a "dot-blot" apparatus.

III. RNA Mismatch

RNaseA mismatch cleavage analysis has also been used for detecting the base substitution responsible for *ras* gene activation (FORRESTER et al. 1987). This technique relies on the ability of RNaseA to recognise and cleave RNA at the position of single base pair mismatches in RNA-RNA duplexes. RNA from the tumour or transformed cell line under investigation is hybridised to a radiolabelled antisense RNA probe complementary to the coding strand of the *ras* proto-oncogene. Analysis by gel electrophoresis of the RNA fragments of defined length that remain after digestion with RNaseA allows the position of the base substitution to be determined.

IV. Restriction Fragment Length Polymorphisms

The point mutations responsible for *ras* gene activation can either remove existing restriction endonuclease sites or create new restriction sites. When this occurs the activating mutations may be conveniently monitored by Southern analysis of cellular DNAs that have been digested with the appropriate restriction endonuclease. Indeed, in some model systems (see below) the analysis of restriction fragment length polymorphism (RFLPs) provides the method of choice for detecting activated *ras* genes (ZARBL et al. 1985; QUINTANILLA et al. 1986). Southern analysis of tumour DNA is also usually the method of choice when the particular genetic alterations under investigation are not conveniently detected by DNA transfection into NIH 3T3 cells. Thus gene rearrangements, which can result from chromosomal translocation or viral integration, and gene amplification are most easily detected by this method (WARREN et al. 1987; SAWEY et al. 1987).

All of the methods described above have their own virtues and limitations. Transfection into NIH 3T3 cells detects a broad spectrum of activated genes and provides direct biological evidence that the altered gene has transforming potential. Oligonucleotide hybridisation does not provide information about the biological consequences of the mutations but is the best method for screening large numbers of tumours for point mutations in selected regions at a predetermined target gene (usually K-*ras*, N-*ras* or H-*ras*). A combination of these techniques provides the maximum amount of information.

C. Oncogene Activation in Chemically Induced Tumours

I. Skin Papillomas and Carcinomas

Papillomas may be induced in mouse skin following the topical application of a single dose of initiating agent, such as 7,12-dimethylbenz[*a*]anthracene (DMBA), and multiple doses of a promoter, such as 12-*O*-tetradecanoylphorbol-13-acetate (TPA) (HECKER et al. 1982). Papillomas are benign lesions that usually regress when promoter treatment is stopped. However, some papillomas become autonomous, persisting in the absence of promoter treatment, while others (usually 5%–7%) progress to form malignant carcinomas. BALMAIN and co-workers

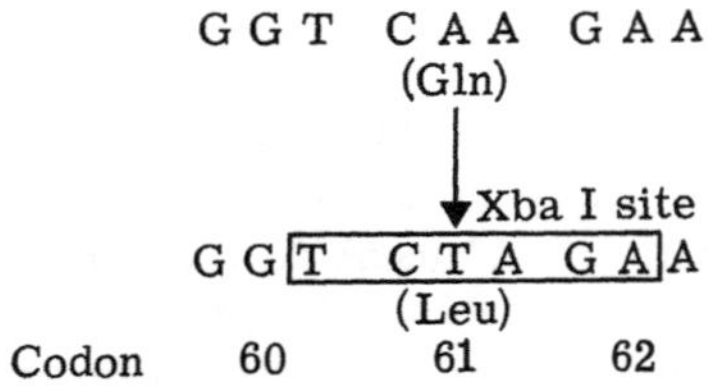

Fig. 1. Mechanism of H-*ras* activation in mouse skin tumours initiated with 7,12-dimethylbenz[*a*]anthracene and promoted with 12-*O*-tetradecanoylphorbol-13-acetate

(Balmain and Pragnell 1983; Balmain et al. 1984) have shown that DNA from both papillomas and carcinomas can transform NIH 3T3 mouse fibroblasts in DNA transfection experiments and that the gene responsible for the observed transformation is an activated version of the mouse cellular H-*ras* gene. The papillomas were usually heterozygous, containing both normal and activated versions of H-*ras*, but the carcinomas had frequently lost the normal H-*ras* allele or had several copies of the activated H-*ras* gene (Quintanilla et al. 1986). These observations suggested (i) that activation of the H-*ras* gene is a relatively early event in tumour development and (ii) that further changes at the H-*ras* locus are a common feature of tumour progression. Additional alterations at the H-*ras* locus were not, however, detected in all carcinomas, indicating that other types of changes may also be involved in the transition to malignancy.

Analysis of the mechanisms of proto-oncogene activation revealed that 90% of the papillomas and carcinomas initiated with DMBA and promoted with TPA contained H-*ras* that was activated by a A:T→T:A transversion mutation at the second position of codon 61 (Fig. 1, Table 1). This particular alteration results in the substitution of leucine for glutamine at amino acid 61 of the p21 *ras* gene product and introduces a *Xba*I site into the H-*ras* allele, allowing its presence in papillomas and carcinomas to be conveniently monitored by Southern analysis of *Xba*I-digested tumour DNA. Using this *Xba*I polymorphism to detect mutated H-*ras*, Quintanilla et al. (1986) examined the effect of using different initiators and promoters on the frequency of the A:T→T:A mutation. Their results showed that the nature of the promoter did not alter the frequency of this mutation. Thus, the *Xba*I restriction fragments diagnostic for the presence of the A:T→T:A transversion were also detected in tumours that were initiated with DMBA and promoted with chrysarobin, a promoter that is structurally unrelated to TPA and that does not bind to the phorbol ester receptor. This particular mutation was not, however, detected in a total of 12 tumours that were initiated with *N*-methyl-*N'*-nitro-*N*-nitrosoguanidine (MNNG). Since MNNG and DMBA are chemically distinct and would not be expected to induce the same types of mumutation, this observation indicates that proto-oncogene activation may be a direct consequence of the mutagenic action of DMBA and MNNG and may occur at, or very close to, the time of application of the single dose of initiating agent.

Brown et al. (1986) argued that if the *ras* mutation is indeed involved in initiation it should be possible to replace the treatment with initiating carcinogen by directly introducing activated *ras* into mouse skin epidermal cells. To achieve this, mouse skin was infected with Harvey murine sarcoma virus (HaSV), which

Table 1. Oncogene activation in chemically induced tumours

Tumour type	Inducing chemical	Gene activated	Codon	Observed mutations	References
Skin papillomas and carcinomas	DMBA/TPA	Mouse H-*ras*	61	CAA(gln)→CTA(leu)	QUINTANILLA et al. (1986)
	DB[c,h]A/TPA	Mouse H-*ras*	61	CAA(gln)→CTA(leu)	BIZUB et al. (1986)
	DMBA	Mouse H-*ras*	61	CAA(gln)→CTA(leu)	BIZUB et al. (1986)
	TPA	Mouse H-*ras*	61	CAA(gln)→CTA(leu)	PELLING et al. (1988)
Skin keratoacanthomas	DMBA	Rabbit H-*ras*	61		LEON et al. (1988)
Mammary tumours	MNU	Rat H-*ras*	12	GGA(gly)→GAA(glu)	ZARBL et al. (1985)
	DMBA	Mouse H-*ras*	61	CAA(gln)→CTA(leu)	DANDEKAR et al. (1986)
Thymomas	MNU	Mouse N-*ras*	12	GGT(gly)→GAT(asp)	GUERRERO and PELLICER (1987)
			61	CAA(gln)→AAA(lys)	GUERRERO and PELLICER (1987)
		Mouse K-*ras*	12	GGT(gly)→GAT(glu)	GUERRERO and PELLICER (1987)
	MNU	Mouse K-*ras*	12	GGT(gly)→GAT(glu)	WARREN et al. (in preparation)
	MNU	Mouse *myc*		Activated by proviral integration	WARREN et al. (1987)
	MNU	Mouse *pim*-1		Activated by proviral integration	WARREN et al. (1987)
Fibrosarcomas	MCA	Mouse K-*ras*			EVA and Trimmer (1986)
	MCA	Mouse K-*ras*			EVA and AARONSON (1983)
Lung	1,8-DNP	Rat K-*ras*	12	GGT(gly)→TGT(cys)	TAHIRA et al. (1986)
	TNM	Rat K-*ras*	12	GGT(gly)→GAT(glu)	STOWERS et al. (1987)
		Mouse K-*ras*	12	GGT(gly)→(GAT(glu)	STOWERS et al. (1987)
Hepatomas	various	Mouse H-*ras*		see Table 2	
Renal mesenchymal	DMN-OMe	Rat K-*ras*			SUKUMAR et al. (1986)
		Rat N-*ras*			SUKUMAR et al. (1986)
Plasmocytomas	Pristane and mineral oil	Mouse *myc*		Rearrangement with Immunglobin heavy chain gene	CORY (1986)
	Pristane and mineral oil	Mouse *mos*		Insertion of IAP sequences	CAANANI et al. (1983); COHEN et al. (1983); KUFF et al. (1985)
Nasal squamous cell carcinoma	MMS	Rat unidentified			GARTE et al. (1985)
Ethesioneuroepithelioma	MMS	Rat unidentified			GARTE et al. (1985)
Neuroblastomas and glioblastomas	ENU	*neu*	664	GTG(val)→GAG(glu)	BARGMANN et al. (1986b)

DMBA, 7,12-dimethylbenz[*a*]anthracene; TPA, 12-*O*-tetradecanylphorbol-13-acetate; MNU, *N*-methyl-*N*-nitrosourea; MCA, 3-methylcholanthrene; 1,8-DNP, 1,8-dinitropyrene; TNM, tetranitromethane; DMN-OMe, methyl(methoxymethyl)nitrosourea; MMS, methylmethanesulphonate; ENU, *N*-ethyl-*N*-nitrosourea; IAP Sequences intracisternal A-particle sequences; DB[c,h]A, Dibenz [c,h]acridine.

contains activated viral H-*ras*. Treatment with HaSV alone failed to induce tumours, but subsequent treatment with TPA either immediately after HaSV treatment or after a latent period of 4 months produced papillomas within 4 weeks. Subsequently, some of the papillomas underwent a rapid transition to form invasive carcinomas. As expected, viral-specific mRNA and expression of viral H-*ras* p21 was detected in both papillomas and carcinomas. Analysis of the distribution of p21 using immunocytochemical techniques revealed that the highest level of p21 were present in the basal epidermal cell layer in which most cell proliferation in the papilloma takes place. Some p21 was also detected in the upper epidermal cell layers, but the levels of p21 decreased as the cells became more differentiated.

PELLING et al. (1988) examined the mechanism of H-*ras* activation in papillomas and carcinomas that arise at low frequency in mice treated with tumour promoter alone; the mice were not exposed to DMBA or other initiating agents. Remarkably, they found that 7 of 9 tumours induced by TPA or mezerein contained an *Xba*I RFLP indicating the presence of A:T→T:A point mutations at the second base of codon 61 (Table 1). In order to account for this specificity, they suggested that spontaneous mutations occur randomly in the H-*ras* gene of skin epithelial cells and proposed that only those cells containing A:T→T:A mutations develop into tumours because of the enhanced sensitivity of cells containing this particular mutation to treatment with promoting agents. Extending this model one would predict that the types of mutations observed in chemically induced tumours would be determined both by the spectrum of mutations brought about by the initiating agent and by the relative susceptibility of cells containing each of the activating mutations to treatment with tumour promoters.

BIZUB et al. (1986) examined the activation of proto-oncogenes in mouse skin tumours induced (i) by the repetitive application of DMBA and (ii) by a single application of either dibenz[*c,h*]acridine (DB[c,h]A) or benzo[*a*]pyrene (B[a]P) followed by promotion with TPA. H-*ras* that had been activated by an A:T→T:A mutation at the second position of codon 61 was detected in the DMBA-induced carcinomas and in the DB[c,h]A-initiated carcinomas and papillomas (Table 1). Activated proto-oncogenes were not, however, detected in B[a]P-initiated carcinomas. The latter result was considered surprising since the human H-*ras* protooncogene can be activated by mutation at the 12th and 61st codon after in vitro modification of the cloned proto-oncogene with activated dervatives of B[a]P (MARSHALL et al. 1984). In addition, using DNA transfection into NIH 3T3 mouse fibroblasts, activated genes that were apparently unrelated to *ras* were detected in one DMBA-induced carcinoma and in one DB[*c,h*]A-initiated carcinoma.

The high frequency of A:T→T:A mutations at the H-*ras* locus in DMBA-induced skin tumours has focussed attention on the adducts formed in reactions of activated metabolites of DMBA with deoxyadenosine in DNA. The metabolic activation of DMBA involves the formation of the 3,4-dihydrodiol derivative that is further metabolised to form two vicinal "bay-region" diol epoxides, namely, the *syn*- and *anti*-DMBA-3,4-dihydrodiol-1,2-oxides (BIGGER et al. 1983). The major carcinogen-DNA adducts detected in mouse skin are formed in reactions of these diol-epoxides with deoxyadenosine, although some deoxyguanosine adducts are also observed. This abundance of deoxyadenosine-hydrocarbon adducts contrasts with the predominance of deoxyguanosine-hydrocarbon adducts formed in reactions of diol-epoxide derivatives of most other polycyclic aromatic hydrocarbons (PAHs), including B[a]P, with DNA. BIGGER et al. (1983) and QUANTILLA et al. (1986) have suggested that the potency, as an initiating agent, of each PAH may be related to its ability to bind to, and therefore induce mutations at, deoxyadenosine. In support of this hypothesis they noted that B[a]P, which is a 30-fold weaker initiator than DMBA, exhibits a corresponding decrease in the extent of reaction with deoxyadenosine, but not with deoxyguanosine.

II. Neuroblastomas and Glioblastomas

The cellular *neu* gene is frequently activated in neuroblastomas and glioblastomas of BDIX rats. A high proportion of the offspring of pregnant BDIX rats that have been treated with a single dose of *N*-ethyl-*N*-nitrosourea (ENU) during the second half of gestation develop central and peripheral nervous system tumours, including neuroblastomas and glioblastomas, after a latency of approximately 250 days (RAJEWSKY 1983; LANTOS 1986). SHIH et al. (1981) demonstrated that DNA from cell lines derived from four independent tumours could transform NIH 3T3 cells in DNA transfection-transformation experiments. The transforming gene transferred in these experiments was unrelated to *ras* and was associated with the expression of a phosphoprotein of relative molecular mass 185000 (p185) (PADHY et al. 1982). Subsequent studies demonstrated that the *neu* gene was related to, but distinct from, the gene that encodes the EGF receptor; an EGF receptor (c-*erb*B) probe hybridised to *neu* when the hybridisations were completed under conditions of relaxed stringency, and antisera raised against the EGF receptor cross-reacted with p185 (SCHECHTER et al. 1984, 1985). The nucleotide sequence of the *neu* cDNA revealed a 1260-amino acid protein that exhibits 50% amino acid homology to the EGF receptor and possesses several features (Fig. 2), including a transmembrane domain and a cytoplasmic protein tyrosine kinase domain, that are the hallmarks of a mem-

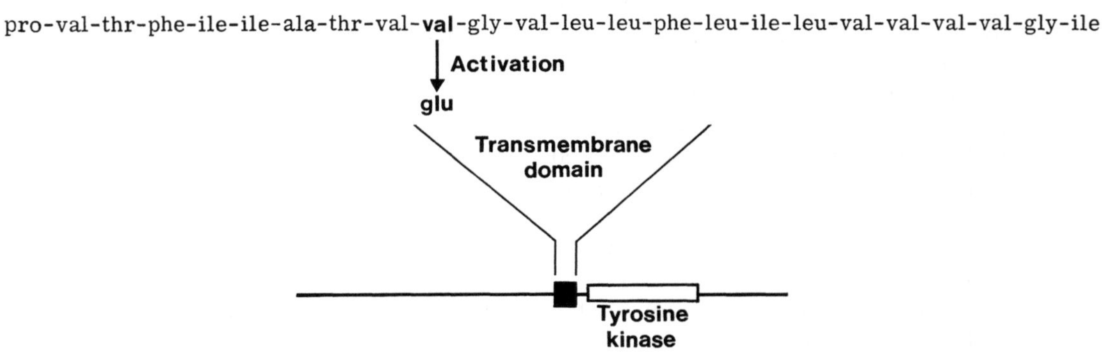

Fig. 2. Mechanism of activation of the *neu* protein in rat neuroblastomas and glioblastomas

brane receptor protein (BARGMANN et al. 1986a). The structure of the product of the *neu* gene is shown (Fig. 2). When considered together, these observations strongly suggest that *neu* encodes a growth factor receptor, although the identity of the ligand that binds to this putative receptor remains to be determined.

Comparisons of the activated and normal versions of the *neu* gene have demonstrated that activated *neu* in four independent cell lines involves the same T:A→A:T transversion mutation (BARGMANN et al. 1986b). Unexpectedly, this alteration, which changes valine to glutamic acid, falls within the putative transmembrane domain (Fig. 2). How does this single amino acid change activate the transforming potential of the *neu* gene? One possibility is that the insertion of a charged amino acid in the transmembrane domain might alter the subcellular localisation of *neu*. However, it has been demonstrated that, like its normal counterpart, at least part of the p185 product of the activated *neu* gene is membrane associated (DREBIN et al. 1984). In addition, this membrane-associated p185 appears to be responsible for transformation because antisera to p185 suppress the transformed phenotype in *neu*-transformed cells (DREBIN et al. 1985). Another possibility is that the single amino acid change may mimic the effect of the binding of ligand to the receptor, conceivably by stabilising the configuration that the receptor adopts when the ligands bind. For example, activation might affect receptor clustering or in some way alter the conformation of the receptor.

Reactions of ENU with the target tissue inflicts many distinct types of damage on cellular DNA, but it is the formation of O^6-ethylguanine (O^6-ethylG) and O^4-ethylthymine (O^4-ethylT) that has received the most attention. These alkylated bases are considered to be the major promutagenic lesions and are believed to act by causing mispairing during DNA replication. Mispairing with O^6-ethylG and O^4-ethylT would be expected to cause, respectively G:C→A:T and T:A→C:G mutations. Indeed, analysis of the types of mutations involved in the xanthine guanine phosphoribosyl transferase gene following exposure of *Escherichia coli* to ENU reveal that the majority of mutations were G:C→A:T (24/33) and T:A→C:G (7/33) transitions. One G:C→C:G transversion and one A:T→C:G transversion were observed, but notably no A:T→T:A transversions were found (RICHARDSON et al. 1987). A:T→T:A transversion mutations have, however, been detected in other studies on ENU mutagenesis: two mutations detected in the globin genes of the progeny of female mice treated with ENU were both A:T→T:A transversions (POPP et al. 1983; LEWIS et al. 1985).

The development of central nervous system (CNS) tumours in the offspring of ENU-treated pregnant BDIX rats appears to be a multistep process (reviewed by LANTOS 1986). It has been proposed that the target cells for ENU-induced transformation are undifferentiated stem cells present in the subependymal plate. In agreement with this idea, focal cellular hyperplasia of the subependymal plate and abnormal clusters of undifferentiated subependymal cells that may represent intermediate stages in the process of tumour development have been observed during histological examination of the brains of ENU-treated rats (LANTOS 1986). Evidence that transformation of brain cells is a multistep process was also obtained from the examination of cultures of cells prepared from brains at various intervals after exposure to the carcinogen (LAERUM and RAJEWSKY 1975; ROSCOE and CLAISSE 1976, 1978). Cultures of brain cells from treated rats, but not from untreated rats, continued to grow and sequentially acquire the ability to pile up, to form colonies in semi-solid media and to form tumours when injected into BDIX rats. The precise role that the activated *neu* gene plays in inducing CNS tumours and the stage at which the *neu* gene is activated to be established, although it is tempting to speculate that *neu* activation is the initiating event in tumour induction.

Studies on the repair of alkylated DNA in the tissues of BDIX rats treated with ENU have focussed attention on O^6-ethylG. This modified nucleoside is removed rapidly from DNA in the liver but much more slowly from the brain. In contrast, ethylated products such as 3-ethylA and 7-ethylG are removed more rapidly than O^6-ethylG, both in the brain and in other rat tissues (RAJEWSKY 1983). These observations have provided support for the idea that the persistance of O^6-ethylG is an important factor in the induction of tumours of the CNS. However, it seems unlikely that this lesion is responsible for the A:T→T:A transversion involved in *neu* activation. To explain how exposure to ENU causes A:T→T:A mutations, it may be necessary to search for a novel promutagenic lesion from amongst the spectrum of different products that result from exposure to ENU.

III. Mammary Tumours

Mammary cancer can be induced rapidly in female rats either by oral administration of PAHs such as DMBA and 3-methylcholanthrene (MCA) (HUGGINS et al. 1961) or by intravenous injection of *N*-methyl-*N*-nitrosourea (MNU) (GULLINO et al. 1975). Small lesions with the histological characteristics of mammary adenocarcinomas are observed 20 days after treatment with DMBA, and it is these lesions, which appear to originate from the epithelial cells of the inner lining of the mammary duct, that are thought to proliferate to form mammary tumours (SINHA and DAO 1975). Hyperplastic alveolar nodules (HAN) that develop in the mammary glands of rats and mice treated with DMBA have also been proposed as intermediates in tumour development since tumours may arise after transplantation of HANs into gland-free fat pads. However, the lack of correlation between the frequency of HAN formation and tumour incidence favours the view that most chemically induced mammary tumours do not involve the intermediate formation of these types of lesion (SINHA and DAO 1975).

A high proportion of MNU-induced mammary cancers contain H-*ras* that is invariably activated by a G:C→A:T transition mutation in the second position of codon 12 (SUKUMAR et al. 1983; ZARBL et al. 1985) (Fig. 3, Table 1). This is

```
         G C T   G G A   G G C
                 (Gly)
                   |
                   ↓
         G C T   G A A   G G C
                 (Glu)
Codon     11      12      13
```

Fig. 3. Mechanism of activation of H-*ras* in *N*-methyl-*N*-nitrosourea-induced rat mammary carcinomas

precisely the type of mutation that would be expected following the formation of O^6-methylG, the principle promutagenic lesion that results from exposure to MNU, and is the predominant type of point mutation observed after exposure of bacteria to MNU (RICHARDSON et al. 1987). As a consequence of alkylation of the O^6-position of guanine, the modified guanine residue can pair with thymidine instead of cytidine, leading to the generation of G:C→A:T transitions. Activated H-*ras* was also detected in DMBA-induced mammary tumours, but the proportion of tumours containing activated H-*ras* (3 of 14) was lower than that observed for MNU-induced tumours, and activation involved point mutations in codon 61 (ZARBL et al. 1985); the G:C→A:T transition in codon 12 was not observed in DMBA-induced tumours. This dramatic dependence of the activating mutation on the particular chemical used for tumour induction can be considered to provide further support for the idea that activation results from direct interaction of the carcinogen with DNA. Since MNU has a short half-life after intravenous injection, most alkylation of DNA must occur within a few hours of administration. It therefore seems probable that *ras* activation is a very early event that is closely associated with tumour initiation.

DANDEKAR et al. (1986) examined transforming gene activation in mouse mammary tumours arising from a transplantable line (DI/UCD HPO) that was derived from HAN. Although 22 weeks after transplantation a 22% incidence of spontaneous tumours was observed, the tumour incidence could be increased to 81% by exposure to DMBA. Activated transforming genes were not detected in spontaneous tumours or in the parent DI/UCD HPO line, but activated H-*ras* was detected in each of four DMBA-induced tumours. Activation invariably involved A:T→T:A transversion mutations in the second position of codon 61, precisely the same mutation that is found in DMBA-initiated skin tumours (Table 1).

IV. Thymomas

Thymomas (lymphocytic lymphomas of the thymus) can be induced in most mouse strains either by administering chemical carcinogens such as MNU, DMBA and MCA or by exposure to γ-radiation (LAWLEY 1976). An insight into the mechanism of thymoma induction has been provided by analyses of the kinetics of thymoma appearance following administration of a single dose of MNU. From the data obtained it was possible to calculate the "hit-number" which can be interpreted as the number of dose-dependent "carcinogen-hits" required to induce tumour formation. For most mouse strains the "hit-number" is 2.5–3.0, indicating that two or three carcinogen-dependent critical events are required for thymoma development (FREI 1980; FREI and LAWLEY 1980).

The results of DNA transfection experiments have revealed that a high proportion of thymomas contain activated *ras* genes, indicating that *ras* activation may correspond to one of these "hits". In their initial study GUERERRO et al. (1984a) found that 5 of 6 thymomas induced by MNU in the AKR/RF mouse strain contained activated N-*ras*, but in a later, more extensive study using different mouse strains and large groups of mice, they discovered that the majority of MNU-induced thymomas contained activated K-*ras* (GUERRERO et al. 1986; DIAMOND et al. 1988). K-*ras* was the only altered *ras* gene detected in two other studies. Thus, WARREN et al. (1987, and unpublished observations) identified activated K-*ras* in 9 of 36 thymomas induced by treating AKR mice with MNU, and EVA and TRIMMER (1986) found activated K-*ras* in 10 of 12 thymomas that developed in RJF mice following skin painting with MCA. The activating mutations observed in the MNU-induced thymomas are less specific than those detected in MNU-induced rat mammary tumours in which H-*ras* is invariably activated by a G:C→A:T mutation at the second position of codon 12 (Table 1). This particular mutation was found in 8 of the 9 activated versions of K-*ras* examined by Warren et al. (personal communication), in 29 of 31 activated K-*ras* genes studied by DIAMOND et al. (1988) but in only 1 of the 5 activated N-*ras* alleles (GUERRERO et al. 1984b, 1986; DIAMOND et al. 1988) (Table 1). From the known chemistry of reaction of DNA with a series of alkylating carcinogens of varying reactivities, evidence was provided that the promutagenic alkylation of DNA at the O^6 atom of guanine is most probably as essential cellular reaction for induction of thymomas in mice (FREI et al. 1978). This lesion could account for the induction of the G:C→A:T transition mutations but not for the induction of a C:G→A:T transversion that was detected in one of the activated N-*ras* gene examined by GUERRERO et al. (1984b). Indeed, the observations that only 1 of the 5 activated N-*ras* genes detected by GUERRERO et al. (1986b); GUERRERO and PELLICER (1987) contained the G:C→A:T transition led these workers to suggest that N-*ras* activation in their particular model system may not be a consequence of the direct mutagenic action of MNU.

Endogenous murine leukaemia viruses (MuLVs) have been implicated in the induction of mouse thymomas, particularly in the AKR strain, which develops thymomas spontaneously after 6 months of age. In some spontaneous AKR thymomas and in some thymomas induced by infecting young mice of other strains with MuLVs, a critical event in tumour development appears to involve modification of the cellular c-*myc* and *pim*-1 genes by proviral insertion of these loci (CORCORAN et al. 1984; CUYPERS et al. 1984). The *pim*-1 gene was in fact originally isolated as a specific site of integration of MuLVs in mouse thymomas and was subsequently found to be a member of a family of genes that encode protein kinases (SELTAN et al. 1986).

Proviral integrations at the *pim*-1 and c-*myc* loci have also been detected in thymomas induced by treating AKR mice with MNU but have never been detected in the thymomas induced by exposing other mouse strains to chemical carcinogens (WARREN et al. 1987) (Table 1). The MNU-induced thymomas from AKR mice are quite distinct from the spontaneous thymomas that develop in this mouse strain because (i) they appear much earlier than the spontaneous thymomas, (ii) they lack a class of recombinant MuLVs (called MCF viruses) that are found in all spontaneous tumours, and (iii) in contrast to spontaneous thymomas, they frequently contain activated *ras* genes (WARREN et al. 1987). Analysis of the kinetics of appearance of thymomas in MNU-treated AKR mice revealed a "hit number" of 1.4, compared with 2.5–3.0 for other mouse strains (FREI 1980). This indicates that one "carcinogenic-hit" might have already occurred in the AKR strain and suggests the possibility that the development of thymomas in MNU-treated AKR mice involves cooperation between the chemical carcinogen and the endogenous MuLVs.

Cytogenetic studies have revealed that a variable proportion of the cells present in both virally and chemically induced thymomas exhibit trisomy of chromosome 15. Although the significance of this non-random chromosomal abnormality remains to be established, it has been proposed that the modest increase in copy number of genes, such as c-*myc* and p53, that are located on this chromosome may facilitate thymoma development.

V. Hepatomas

Male C57BL/6J × C3H/HeJ F1 (B6C3F1) mice exhibit an approximately 25% incidence of spontaneous hepatomas by the age of 24 months. However, hepatomas develop much earlier and in a higher proportion of animals when 12-day-old B6C3F1 mice are treated with a single dose of one of a wide variety of chemical carcinogens (FENNELL et al. 1985; LAI et al. 1985). The interpretation of the effect of chemical treatment on tumour incidence is controversial. One possibility is that the chemical treatment simply accelerates the process of spontaneous tumour development, possibly by a cytotoxic mechanism. Alternatively, chemical exposure might have a more direct role in tumour induction, perhaps by inducing somatic mutations. This problem has received particular attention because B6C3F1 mice have been used extensively in the National Toxicity Program/National Cancer Institute's long-term bioassay for carcinogenicity. The precise interpretation of the bioassay results may therefore be important for the assessment of potential human risk.

To help to resolve this issue DNA transfection into NIH 3T3 cells has been used to examine the spectrum of transforming genes activated in both spontaneous and chemically induced hepatomas. FOX and WATANABE (1985) demonstrated that DNA from a high proportion of hepatomas (9 of 11) could transform NIH 3T3 cells but did not identify the genes transformed in their experiments. In subsequent analyses REYNOLDS et al. (1986) found that H-*ras* was the gene most frequently activated both in spontaneous hepatocellular carcinomas (14 of 17) and in spontaneous adenomas; carcinomas and adenomas represent different stages of tumour progression, and it is generally believed that adenomas have the potential to progress to form carcinomas.

H-*ras* was also the gene most frequently activated in chemically induced tumours. Thus, WISEMAN et al. (1986) found H-*ras* activation in the majority of tumours induced by *N*-hydroxy-2-acetylaminofluorene (N-OH-AAF), vinyl carbamate (VC) or 1′-hydroxy-2′,3′-dihydroestragole (OH-DHE). Analysis of the activating mutations showed that, in N-OH-AAF-induced tumours, H-*ras* was invariably activated by C:G→A:T transversion mutations at the first position of codon 61 while, in VC- and OH-DHE-induced tumours, activation involved A:T→T:A transversions and A:T→G:C transitions at the second position of codon 61 (Table 2). This dependence of the observed mutation on the particular chemical used for tumour induction initially suggested that the mutation may be a direct result of chemical modification of the H-*ras* gene and supported the idea that the chemical has a direct role in tumour induction. However, the interpretation of these results became more difficult following the discovery that each of the activated H-*ras* genes found in spontaneous tumours (REYNOLDS et al. 1987)

Table 2. Gene activation in mouse hepatomas

Treatment	Transfection frequency (tumours positive/tumours examined)	Activated oncogenes							K-*ras*	*raf*	Un-known
		H-*ras*, codon 61 CAA(Gln) mutated to			H-*ras*, codon 13 GGC(Gly) mutated to		H-*ras*, codon 117 AAG(Lys) mutated to				
		AAA (Lys)	CTA (Leu)	CGA (Arg)	GTC (Val)	CGC (Arg)	AAC (Asn)	AAT (Asn)			
N–OH–AFF	7/7	7	0	0							
VC	7/7	0	6	1							
HO–DHE	11/11	0	5	5					1		
DEN	14/33	7	4	3							
Furan	13/29	4	0	1			2	2	1	1	0
Furfural	13/16	5	0	1	1	1	0	1	1	0	3
None	17/27	9	3	3						1	1

N–OH–AAF, *N*-hydroxy-2-acetylaminofluorene; VC, vinyl carbamate; HO–HDE, 1′-hydroxy-2′,3′-dihydroestragole; DEN, *N*-nitrosodiethylamine and in tumours induced by the structurally unrelated chemical *N*-nitrosodiethylamine (STOWERS et al. 1988) contained one of these same three mutations in codon 61. Additional support for the idea that the mechanisms of development of spontaneous and chemically induced tumours are distinct was, however, provided by analyses of the mutations responsible for H-*ras* activation in furfural- and furan-induced adenomas and carcinomas. Thus, in addition to the codon 61 mutations, a significant proportion of the activated genes had mutations in codon 13(2/19) and codon 117 (5/19), two weakly transforming mutations that had not previously been detected in tumour DNAs. Mutations at these positions were not detected in spontaneous tumours (REYNOLDS et al. 1987). The activation of transforming genes other than H-*ras* was also observed in these transfection experiments. Activated K-*ras* was detected in four chemically induced carcinomas, evidence for the presence of activated *raf* was obtained for one chemically induced and one spontaneous carcinoma, and currently unidentified transforming genes were found in one spontaneous carcinoma and three furfural-induced carcinomas (WISEMAN et al. 1986; REYNOLDS et al. 1987).

In an independent study MCMAHON et al. (1986) provided evidence for transforming gene activation in a high proportion of hepatocellular carcinomas induced in Fischer rats by repeated intraperitoneal injections of aflatoxin B1. Activated K-*ras* was detected in two carcinomas, while evidence that 8 of the 11 tumours contained activated genes that were unrelated to *ras* was also given. The differences between the results obtained in studies on liver tumours from B6C3F1 mice and results obtained in studies on aflatoxin-induced rat liver tumours are striking and worthy of further investigation.

VI. Fibrosarcomas

Activated oncogenes have been detected in fibrosarcomas induced by treatment either with MCA or with 1,8-dinitropyrene (1,8-DNP), a potent mutagen that is present in engine exhaust (Table 1). The observation that activated genes are present in two of four cell lines prepared from MCA-induced mouse fibrosarcomas was of particular historical significance since it provided the first demonstration that activated transforming genes are present in chemically induced tumours (EVA and AARONSON 1983). In later studies OCHAI et al. (1985) and TAHIRA et al. (1986) found evidence that four rat fibrosarcomas induced by 1,8-DNP also contained activated transforming genes; one tumour contained activated K-*ras*, while three others were believed to contain genes unrelated to *ras*.

Also of historical significance is the observation that the protein designated p53 is overexpressed in many types of chemically transformed cells (DELEO et al. 1979; ROTTER 1983). This protein was originally identified both as a tumour-specific transplantation antigen in MCA-induced mouse fibrosarcomas (DELEO et al. 1979) and as a protein that bound tightly to the product of a DNA tumour virus oncogene, the large T protein (LANE and CRAWFORD 1979). Subsequently, the p53 gene was recognised as an oncogene in its own right that can immortalise cultures of primary cells and complement *ras* to induce full transformation of primary rodent fibroblasts (for a review see LANE 1987). In initial experiments, the biological effects were achieved by overexpression of the normal p53 gene but

JENKINS et al. (1985) subsequently demonstrated that p53 can immortalise primary cells by a second distinct mechanism involving alterations of the coding sequences that generate a stable protein product. Looking to the future, analysis of p53 genes present in chemically transformed cells to determine whether they contain alterations that result in overexpression and/or increased protein stability may provide important insights into the mechanism of chemical carcinogenesis.

VII. Plasmocytomas

Plasmocytomas (a B-lymphocyte neoplasm) may be induced in a few selected mouse strains, including the BALB/c and NZB strains, by intraperitoneal injection of either mineral oils or pure alkanes such as pristane (2,6,10,14-tetramethylpentadecane). The mechanism of plasmocytomagenesis is obscure (POTTER 1984). These agents are apparently not capable of attacking DNA directly but induce a severe inflammatory response at the site of injection. Migrating macrophages and neutrophils phagocytose oil droplets and adhere to the peritoneal surface where they eventually form an oil granuloma that becomes infiltrated by lymphocytes. The plasmacytoma is detected as free cells after a latent period of at least 130 days.

Cytogenetic analyses have revealed that the majority of plasmocytomas possess specific chromosomal abnormalities. The most frequent alteration is a translocation between chromosomes 15 and 12 that, at the molecular level, corresponds to a translocation between the c-*myc* locus on chromosome 15 and the immunoglobin heavy chain (IgH) locus on chromosome 12 (Table 1). Translocations can also occur between the immunoglobin κ light chain gene on chromosome 6 and a locus designated *pvt*-1 that maps cytogenetically to the same band on chromosome 15 as c-*myc* (for review, see CORY 1986).

Immunoglobin heavy and light chains undergo interchromosomal rearrangements during normal B-cell development, allowing the mature B cells to express immunoglobins. The normal mechanism of IgH rearrangement is shown in Fig. 4a. First, the portion of the gene encoding the immunoglobin variable region is assembled by a series of recombinants between variable (V), diversity (D) and joining (J) elements, leading initially to the production of a gene encoding the μ heavy chain. Recombination may then occur between the switch regions (Sμ, Sα, etc.), allowing the synthesis of other classes of immunoglobins. For example, in the rearrangement illustrated in Fig. 4a, a switch occurs between Sμ and Sα, allowing the production of immunoglobin containing the α chain constant region. Within the context of this model of immunoglobin gene construction, it is generally supposed that the translocation detected in plasmocytomas results from rare aberrant interchromosomal recombinations that occur during B-cell maturation. The precise role that exposure to mineral oil or pristane plays in generating these aberrant rearrangements is, however, not known.

The c-*myc* gene is composed of three exons (Fig. 4b) and the majority of the translocations occur (i) upstream (5′) of exon 1, (ii) within the first exon, or (iii) between the first and second exons. Within the IgH locus the prominent targets for translocation are the germline and switched Sα elements. For example, in

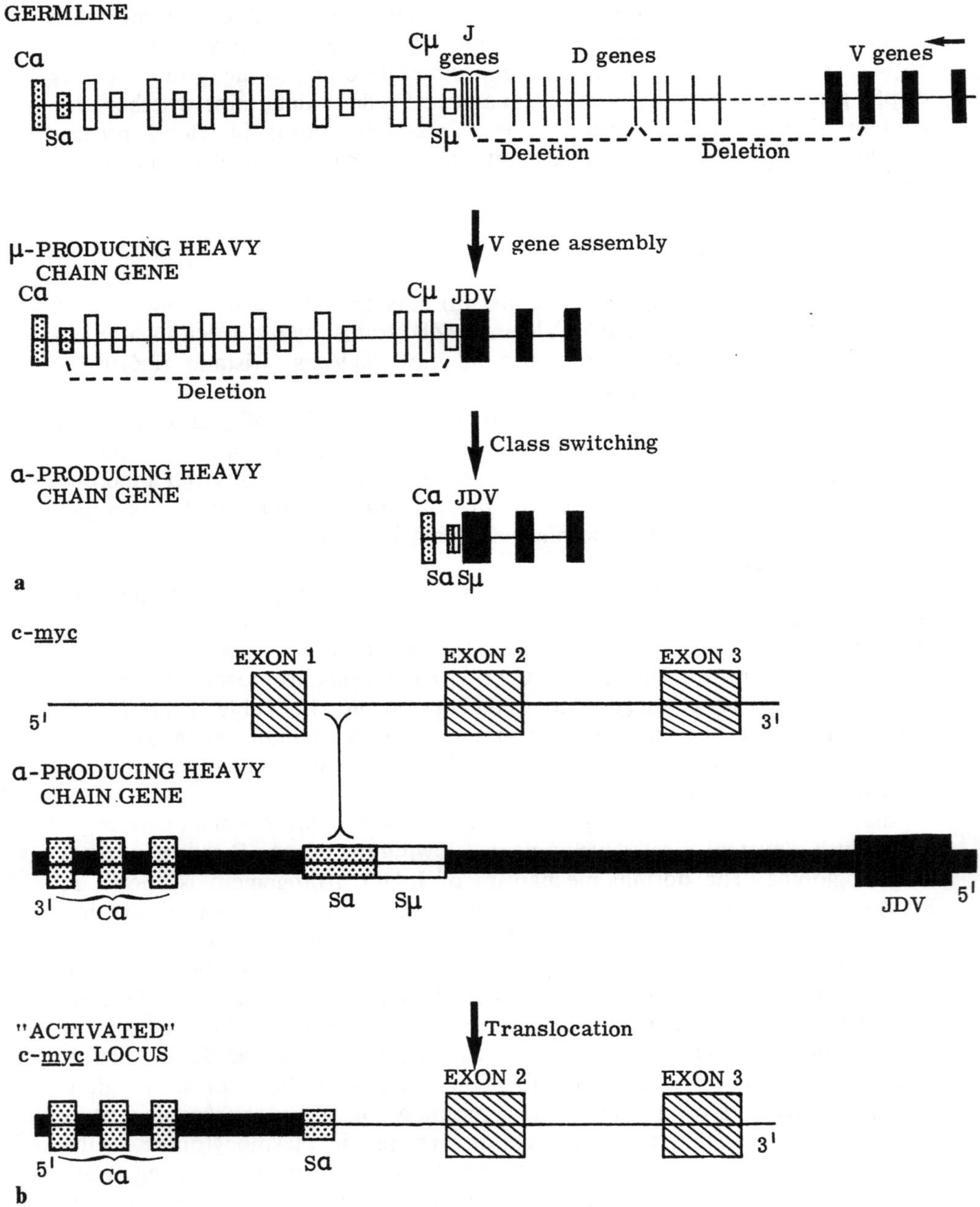

Fig. 4. a Normal mechanism of rearrangement of the immunoglobin heavy chain. **b** Mechanism of activation of c-*myc* gene in pristane- and mineral oil-induced mouse plasmocytomas involves invasion of c-*myc* by the immunoglobin heavy chain gene. Immunoglobin variable (*V*), diversity (*D*), joining (*J*), constant (*C*), and switch (*S*) regions are shown. The *arrow* indicates the normal direction of transcription of the immunoglobin genes

Fig. 4b the target for recombination is a Sμ element that has already become joined to a Sα element. This translocation brings together the altered c-*myc* locus and the 3′-end of the IgH locus in a "head-to-head" configuration. The most dramatic consequence of IgH invasion of the c-*myc* locus is the deregulation of *myc* expression. This is most clearly illustrated by the observation that untranslocated c-*myc* is usually transcriptionally silent in plasmocytomas while the translocated allele is actively transcribed (BERNARD et al. 1983). Indeed it is the constitutive expression of the rearranged c-*myc* locus that is believed to play a major role in the induction of plasmocytomas.

Activated version of the cellular *mos* gene have been detected in two plasmocytoma cell lines designated XRPC24 and NSI (RECHAVI et al. 1982; CANAANI et al. 1983; COHEN et al. 1983; GATTONI-CELLI et al. 1983; KUFF et al. 1983). *Mos* was originally identified as the transforming sequence of Moloney murine sarcoma virus and encodes a protein serine kinase that has a cytoplasmic location. The cellular *mos* gene is usually expressed at only very low levels and its transforming potential can be activated simply by elevating its level of expression (BLAIR et al. 1981).

The aberrant *mos* genes present in the XRPC24 and NSI cell lines were originally detected as gene rearrangements by Southern analysis and are referred to as rc-mos^{X24} and rc-mos^{NSI} (rc-*mos* stands for rearranged c-*mos*). The rearrangement present in each cell line results from integration of a intracisternal A-particle (IAP) genome at the 5′ end of the *mos* gene (CANNANI et al. 1983; COHEN et al. 1983; KUFF et al. 1983). IAPs are non-infectious, retroviral-like entities that are abundantly expressed in many murine tumours, including plasmocytomas. IAP sequences homologous to the RNA present in IAP particles are located at

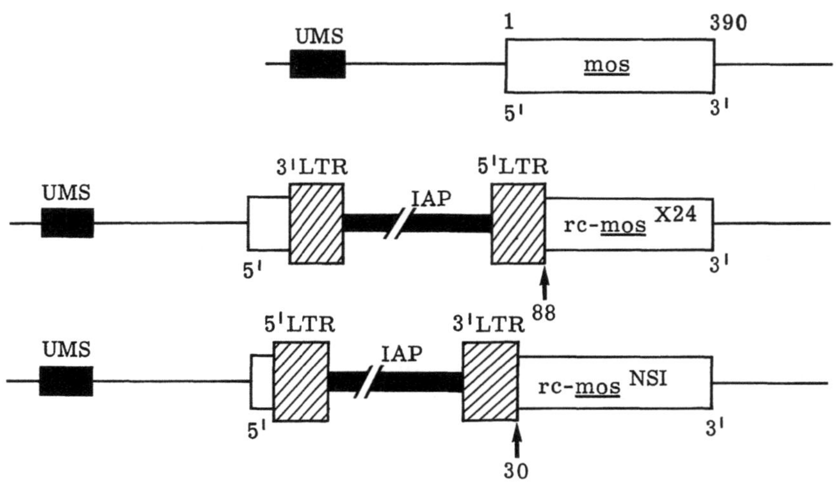

Fig. 5. Activation of the *mos* gene in mouse plasmocytomas involves insertion of an intracisternal A-particle (*IAP*) genome at the 5′ end of the *mos* gene. The IAP long terminal repeats (*LTR*) and the inhibitory upstream mouse *mos* sequence (*UMS*) are shown

1000 or more sites per haploid genome and are generally considered to represent a class of movable genetic elements. The IAP genomes present in rc-*mos*^{X24} and rc-*mos*^{NSI} are ocated within the coding region of the *mos* gene, at codons 88 and 30, respectively (Fig. 5). The orientation of the IAP genome relative to c-*mos* is head-to head for rc-*mos*^{X24} and tail-to-head for rc-*mos*^{NSI}. Integration of the IAPs enhances transcription of *mos*, and in each case the transcriptional activation is believed to result from the juxtaposition of the *mos* sequences and transcriptional control elements present in the long terminal repeats (LTRs) of the IAP genome. It is, however, worthy of note that the integrated IAP gene also separates *mos* from the *cis*-acting negative element normally located around 1 kb upstream from the *mos* coding region (BLAIR et al. 1984).

The integration of IAP genetic elements into the κ light chain gene in two mutant hybridoma sub-lines has also been reported (HAWLEY et al. 1982). It therefore seems likely that the relocation of IAP sequences within the mouse genome may be a relatively frequent event, and it is possible that IAP genes may occasionally have a role in the transformation of mouse cells by activating cellular proto-oncogenes.

VIII. Other Animal Models

Oncogene activation has also been found in lung tumours induced by tetranitromethane (TNM) in rats and mice (STOWERS et al. 1987), nasal squamous carcinomas induced by methylmethanesulphonate (MMS) in rats (GARTE et al. 1985), renal mesenchymal tumours induced by methyl(methoxymethyl)nitrosamine (DMN-OMe) in rats (SUKUMAR et al. 1986) and keratoacanthomas induced by DMBA in rabbits.

Lung tumours classified as adenocarcinomas, squamous cell carcinomas or adenosquamous carcinomas developed in Fischer 344 rats that were chronically exposed to TNM by inhalation. Similar treatment of B6C3F1 mice gave rise to lung tumours that were classified as papillary adenocarcinomas or adenomas. Using a combination of DNA transfection into NIH 3T3 cells and oligonucleotide hybridisation, cellular K-*ras* that was invariably activated by a G→A mutation at the second base of codon 12 was found in 18 of 19 rat lung tumours and 10 of 10 mouse lung tumours (STOWERS et al. 1987) (Table 1). Oncogene activation was believed to represent an early event in carcinogenesis since activated K-*ras* was detected in both adenocarcinomas and adenomas. Loss of the normal K-*ras* allele was observed in one rat lung tumour, suggesting that loss of the normal allele may, in some cases, play a role in tumour progression. Although possible mechanisms of interaction of TNM and DNA have currently not been defined, it is possible that the ability of TNM to induce lung tumours containing activated K-*ras* and to induce point mutations inbetween may be related to its ability to act as a nitrating agent under physiological conditions.

Treatment of newborn Fischer 344 rats with DMN-OMe results in the development of renal mesenchymal tumours within 40 weeks. The chemistry of interaction of DMN-OMe with DNA is still under investigation, but from a comparison of its structure with that of nitrosamines whose chemistry is well-defined, it would be expected to act as a simple methylating agent. DNA from 11 of 25 tumours was capable of transforming NIH 3T3 cells, and the activated genes were identified by SUKUMAR et al. (1986) as N-*ras* (1 tumour) and K-*ras* (10 tumours).

GARTE et al. (1985) found that DNA from 7 nasal squamous cell carcinomas and 1 ethesioneuroepithioma that were induced in rats by inhalation of MMS efficiently transformed NIH 3T3 cells (Table 1). MMS is a direct alkylating agent that only produces low levels of the O-alkylated bases that have been implicated in the genesis of point mutations and would be expected to be only a poor inducer of the point mutations that are required for *ras* gene activation. Accordingly the genes detected in the MMS-induced tumours were not closely related to H-, K- and N-*ras*.

Keratoacanthomas are benign and self-regressing skin tumours that can be induced by the repeated application of DMBA to rabbits' ears. LEON et al. (1988) have demonstrated that a high proportion of these chemically induced tumours contain activated H-*ras*, and preliminary evidence was provided indicating that activation involved a mutation at codon 61. The finding of H-*ras* activation in benign tumours that self-regress indicates that activated *ras* genes alone are not sufficient to maintain the presence of a tumour.

D. Oncogene Activation in Cells Transformed in Culture

I. Rodent Fibroblasts

DiPaolo and co-workers have examined transforming gene activation in lines of fetal guinea pig fibroblast-like cells that have been transformed by treatment with a single dose of carcinogen. Several distinct stages can be identified as the chemically treated cultures progress to form tumorigenic cell lines. Within 8 days of treatment the first morphological alterations are observed, but full morphological transformation may not occur until the cells have been passaged for several months. The potential to grow in soft agar and to form tumours in syngeneic guinea pigs arise together and may appear either at the same time or subsequent to full morphological transformation (EVANS and DIPAOLO 1975). Activated N-*ras* was found in five independent lines of tumorigenic guina pig cells that were derived from cultures treated with MNNG (2 lines), diethylnitrosamine (DENA), MCA or BP (SUKUMAR et al. 1984; DONIGER et al. 1987). Detection of the transforming genes occurred concomitantly with the appearance of tumorigenic properties, indicating that *ras* activation is associated with the acquisition of the fully transformed phenotype and may be a late stage in carcinogenesis. Remarkably, the same point mutation, an A:T→T:A transversion at the third position of codon 61 (Table 3), was found in all cell lines regardless of the carcinogen used to induce transformation (DONIGER et al. 1987). This observation, together with the late appearance of cells containing activated N-*ras*, prompted the authors to suggest that activation of the N-*ras* gene in this particular model system was independent of the direct mutagenic activity of the initiating carcinogen.

C3H/10T1/2 mouse fibroblasts that had been transformed by MCA were among some of the first cell lines to be examined using the NIH 3T3 transfection

Table 3. Oncogene activation in cells transformed in culture

Cell type	Transforming chemical	Gene activated	Observed mutations		References
			Codon	Mutation	
Fetal fibroblasts	DENA	Guinea pig N-*ras*	61	CAA(gly)→CAT(his)	DONIGER et al. (1987)
	MNNG	Guinea pig N-*ras*	61	CAA(gly)→CAT(his)	DONIGER et al. (1987)
	B[a]P	Guinea pig N-*ras*	61	CAA(gly)→CAT(his)	DONIGER et al. (1987)
	MCA	Guinea pig N-*ras*	61	CAA(gly)→CAT(his)	DONIGER et al. (1987)
C3H/10T1/2 fibroblasts	MCA	Mouse K-*ras*			PARADA and WEINBERG (1983)
4DH2 fibroblasts	MNU	Hamster, not *ras*			SHINER and COOPER (in preparation)
	ENU	Hamster, not *ras*			SHINER and COOPER (in preparation)
	DMS	Hamster, not *ras*			SHINER and COOPER (in preparation)
Bladder epithelial cells	DMBA	Mouse K-*ras*	12	GGT(gly)→AGT(ser)	BROOKES et al. (1988)
	None (spontaneous)	Mouse K-*ras*	12	GGT(gly)→AGT(ser)	BROOKES et al. (1988)
HOS cells	MNNG	Human *met*		*met* tyrosine kinase domain fused to *trp* gene	CHAN et al. (1987)
312H cells	MCA	Human H-*ras*	61		RHIM et al. (1987)

DENA, diethylnitrosamine; MNNG, *N*-methyl-*N'*- nitro-*N*- nitrosoguanidine; B[a]P, benzo[*a*]pyrene; MCA, 3-methylcholanthrene; MNU, *N*-methyl-*N*-nitrosourea; ENU, *N*-ethyl-*N*-nitrosourea; DMS, dimethylsulphate; DMBA, 7,12-dimethylbenz[*a*]anthracene.

assay (SHIH et al. 1979). A high proportion of MCA-transformed cell lines examined in these early studies contained transforming genes that were subsequently identified as activated homologues of mouse K-*ras* (PARADA and WEINBERG 1983) (Table 3). Transforming genes have also been identified in immortalised hamster dermal fibroblasts ($4DH_2$ cells) that had been transformed by exposure to MNU, ENU or DMS. The identity of the gene activated in DMS-transformed fibroblasts was of particular interest because DMS is only a weak point mutagen and would not be expected to be an efficient inducer of the mutations required for *ras* gene activation. As expected, the genes activated in DMS-transformed cells were not closely related to *ras*, but, perhaps surprisingly, the genes activated in cells transformed by the potent point mutagens ENU and MNU also contained genes that showed no detectable homology to *ras* (Shiner and Cooper, unpublished observations).

II. Mouse Bladder Epithelium

SUMMERHAYES and FRANKS (1979) examined the ability of DMBA to induce foci of morphologically transformed cells in cultures of urothelial cells from young (5–6 months) and old (28–30 months) C57BL/Icrf-a^1 mice, a mouse strain that does not exhibit a significant incidence of spontaneous bladder cancer. In experiments with urothelial cells from young mice, no transformation was observed in control cultures, and only a low level of transformation (1 foci/180 dishes) was observed in DMBA-treated cultures. In contrast, in experiments with urothelial cells from old mice, some transformation was detected in control cultures (4 foci/80 dishes), but a significantly higher (fivefold) frequency of transformation was detected after DMBA treatment (21 foci/80 dishes). BROOKES et al. (1988) detected activated K-*ras* genes in 4 of 5 transformed cell lines that arose in DMBA-treated cultures of urothelial cells from old mice and in a single transformed cell line derived from control cultures. Unexpectedly, the *ras* genes in cell lines derived from both DMBA-treated and control cultures were activated by a G:C→A:T transition mutation at the first base of codon 12 (Table 3). This was particularly surprising considering that most of the activating *ras* mutations found in DMBA-induced tumours are A:T→T:A transversions at the second base of codon 61 (DANDEKAR et al. 1986; QUINTANILLA et al. 1986). One interesting interpretation of these results was that some cells present in the urothelium of old mice may already contain activated K-*ras*, presumably as a result of mutations that accumulate throughout the lifetime of the animal. Disaggregation of the tissue and maintenance in tissue culture may, in some cases, allow cells containing activated *ras* to develop into foci of morphologically transformed cells. Within the context of this model, DMBA may simply act as a "promoter" by facilitating the development of foci of morphologically transformed cells from individual cells that already contain activated *ras*.

III. Human Cells

Cultures of human cells are in general much more resistant to chemical transformation than equivalent cultures of rodent cells. However, transformation has been reported following treatment of immortal human cell lines with chemical carcinogens and, at least in some cases, transformation is accompanied by the activation of transforming genes. The activated *met* gene was detected by transfection of DNA from a transformed human cell line called MNNG-HOS (COOPER et al. 1984a, b) that was derived by treating HOS cells with MNNG. The HOS cell line, as its name implies, was originally derived from a human osteosarcoma. HOS cells exhibit a relatively flat morphology when grown in tissue culture and do not form tumours when injected into nude mice. They can, however, be converted into morphologically transformed cells that form tumours when injected into nude mice after treatment either with chemical carcinogens such as MNNG and DMBA or with transforming viruses such as Kirsten sarcoma virus (RHIM et al. 1975a, b). The activated *met* gene was detected in MNNG-HOS cells but not in the parental HOS cell line, indicating that treatment of HOS cells with MNNG gave rise to an activated transforming gene (COOPER et al. 1984a).

The *met* proto-oncogene encodes a growth factor receptor-related protein that contains a cytoplasmic tyrosine kinase domain, a transmembrane domain and a large extracellular ligand binding domain (PARK et al. 1987; CHAN et al. 1988). The activation of *met*, shown in Fig. 6, involves a chromosomal rearrangement in which the region of the *met* gene encoding the extracellular and transmembrane domain is replaced by the 5′ region from an unrelated gene that has

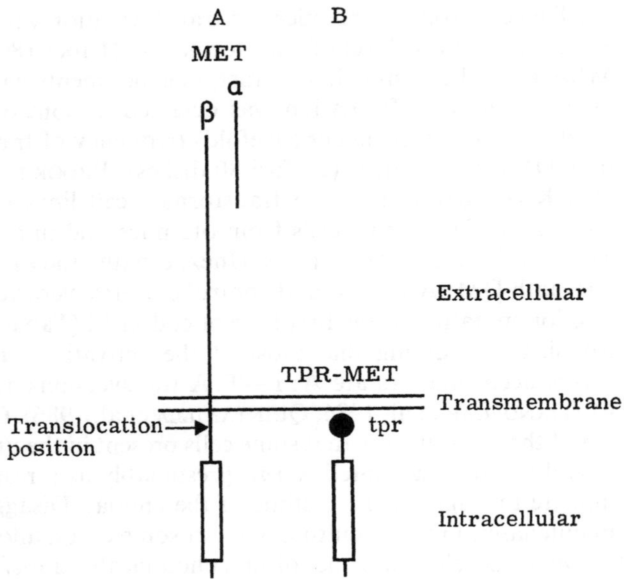

Fig. 6. Mechanism of *met* gene activation. The structures of the products encoded (*A*) by the *met* proto-oncogene and (*B*) by the *tpr-met* fusion gene found in the MNNG-HOS chemically transformed human cells are shown. The β subunit structure at the mature *met* protein proposed by TEMPEST et al. (1988) is illustrated

been designated *tpr* (translocated promoter region) (PARK et al. 1986; TEMPEST et al. 1986). The chimaeric gene is transcribed to produce a unique 5.0-kb hybrid *tpr-met* message that is in turn translated to form a protein in which the protein tyrosine kinase domain of *met* is fused to a *tpr*-encoded amino acid sequence that exhibits 35% homology to the extracellular matrix protein laminin B1 (CHAN et al. 1987). This weak homology between *tpr* and laminin B1 inicates that the unrearranged *tpr* gene encodes a structural protein. Although the identity and subcellular location of this putative structural protein has not been determined, it is interesting to speculate that the generation of the fusion protein may confer transforming potential on *met* by redirecting its subcellular location. In addition, the modification to the structure of *met* may alter its response to normal cellular control mechanisms.

The mechanism of activation of *met* is similar to that observed for the *trk* and *abl* genes. During activation of *trk*, which has occurred in a human colon carcinoma cell line, the carboxyl-terminal tyrosine kinase domain of a putative transmembrane receptor becomes attached to the amino-terminal 221 amino acids of a non-muscle tropomyosin molecule (MARTIN-ZANCA et al. 1986). Activation of c-*abl* occurs in chronic myelogenous leukaemia, in which the Philadelphia translocation results in the substitution of the 5′ sequences of the c-*abl* gene with *bcr* gene sequences. The activated gene retains the PTK domain of c-*abl* and exhibits enhanced PTK activity when compared with that of the normal c-*abl* protein (KONOPKA et al. 1984). Thus in each case the 3′ end of the activated gene encodes a PTK while initiation of transcription occurs in a separate DNA domain that composes the 5′-end of the gene.

Recently, RHIM et al. (1987) identified an activated H-*ras* gene in a morphologically transformed cell line derived by treating 312H human cells, a derivative of the HOS cell line, with MCA.

E. Mechanisms of Oncogene Activation

The proposed mechanisms of gene activation following treatment of animals or cultured cells with chemical carcinogens can be divided conveniently into two categories. Interaction of the carcinogen with the target gene may directly induce mutations by causing errors in DNA replication. Alternatively, gene activation may be an indirect consequence of treatment with the chemical. Thus, reaction of the carcinogen with DNA may cause mutations indirectly following the induction of, for example, error prone (SOS) repair mechanisms or may be completely unrelated to the ability of the carcinogen to interact with DNA.

I. Evidence of Direct Activation by Chemicals

In several of the model systems described in this chapter, the nature of the mutation responsible for *ras* gene activation is dependent on the particular class of chemical used for tomour induction, an observation that is considered to provide evidence that gene activation is a direct consequence of the interaction of the carcinogen with the target gene. For example in the mammary carcinogenesis model, MNU-induced tumours contain H-*ras* that is invariably activated by a G:C→A:T mutation of codon 12, while DMBA-induced tumours contain mutations in codon 61 (ZARBL et al. 1985). The argument for direct interaction of a particular chemical carcinogen with the target gene becomes stronger when the mutations responsible for *ras* activation are precisely those expected from the known chemistry of interaction of the carcinogen with DNA and from analysis of the point mutations induced in bacteria. Again, a good example is the G:C→A:T mutation responsible for H-*ras* activation in MNU-induced mammary tumours; this is exactly the mutation that would be expected to result from alkylation of the O^6-position of guanine by MNU (ZARBL et al. 1985). Similar arguments have been used in support of the idea that H-*ras* activation in DMBA-induced mouse skin papillomas and carcinomas and in N-OH-AAF-induced mouse hepatomas is a direct consequence of interaction of the carcinogen with DNA (WISEMAN et al. 1986).

Although the evidence for direct involvement of MNU during mammary carcinogenesis in rats is compelling, it is curious that exactly the same mutation (a G:C→A:T transition at the second position of codon 12) is almost invariably found in H-*ras* activated in MNU-induced mammary tumours (ZARBL et al. 1985), in K-*ras* activated in TNM-induced lung tumours (STOWERS et al. 1987) and in N-*ras* and K-*ras* activated in X-ray-induced mouse thymic lymphomas (GUERRERO and PELLICER 1987; DIAMOND et al. 1988). Unfortunately, the mechanism of metabolic activation of TNM has not been defined, but in the case of X-ray-induced lymphomas, it is not at all clear how the major lesions produced in DNA after exposure to radiation could lead exclusively to the induction of this particular mutation (Ames and Saul 1986). It may therefore not be heretical to consider the possibility that induction of G:C→A:T mutations following treatment with MNU, TNM or X-rays involves a common, but indirect, mechanism.

Another remarkable feature of MNU-induced mammary tumours is that the activating G:C→A:T mutation is always at the second position of codon 12 (codon 12 in rat H-*ras* is GGA), although G→A mutations of both the first and second positions should generate potent transforming genes (Zarbl et al. 1985; Seeburg et al. 1984). A remarkably similar specificity for mutation at particular G residues was observed in the bacterial xanthine-guanine phosphoribosyltransferase gene (*gpt*) following exposure to MNU (Richardson et al. 1987). Thus, in 32 of 39 of the G→A transition mutations detected, a *gpt* occurred in the middle G residue of the sequence 5′-GG(A or T)3′ while, at least for this limited series of mutations, when two or more G residues were found together a mutation was never observed in the 5′ G residue. Applying the rules derived from these studies on mutations in bacteria to the sequence around codon 12 in rat H-*ras* (the sequence at codons 11–13 is GCT GGA GGC), one would predict that the predominant activating mutation would occur at the second position of codon 12. The precise reason for this sequence specificity, both in the bacterial mutation studies and during activation of H-*ras*, remains to be established, but it could, in theory, be caused by variations in the local environment of G residues that result in preferential methylation, sequence-specific repair and/or variations in miscoding potential.

Topal et al. (1986) examined the distribution of mutations induced in the ampicillinase gene after incorporation of O^6-methyl GTP opposite T during *E. coli* DNA polymerase I-catalysed replication in vitro and transfer of the modified gene to *E. coli*. In this system, most mutations are a consequence of enzyme-catalysed removal of the methyl group from O^6-methylG:T base pairs forming G:T intermediates that, when replicated in *E. coli*, generate mutations. Their results showed that one-third of the A:T base pairs (bps) were resistant to mutation, and this was attributed, at least in part, to the inefficient repair of O^6-methylG:T bps at particular sequences after transfer of the modified gene to *E. coli*. One of the most notable features of this study was the discovery of a high (75%) homology between the consensus sequence of the regions surrounding the mutation-resistant A:T bps and the sequence surrounding the middle base at codon 12 in human and rat H-*ras* genes (Fig. 7). This striking homology may indicate that an O^6-methylG:C bps formed at the second base of codon 12 of H-*ras* would also be inefficiently repaired, leading to a high frequency of G:C→A:T transition mutations at this position.

Direct evidence that *ras* gene activation can result from the reaction of carcinogens with DNA was provided by experiments in which in vitro modification of plasmids containing the human H-*ras* proto-oncogene with activated derivatives of chemical carcinogens gave rise to transforming genes when the modified DNA was introduced into NIH 3T3 mouse fibroblasts (Marshall et al. 1984; Vousden et al. 1986). Activation of H-*ras* was observed after in vitro modifica-

			Codon 12	
rat H-ras :	G G T G G G C G C T -	G	G	A G G
Consensus :	G C T G G T C G C C A	G	G Me	A G G

Fig. 7. Comparison of the sequence around codon 12 of the *rat H-ras* gene and the *consensus* sequence for resistance to repair of O^6-methylG:T base pairs as determined by TOPAL et al. (1986)

tion with *r*-7,*t*-8-dihydroxy-*t*-9,10-epoxy-7,8,9,10-tetrahydrobenzo[*a*]pyrene (*anti*-BPDE), with *N*-acetoxy-2-acetylaminofluorene (N-Aco-AFF), with 1′-acetoxysafrole (Aco-S) or with the 3-*N*-,*N*-acetoxyacetyl derivative (N-Aco-AGlu-P-3) of the mutagenic L-glutamic acid pyrolysis product, 3-amino-4,6-dimethyldipyrido[1,2-a:3′,2′-d]imidazole (Table 4). In addition H-*ras* activation was also observed after the generation of apurinic sites by heating the plasmid to 70 °C at pH 4.0 (Table 4). In contrast to the results obtained with many of the animal models, a broad spectrum of activating mutations was observed. For example, with *anti*-BPDE, five different mutations were found at codon 61 in addition to mutations at codons 12 and 13 (VOUSDEN et al. 1986). In general, the types of mutation were similar to those observed in bacterial mutation studies, perhaps with the exception that G:C→C:G transversions were detected after modification with *anti*-BPDE. Notably, most of the mutations induced by N-Aco-AGlu-P-3 and by N-Aco-AAF were G:C→T:A transversions. This is in agreement with the known preference of reactions of N-Aco-AGlu-P-3 and N-Aco-AAF with G residues and with the observation that 7 of 7 N-OH-AAF-induced mouse hepatomas contain H-*ras* activated by this transversion mutation (WISEMAN et al. 1986). In mouse hepatomas, all of the mutations were at the first position of codon 61, but, after modification of plasmid in vitro, mutations occurred at both the first and third positions of this codon, again demonstrating that the activation occurring after in vitro modification is less specific than that observed in vivo.

II. Evidence for an Indirect Role of Chemical Exposure

In several of the model systems described in the previous sections oncogene activation is believed to be an indirect consequence of carcinogen treatment. One example is provided by the c-*myc* rearrangement found in most pristane-induced plasmocytomas. The precise mechanism of pristane-induced carcinogenesis still remains to be established, but as mentioned above pristane is a relatively inert chemical that is not known to interact with DNA directly. Analysis of the mutations responsible for *ras* gene activation have, in some cases, also supported the notion that activation involves an indirect mechanism. This is particularly true when the same activating mutation is invariably found, regardless of the chemical used to induce transformation, and when the mutations observed are not those expected from the known chemistry of the carcinogen and from studies of muta-

Table 4. Activating mutations in H-*ras* induced by chemicals and depurination[a]

Treatment	No. of mutants analysed	Codon 12		Codon 13		Codon 61				
		Base 1	Base 2	Base 1	Base 2	Base 1	Base 2		Base 3	
		G→T (cys)	G→A (asp)	G→C (arg)	G→A (asp)	G→T (lys)	A→T (leu)	A→C (pro)	G→C (his)	G→T (his)
anti-BPDE	13	1			1	4	3	1	2	1
N-Aco-AAF	6	1		1		2				2
Aco-S	3	1				1				
N-Aco-AGlu-P-3	6					3				3
Depurination	3		1	1			1			

anti-BPDE, *r*-7,*t*-8-dihydroxy-*t*-9,10-epoxy-7,8,9,10-tetrahydrobenzo[*a*]pyrene; N-Aco-AAF, *N*-acetoxy-2-acetylaminofluorene; Aco-S, 1′-acetoxysafrole; N-Aco-AGlu-P-3, 3-N,N-acetoxyacetyl derivate of 3-amino-4,6-dimethyldipyrido[1,2-a: 3′,2′-d]imidazole.

[a] Data from VOUSDEN et al. (1986) and IRELAND et al. (1988).

tions induced in bacteria. For example, the observations (i) that the same activating mutation is found in lines of fetal guinea pig cells transformed by DMBA, BP, MNNG or DENA and (ii) that the mutations responsible for *ras* gene activation in thymomas induced by MNU in AKR/RF mice were not the expected G:C→A:T transition mutations suggest indirect mechanisms of *ras* gene activation (GUERRERO et al. 1986; DONIGER et al. 1987).

III. Spontaneous Gene Activation

Activation of *ras* may also occur spontaneously during carcinogenesis. Thus, activated H-*ras* was detected in spontaneous hepatomas (FOX and WATANABE 1985; REYNOLDS et al. 1986) from mice while activated K-*ras* was detected in a transformed cell line that arose in cultures of urothelium from old mice (BROOKES et al. 1988). The mechanism of spontaneous *ras* gene activation is unknown. Activation of cellular genes might result from background exposure to carcinogens and ionising radiation of an animal kept under normal laboratory conditions. Oxidative damage to DNA, including the formation of thymidine glycol and 5-hydroxymethyluracil, may result from exposure to ionising radiation (AMES and SAUL 1986). In addition, experiments using the very sensitive ^{32}P-post-labelling procedure have detected low levels of nucleic acid derivatives with the chromatographic properties of aromatic hydrocarbon-DNA adducts in DNA from mice. Alternatively, damage to DNA, including base loss, deamination and base oxidation, may be caused by reactive oxygen species that arise through oxidative metabolism. DNA damage may also be caused by the endogenous methylating agent *S*-adenosylmethionine or by exposure to normal metabolites, such as formaldehyde and methylglyoxal (AMES and SAUL 1986). Another possibility is that mutations arise through inherent deficiencies in the fidelity of DNA replication that might, for example, be a consequence of base tautomerisation, base rotation or base ionisation.

F. Specificity of Gene Activation

An interesting feature of most of the model systems described above is the strict tissue specificity of gene activation. For example, H-*ras* is the gene invariably detected in chemically induced skin papillomas and rat mammary tumours (SUKUMAR et al. 1983; BALMAIN et al. 1984), whilst N-*ras* is the gene preferentially activated following chemical treatment of guinea pig embryo cells (DONIGER et al. 1987). The specificity for activation of a particular gene in chemically induced thymomas, lymphomas and renal mesenchymal tumours is less marked since both K-*ras* or N-*ras* are activated in each of these model systems. However, activation of H-*ras* is not observed, indicating that there is still some specificity for activation of particular *ras* genes.

Preferential activation of particular genes is also found in certain types of human tumours, although the degree of specificity is usually less dramatic than that observed in animal models. Thus, N-*ras* is the gene most frequently activated in acute non-lymphocytic leukaemia (BOS et al. 1985), K-*ras* is frequently activated in colon tumours (FORRESTER et al. 1987; BOS et al. 1987), and H-*ras* is preferentially activated in bladder cancer (FUJITA et al. 1985). This specificity can be explained, at least in part, by the differential expression of the target *ras* gene. LEON et al. (1987) have shown that, of the three *ras* genes, H-*ras* is expressed at high levels in mouse skin. Conversely, mouse thymus contains high levels of transcripts from the K-*ras* and N-*ras* genes but only low levels of H-*ras* transcripts. These results are consistent with the preferential activation of particular *ras* genes in skin papillomas and in mouse thymomas. The correlation between expression and gene activation is perhaps not surprising since a point mutation in a gene that is not expressed would be unlikely to have any effect on cell proliferation.

G. Prospectives

Several distinct types of genetic alteration have been implicated in the genesis of human cancers. These include the point mutations that are associated with *ras* gene activation, gene rearrangements, gene amplification and gene deletion. Studies on the genetic changes present in chemically induced tumours have focussed attention on point mutations and gene rearrangements, although, in a recent study on radiation-induced mouse skin tumours, amplification of c-*myc* was observed (SAWEY et al. 1987). In addition WONG (1987) has detected amplification of the gene encoding the epidermal growth factor receptor (c-*erb*B1) in oral carcinomas induced by treating hamsters with DMBA. This lack of information on the amplification of specific genes in chemically induced tumours is curious since certain types of chemically induced tumours are known to contain double minute chromosomes (COWELL 1981), the morphological hallmark of gene amplification. A second, potentially exciting area for future investigation is the identification of specific genes that may be deleted or inactivated during chemical carcinogenesis. Such genes may correspond to the so-called "suppressor genes" or "anti-oncogenes" that need to be removed to allow cell transformation (reviewed by KLEIN 1987). Since the loss of specific chromosomal material is a common feature of the development of many types of human cancer, it would be very useful to have an animal model allowing the mechanism of gene loss and its role in carcinogenesis to be studied in more detail.

The observation that *ras* gene activation is found in a high proportion of certain types of chemically induced tumours and in a significant proportion of some human tumours raises the provocative question: is the activation of *ras* genes in human cancer a consequence of exposure to chemical carcinogens? In an attempt to answer this question, it is worthy of note that *ras* gene activation is found in classes of human cancer, such as lung and bladder cancer and acute non-lymphatic leukaemia (PULCIANI et al. 1982; BOS 1985; FUJITA et al. 1985), whose incidence is known to be increased as a result of environmental and occupational exposure to chemicals. It may, however, be difficult to provide a definitive answer to this question, because analysis of animal models has revealed that *ras* gene activation can also occur in tumours that arise "spontaneously" or following exposure to radiation (GUERRERO et al. 1984a; FOX and WATANABE 1985; REYNOLDS et al. 1986; SAWEY et al. 1987).

Carcinogenesis is generally considered to be a multistep process. Evidence for this is provided by studies on the kinetics of appearance of cancer (NORDLING 1953), by analysis of the pathology of cancer development (FOULDS 1969) and by studying the transformation of cells in tissue culture. Transformation of primary cells in tissue culture can, in some cases, be achieved by cooperation between activated genes. For example, in a classic study, LAND et al. (1983) demonstrated that primary embryonic fibroblasts can be efficiently transformed by cooperation between activated forms of *ras* and *myc* genes. This type of experiment suggests that several genetic changes may be required to achieve full transformation. In most of the model systems described in this chapter, only a single consistent genetic change (usually *ras* gene activation) has been defined. The identification of other genetic, or even epigenetic, changes that may cooperate with *ras* during chemical carcinogenesis provides a fruitful area for future studies.

References

Ames BN, Saul RL (1986) Oxidative DNA damage as related to cancer and ageing. Prog Clin Biol Res 209A:11–26

Balmain A, Pragnell IB (1983) Mouse skin carcinomas induced in vivo by chemical carcinogens have a transforming Harvey-*ras* oncogene. Nature 303:72–74

Balmain A, Ramsden M, Bowden GT, Smith J (1984) Activation of mouse cellular Harvey-*ras* in chemically induced benign skin papillomas. Nature 307:658–660

Barbacid M (1987) *ras* genes. Ann Rev Biochem 56:779–827

Bargmann CI, Hung M-C, Weinberg RA (1986a) The *neu* oncogene encodes an epidermal growth factor-related protein. Nature 319:226–230

Bargmann CI, Hung C-M, Weinberg RA (1986b) Multiple independent activations at the *neu* oncogene by a point mutation altering the transmembrane domain of p185. Cell 45:649–657

Bernard O, Cory S, Gerondakis S, Webb E, Adams JM (1983) Sequence of murine and human cellular *myc* oncogenes and two models of *myc* transcription resulting from chromosome translocation in B lymphoid tumours. EMBO J 2:2375–2383

Bigger CAH, Sawicki JT, Blake DM, Raymond LG, Dipple A (1983) Products of binding of 7,12-dimethylbenz[*a*]anthracene to DNA in mouse skin. Cancer Res 43:5647–5651

Bizub D, Wood AW, Skalka AM (1986) Mutagenesis of the Ha-*ras* oncogene in mouse skin tumours induced by polycyclic aromatic hydrocarbons. Proc Natl Acad Sci USA 83:6048–6052

Blair DG, Oskarsson MK, Wood TG, McClements WL, Fischinger PJ, Vande Woude GF (1981) Activation of the transforming potential of a normal cell sequence: a model for oncogenesis. Science 212:941–943

Blair DG, Cooper CS, Oskarsson MK, Eader LA, Vande Woude GF (1982) New method for detecting cellular transforming genes. Science 218:1122–1125

Blair DG, Wood TG, Woodworth AM, McGeady ML, Oskarsson MK, Propst F, Tainsky MA, Cooper CS, Watson R, Baroudy BM, Vande Woude GF (1984) Properties of the mouse and human *mos* oncogene loci. In: Vande Woude GF, Levine AJ, Topp WC, Watson JD (eds) Cancer cells: oncogene and viral genes, vol 2. Cold Spring Harbor Laboratory Press, New York, pp 281–289

Bos JL, Verlaan-de Vries M, Jansen AM, Veeneman GH, van Boom JH, van der Eb AJ (1984) Three different mutants in codon 61 of the human N-*ras* gene detected by synthetic oligonucleotide hybridization. Nucleic Acid Res 12:9155–9162

Bos JL, Toksoz D, Marshall CJ, Verlaan-de Vries M, Veeneman GH, van der Eb AJ, van Boom JH, Janssen JWG, Steenvoorden ACM (1985) Amino-acid substitutions at codon 13 of the N-*ras* oncogene in human acute myeloid leukaemia. Nature 315: 726–730

Bos JL, Fearon ER, Hamilton SR, Verlaan-de Vries M, van Boom JH, van der Eb A, Vogelstein B (1987) Prevalence of *ras* gene mutations in human colorectal cancers. Nature 327:293–297

Brookes P, Lawley PD (1964) Evidence for the binding of polynuclear aromatic hydrocarbons to the nucleic acids of mouse skin: relation between carcinogenic powder of hydrocarbons and their binding to deoxyribonucleic acid. Nature 202:781–784

Brookes P, Cooper CS, Ellis MV, Warren W, Gardner E, Summerhayes IC (1988) Activated K-*ras* genes in bladder epithelial cell lines transformed by treatment of primary mouse bladder explant cultures with 7,12-dimethylbenz[*a*]anthracene. Mol Carcinogenesis 1:82–88

Brown K, Quintanilla M, Ramsden M, Kerr IB, Young S, Balmain A (1986) v-*ras* genes from Harvey and BALB murine sarcoma viruses can act as initiators of two-stage mouse skin carcinogenesis. Cell 46:447–456

Canaani E, Dreazen O, Klar A, Rechav G, Ram D, Cohen JB, Givol D (1983) Activation of the c-*mos* oncogene in mouse plasmocytoma by insertion of an endogenous intracisternal A-particle genome. Proc Natl Acad Sci USA 80:7118–7122

Chan AM-L, King HWS, Tempest PR, Deakin EA, Cooper CS, Brookes P (1987) Primary structure of the *met* protein tyrosine kinase domain. Oncogene 1:229–233

Chan AM-L, King HWS, Deakin EA, Tempest PR, Hilkens J, Kroezen V, Edwards DR, Wills AJ, Brookes P, Cooper CS (1988) Characterisation of the mouse *met* protooncogene. Oncogene 2:593–599

Cohen JB, Unger T, Rechavi G, Canaani E, Givol D (1983) Rearrangement of the oncogene c-*mos* in mouse myeloma NSI and hybridomas. Nature 306:797–799

Cooper CS, Blair DG, Oskarsson MK, Tainsky MA, Eader LA, Vande Woude GF (1984 a) Characterization of human transforming genes from chemically transformed teratocarcinomas and pancreatic carcinoma cell lines. Cancer Res 44:1–10

Cooper CS, Park M, Blair DG, Tainsky MA, Huebner K, Croce CM, Vande Woude GF (1984 b) Molecular cloning of a new transforming gene from a chemically transformed human cell line. Nature 311:29–33

Cooper GM, Okerquist S, Silverman L (1980) Transforming activity of DNA of chemically transformed and normal cells. Nature 284:418–421

Corcoran LM, Adams JM, Dunn AR, Cory S (1984) Murine T lymphomas in which the cellular *myc* oncogene has been activated by retroviral insertion. Cell 37: 113–122

Cory S (1986) Activation of cellular oncogenes in hemopoietic cells by chromosome translocation. Adv Cancer Res 47:189–234

Cowell JK (1981) Chromosome abnormalities associated with salivary gland epithelial cell lines transformed in vitro and in vivo with evidence of a role for genetic imbalance in transformation. Cancer Res 41:1508–1517

Cuypers HT, Selten G, Quint W, Zijlstra M, Robanus-Maandag E, Boelens W, van Wezenbeek P, Melief C, Berns A (1984) Murine leukaemia virus-induced T-cell lymphomagenesis: integration of provirus in a distinct chromosomal region. Cell 37:141–150

Dandekar S, Sukumar S, Zarbl H, Young LJT, Cardiff RD (1986) Specific activation of the cellular Harvey-*ras* oncogene in dimethylbenzanthracene-induced mouse mammary tumours. Mol Cell Biol 6:4104–4108

DeLeo AB, Joy G, Appella E, Dubois GC, Law LW, Old LJ (1979) Detection of a transformation-related antigen in chemically induced sarcomas and other transformed cells of the mouse. Proc Natl Acad Sci USA 76:2420–2424

Diamond LE, Guerrero I, Pellicer A (1988) Concomitant K- and N-*ras* gene point mutations in cloncal murine lymphomas. Mol Cell Biol 8:2233–2236

Doniger J, Notario V, DiPaolo JA (1987) Carcinogens with diverse mutagenic activities initiate neoplastic guinea pig cells that acquire the same N-*ras* point mutation. J Biol Chem 262:3813–3819

Drebin JA, Stern DF, Link VL, Weinberg RA, Greene MI (1984) Monoclonal antibodies identify a cell-surface antigen associated with an activated cellular oncogene. Nature 312:545–548

Drebin JA, Link VC, Stern DF, Weinberg RA, Green MI (1985) Down-modulation of an oncogene protein product and reversion of the transformed phenotype by monoclonal antibodies. Cell 41:695–706

Eva A, Aaronson SA (1983) Frequent activation of c-*kis* as a transforming gene in fibrosarcomas induced by methylcholanthrene. Science 220:955–95

Eva A, Trimmer W (1986) High frequency of c-K-*ras* activation in 3-methylcholanthrene-induced mouse thymomas. Carcinogenesis 7:1931–1933

Evans CH, DiPaolo JA (1975) Neoplastic transformation of guinea pig fetal cells in culture induced by chemical carcinogens. Cancer Res 35:1035–1044

Fennell TR, Wiseman RW, Miller JA, Miller EC (1985) Major role of hepatic sulfotransferase activity in the metabolic activation, DNA adduct formation, and carcinogenicity of 1′-hydroxy-2′,3′-dehydroestragole in infant male C57BL/6J × C3H/HeJF_1 mice. Cancer Res 45:5310–5320

Forrester K, Almoguera C, Han K, Grizzle WE, Perucho M (1987) Detection of high incidence of K-*ras* oncogenes during human colon tumorigenesis. Nature 327:298–303

Foulds L (1969) Neoplastic development, vol 1. Academic, London

Fox TR, Watanabe PG (1985) Detection of a cellular oncogene in spontaneous liver tumours of B6C3F1 mice. Science 228:596–597

Frei JV (1980) Methylnitrosourea induction of thymomas in AKR mice requires one or two "hits" only. Carcinogenesis 1:721–723

Frei JV, Lawley PD (1980) Thymomas induced by simple alkylating agents in C57BL/Cbi mice: kinetics of the dose response. JNCI 64:845–856

Frei JV, Swenson DH, Warren W, Lawley PD (1978) Alkylation of DNA in vivo in various organs of C57BL mice by the carcinogens *N*-methyl-*N*-nitrosourea and ethylmethanesulphonate in relation to induction of thymic lymphomas: some applications of high-pressure liquid chromatography. Biochem J 174:1031–1044

Fujita J, Srivastava SK, Kraus MH, Rhim JS, Tronick SR, Aaronson SA (1985) Frequency of molecular alterations affecting *ras* protooncogene in human urinary tract tumour. Proc Natl Acad Sci USA 82:3849–3853

Garte SJ, Hood AT, Hochwalt AE, D'Eustachio P, Snyder CA, Segal A, Albert RE (1985) Carcinogen specificity in the activation of transforming genes by direct-acting alkylating agents. Carcinogenesis 6:1709–1712

Gattoni-Celli S, Hsiao WW-L, Weinstein IB (1983) Rearrangement of c-*mos* locus in MOPC21 murine myeloma cell line and its persistence in hybridomas. Nature 306:795–796

Guerrero I, Pellicer A (1987) Mutational activation of oncogenes in animal model systems of carcinogenesis. Mutat Res 135:293–308

Guerrero I, Calzada P, Mayer A, Pellicer A (1984a) A molecular approach to leukaemogenesis: mouse lymphomas contain an activated c-*ras* oncogene. Proc Acad Natl Sci USA 81:202–205

Guerrero I, Villasante A, D'Eustachio P, Pellicer A (1984b) Isolation, characterization and chromosomal assignment of mouse N-*ras* gene from carcinogen-induced thymic lymphoma. Science 225:1041–1043

Guerrero I, Villasante A, Diamond L, Berman JW, Newcomb EW, Steinberg JJ, Lake R, Pellicer A (1986) Oncogene activation and surface markers in mouse lymphomas induced by radiation and nitrosomethylurea. Leukaemia Res 10:851–858

Gullino PM, Pettigrew HM, Grantham FH (1975) *N*-Nitrosomethylurea as a mammary gland carcinogen in rats. JNCI 54:401–409

Hawley RG, Shulman MJ, Murialdo H, Gibson DM, Hozumi N (1982) Mutant immunoglobin genes have repetitive DNA elements inserted into their intervening sequences. Proc Natl Acad Sci USA 79:7425–7429

Hecker E, Fusenig NE, Kinz W, Marks F, Thielmann HW (eds) (1982) Carcinogenesis, vol 7. Raven, New York

Huggins C, Grand LC, Brillantes FP (1961) Mammary cancer induced by a single feeding of polynuclear hydrocarbons and its suppression. Nature 189:204–207

Ireland CM, Cooper CS, Marshall CJ, Hebert E, Phillips DH (1988) Activating mutations in human c-Ha-*ras*-1 gene induced by reactive derivatives of safrole and the glutamic acid pyrolysate, Glu-P-3. Mutagenesis 3:429–435

Jenkins JR, Rudge K, Chumakov P, Currie GA (1985) The cellular oncogene p53 can be activated by mutagenesis. Nature 317:816–818
Klein G (1987) The approaching era of the tumour suppressor genes. Science 238: 1539–1545
Konopka JB, Watanabe SM, Witte ON (1984) An alteration of the human c-*abl* protein in K562 leukaemia cells unmasks associated tyrosine kinase activity. Cell 37:1035–1045
Kuff EL, Feenstra A, Lueders K, Rechavi G, Givol D, Canaani E (1983) Homology between an endogenous viral LTR and sequence inserted in an activated cellular oncogene. Nature 302:547–548
Laerum OD, Rajewsky MF (1975) Neoplastic transformation of fetal rat brain cells in culture after exposure to ethyl nitrosourea in vivo. JNCI 55:1177–1184
Lai C-C, Miller JA, Miller EC, Liem A (1985) *N*-Sulfoxy-2-aminofluorene is the major ultimate electrophilic and carcinogenic metabolite of *N*-hydroxy-2-acetylaminofluorene in the liver of infant male C57BL/6J × C3H/HeJF_1 (B6C3F_1) mice. Carcinogenesis 6:1037–1045
Land H, Parada LF, Weinberg RA (1983) Tumorigenic conversion of primary embryo fibroblasts requires at least two cooperating oncogenes. Nature 304:596–602
Lane DP (1987) p53 in Paris, an oncogene comes of age. Oncogene 1:241–242
Lane DP, Crawford LV (1979) T antigen is bound to a host protein in SV40-transformed cells. Nature 278:261–263
Lantos PL (1986) Development of nitrosourea-induced brain tumours with a special note on changes occurring during latency. Chem Toxicol 24:121–127
Lawley PD (1976) Comparison of alkylating agents and radiation carcinogenesis: some aspects of the possible involvement of effects of DNA. In: Yunkas JM, Tennant RW, Regan JD (eds) Biology of radiation carcinogenesis. Raven, New York, pp 165–174
Leon J, Guerrero I, Pellicer A (1987) Differential expression of the *ras* gene family in mice. Mol Cell Biol 7:1535–1540
Leon J, Kamino H, Steinberg JJ, Pellicer A (1988) H-*ras* activation in benign and self-regressing skin tumours (keratoacanthomas) in both humans and on animal model system. Mol Cell Biol 8:786–793
Lewis SE, Johnson FM, Skow LC, Popp D, Barnett LB, Popp RA (1985) A mutation in the *β*-globin gene detected in the progeny of a female mouse treated with ethylnitrosourea. Proc Natl Acad Sci USA 82:5829–5831
Marshall CJ, Vousden KH, Phillips DH (1984) Activation of c-Ha-*ras*-1 proto-oncogene by in vitro modification with a chemical carcinogen, benzo[*a*]pyrene diol-epoxide. Nature 310:586–589
Martin-Zanca D, Hughes SH, Barbacid M (1986) A human oncogene formed by the fusion of truncated tropomyosin and protein kinase sequences. Nature 319:743–748
McCann J, Ames BN (1976) Detection of carcinogens as mutagens in the *Salmonella*/microsome test: assay of 300 chemicals: Discussion. Proc Natl Acad Sci USA 73:950–954
McCann J, Choie E, Yamasaki E, Ames BN (1975) Detection of carcinogens as mutagens in the *Salmonella*/microsome test: assay of 300 chemicals. Proc Natl Acad Sci USA 72:5135–5139
McMahon G, Hanson L, Lee J-J, Wogan GN (1986) Identification of an activated c-Ki-*ras* oncogene in rat liver tumours induced by aflatoxin B_1. Proc Natl Acad Sci USA 83:9418–9422
Nordling CO (1953) A new theory on the cancer-inducing mechanism. Br J Cancer 7:68–72
Ochai M, Nagao M, Tahira T, Ishikawa F, Hayash K, Ohgaki H, Terada M, Tsuchida N, Sugimura T (1985) Activation of K-*ras* and oncogenes other than *ras* family in rat fibrosarcoma induced by 1,8-dinitropyrene. Cancer Lett 29:119–125
Padhy LC, Shih C, Cowing D, Finkelstein R, Weinberg RA (1982) Identification of a phosphoprotein specifically induced by the transforming DNA of rat neuroblastomas. Cell 28:865–871
Parada LF, Weinberg RA (1983) Presence of a Kirsten murine sarcoma virus *ras* oncogene in cells transformed by 3-methylcholanthrene. Mol Cell Biol 3:2298–2301
Park M, Dean M, Cooper CS, Schmidt M, O'Brien SJ, Blair DG, Vande Woude GF (1986) Mechanism of *met* oncogene activation. Cell 45:895–904

Park M, Dean M, Karl K, Braun MJ, Gonda MA, Vande Woude GF (1987) Sequence of MET protooncogene cDNA has features characteristic of the tyrosine kinase family of growth-factor receptors. Proc Natl Acad Sci USA 84:6379–6383

Pelling JC, Neodes R, Strawhecker J (1988) Epidermal papillomas and carcinomas induced in uninitiated mouse skin by tumour promoters alone containt point mutations in the 61st codon of the Ha-*ras* oncogene. Carcinogenesis 9:665–667

Popp RA, Baliff EG, Skow LC, Johnson FM, Lewis SE (1983) Analysis of a mouse α-globin gene mutation induced by ethylnitrosourea. Genetics 105:157–167

Potter M (1984) Genetics of susceptibility to plasmocytoma development in BALB/c mice. Cancer Surv 3:247–264

Pulciani B, Santos E, Lauver AV, Long LK, Aaronson SA, Barbacid M (1982) Oncogenes in solid human tumours. Nature 300:539–542

Quintanilla M, Brown K, Ramsden M, Balmain A (1986) Carcinogen-specific mutation and amplification of H-*ras* during mouse skin carcinogenesis. Nature 322:78–80

Rajewsky MF (1983) Structural modifications and repair of DNA in neuro-oncogenesis by *N*-ethyl-*N*-nitrosourea. Recent Results Cancer Res 84:63–75

Rechavi G, Givol D, Canaani E (1982) Activation of a cellular oncogene by DNA rearrangement: possible involvement of an IS-like element. Nature 300:607–610

Reddy EP, Reynolds PK, Santos E, Barbacid M (1982) A point mutation is responsible for the acquisition of transforming properties by the T24 human bladder carcinoma oncogene. Nature 300:149–152

Reynolds SH, Stowers SJ, Maronpot RR, Anderson MW, Aaronson SA (1986) Detection and identification of activated oncogenes, in spontaneously occurring benign and malignant hepatocellular tumours of the B6C3F1 mouse. Proc Natl Acad Sci USA 83:33–37

Reynolds SH, Stowers SJ, Patterson RM, Maronpot RR, Aaronson SA, Anderson MW (1987) Activated oncogenes in B6C3F1 mouse liver tumours: implications for risk assessment. Science 237:1309–1316

Rhim JS, Kin CM, Arnstein OP, Huebner RJ, Weisburger EK, Nelson-Rees WA (1975a) Transformation of human osteosarcoma cells by a chemical carcinogen. JNCI 55: 1291–1294

Rhim JS, Park DK, Arnstein P, Huebner RJ, Weisburger EK (1975b) Transformation of human cells in culture by *N*-methyl-*N'*-nitro-*N*-nitrosoguanidine. Nature 256: 751–753

Rhim JS, Fujita J, Park JB (1987) Activation of H-*ras* oncogene in 3-methylcholanthrene-transformed human cell line. Carcinogenesis 8:1165–1167

Richardson KK, Richardson FC, Crosby RM, Swenberg JA, Stopek TR (1987) DNA base charges and alkylation following in vivo exposure of *Escherichia coli* to *N*-methyl-*N*-nitrosourea or *N*-ethyl-*N*-nitrosourea. Proc Natl Acad Sci USA 84:344–348

Roscoe JP, Claisse PJ (1976) A sequential in vivo – in vitro study of carcinogenesis induced in the rat brain by ethylnitrosourea. Nature 262:314–316

Roscoe JP, Claisse PJ (1978) Analysis of *N*-ethyl-*N*-nitrosourea-induced brain carcinogenesis by sequential culturing during the latent period. I. Morphology and tumorigenicity of the cultured cells and their growth in agar. JNCI 61:381–386

Rotter V (1983) p53, a transformation-related cellular-encoded protein, can be used as a biochemical marker for the detection of primary mouse tumour cells. Proc Natl Acad Sci USA 80:2613–2617

Sawey MS, Hood AT, Burns FJ, Garte SV (1987) Activation of c-*myc* and c-K-*ras* oncogenes in primary rat tumours induced by ionizing radiation. Mol Cell Biol 7:932–935

Schechter AL, Stern DF, Vaidyanathan L, Decker SJ, Drebin JA, Green MI, Weinberg RA (1984) The *neu* oncogene: as *erb*B-related gene encoding a 185000-M, tumour antigen. Nature 312:512–516

Schechter AL, Hung M-C, Vaidyanathan L, Weinberg RA, Yang-Feng T, Francke U, Ullrich A, Coussens L (1985) The *neu* gene: an *erb*B-homologous gene distinct from and unlinked to the gene encoding the EGF receptor. Science 229:976–978

Seeburg PH, Colby WW, Capon DJ, Goeddel DV, Levinson AD (1984) Biological properties of human c-Ha-*ras*-1 genes mutated at codon 12. Nature 312:71–75

Seltan G, Cuypers HT, Boelens W, Robanus-Maandog E, Verbeek J, Domen J, van-Beveren C, Berns A (1986) The primary structure of the putative oncogene *pim*-1 shows extensive homology with protein kinases. Cell 46:603–611

Shih C, Shilo B-Z, Goldfarb MP, Dannenberg A, Weinberg RA (1979) Passage of phenotypes of chemically transformed cells via transfection of DNA and chromatin. Proc Natl Acad Sci USA 76:5714–5718

Shih C, Padhy LC, Murray M, Weinberg RA (1981) Transforming genes of carcinomas and neuroblastomas introduced into mouse fibroblasts. Nature 290:261–264

Sinha D, Dao TL (1975) Site of origin of mammary tumours induced by 7,12-dimethylbenz[*a*]anthracene in the rat. JNCI 54:1007–1008

Stowers SV, Glover PL, Reynolds SH, Boone LR, Maronpot RR, Anderson MW (1987) Activation of the K-*ras* protooncogene in lung tumours from rats and mice chronically exposed to tetranitromethane. Cancer Res 47:3212–3219

Stowers SJ, Wiseman RW, Ward JM, Miller EC, Miller JA, Anderson MW, Eva A (1988) Detection of activated proto-oncogenes in *N*-nirosodiethylamine-induced liver tumours: a comparison between $B6C3F_1$ mice and Fischer 344 rats. Carcinogenesis 9:271–276

Strong LC (1949) The induction of mutations by a carcinogen. Br J Cancer 3:97–108

Sukumar S, Notario J, Martin-Zanca D, Barbacid M (1983) Induction of mammary carcinomas in rats by nitrosomethylurea involves malignant activation of H-*ras*-1 locus by single point mutations. Nature 306:658–661

Sukumar S, Pulciani S, Doniger J, DiPaolo JA, Evans CH, Zbar B, Barbacid M (1984) A transforming *ras* gene in tumorigenic guinea pig cell lines initiated by diverse chemical carcinogens. Science 223:1197–1199

Sukumar S, Perantoni A, Reed C, Rice JM, Wenk ML (1986) Activated K-*ras* and N-*ras* oncogenes in primary renal mesenchymal tumours induced in F344 rats by methyl(methoxymethanol)nitrosamine. Mol Cell Biol 6:2716–2720

Summerhayes IC, Franks LM (1979) Effect of donor age on neoplastic transformation of adult mouse bladder epithelium in vitro. JNCI 62:1017–1023

Tabin CJ, Bradley SM, Bargmann CI, Weinberg RA, Papageorge AG, Scolnick EM, Dhar R, Lowy DR, Chang EH (1982) Mechanism of activation of a human oncogene. Nature 300:143–149

Tahira T, Hayashi K, Ochiai M, Tsuchida N, Nagao M, Sugimura T (1986) Structure of the c-Ki-*ras* gene in a rat fibrosarcoma induced by 1,8-dinitropyrene. Mol Cell Biol 6:1349–1351

Tempest PR, Reeves BR, Spurr NK, Rance AJ, Chan AM-L, Brookes P (1986) Activation of the *met* oncogene in the human MNNG-HOS cell line involves a chromosomal rearrangement. Carcinogenesis 7:2051–2057

Tempest PR, Sratton MR, Cooper CS (1988) Structure of the *met* protein and variation of *met* protein kinase activity among human tumour cell lines. Br J Cancer (in press)

Topal MD, Eadie JS, Conrad M (1986) O^6-Methylguanine mutation and repair is non-uniform. J Biol Chem 261:9879–9885

Vousden KH, Bos JL, Marshall CJ, Phillips DH (1986) Mutations activating human c-Ha-*ras*-1 protooncogene (HRASI) induced by chemical carcinogens and depurination. Proc Natl Acad Sci USA 83:1222–1226

Warren W, Lawley PD, Gardner E, Harris G, Ball JK, Cooper CS (1987) Induction of thymomas by *N*-methyl-*N*-nitrosourea in AKR mice: interaction between the chemical carcinogen and endogenous murine leukaemia viruses. Carcinogenesis 8:163–172

Wiseman RW, Stowers SV, Miller EC, Anderson MW, Miller JA (1986) Activating mutations of the c-Ha-*ras* protooncogenes in chemically induced hepatomas of the male B6C3F1 mouse. Proc Natl Acad Sci USA 83:5825–5829

Wong DT (1987) Amplification of the c-*erb* B1 oncogene in chemically induced oral carcinomas. Carcinogenesis 8:1963–1965

Zarbl H, Sukumar S, Arthur AV, Martin-Zanca D, Barbacid M (1985) Direct mutagenesis of Ha-*ras*-1 oncogenes by *N*-nitroso-*N*-methylurea during initiation of mammary carcinogenesis in rats. Nature 315:382–385

CHAPTER 13

Growth Factors and Their Receptors

C.-H. HELDIN and B. WESTERMARK

A. Introduction

The growth of cells in culture is controlled by polypeptide hormones that stimulate or inhibit proliferation. More than 20 different growth factors have been extensively characterised, and the corresponding cDNAs have been cloned (Table 1). Several additional growth factors are known from their biological activities but have not yet been structurally characterised. The in vivo functions of growth factors are assumed to be the stimulation of fetal and placental growth during development, the regulation of the growth and differentiation of continuously regenerating tissues, e.g. in haematopoiesis, and the stimulation of tissue repair processes. Specific growth inhibitory polypeptides have also recently been identified. Transforming growth factor (TGF)-β and the interferons are the most well studied, but it is anticipated that additional growth inhibitors will be found. The functions of growth inhibitors in vivo remain to be elucidated; it is possible that they have a role in the feedback inhibition of cell growth and regulation of cell differentiation.

I. Mechanism of Action of Growth Factors

Growth factors exert their mitogenic action by interaction with specific cell surface receptors on responsive cells. Many growth factors form families of structurally related peptides that bind to a common receptor (Table 1). Several receptors are equipped with a ligand-activated protein tyrosine kinase activity (Table 1), indicating that tyrosine phosphorylation of specific substrates is important in growth stimulation. In spite of intensive investigations (reviewed in HUNTER and COOPER 1985), however, no such substrate has been identified with a proven function in the mitogenic pathway.

Certain growth factors have been found to stimulate turnover of phosphatidylinositol, leading to the formation of diacylglycerol and inositol-trisphosphate, two second messengers of potential interest in growth stimulation. Diacylglycerol stimulates the activity of protein kinase C, an enzyme that can also be activated by certain tumour promoters (reviewed in NISHIZUKA 1984). Inositol-trisphosphate mobilises Ca^{2+} from internal stores (reviewed in BERRIDGE and IRVINE 1984). The importance of the phosphatidylinositol cycle in the regulation of cell growth remains to be established, however, since there is a poor correlation between the ability of different factors to stimulate phosphatidylino-

Table 1. Characterisation of growth factors and their receptors

Growth factor	Structure of mature product[a]	Receptor[a]	Target cells
Platelet-derived growth factor family PDGF-AA PDGF-AB PDGF-BB	Dimers of A chains (17 K) and B chains (c-*sis*) (16 K) in different combinations	Type A binds PDGF-AA, PDGF-AB and PDGF-BB, 170 K; Type B binds PDGF-BB, and at lower affinity PDGF-AB, 170–185 K, protein tyrosine kinase	Mesenchymal cells, glial cells
Epidermal growth factor family EGF Transforming growth factor (TGF)-α vaccinia virus growth factor	Monomers of 6–9 K displaying amino acid sequence homology to each other and formed from membrane-bound precursors	c-*erb*B, 175 K, protein tyrosine kinase	Epithelial cells, mesenchymal cells
Insulin-like growth factor family Insulin	Two chains of 5700 DA formed from proinsulin by proteolysis	$(130\ K, 90\ K)_2$, protein tyrosine kinase	Various cell types
Insulin-like growth factor-I (IGF-I)	Monomer of 7 K homologous to proinsulin	$(130\ K, 90\ K)_2$, protein tyrosine kinase	Various cell types
IGF-II	Monomer of 7 K homologous to proinsulin	250 K	Various cell types
Fibroblast growth factor family acidic FGF basic FGF	Monomers of about 17 K displaying amino acid sequence homology to each other and to the IL-1 family	160 K, protein tyrosine kinase	Endothelial mesenchymal cells, endocrine epithelial cells
Bombesin family GRP Neuromedin B and C Litorin	Monomers of short peptides with a homologous carboxy-terminal hepta-peptide	75–85 K	Fibroblasts, bronchial epithelial cells

Interleukin-1 family IL-1α IL-1β	Monomers of 15–30 K displaying amino acid sequence homology to each other and to the FGF family	80 K	Various cell types
IL-2	Monomer of 15 K	55 K and 70 K	T cells, B cells
IL-3 (multi colony-stimulating factor)	Monomer of glycosylated 15 K peptide	50–70 K	Haemopoietic cells
IL-4	Monomer of 20 K	60 K	B cells, T cells, mast cells
IL-5	Monomer of 13 K	?	B cells, eosinophilic granulocytes
IL-6	Monomer of 21 K	?	B cells, B cell hybridomas
G-CSF	Monomer of glycosylated 19 K peptide	150 K	Granulocyte lineage of haemopoietic cells
M-CSF (CSF-1)	Two variants due to differential splicing: homodimers of glycosylated 14 or 21 K chains	c-*fms*, 150 K, protein tyrosine kinase	Macrophage lineage of haemopoietic cells
GM-CSF	Monomer of glycosylated 14 K peptide	51 K	Granulocyte/macrophage lineage of haemopoietic cells
Erythropoietin	Monomer of glycosylated 18 K peptide	?	Erythroid lineage of haemopoietic cells
?	?	c-*kit*, 145 K, protein tyrosine kinase	?
?	?	c-*neu*, 185 K, protein tyrosine kinase	?
?	?	c-*met*, 145 K, protein tyrosine kinase	?
?	?	c-*trk*, protein tyrosine kinase	?
?	?	c-*ret*, protein tyrosine kinase	?
?	?	c-*ros*, protein tyrosine kinase	?
?	?	c-*eph*, protein tyrosine kinase	?

[a] Molecular sizes are given as determined from migrations in SDS-gel electrophoresis.

sitol turnover and their ability to stimulate growth, especially since PDGF also stimulates the growth of cells depleted of protein kinase C (COUGHLIN et al. 1985).

Somewhat later in the mitogenic pathway, 10 min to a few hours after addition of the growth factor, specific genes are induced (FERRARI and BASERGA 1987). Although the functions of the corresponding gene products are as yet unknown, they most likely perform important functions in the machinery that drives the cell cycle.

II. Subversion of the Mitogenic Pathway of Growth Factors in Cell Transformation

Recent studies on oncogenes and their products have linked their functional activities to the mitogenic pathway of growth factors (reviewed in HELDIN et al. 1987; WESTERMARK and HELDIN 1988). There are examples of transforming proteins that act as growth factors, others represent structurally altered versions of growth factor receptors, and yet others are activated components of the intracellular messenger system. Available data support the hypothesis that oncogene products act in cell transformation by subverting the normal mitogenic pathway, thereby giving the cell a constitutive growth stimulus leading to uncontrolled growth. The purpose of the present article is to review data on the role of perturbations of growth factors and growth factor receptors in cell transformation.

B. Autocrine Growth Factors in Cell Transformation

One of the first to suggest that tumour cell growth may be driven by endogeneously produced growth factors was TEMIN (1966). This idea was later developed into the autocrine concept of cell growth (reviewed in SPORN and TODARO 1980; SPORN and ROBERTS 1985). A prerequisite for autocrine growth stimulation is that the same cell both produces a growth factor and expresses the corresponding receptor. The autocrine concept has received considerable interest, and there are several examples of growth-factor-producing cells.

Only in a few cases, however, has it been possible to demonstrate that a cell is dependent on an autocrine factor for its growth.

I. Platelet-Derived Growth Factor

Platelet-derived growth factor (PDGF) is a major mitogen in the serum for connective tissue cells in vitro (for review, see HELDIN et al. 1985; ROSS et al. 1986). Structurally, it is a dimer composed of two homologous disulphide-bonded polypeptide chains, denoted A and B. All three possible dimeric forms of PDGF have been found: a major part of PDGF purified from human platelets is a heterodimer (HAMMACHER et al. 1988a); PDGF-AA has been purified from conditioned media of human osteosarcoma (HELDIN et al. 1986), melanoma (WESTERMARK et al. 1986) and glioma (HAMMACHER et al. 1988b) cell lines; the transforming product of simian sarcoma virus (SSV) (WATERFIELD et al. 1983; DOOLITTLE et al. 1983; ROBBINS et al. 1983) and PDGF purified from porcine platelets (STROOBANT and WATERFIELD 1984) have been identified as PDGF-BB.

Recent studies have indicated that the dimeric forms of PDGF have different functional activities. Thus, PDGF-AA was found to have a lower mitogenic activity compared with PDGF-AB (NISTÉR et al. 1988a). Furthermore, in contrast to PDGF-AB, PDGF-AA has no chemotactic activity and did not stimulate actin reorganisation and membrane ruffling of human fibroblasts (NISTÉR et al. 1988a). The functional differences are most likely due to the fact that the three dimers interact with different specificities to two distinct receptor species, denoted type A and type B, on the fibroblast cell surface (HELDIN et al. 1988; HART et al. 1988). The B type receptor is identical to the previously identified PDGF receptor with protein tyrosine kinase activity (reviewed in HELDIN and RÖNNSTRAND 1989), which has been molecularly cloned (YARDEN et al. 1987; CLAESSON-WELSH et al. 1988). All dimeric PDGF forms bind to the A type receptor, whereas the B type receptor binds only PDGF-BB and PDGF-AB, but the latter at lower affinity (Table 1).

The transforming potential of the B-chain gene of PDGF is illustrated by the fact that it has been found to be transduced by two separate isolates of acutely transforming retroviruses, SSV (DEVARE et al. 1983) and Parodi-Irgens feline sarcoma virus (BESMER et al. 1983). There is strong evidence to support the hypothesis that SSV transformation is exerted by an autocrine PDGF-like growth factor (reviewed in WESTERMARK et al. 1987). In short, SSV transforms only cell types that are responsive to PDGF, i.e. that have PDGF receptors (LEAL et al. 1985). Furthermore, SSV-transformed cells in vitro produce a growth factor that binds to PDGF receptors and is neutralised by PDGF antibodies (references in WESTERMARK et al. 1987), and SSV transformation can be inhibited by agents that prevent PDGF from binding to its receptor, e.g. PDGF antibodies (JOHNSSON et al. 1985) and suramin (BETSHOLTZ et al. 1986b). Moreover, the transforming activity of *sis* is lost following deletion of the hydrophobic leader sequence, which causes transfer into the endoplasmic reticulum, secretion and the possibility for the *sis*-product to interact with the ligand-binding domain of the PDGF receptor (HANNINK and DONOGHUE 1984; KING et al. 1985a).

Constructs using genomic (GAZIT et al. 1984) or cDNA (CLARKE et al. 1984; JOSEPHS et al. 1984) c-*sis* sequences placed under the control of viral promoters have been shown to cause cell transformation in transfection experiments. Similar experiments with constructs of the A-chain gene have revealed that it too has transforming activity, albeit less potent than that of the B-chain gene (BECKMANN et al. 1988; BYWATER et al. 1988). These findings raise the important question as to whether unscheduled expression of the c-*sis*/PDGF B-chain gene or PDGF A-chain gene is involved in the genesis of "spontaneous" human tumours. The expression of PDGF A- or B-chain mRNA in several human tumour cell lines, which presumably are derived from PDGF-responsive tissues, e.g. gliomas and sarcomas (BETSHOLTZ et al. 1986a), is compatible with such a possibility.

Human glioma cell lines have been shown to have PDGF-like activity (NISTÉR et al. 1984; PANTAZIS et al. 1985), and all three dimeric forms of PDGF were identified in the conditioned medium of one such cell line, PDGF-AA being the predominant species (HAMMACHER et al. 1988 b). PDGF production seems to be very common in human glioma cell lines, 23 and 17 of 23 investigated lines showing expression of A- and B-chain mRNA, respectively (NISTÉR et al. 1988 b).

The actual role of PDGF in the pathogenesis of human malignancies, however, remains to be elucidated. Addition of neutralising PDGF antibodies to PDGF-producing osteosarcoma (BETSHOLTZ et al. 1984) or glioma (NISTÉR et al., unpublished data) cell lines had no effect on their growth rate in serum-free medium, in contrast to the effect on SSV-transformed fibroblasts. These findings do not invalidate the possibility that autocrine growth stimulation by PDGF or other factors can be part of the pathogenesis of sarcomas and gliomas. It is possible that autocrine stimulation is an early event in tumour development and that subsequent changes in the cell genome abolish the requirement for the endogenously produced factors. Notably, PDGF antibodies have no, or only a marginal, effect on the growth rate of serially passaged, SSV-transformed, immortalised cell lines (HUANG et al. 1984), in contrast to the marked effect on acutely transformed human diploid fibroblasts (JOHNSSON et al. 1985).

There are several examples of tumour cell types that do not have PDGF receptors but still produce PDGF, e.g. mammary carcinoma cells (ROZENGURT et al. 1985; PEREZ et al. 1987; BRONZERT et al. 1987), melanoma cells (WESTERMARK et al. 1986) and lung carcinoma cells (BETSHOLTZ et al. 1987; SÖDERDAHL et al. 1988). In these cases it is highly unlikely that the endogenously produced PDGF has an autocrine function; it is possible, however, that it has a function in vivo in paracrine stimulation of connective tissue cell stroma.

II. Transforming Growth Factor α

TGF-α is synthesised as a membrane-bound precursor (BRINGMAN et al. 1987; TEIXIDÓ et al. 1987; GENTRY et al. 1987), a property which is shared by the other known members of this family of growth factors, epidermal growth factor (EGF) (GRAY et al. 1983; SCOTT et al. 1983) and the vaccinia virus growth factor (STROOBANT et al. 1985; TWARDZIK et al. 1985); these active molecules show about 40% amino acid similarity and bind to a common receptor (CARPENTER 1987). TGF-α is synthesised by cells transformed by certain retroviruses; it is not the product of any oncogene per se, but its synthesis is regulated by a number of different oncogenes, e.g. *abl*, *mos*, *fes*, *fms*, and *ras* (DE LARCO and TODARO 1978; TODARO et al. 1979; OZANNE et al. 1980; TWARDZIK et al. 1982, 1983; MARQUARDT et al. 1983). TGF-α has also been found to be produced by cells transformed by DNA tumour viruses (OZANNE et al. 1980; KAPLAN and OZANNE 1982) and by cell lines established from human tumours (TODARO et al. 1980; DERYNCK et al. 1984, 1987; NISTÉR et al. 1988 b). In contrast, expression of EGF in tumour cell lines is less common; one example found, a human salivary gland adenocarcinoma cell line, may represent a normal function of this cell type (SATO et al. 1985).

The designation "transforming" growth factor-α is somewhat of a misnomer and has been given because of its ability to stimulate growth in soft agar of certain cells in synergy with TGF-β (ANZANO et al. 1983), an effect which TGF-α shares with other growth factors.

The question whether TGF-α production serves an autocrine function in cell transformation was addressed by TODARO et al. (1979) and KAPLAN et al. (1982) using temperature-sensitive mutants of Moloney and Kirsten sarcoma viruses. At a permissive temperature, when TGF-α production occurs, the cells have a low serum requirement, down-regulated EGF receptors and do not respond to exogenous EGF, whereas, at non-permissive temperatures, the cells have EGF receptors and respond to EGF. These findings indicate that oncogene expression regulates the production of functionally active TGF-α but cannot be taken as a proof that autocrine stimulation by TGF-α is important in cell transformation. Rather, data have been presented indicating that *ras* transformation occurs equally well in cells that are devoid of EGF receptors and thereby are not responsive to TGF-α (MCKAY et al. 1986).

Artificial constructs containing gene sequences of TGF-α or EGF under the control of viral promoters have been shown to have transforming effects on cells in vitro and in vivo; the transforming effects were, however, considerably weaker than, for example, corresponding constructs with the *sis* oncogene/PDGF B chain (ROSENTHAL et al. 1986; WATANABE et al. 1987; STERN et al. 1987; FINZI et al. 1987). The frequent expression of TGF-α in EGF receptor-containing cell lines of human tumours (DERYNCK et al. 1987; SMITH et al. 1987; NISTÉR et al. 1988 b) suggests that autocrine stimulation of cell growth involving TGF-α may operate in the development of human neoplasia. This possibility is further supported by the fact that factors interacting with the EGF receptor have been identified in the urine of cancer patients (SHERWIN et al. 1983).

III. Fibroblast Growth Factor and Related Factors

Acidic and basic fibroblast growth factors (FGFs) are two homologous endothelial cell mitogens that bind to the same receptor (reviewed in GOSPODAROWICZ et al. 1986). The two FGFs also share the property of having a strong affinity for heparin, which has simplified their identification, characterisation and purification from various sources. Since FGFs have been shown to stimulate the formation of new vessels in vivo, factors in this family have attracted considerable interest as potential tumour-derived angiogenic factors (FOLKMAN and KLAGSBRUN 1987).

Acidic and basic FGFs were originally purified from neuronal tissue but have subsequently been found in a variety of normal (MOSCATELLI et al. 1986; VLODAVSKY et al. 1987; FERRARA et al. 1987; WINKLES et al. 1987) and transformed (MOSCATELLI et al. 1986; SCHWEIGERER et al. 1987; LIBERMANN et al. 1987; KLAGSBRUN et al. 1986) cells.

Several of the malignant producer cell types also respond to FGF, e.g. human glioma (LIBERMANN et al. 1987) and rhabdomyosarcoma (SCHWEIGERER et al. 1987); FGF may thus potentially have the dual function of both stimulating the producer cells by an autocrine mechanism and inducing neovascularisation of the tumour by a paracrine route. Both these mechanisms, however, require that FGF is secreted from the producer cell; how that can occur remains to be elucidated since cloning of cDNAs for acidic (JAYE et al. 1986) and basic (ABRAHAM et al. 1986) FGFs revealed no signal sequences. That the gene for basic FGF is indeed a potential oncogene was recently demonstrated (ROGELJ et al. 1988). Transfection of NIH 3T3 cells with bFGF cDNA linked to an immunoglobulin signal sequence resulted in focus formation in vitro and tumour formation in nude mice, whereas only a slight change in growth behaviour was recorded when a construct with the bFGF cDNA alone was used for transfection.

Two oncogenes were recently found to be related to FGF; the products of *hst* and *int*-2 show about 40% amino acid similarity to each other and to acidic or basic FGF. The *hst* oncogene was isolated from a human gastric carcinoma (SAKAMOTO et al. 1986; TAIRA et al. 1987) but has also been isolated from Kaposi's sarcoma (DELLI BOVI et al. 1987), whereas *int*-2 is implicated in murine breast cancer induced by mouse mammary tumor virus (NUSSE 1986; DICKSON and PETERS 1987; MOORE et al. 1986). The *hst* and *int*-2 products have, in contrast to FGFs, hydrophobic signal sequences in the N-termini, which indicates that they are secretory products. The product of the *hst* gene, which was isolated from a Kaposi's sarcoma, hat growth-factor activity (DELLI BOVI et al. 1987), and the *int*-2 product may also be a growth factor. The mechanism for the transforming activities of these oncogenes is most likely analogous to that of the *sis* oncogene, i.e. autocrine growth stimulation exerted by their products.

IV. Bombesin-like Peptides

There are several human peptides which share an identical C-terminal heptapeptide with bombesin and interact with a common receptor, thereby inducing a mitogenic effect in certain cell types (ZACCHARY and ROZENGURT 1985). An involvement of bombesin-like peptides as autocrine growth factors in human small cell lung carcinoma was recently suggested (CUTTITTA et al. 1985). A monoclonal antibody against the C-terminus of bombesin was found to inhibit growth of small cell lung carcinoma cells in vivo. Remarkably, several of the xenografted animals developed progressively growing tumours in the presence of antibodies (CUTTITTA et al. 1985), which may reflect a progression into a non-autocrine malignant phenotype.

V. Interleukins and Colony-Stimulating Factors

Growth factors acting on cells in the lymphoid system and bone marrow have been designated interleukins (IL) and colony-stimulating factors (CSF). The number of factors in this group increases continuously; at present ten factors have been cloned (IL-1 to IL-6, M-CSF, G-CSF, GM-CSF and erythropoietin; IL-3 is also called multi-CSF). There is, as yet, no example of an oncogene that encodes an interleukin or a colony-stimulating factor, but it is becoming increasingly clear that aberrant expression of such factors plays a role in the genesis of leukaemias and lymphomas.

IL-1 is derived from macrophages and mediates a wide variety of immunological and inflammatory responses (DURUM et al. 1985); two related forms exist, IL-1α and IL-1β, that bind to the same receptor (MARCH et al. 1985). Recently IL-1 has been found to synergise with CSFs to stimulate the proliferation of immature haemopoietic cells (MOCHIZUKI et al. 1987).

IL-2 stimulates the proliferation of T cells (Cantrell and Smith 1984; Taniguchi et al. 1986) and B cells (Mingari et al. 1984). A transient co-expression of IL-2 and its receptor occurs after antigen stimulation of T cells. Exogenously added IL-2 promotes long-term growth of normal T cells in vitro, as well as of T cells transformed by human T-cell leukaemia type I virus (HTLV-I) (Morgan et al. 1976; Gootenberg et al. 1981). HTLV-I-transformed cell lines may progress to IL-2 independence upon long-term culture, most likely as a function of an autocrine production of IL-2 (Gootenberg et al. 1981). Non-autocrine mechanisms of growth factor independence must also exist, however, since some IL-2-independent HTLV-I-transformed T-cell lines lack IL-2 production (Arya et al. 1984). Autocrine mechanisms were directly demonstrated in the human T-cell lymphoma cell line IARC 301 (Duprez et al. 1985) and in T-cell leukaemia cells in primary culture (Arima et al. 1986), since antibodies against IL-2 or its receptor retarded the growth of the cells.

IL-4, previously called B-cell stimulatory factor-1 (Lee et al. 1986; Noma et al. 1986), acts on B cells at several phases of their differentiation and also on mast cells and T cells (Ohara and Paul 1987). Antigen stimulation of certain T helper cell lines in the presence of adherent cells leads to release of IL-4, which apparently serves an autocrine function, since addition of a monoclonal antibody against IL-4 inhibits cell growth (Fernandez-Botran et al. 1986). Also, a majority of transformed mast cell lines and all IL-3-dependent non-transformed mast cells investigated were found to express IL-4 mRNA (Brown et al. 1987), suggesting an autocrine mechanism of cell growth.

IL-5 is also known as T-cell replacing factor, B-cell growth factor II and eosinophil differentiation factor (Kinashi et al. 1986; Campbell et al. 1987); as is indicated by the various designations, it has a multifunctional role in the stimulation of growth and maturation of B cells and eosinophilic granulocytes.

Another multifunctional molecule is IL-6, which is also known as interferon-β_2 (Haegeman et al. 1986; Zilberstein et al. 1986; May et al. 1986), B-cell differentiation factor (BCDF or BSF-2) (Hirano et al. 1986) and hybridoma growth factor (van Snick et al. 1986). Interestingly, this molecule acts both as a growth factor, e.g. for hybridoma cells (van Snick et al. 1986), and as a growth inhibitor for fibroblasts (Kohase et al. 1986). Human myeloma cells in primary culture have been found to produce IL-6 and express its receptors (Kawano et al. 1988). An autocrine effect of IL-6 in these cells was suggested by a growth-inhibitory effect of IL-6 antibodies.

CSFs regulate normal haematopoiesis and have been identified by their ability to support growth of bone marrow cells in semi-solid media (Metcalf 1986; Clark and Kamen 1987).

G-CSF and M-CSF stimulate the proliferation only of relatively late progenitors committed to the granulocyte and macrophage lineages, respectively. GM-CSF interact, in addition, with more immature progenitor cells, and IL-3 (multi-CSF) has the ability to support the development of cells from relatively early pluripotent progenitors to mature cells of many different lineages (references in Clark and Kamen 1987). Erythropoietin stimulates growth and differentiation of cells in the erythrocyte lineage of haemopoietic cells (Jacobs et al. 1985) (Table 1).

Studies on myeloid leukaemia cells in culture have shown that they are, in general, as dependent on the addition of exogenous CSFs as are normal myeloid cells, suggesting that autocrine stimulation by CSFs is rarely involved in the pathogenesis of "spontaneous" leukaemia (METCALF 1985; GOUGH 1986). There are, however, exceptions: GM-CSF induces an autocrine response in acute myeloblastic leukaemia (YOUNG and GRIFFIN 1986).

Non-autocrine, non-tumorigenic myleoid cells in culture may be converted to an autocrine tumorigenic state by spontaneous mutation (SCHRADER and CRAPPER 1983) or experimentally by introducing suitable CSF cDNA expression vectors into immortalised cell lines (LANG et al. 1985; LAKER et al. 1987). The development of a growth-factor-independent state probably occurs in two steps, the first through a true autocrine stimulation (which can be blocked by antibodies), and the second through secondary mutational events which cause abolition of growth factor dependence by a non-autocrine mechanism (COOK et al. 1985; PIERCE et al. 1985; LAKER et al. 1987).

VI. Mechanism of Activation of Growth Factor Expression in Cell Transformation

There are several examples of normal cells that produce autocrine factors, indicating that autocrine stimulation is not just connected with cell transformation. Presumably, autocrine loops in normal cells are tightly controlled and occur only transiently, as a response to external stimuli or in a specific phase of development. Examples include the production of PDGF by vascular smooth muscle cells (SEIFERT et al. 1984; NILSSON et al. 1985b; SEJERSEN et al. 1986), placental cytotrophoblasts (GOUSTIN et al. 1985) or mitogen-stimulated fibroblasts (PAULSSON et al. 1987). The finding of an autoinduction of PDGF in fibroblasts is interesting and may suggest the presence of a positive autocrine feedback loop in growth stimulation. Analogously, TGF-α has recently been found to induce TGF-α production in keratinocytes (COFFEY et al. 1987). Like PDGF, TGF-α may also have a function during development; it has been found to be expressed in embryonic tissue (LEE et al. 1985; TWARDZIK 1985). Additional examples of autocrine stimulation of normal cells are found in the immune system, in which autocrine loops are involved in the expansion of antigen-stimulated lymphocytes.

It is possible that the establishment of an autocrine loop that drives cell division is an important step in the development of the malignant phenotype. Given that autocrine loops also occur under certain conditions in normal cells, it is, on the other hand, possible that it reflects a normal feature of a cell at a certain developmental stage and thus is a result of transformation rather than a cause of it.

In only a few cases have the mechanisms behind unscheduled growth factor synthesis been elucidated. The PDGF B-chain gene has been transduced by two retroviruses (DEVARE et al. 1983; BESMER et al. 1983) as the v-*sis* oncogene. In addition, there are some examples of nearby insertion of retroviral DNA – the constitutive expression of the IL-3 gene in WEHI-3B leukaemia cells (YMER et al. 1985), the activation of the IL-2 gene in a T-cell lymphoma cell line (CHEN et al. 1985) and the activation of the putative growth-factor gene *int*-2 in mammary carcinomas (NUSSE 1986). *int*-1 is another gene activated by insertion of proviral DNA and implicated in the development of mammary carcinoma. Since the structure of *int*-1 indicates that the protein product is secreted, it is possible that the *int*-1 product acts as an inter-cellular signal substance (VAN OOYEN and NUSSE 1984). Interestingly, it was recently found that *int*-1 corresponds to the developmental gene *wingless* in *Drosophila* (RIJSEWIK et al. 1987), indicating an important role of this gene product in development. In all these cases, the normal regulatory functions are lost, and growth factor gene expression is driven by *cis*-acting viral elements. *Trans*-activating mechanisms are also likely to occur, e.g. in the expression of TGF-α in cells transformed by *abl, mos, fes, fms* and *ras* (see above) or myelomonocytic growth factor in cells transformed by v-*src* or v-*mil* (ADKINS et al. 1984; GRAF et al. 1986; VON WEIZÄCKER et al. 1986).

C. Perturbations of Growth Factor Receptors in Cell Transformation

Growth factor receptors are transmembrane proteins with an external ligand-binding domain and an internal effector domain. cDNAs for several growth factor receptors have been cloned; with few exceptions their sequences contain regions with homology to protein tyrosine kinases (ULLRICH et al. 1984, 1985, 1986; YARDEN et a. 1986; COUSSENS et al. 1986) (Table 1). Tyrosine phosphorylation thus seems to be of major importance in growth stimulation. There are, however, examples of growth factor receptors without protein tyrosine kinase activity. For example, the IGF-II receptor which has a large extracellular part made up of fifteen repeat sequences and a small region homologous to the collagen binding domain of fibronectin; its intracellular domain is short, only 164 residues, and without homology to kinases (MORGAN et al. 1987). Similarly, the intracellular part of the IL-2 receptor is short, only 13 residues, and without homology to kinases (LEONARD et al. 1983); it is clear, however, that the IL-2 receptor contains additional components (SHARON et al. 1986) that have not yet been characterised. The receptor for substance K, a neuropeptide with a mitogenic effect in certain cell systems (NILSSON et al. 1985a), and muscarinic, cholinergic and adrenergic receptors form a family of receptors in which the peptide traverses the cell membrane seven times and may form a membrane channel (reviewed in SIBLEY et al. 1987); these receptors have no tyrosine kinase activity, but there are indications that they interact with coupling factors (G-proteins).

It should be emphasised, however, that there are examples of protein tyrosine kinases which are not transmembrane growth factor receptors, e.g. the products of *abl, fes/fps,* and the *src* family (reviewed in HUNTER 1987). For instance, the *src* family consists of a group of homologous proteins (*src, yes, fyn, hck, lyn, fyn, lck*) which show different tissue distributions. Since the *src* product in normal cells is mainly found in non-proliferating cell types, e.g. platelets (GOLDEN et al. 1986) and neurons (COTTON and BRUGGE 1983), it is possible that the normal functions of the members of the *src* family are related to cell differentiation rather than to cell proliferation.

Some oncogene products are clearly mutated versions of growth factor receptors, like *erbB* and *fms,* which correspond to the EGF receptor and M-CSF receptor, respectively. Others, like *neu, trk, ros, met, kit, ret,* and *eph* are most likely also derived from growth factor receptors, since they are transmembrane protein tyrosine kinases with receptor-like structures (Table 1). The mechanism of perturbations of these components in cell transformation is reviewed below.

I. Epidermal Growth Factor Receptor

The EGF receptor is one of the most well-characterised growth factor receptors (reviewed in CARPENTER 1987) and the first to be cloned (ULLRICH et al. 1984). The receptor consists of an external ligand-binding domain comprising 621 amino acids with two cysteine-rich areas, one transmembrane stretch of hydrophobic amino acids and an internal effector domain of 542 amino acids. The internal domain contains a region of homology to other tyrosine kinases, as well as a C-terminal tail with the three major autophosphorylation sites (DOWNWARD et al. 1984a).

There are several exmples in which the EGF receptor, or mutated versions of it, is involved in cell transformation. Firstly, v-*erbB,* one of two oncogenes of avian erythroblastosis virus, corresponds to a truncated EGF receptor (DOWNWARD et al. 1984b). The v-*erbB* product, which transforms erythroblasts and fibroblasts, has lost the ligand binding part of the EGF receptor, as well as 32 amino acids in the C-terminus comprising the major autophosphorylation site (YAMAMOTO et al. 1983; ULLRICH et al. 1984). The importance of the loss of the autophosphorylation site is illustrated by the fact that a chimaera of an intact extracellular domain of the EGF receptor fused with the intracellular part of the v-*erbB* product retains transforming potential (RIEDEL et al. 1987). The v-*erbB* product has a constitutively active protein tyrosine kinase (KRIS et al. 1985; GILMORE et al. 1985), which may be involved in a continuous and uncontrolled stimulation of cell growth.

Secondly, the EGF receptor gene can be activated by proviral insertion. In a large number of cases of avian leukosis virus-induced erythroblastosis that have been examined, most of the proviral integration sites were clustered within a few hundred bases in the middle of the EGF receptor gene, leading to the expression of a truncated receptor of a size similar to that of the v-*erbB* product (RAINES et al. 1985; NILSEN et al. 1985; MILES and ROBINSON 1986). These viruses were inefficient in transforming fibroblasts (BEUG et al. 1986). As they seem to contain intact C-termini (NILSEN et al. 1985; LAX et al. 1985), it is possible that a C-terminal truncation is important for fibroblast transformation but not required for transformation of erythroblasts.

Thirdly, there are several examples of EGF receptor amplification in human tumours. Thus, about 30% of human glioblastomas show over-expression of EGF receptors and amplification of the EGF receptor gene (LIBERMANN et al. 1985; BIGNER et al. 1987; WONG et al. 1987). No amplification was found in astrocytomas of higher differentiation grade. EGF receptor over-expression was also found in certain carcinomas (COWLEY et al. 1984; HENDLER and OZANNE 1984; YAMAMOTO et al. 1986b; GULLICK et al. 1986) and sarcomas (GUSTERSON et al. 1986).

II. Macrophage Colony-Stimulating Factor Receptor

M-CSF (CSF-1) stimulates growth and maturation of the macrophage lineage of haematopoietic stem cells (STANLEY et al. 1983) and may also have a role in placental development (POLLARD et al. 1987).

The M-CSF receptor is a transmembrane protein with a ligand-activated protein tyrosine kinase. However, it differs slightly from the EGF receptor in its overall organisation (COUSSENS et al. 1986). The external part lacks clusters of cysteine residues; rather, it contains ten evenly distributed cysteine residues, the spacing of which predicts that the external part contains five immunoglobulin-like domains. In addition, the tyrosine kinase domain is split into two parts by a stretch of amino acids without homology to other kinases. These features are also found in the PDGF receptor (YARDEN et al. 1986; CLAESSON-WELSH et al. 1988) and the c-*kit* product (YARDEN et al. 1987).

By analogy with the EGF receptor, a mutated version of the M-CSF receptor has been transduced by a retrovirus (SHERR et al. 1985). The corresponding oncogene has been designated v-*fms*. Like the v-*erbB* product, it has lost its C-terminal tail, comprising a potential regulatory autophosphorylation site (VERBEEK et al. 1985; COUSSENS et al. 1986; WHEELER et al. 1986a). Replacement of the C-terminal end of v-*fms* with normal c-*fms* sequences reduces the transforming potential (BROWNING et al. 1986), whereas mutation of Tyr_{969} in c-*fms* to a phenylalanine residue increases the transforming activity (ROUSSEL et al. 1987). Taken together, this indicates that autophosphorylation at Tyr_{969} is involved in the regulation of the M-CSF receptor and that loss of this residue is part of the activation of the receptor gene to an oncogene.

Proviral integration in the c-*fms* gene (at the *fms*-2 site) was found to cause a high expression of a normal sized c-*fms* mRNA; this event has been found to be linked to the development of murine myeloblastic leukaemias (GISSELBRECHT et al. 1987).

In contrast to the v-*erbB* product, the v-*fms* product contains an intact ligand-binding domain. In spite of this, the tyrosine kinase of the v-*fms* product seems to be active in the absence of ligand (SACCA et al. 1986; WHEELER et al. 1986b).

III. Receptors with Unknown Ligands (c-*neu*, c-*met*, c-*ros*, c-*kit*, c-*trk*, c-*ret*, c-*eph*)

The proto-oncogenes described in this chapter have all been identified by the transforming capacity of mutated versions of them. Their products are all transmembrane proteins with protein tyrosine kinase activities. In addition, they also share other structural features with known growth factor receptors. It is thus likely that they are growth factor receptors; however, their ligands remain to be identified.

The c-*neu* (c-*erbB*-2) product is a 185 K protein structurally related to the EGF receptor (COUSSENS et al. 1985; BARGMANN et al. 1986a; YAMAMOTO et al. 1986a). Interestingly, c-*neu* was converted to an oncogene by a chemically induced point mutation in the transmembrane domain, changing a valine residue to a glutamic acid residue (BARGMANN et al. 1986b). This implies that the transmembrane region has a critical role in the transduction of the mitogenic signal; it is possible that the introduction of a charged residue in this domain locks it in a constitutively activated form.

Another mechanism for activation of c-*neu* involves amplification of the gene. As many as 20% of mammary carcinoma cell lines show amplification of c-*neu* (KING et al. 1985b; SLAMON et al. 1986; KRAUS et al. 1987; VAN DE VIJVER et al. 1987; ZHOU et al. 1987). Amplification of c-*neu* has also been found in a salivary gland adenocarcinoma (SEMBA et al. 1985) and in a gastric cancer cell line (FUKUSHIGE et al. 1986). Over-expression of the normal c-*neu* product in NIH 3T3 cells has been shown to cause a transformed cellular phenotype (HUDZIAK et al. 1987).

The nucleotide sequence of c-*met* predicts a product with a large extracellular domain of 926 amino acid residues with potential ligand-binding properties and an intracellular protein tyrosine kinase domain which shows 44% amino acid similarity with the insulin receptor (PARK et al. 1987). The oncogene *met* was identified in human osteosarcoma cells treated with the carcinogen methylnitrosourea (COOPER et al. 1984). Activation in this cell line involved gene rearrangement such that the 5′ end of the transforming gene comprises gene sequences derived from chromosome 1 fused to the 3′ end of a truncated c-*met* gene (PARK et al. 1986). Amplification of c-*met* may also cause cell transformation; a high proportion of "spontaneously" occurring foci were found in DNA transfection experiments using NIH 3T3 cells as target cells to contain a four- to eightfold amplification of c-*met* and at least a 20-fold over-expression of an apparently normal c-*met* transcript (COOPER et al. 1986). No cases of rearranged or amplified c-*met* in spontaneously occurring tumours have so far been found.

c-*ros* is expressed at high levels in the kidney (SHIBUYA et al. 1982; NECKAMEYER et al. 1986). A truncated version of c-*ros* was transduced by avian sarcomavirus UR2 (BALDUZZI et al. 1981).

c-*kit* encodes a protein similar to the receptors for PDGF and M-CSF (YARDEN et al. 1987); an N- and C-terminally truncated version was transduced by a feline sarcomavirus (BESMER et al. 1986).

The oncogene *trk* was isolated from a human colon carcinoma, as a hybrid between a protein tyrosine kinase with a transmembrane domain and a non-muscle tropomyosin molecule (MARTIN-ZANCA et al. 1986). It is conceivable that the extracellular ligand-binding domain of a receptor of an unidentified growth factor has been replaced in this case by the tropomyosin molecule by somatic rearrangement. Analysis of a transforming gene sequence from a human breast carcinoma revealed that it analogously consisted of a truncated *trk* gene fused in its 5′ end with another gene which had not previously been characterised (KOZMA et al. 1988).

Activation of the *ret* oncogene involved a recombination of two unlinked sequences of human DNA, which most likely occurred during transfection of NIH 3T3 cells (TAKAHASHI and COOPER 1987). Similar to the case with *trk*, a truncated protein tyrosine kinase was fused at its N-terminal end with an unrelated sequence, presumably causing a constitutively activated kinase.

The putative protein tyrosine kinase receptor c-*eph* was identified by low-stringency hybridisation using a v-*fps* probe; its sequence predicts a receptor with a cysteine-rich domain in the extracellular part and a protein tyrosine kinase domain in the intracellular part (HIRAI et al. 1987). The *eph* gene is over-expressed in several human carcinomas, suggesting that it may be involved in neoplastic transformation (HIRAI et al. 1987).

IV. Mechanism for Perturbations at the Growth Factor Receptor Level in Cell Transformation

The production of a growth factor in a cell that carries the cognate receptor will cause an autocrine loop which drives cell multiplication (see Sect. B). In theory, the expression of a growth factor receptor in a cell that has, as part of its normal function, to produce the corresponding growth factor, would similarly result in autocrine stimulation of growth. No such examples have yet been found, but, as exemplified above, there are several other examples of perturbations of growth factor receptors.

Amplification of apparently normal growth factor receptor genes have been connected with malignancies in the cases of the EGF receptors c-*erbB* and c-*neu*. Under experimental conditions, amplification of c-*met* has also been observed. Over-expression of a seemingly normal c-*fms* mRNA was also seen after proviral insertion (GISSELBRECHT et al. 1987). It is not clear how over-expression of a normal growth factor receptor can contribute to excessive cell growth. It is possible that each receptor has a low background activity in the absence of ligand; an excessive number of receptors could, in such a case, cause a constitutive growth stimulation. In addition, a cell with an excess of growth factor receptors would be more sensitive to minute amounts of growth factor in the environment or to endogenously produced growth factor.

Truncation is commonly associated with activation of growth factor receptor genes, involving loss of the ligand-binding domain (v-*erbB*, v-*kit*, *trk*, *ret*) or loss of a C-terminal regulatory autophosphorylation site (v-*erbB*, v-*fms*). There are several mechanisms for truncation, e.g. retroviral transduction (v-*erbB*, v-*kit*, v-*fms*, v-*ros*), insertional mutagenesis (*erbB*) and somatic rearrangements resulting in chimaeric protein products (*met*, *trk*, *ret*). In the case of *trk*, the N-terminal fusion to tropomyosin results in a cytoplasmic localisation of the hybrid molecule, since tropomyosin lacks a signal sequence; the altered cellular localisation may be of importance in cell transformation (Barbacid, personal communication).

The examples given above illustrate that truncation in the N-terminus (the ligand-binding domain) or the C-terminus (with removal or regulatory autophosphorylation sites) may cause constitutive activation of a growth factor receptor. There is, however, also one example of a perturbation in the communication between the extracellular ligand-binding domain and the intracellular effector domain caused activation, i.e. the point mutation in the transmembrane part of c-*neu*.

D. Subversion of Growth Regulatory Pathways in Multistep Carcinogenesis

The fully malignant phenotype includes, in addition to uncontrolled growth, also invasive growth, capacity to form metastases and immortalisation. Subversion of growth regulatory pathways by unscheduled production of growth factors, by perturbation of growth factor receptors or by activation of regulatory components in the intracellular messenger system may account for uncontrolled growth and may possibly also affect invasiveness and the ability to form metastases but can hardly cause immortalisation (cf. JOHNSSON et al. 1986). In terminally differentiating systems, as in haematopoieses, subversion of growth-regulatory pathways must be complemented by a differentiation block in order to cause malignancy. It is notable, for instance, that efficient v-*erbB* transformation of erythrocytes requires the cooperation of v-*erbA*, the role of which is probably to block differentiation (KAHN et al. 1986). Clearly, the development of a fully malignant phenotype is a multihit event, in which the subversion of growth regulatory pathways may be an important and perhaps necessary, but not sufficient, part.

Acknowledgements. We thank Linda Baltell for invaluable secretarial assistance and Lena Claesson-Welsh for critically reading the manuscript.

References

Abraham JA, Whang JL, Tumolo A, Mergia A, Friedman J, Gospodarowicz D, Fiddes JC (1986) Human basic fibroblast growth factor: nucleotide sequence and genomic organization. EMBO J 5:2523–2528

Adkins B, Leutz A, Graf T (1984) Autocrine growth induced by *src*-related oncogenes in transformed chicken myeloid cells. Cell 39:439–445

Anzano MA, Roberts AB, Smith JM, Sporn MB, De Larco JE (1983) Sarcoma growth factor from conditioned medium of virally transformed cells is composed of both type α and type β transforming growth factors. Proc Natl Acad Sci USA 80:6264–6268

Arima N, Daitoku Y, Oghaki S, Fukumori J, Tanaka H, Yamamoto Y, Fujimoto K, Onoue K (1986) Autocrine growth of interleukin 2-producing leukemic cells in a patient with adult T cell leukemia. Blood 68:779–782

Arya SL, Wong-Staal F, Gallo RC (1984) T-cell growth factor gene: lack of expression in human T-cell leukemia-lymphoma virus-infected cells. Science 223:1086–1087

Balduzzi PC, Notter MDF, Morgan HR, Shibuya M (1981) Some biological properties of two new avian sarcoma viruses. J Virol 40:268–275

Bargmann CI, Hung M-C, Weinberg RA (1986a) The *neu* oncogene encodes an epidermal growth factor receptor related protein. Nature 319:226–230

Bargmann CI, Hung M-C, Weinberg RA (1986b) Multiple independent activations of the *neu* oncogene by a point mutation altering the transmembrane domain of p185. Cell 45:649–657

Beckmann MP, Betsholtz C, Heldin C-H, Westermark B, Di Marco E, Di Fiore PP, Robbins KC, Aaronson SA (1988) Human PDGF-A and -B chains differ in their biological properties and transforming potential. Science 241:1346–1349

Berridge MJ, Irvine RF (1984) Inositol trisphosphate, a novel second messenger in intracellular signal transduction. Nature 312:315–321

Besmer P, Snyder HW Jr, Murphy JE, Hardy WD Jr, Parodi A (1983) The Parodi-Irgens feline sarcoma virus and simian sarcoma virus have homologous oncogenes but in different contexts of the viral genomes. J Virol 46:606–613

Besmer P, Murphy JE, George PC, Qui F, Bergold PJ, Lederman L, Synder HW Jr, Brodeur D, Zuckerman EE, Hardy WD (1986) A new acute transforming feline retrovirus and relationship of its oncogene v-*kit* with the protein kinase gene family. Nature 320:415–421

Betsholtz C, Westermark B, Ek B, Heldin C-H (1984) Coexpression of a PDGF-like growth factor and PDGF receptors in a human osteosarcoma cell line. Implications for autocrine receptor activation. Cell 39:447–457

Betsholtz C, Johnsson A, Heldin C-H, Westermark B, Lind P, Urdea MS, Eddy R, Shows TB, Philpott K, Mellor AL, Knott TJ, Scott J (1986a) cDNA sequence and chromosomal localization of human platelet-derived growth factor A-chain and its expression in tumor cell lines. Nature 320:695–699

Betsholtz C, Johnsson A, Heldin C-H, Westermark B (1986b) Efficient reversion of SSV-transformation and inhibition of growth factor-induced mitogenesis by suramin. Proc Natl Acad Sci USA 83:6440–6444

Betsholtz C, Bergh J, Bywater-Ekegärd M, Pettersson M, Johnsson A, Heldin C-H, Ohlsson R, Scott J, Bell G, Westermark B (1987) PDGF production in human lung cancer cells suggestes a role in tumor stroma formation. Int J Cancer 39:502–507

Beug H, Hayman MJ, Raines MB, Kung HJ, Vennström B (1986) Rous-associated virus 1-induced erythroleukemic cells exhibit a weakly transformed phenotype in vitro and release c-*erbB*-containing retroviruses unable to transform fibroblasts. J Virol 57:1127–1138

Bigner SH, Wong AJ, Mark J, Kinzler KW, Vogelstein B, Bigner DD (1987) Relationships between gene amplification and chromosomal deviations in malignant human gliomas. Cancer Genet Cytogenet 29:165–170

Bringman TS, Lindquist PB, Derynck R (1987) Different transforming growth factor-α species are derived from a glycosylated and palmitoylated transmembrane precursor. Cell 48:429–440

Bronzert DA, Pantazis P, Antoniades HN, Kasid A, Davidson N, Dickson RB, Lippman ME (1987) Synthesis and secretion of platelet-derived growth factor by human breast cancer cell lines. Proc Natl Acad Sci USA 84:5763–5767

Brown MA, Pierce JH, Watson CJ, Falco J, Ihle JN, Paul WE (1987) B cell stimulatory factor-1/interleukin-4 mRNA is expressed by normal and transformed mast cells. Cell 50:809–818

Browning PJ, Bunn HF, Cline A, Shuman M, Nienhius AW (1986) "Replacement" of COOH-terminal truncation of v-*fms* with c-*fms* sequences markedly reduces transformation potential. Proc Natl Acad Sci USA 83:7800–7804

Bywater M, Rorsman F, Bongcam-Rudloff E, Mark G, Hammacher A, Heldin C-H, Westermark B, Betsholtz C (1988) Expression of recombinant PDGF A- and B-chain homodimers in rat-1 cells and human fibroblasts reveals differences in protein processing and autocrine effects. Mol Cell Biol 8:2753–2762

Campbell HD, Tucker WQJ, Hort Y, Martinson ME, Mayo G, Clutterbuck EJ, Sanderson CJ, Young IG (1987) Molecular cloning, nucleotide sequence, and expression of the gene encoding human eosinophil differentiation factor (interleukin 5). Proc Natl Acad Sci USA 84:6629–6633

Cantrell DA, Smith KA (1984) The interleukin-2 T-cell system: a new cell growth model. Science 224:1312–1316

Carpenter G (1987) Receptors for epidermal growth factor and other polypeptide mitogens. Ann Rev Biochem 56:881–914

Chen SJ, Holbrook NJ, Mitchell KF, Vallone CA, Freegard JS, Crabtree GR, Lin Y (1985) A viral long terminal repeat in the interleukin 2 gene of a cell line that constitutively produces interleukin 2. Proc Natl Acad Sci USA 82:7284–7288

Claesson-Welsh L, Eriksson A, Morén A, Severinsson L, Ek B, Östman A, Betsholtz C, Heldin C-H (1988) cDNA cloning and expression of a human PDGF receptor specific for B chain containing molecules. Mol Cell Biol 8:3476–3486

Clark SC, Kamen R (1987) The human hematopoietic colony-stimulating factors. Science 236:1229–1237

Clarke MF, Westin E, Schmidt D, Josephs SF, Ratner L, Wong-Staal F, Gallo RC, Reitz MS (1984) Transformation of NIH 3T3 cells by a human c-*sis* cDNA clone. Nature 308:464–467

Coffey RJ Jr, Derynck R, Wilcox JN, Bringman TS, Goustin AS, Moses HL, Pittelkow MR (1987) Production and auto-induction of transforming growth factor-α in human keratinocytes. Nature 328:817–820

Cook WD, Metcalf D, Nicola NA, Burgess AW, Walker F (1985) Malignant transformation of a growth factor-dependent myeloid cell line by Abelson virus without evidence of an autocrine mechanism. Cell 41:677–683

Cooper CS, Park M, Blair DG, Tainsky MA, Huebner K, Croce CM, Vande Woude GF (1984) Molecular cloning of a new transforming gene from a chemically transformed human cell line. Nature 311:29–33
Cooper CS, Tempest PR, Beckmann MP, Heldin C-H, Brookes P (1986) Amplification and overexpression of the *met* gene in spontaneously transformed NIH 3T3 mouse fibroblasts. EMBO J 5:2623–2628
Cotton PC, Brugge JS (1983) Neural tissues express high levels of the cellular *src* gene product, $pp60^{c\text{-}src}$. Mol Cell Biol 3:1157–1162
Coughlin SR, Lee WMF, Williams PW, Giels GM, Williams LT (1985) c-*myc* gene expression is stimulated by agents that activate protein kinase C and does not account for the mitogenic effect of PDGF. Cell 43:243–251
Coussens L, Yang-Feng TL, Liao Y-C, Chen E, Gray A, McGrath J, Seeburg PH, Libermann TA, Schessinger J, Francke U, Levinson A, Ullrich A (1985) Tyrosine kinase receptor with extensive homology to EGF receptor shares chromosomal location with *neu* oncogene. Science 230:1132–1139
Coussens L, Van Beveren C, Smith D, Chen E, Mitchell RL, Isacke CM, Verma IM, Ullrich A (1986) Structural alteration of viral homologue of receptor protooncogene *fms* at carboxyl terminus. Nature 320:277–280
Cowley G, Smith JA, Gusterson B, Hendler F, Ozanne B (1984) The amount of EGF receptor is elevated in squamous cell carcinomas. Cancer Cells 1:5–10
Cuttitta F, Carney DN, Mulshine J, Moody TW, Fedorko J, Fischler A, Minna JND (1985) Bombesin-like peptides can function as autocrine growth factors in human small-cell lung cancer. Nature 316:823–826
De Larco JE, Todaro GJ (1978) Growth factors from murine sarcoma virus-transformed cells. Proc Natl Acad Sci USA 75:4001–4005
Delli Bovi P, Curatola AM, Kern FG, Greco A, Ittmann M, Basilico C (1987) An oncogene isolated by transfection of Kaposi's sarcoma DNA encodes a growth factor that is a member of the FGF family. Cell 50:729–737
Derynck R, Roberts AB, Winkler ME, Chen EY, Goeddel DV (1984) Human transforming growth factor-α: precursor structure and expression in *E. coli*. Cell 38:287–297
Derynck R, Goeddel DV, Ullrich A, Gutterman JU, Williams RD, Bringman TS, Berger WH (1987) Synthesis of messenger RNA for transforming growth factors α and β and the epidermal growth factor receptor by human tumors. Cancer Res 47:707–712
Devare SG, Reddy EP, Law JD, Robbins KC, Aaronson SA (1983) Nucleotide sequence of the simian sarcoma virus genome: demonstration that its acquired cellular sequences encode the transforming gene product $p28^{sis}$. Proc Natl Acad Sci USA 80:731–735
Dickson C, Peters G (1987) Potential oncogene product related to growth factors. Nature 326:833
Doolittle RF, Hunkapiller MW, Hood LE, Devare SG, Robbins KC, Aaronson SA, Antoniades HN (1983) Simian sarcoma virus oncogene, v-*sis*, is derived from the gene (or genes) encoding a platelet-derived growth factor. Science 221:275–277
Downward J, Parker P, Waterfield MD (1984a) Autophosphorylation sites on the epidermal growth factor receptor. Nature 311:483–485
Downward J, Yarden Y, Mayes E, Scrace G, Totty N, Stockwell P, Ullrich A, Schlessinger J, Waterfield MD (1984b) Close similarity of epidermal growth factor receptor and v-*erb-B* oncogene protein sequences. Nature 307:521–527
Duprez V, Lenoir G, Dautry-Varsat A (1985) Autocrine growth stimulation of a human T-cell lymphoma line by interleukin 2. Proc Natl Acad Sci USA 82:6932–6936
Durum SV, Schmidt TA, Opperheim TT (1985) Interleukin 1: an immunological perspective. Ann Rev Immunol 3:263–287
Fernandez-Botran R, Sanders VM, Olivier KG, Chen Y-W, Krammer PH, Uhr JW, Vitetta ES (1986) Interleukin 4 mediates autocrine growth of helper T cells after antigenic stimulation. Proc Natl Acad Sci USA 83:9689–9693
Ferrara N, Schweigerer L, Neufeld G, Mitchell R, Gospodarowicz D (1987) Pituitary follicular cells produce basic fibroblast growth factor. Proc Natl Acad Sci USA 84:5773–5777

Ferrari S, Basegra R (1987) Oncogenes and cell cycle genes. Bioessays 7:9–13
Finzi E, Fleming T, Segatto O, Pennington CY, Bringman TS, Derynck R, Aaronson SA (1987) The human transforming growth factor type α coding sequences is not a direct-acting oncogene when overexpressed in NIH 3T3 cells. Proc Natl Acad Sci USA 84:3733–3737
Folkman J, Klagsbrun M (1987) Angiogenic factors. Science 235:442–447
Fukushige S-I, Matsubara K-I, Yoshida M, Sasaki M, Suzuki T, Semba K, Toyoshima K, Yamamoto T (1986) Localization of a novel v-*erbB*-related gene, c-*erbB*-2, on human chromosome 17 and its amplification in a gastric cancer cell line. Mol Cell Biol 6:955–958
Gazit A, Igarashi H, Chiu I-M, Srinivasan A, Yaniv A, Tronick SR, Robbins KC, Aaronson SA (1984) Expression of the normal human sis/PDGF-2 coding sequence induces cellular transformation. Cell 39:80–97
Gentry LE, Twardzik DR, Lim GJ, Ranchalis JE, Lee DC (1987) Expression and characterization of transforming growth factor α precursor protein in transfected mammalian cells. Mol Cell Biol 7:1585–1591
Gilmore T, DeClue JE, Martin GS (1985) Protein phosphorylation at tyrosine is induced by the v-*erbB* gene product in vivo and in vitro. Cell 40:609–618
Gisselbrecht S, Fichelson S, Sola B, Bordereaux D, Hampe A, André C, Galibert F, Tambourin P (1987) Frequent c-*fms* activation by proviral insertion in mouse myeloblastic leukemias. Nature 329:259–261
Golden A, Nemeth SP, Brugge JS (1986) Blood platelets express high levels of the $pp60^{c\text{-}src}$-specific tyrosine kinase activity. Proc Natl Acad Sci USA 83:852–856
Gootenberg JE, Ruscetti FW, Mier JW, Gazdar A, Gallo RC (1981) T-cell lymphoma and leukemia cell lines produce and respond to T-cell growth factor. J Exp Med 154:1403–1418
Gospodarowicz D, Neufeld D, Schweigerer L (1986) Fibroblast growth factor. Mol Cell Endocrinol 46:187–204
Gough NM (1986) The granulocyte-macrophage colony-stimulating factors. In: Kahn P, Graf T (eds) Oncogenes and growth control. Springer, Berlin Heidelberg New York, pp 35–42
Goustin AS, Betsholtz C, Pfeifer-Ohlsson S, Persson H, Rydnert J, Bywater M, Holmgren G, Heldin C-H, Westermark B, Ohlsson R (1985) Co-expression of the *sis* and *myc* proto-oncogenes in human placenta suggest autocrine control of trophoblast growth. Cell 41:301–312
Graf T, v. Weizsäcker F, Grieser S, Coll J, Stehelin D, Patschinsky T, Bister K, Bechade C, Calothy G, Leutz A (1986) v-*mil* induces autocrine growth and enhanced tumorigenicity in v-*myc* transformed avian macrophages. Cell 45:357–364
Gray A, Dull TJ, Ullrich A (1983) Nucleotide sequence of epidermal growth factor cDNA predicts a 128 000-molecular weight protein precursor. Nature 303:722–725
Gullick WJ, Marsden JJ, Whittle N, Ward B, Bobrow L, Waterfield MD (1986) Expression of epidermal growth factor receptors on human cervical, ovarian and vulval carcinomas. Cancer Res 46:285–292
Gusterson B, Cowley G, McIlhinney J, Ozanne B, Fisher C, Reeves B (1985) Evidence for increased epidermal growth factor receptors in human sarcomas. Int J Cancer 36:689–693
Haegeman G, Content J, Volckaert G, Derynck R, Tavernier J, Fiers W (1986) Structural analysis of the sequence coding for an inducible 26-kDa protein in human fibroblasts. Eur J Biochem 159:625–632
Hammacher A, Hellman U, Johnsson A, Östman A, Gunnarsson K, Westermark B, Wasteson Å, Heldin C-H (1988a) A major part of PDGF purified from human platelets is a heterodimer of one A and one B chain. J Biol Chem 263:16493–16498
Hammacher A, Nistér M, Westermark B, Heldin C-H (1988b) A human glioma cell line secretes three structurally and functionally different dimeric forms of PDGF. Eur J Biochem 176:179–186
Hannink M, Donoghue DJ (1984) Requirement for a signal sequence in biological expression of the v-*sis* oncogene. Science 230:1197–1199

Hart CE, Forstrom JW, Kelly JD, Seifert RA, Smith RA, Ross R, Murray MJ, Bowen-Pope DF (1988) Two classes of PDGF receptors recognize different isoforms of PDGF. Science 240:1529–1531

Heldin C-H, Rönnstrand L (1989) The platelet-derived growth factor receptor. In: Moudgil VK (ed) Receptor phosphorylation. CRC Press, Boca Raton, FL, pp 149–162

Heldin C-H, Wasteson Å, Westermark B (1985) Platelet-derived growth factor. Mol Cell Endocrinol 39:169–187

Heldin C-H, Johnsson A, Wennergren S, Wernstedt C, Betsholtz C, Westermark B (1986) A human osteosarcoma cell line secretes a growth factor structurally related to a homodimer of PDGF A chains. Nature 319:511–514

Heldin C-H, Betsholtz C, Claesson-Welsh L, Westermark B (1987) Subversion of growth regulatory pathways in malignant transformation. Biochim Biophys Acta 907:219–244

Heldin C-H, Bäckström G, Östman A, Hammacher A, Rönnstrand L, Rubin K, Nistér M, Westermark B (1988) Binding of different dimeric forms of PDGF to human fibroblasts: evidence for two separate receptor types. EMBO J 7:1387–1394

Hendler FJ, Ozanne BW (1984) Human squamous cell lung cancers express increased epidermal growth factor receptors. J Clin Invest 74:647–651

Hirai H, Maru Y, Hagiwara K, Nishida J, Takaku F (1987) A novel putative tyrosine kinase receptor encoded by the *eph* gene. Science 238:1717–1720

Hirano T, Yasukawa K, Harada H, Taga T, Watanabe Y, Matsuda T, Kashiwamura S, Nakajima K, Koyama K, Iwamatsu A, Tsunasawa S, Sakiyama F, Matsui H, Takahara Y, Taniguchi T, Kishimoto T (1986) Complementary DNA for a novel human interleukin (BSF-2) that induces B lymphocytes to produce immunoglobulin. Nature 324:73–76

Huang JS, Huang SS, Deuel TF (1984) Transforming protein of simian sarcoma virus stimulates autocrine growth of SSV-transformed cells through PDGF cell-surface receptors. Cell 39:79–87

Hudziak RM, Schlessinger J, Ullrich A (1987) Increased expression of the putative growth factor receptor $p185^{HER2}$ causes transformation and tumorigenesis of NIH 3T3 cells. Proc Natl Acad Sci USA 84:7159–7163

Hunter T (1987) A thousand and one protein kinases. Cell 50:823–829

Hunter T, Cooper JA (1985) Protein tyrosine kinases. Ann Rev Biochem 54:897–930

Jacobs K, Shoemaker C, Rudersdorf R, Neill SD, Kaufman RJ, Mufson A, Seehra J, Jones SS, Hewick R, Fritsch EF, Kawakita M, Shimizu T, Miyake T (1985) Isolation and characterization of genomic and cDNA clones of human erythropoietin. Nature 313:806–809

Jaye M, Howk R, Burgess W, Ricca GA, Chiu I-M, Ravera MW, O'Brian SJ, Modi WS, Maciag T, Drohan WN (1986) Human endothelial cell growth factor: cloning, nucleotide sequence, and chromosome localization. Science 233:541–545

Johnsson A, Betsholtz C, Heldin C-H, Westermark B (1985) Antibodies against platelet-derived growth factor inhibit acute transformation by simian sarcoma virus. Nature 317:438–440

Johnsson A, Betsholtz C, Heldin C-H, Westermark B (1986) The phenotypic characteristics of simian sarcoma virus-transformed human fibroblasts suggest that the v-*sis* gene product acts solely as a PDGF receptor agonist in cell transformation. EMBO J 5:1535–1541

Josephs SF, Ratner L, Clarke MF, Westin EH, Reitz MS, Wong-Staal F (1984) Transforming potential of human c-*sis* nucleotide sequences encoding platelet-derived growth factor. Science 225:636–639

Kahn P, Frykberg L, Bradly C, Stanley I, Beug H, Vennström B, Graf T (1986) v-*erbA* cooperates with sarcoma oncogenes in leukemic cell transformation. Cell 45:349–356

Kaplan PL, Ozanne B (1982) Polyoma virus-transformed cells produce transforming growth factor(s) and grow in serum-free medium. Virology 123:372–380

Kaplan PL, Andersson M, Ozanne B (1982) Transforming growth factor(s) production enables cells to grow in the absence of serum: an autocrine system. Proc Natl Acad Sci USA 79:485–489

Kawano M, Hirano T, Matsuda T, Taga T, Horii Y, Iwato K, Asaoku H, Tang B, Tanabe O, Tanaka H, Kuramoto A, Kishimoto T (1988) Autocrine generation and requirement of BSF-2/IL-6 for human multiple myelomas. Nature 332:83–85

Kinashi T, Harada N, Severinson E, Tanabe T, Sideras P, Konishi M, Azuma C, Tominaga A, Bergstedt-Lindquist S, Takahashi M, Matsuda F, Yaoita Y, Takatsu K, Honjo T (1986) Cloning of complementary DNA encoding T-cell replacing factor and identity with B-cell growth factor II. Nature 324:70–73

King CR, Giese NA, Robbins KC, Aaronson SA (1985a) In vitro mutagenesis of the v-*sis* transforming gene defines functional domains of its growth factor-related product. Proc Natl Acad Sci USA 82:5295–5299

King CR, Kraus MH, Aaronson SA (1985b) Amplification of a novel v-*erbB*-related gene in a human mammary carcinoma. Science 229:974–976

Klagsbrun M, Sasse J, Sullivan R, Smith JA (1986) Human tumor cells synthesize an endothelial cell growth factor that is structurally related to basic fibroblast growth factor. Proc Natl Acad Sci USA 83:2448–2452

Kohase M, Henriksen-DeStefano D, May LT, Vilcek J, Sehgal PB (1986) Induction of β_2-interferon by tumor necrosis factor: a homeostatic mechanism in the control of cell proliferation. Cell 45:659–666

Kozma SC, Redmond SMS, Xiao-Chang F, Saurer SM, Groner B, Hynes N (1988) Activation of the tyrosine kinase domain of the *trk* oncogene by recombination with two different cellular sequences. EMBO J 7:147–154

Kraus MH, Popescu NC, Amsbaugh SC, King CR (1987) Overexpression of the EGF receptor-related proto-oncogene *erbB*-2 in human mammary tumor cell lines by different molecular mechanisms. EMBO J 6:605–610

Kris RM, Lax I, Gullick W, Waterfield MD, Ullrich A, Fridkin M, Schlessinger J (1985) Antibodies against a synthetic peptide as a probe for the kinase activity of the avian EGF receptor and v-*erbB* proteins. Cell 40:619–625

Laker C, Stocking C, Bergholz U, Hess N, De Lamarter JF, Ostertag W (1987) Autocrine growth stimulation after viral transfer of the granulocyte-macrophage colony stimulating factor gene and genuine factor independent growth are two distinct but interdependent steps in the oncogenic pathway. Proc Natl Acad Sci USA 84:8458–8462

Lang RA, Metcalf D, Gough NM, Dunn AR, Gonda TJ (1985) Expression of a hemopoietic growth factor cDNA in a factor-dependent cell line results in autonomous growth and tumorigenicity. Cell 43:531–542

Lax I, Kris R, Sasson I, Ullrich A, Hayman MJ, Beug H, Schlessinger J (1985) Activation of c-*erbB* in avian leukosis virus-induced erythroblastosis leads to the expression of a truncated EGF receptor kinase. EMBO J 4:3179–3182

Leal F, Williams LT, Robbins KC, Aaronson SA (1985) Evidence that the v-*sis* gene product transforms by interaction with the receptor for platelet-derived growth factor. Science 230:327–330

Lee DC, Rochford R, Todaro GJ, Villarreal LP (1985) Developmental expression of rat transforming growth factor-α mRNA. Mol Cell Biol 5:3644–3646

Lee F, Yokota T, Otsuka T, Meyerson P, Villaret D, Coffman R, Mosmann T, Rennick D, Roehm N, Smith C, Zlotnik A, Arai K (1986) Isolation and characterization of a mouse interleukin cDNA clone that expresses B-cell stimulatory factor-1 activities and T-cell- and mast-cell-stimulating activities. Proc Natl Acad Sci USA 83: 2061–2065

Leonard WJ, Depper JM, Robb RJ, Waldmann TA, Greene WC (1983) Characterization of the human receptor for T-cell growth factor. Proc Natl Acad Sci USA 80: 6957–6961

Libermann TA, Nusbaum HR, Razon N, Kris R, Lax I, Soreq H, Whittle N, Waterfield MD, Ullrich A, Schlessinger J (1985) Amplification, enhanced expression and possible rearrangement of EGF receptor gene in primary brain tumours of glial origin. Nature 313:144–147

Libermann TA, Friesel R, Jaye M, Lyall RM, Westermark B, Drohan W, Schmidt A, Maciag T, Schlessinger J (1987) An angiogenic growth factor is expressed in human glioma cells. EMBO J 6:1627–1632

March CJ, Mosley B, Larsen A, Cerretti DP, Braedt G, Price V, Gillis S, Henney CS, Kronheim SR, Grabstein K, Conlon PJ, Hopp TP, Cosman D (1985) Cloning, sequence and expression of two distinct human interleukin-1 complementary DNAs. Nature 315:641–647

Marquardt H, Hunkapiller MW, Hood LE, Twardzik DR, De Larco JE, Stephensson JR, Torado GJ (1983) Transforming growth factors produced by retrovirus transformed rodent fibroblasts and human melanoma cells: amino acid sequence homology with epidermal growth factor. Proc Natl Acad Sci USA 80:4684–4688

Martin-Zanca D, Hughes SH, Barbacid M (1986) A human oncogene formed by the fusion of truncated tropomyosin and protein tyrosine kinase sequences. Nature 319:743–748

May LT, Helgott DC, Sehgal PB (1986) Anti-β-interferon antibodies inhibit the increased expression of HLA-B7 mRNA in tumor necrosis factor-treated human fibroblasts: structural studies of the β2 interferon involved. Proc Natl Acad Sci USA 83:8957–8961

McKay I, Malone P, Marshall CJ, Hall A (1986) Malignant transformation of murine fibroblasts by a human c-Ha-*ras*-1 oncogene does not require a functional epidermal growth factor receptor. Mol Cell Biol 6:3382–3387

Metcalf D (1985) The granulocyte-macrophage colony-stimulating factors. Science 229:16–22

Miles BD, Robinson HL (1986) High-frequency transduction of c-*erbB* in avian leukosis virus-induced erythroblastosis. J Virol 54:295–303

Mingari MC, Gerosa F, Carra G, Accolla RS, Moretta A, Zubler RH, Waldmann TA, Moretta L (1984) Human interleukin-2 promotes proliferation of activated B cells via surface receptors similar to those of activated T cells. Nature 312:641–643

Mochizuki DY, Eisenman JR, Conlon PJ, Larsen AD, Tushinski RJ (1987) Interleukin-1 regulates hematopoietic activity, a role previously ascribed to hemopoietin 1. Proc Natl Acad Sci USA 84:5267–5271

Moore R, Casey G, Brookes S, Dixon M, Peters G, Dickson C (1986) Sequence, topography and protein coding potential of mouse *int*-2: a putative oncogene activated by mouse mammary tumour virus. EMBO J 5:919–924

Morgan DA, Ruscetti FW, Gallo RC (1976) Selective in vitro growth of T lymphocytes from normal human bone marrows. Science 293:1007–1008

Morgan DO, Edman JC, Standring DN, Fied FA, Smith MC, Roth RA, Rutter WJ (1987) Insulin-like growth factor II receptor as a multifunctional binding protein. Nature 329:301–307

Moscatelli D, Presta M, Joseph-Silverstein J, Rifkin DB (1986) Both normal and tumor cells produce basic fibroblast growth factor. J Cell Physiol 129:273–276

Neckameyer WS, Shibuya M, Hsu M-T, Wang L-H (1986) Proto-oncogene c-*ros* codes for a molecule with structural features common to those of growth factor receptors and displays tissue-specific and developmentally regulated expression. Mol Cell Biol 6:1478–1486

Nilsen TW, Maroney PA, Goodwin RG, Rottman FM, Crittenden LB, Raines MA, Kung H-J (1985) c-*erbB* activation in ALV-induced erythroblastosis: novel RNA processing and promoter insertion result in expression of an amino-truncated EGF receptor. Cell 41:719–726

Nilsson J, Euler AM v, Dalsgaard C-J (1985 a) Stimulation of connective tissue cell growth by substance P and substance K. Nature 315:61–63

Nilsson J, Sjölund M, Palmberg L, Thyberg J, Heldin C-H (1985 b) Arterial smooth muscle cells in primary culture produce a platelet-derived growth factor-like protein. Proc Natl Acad Sci USA 82:4418–4422

Nishizuka Y (1984) The role of protein kinase C in cell surface signal transduction and tumor promotion. Nature 308:693–698

Nistér M, Heldin C-H, Wasteson Å, Westermark B (1984) A glioma-derived analog to platelet-derived growth factor: demonstration of receptor competing activity and immunological crossreactivity. Proc Natl Acad Sci USA 81:926–930

Nistér M, Hammacher A, Mellström K, Siegbahn A, Rönnstrand L, Westermark B, Heldin C-H (1988 a) A glioma-derived PDGF A chain homodimer has different functional activities than a PDGF AB heterodimer purified from human platelets. Cell 52:791–799

Nistér M, Libermann T, Betsholtz C, Pettersson M, Claesson-Welsh L, Heldin C-H, Schlessinger J, Westermark B (1988b) Expression of messenger RNA for platelet-derived growth factor and transforming growth factor-α and their receptors in human malignant glioma cell lines. Cancer Res 48:3910–3918

Noma Y, Sideras P, Naito T, Bergstedt-Lindquist S, Azuma C, Severinson E, Tanabe T, Kinashi T, Matsuda F, Yaoita Y, Honjo T (1986) Cloning of cDNA encoding the murine IgG1 induction factor by a novel strategy using SP6 promoter. Nature 319:640–646

Nusse R (1986) The activation of cellular oncogenes by retroviral insertion. Trends Genet 244–247

Ohara J, Paul WE (1987) Receptors for B-cell stimulatory factor-1 expressed on cells of haematopoietic lineage. Nature 325:537–540

Ozanne B, Fulton RJ, Kaplan PL (1980) Kirsten murine sarcoma virus transformed cell lines and a spontaneously transformed rat cell line produce transforming factors. J Cell Phys 105:163–180

Pantazis P, Pelicci PG, Dalla-Favera R, Antoniades HN (1985) Synthesis and secretion of protein resembling platelet-derived growth factor by human glioblastoma and fibrosarcoma cells in culture. Proc Natl Acad Sci USA 82:2404–2408

Park M, Dean M, Cooper CS, Schmidt M, O'Brien SJ, Blair DG, Vande Woude GF (1986) Mechanism of *met* oncogene activation. Cell 45:895–904

Park M, Dean M, Kaul K, Braun MJ, Gonda MA, Vande Woude G (1987) Sequence of MET protooncogene cDNA has features characteristic of the tyrosine kinase family of growth factor receptors. Proc Natl Acad Sci USA 84:6379–6383

Paulsson Y, Hammacher A, Heldin C-H, Westermark B (1987) Possible positive autocrine feed back in the prereplicative phase of human fibroblasts. Nature 318:715–717

Perez R, Betsholtz C, Westermark B, Heldin C-H (1987) Frequent expression of growth factors for mesenchymal cells in human mammary carcinoma cell lines. Cancer Res 47:3425–3429

Pierce JH, Di Fiore PP, Aaronson SA, Potter M, Pumphrey J, Scott A, Ihle JN (1985) Neoplastic transformation of mast cells by Abelson-MuLV: abrogation of IL-3 dependence by a nonautocrine mechanism. Cell 41:685–693

Pollard JW, Bartocci A, Arceci R, Orlofsky A, Ladner MB, Stanley ER (1987) Apparent role of the macrophage growth factor, CSF-1, in placental development. Nature 330:484–487

Raines MA, Lewis WG, Crittenden LB, Kung H-J (1985) c-*erbB* activation in avian leukosis virus-induced erythroblastosis: clustered integration sites and the arrangement of ALV provirus in the c-*erbB* alleles. Proc Natl Acad Sci USA 82:2287–2291

Riedel H, Schlessinger J, Ullrich A (1987) A chimeric, ligand-binding v-*erbB*/EGF receptor retains transforming potential. Science 236:197–200

Rijsewik F, Schuermann M, Wagenaar E, Parren P, Weigel D, Nusse R (1987) The *Drosophila* homolog of the mouse mammary oncogene *int*-1 is identical to the segment polarity gene *wingless*. Cell 50:649–657

Robbins KC, Antoniades HN, Devare SG, Hunkapiller MW, Aaronson SA (1983) Structural and immunological similarities between simian sarcoma virus gene product(s) and human platelet-derived growth factor. Nature 305:605–608

Rogelj S, Weinberg RA, Fanning P, Klagsbrun M (1988) Basic fibroblast growth factor fused to a signal peptide transforms cells. Nature 331:173–175

Rosenthal A, Lindquist PB, Bringman TS, Goeddel DV, Derynck R (1986) Expression in rat fibroblasts of a human transforming growth factor-α cDNA results in transformation. Cell 46:301–309

Ross R, Raines EW, Bowen-Pope DF (1986) The biology of platelet-derived growth factor. Cell 46:155–169

Roussel MF, Dull TJ, Rettenmier CW, Ralph P, Ullrich A, Sherr CJ (1987) Transforming potential of the c-*fms* proto-oncogene (CSF-1 receptor). Nature 325:549–552

Rozengurt E, Sinnett-Smith J, Taylor-Papadimitriou J (1985) Production of PDGF-like growth factor by breast cancer cell lines. Int J Cancer 36:247–252

Sacca R, Stanley ER, Sherr CJ, Rettenmier CW (1986) Specific binding of the mononuclear phagocyte colony-stimulating factor CSF-1 to the product of the v-*fms* oncogene. Proc Natl Acad Sci USA 83:3331–3335

Sakamoto H, Mori M, Taira M, Yoshida T, Matsukawa S, Shimizu K, Sekiguchi M, Terada M, Sugimura T (1986) Transforming gene from human stomach cancers and a noncancerous portion of stomach mucosa. Proc Natl Acad Sci USA 83:3997–4001
Sato M, Yoshida H, Hayashi Y, Miyakami K, Bando T, Yanagawa T, Yuna Y, Azuma M, Ueno A (1985) Expression of epidermal growth factor and transforming growth factor-β in a human salivary gland adenocarcinoma cell line. Cancer Res 45:6160–6167
Schrader JW, Crapper RM (1983) Autogenous production of a hemopoietic growth factor, persisting cell-stimulating factor as a mechanism for transformation of bone marrow-derived cells. Proc Natl Acad Sci USA 80:6892–6896
Schweigerer L, Neufeld G, Mergia A, Abraham JA, Fiddes FC, Gospodarowicz D (1987) Basic fibroblast growth factor in human rhabdomyosarcoma cells: implications for the proliferation and neovascularization of myoblast-derived tumors. Proc Natl Acad Sci USA 84:842–846
Scott J, Urdea M, Quiroga M, Sanchez-Pescador R, Fong N, Selby M, Rutter WJ, Bell GI (1983) Structure of a mouse submaxillar messenger RNA encoding epidermal growth factor and seven related proteins. Science 221:236–240
Seifert RA, Schwartz SM, Bowen-Pope DF (1984) Developmentally regulated production of platelet-derived growth factor-like molecules. Nature 311:669–671
Sejersen T, Betsholtz C, Sjölund M, Heldin C-H, Westermark B, Thyberg J (1986) Rat skeletal myoblasts and arterial smooth muscle cells express the gene for the A chain but not the B chain (c-*sis*) of platelet-derived growth factor (PDGF) and produce a PDGF-like protein. Proc Natl Acad Sci USA 83:6844–6848
Semba K, Kamata N, Toyoshima K, Yamamoto T (1985) A v-*erbB* related protooncogene, c-*erbB*-2, is distinct from the c-*erbB*-1/epidermal growth factor-receptor gene and is amplified in a human salivary gland adenocarcinoma. Proc Natl Acad Sci USA 82:6497–6501
Sharon M, Klausner RD, Cullen BR, Chizzonite R, Leonard WJ (1986) Novel interleukin-2 receptor subunit detected by cross-linking under high-affinity conditions. Science 234:859–863
Sherr CJ, Rettenmier CW, Sacca R, Roussel MF, Look AT, Stanley ER (1985) The c-*fms* proto-oncogene product is related to the receptor for the mononuclear phagocyte growth factor, CSF-1. Cell 41:665–676
Sherwin SA, Twardzik DR, Bohn WH, Cockley KD, Todaro GJ (1983) High molecular-weight transforming growth factor activity in the urine of patients with disseminating cancer. Cancer Res 43:403–407
Shibuya M, Hanafusa H, Balduzzi PC (1982) Cellular sequences related to three new *onc* genes of avian sarcoma (*fps, yes,* and *ros*) and their expression in normal and transformed cells. J Virol 42:143–152
Sibley DR, Benovic JL, Caron MG, Lefkowitz RJ (1987) Regulation of transmembrane signaling by receptor phosphorylation. Cell 48:913–922
Slamon DJ, Clark GM, Wong SG, Levin WJ, Ullrich A, McGuire WL (1987) Human breast cancer: correlation of relapse and survival with amplification of the HER-2/*neu* oncogene. Science 235:177–182
Smith JJ, Derynck R, Korc M (1987) Production of transforming growth factor α in human pancreatic cancer cells: evidence for a superagonist autocrine cycle. Proc Natl Acad Sci USA 84:7567–7570
Söderdahl G, Betsholtz C, Johansson A, Nilsson K, Berg J (1988) Differential expression of platelet-derived growth factor and transforming growth factor genes in small- and non-small cell human lung carcinoma cell lines. Int J Cancer 41:636–641
Sporn MB, Roberts AB (1985) Autocrine growth factors and cancer. Nature 313:745–747
Sporn MB, Todaro GJ (1980) Autocrine secretion and malignant transformation of cells. N Engl J Med 303:878–880
Stanley ER, Guilbert LJ, Tushinski RJ, Bartelmez SH (1983) CSF-1 a mononuclear phagocyte lineage-specific hemopoietic growth factor. J Cell Biochem 21:151–159
Stern DF, Hare DL, Cecchini MA, Weinberg RA (1987) Construction of a novel oncogene based on synthetic sequences encoding epidermal growth factor. Science 235:321–324
Stroobant P, Waterfield MD (1984) Purification and properties of porcine platelet-derived growth factor. EMBO J 3:2963–2967

Stroobant P, Rice AP, Gullick WJ, Chen DJ, Kerr IM, Waterfield MD (1985) Purification and characterization of vaccinia virus growth factor. Cell 42:383–393

Taira M, Yoshida T, Miyagawa K, Sakamoto H, Terada M, Sugimura T (1987) cDNA sequence of human transforming gene *hst* and identification of the coding sequence required for transforming activity. Proc Natl Acad Sci USA 84:2980–2984

Takahashi M, Cooper GM (1987) *ret* transforming gene encodes a fusion protein homologous to tyrosine kinases. Mol Cell Biol 7:1378–1385

Taniguchi T, Matsui H, Fujita Y, Hatakeymana M, Kashima N, Fuse A, Hamuro J, Hishi-Takaoka C, Yamada G (1986) Molecular analysis of the interleukin-2 system. Immunol Rev 92:121–133

Teixidó J, Gilmore R, Lee DC, Massagué J (1987) Integral membrane glycoprotein properties of the prohormone pro-transforming growth factor-α. Nature 326:883–885

Temin HM (1966) Control by factors in serum of multiplication of uninfected cells and cells infected and converted by avian sarcoma viruses. In: Defendi V, Stoker M (eds) Growth regulating substances for animal cells in cultures. Wistar symposium monograph no. 7, Wistar Institute Press, Philadelphia, PA, pp 103–116

Todaro GJ, DeLarco JE, Marquardt H, Bryant ML, Sherwin SA, Sliski AH (1979) Polypeptide growth factors produced by tumor cells and virus-transformed cells: a possible growth advantage for the producer cells (1979). In: Sato GH, Ross R (eds) Hormones and cell culture, book A, Cold Spring Harbor conferences on cell proliferation, 6. Cold Spring Harbor, New York, pp 113–127

Todaro GJ, Fryling C, DeLarco JE (1980) Transforming growth factors produced by certain human tumor cells: polypeptides that interact with epidermal growth factor receptors. Proc Natl Acad Sci USA 77:5258–5262

Twardzik DR (1985) Differential expression of transforming growth factor-α during prenatal development of the mouse. Cancer Res 45:5413–5416

Twardzik DR, Todaro GJ, Marquardt H, Reynolds FH Jr, Stephenson JR (1982) Transformation induced by Abelson murine leukemia virus involves production of a polypeptide growth factor. Science 216:894–896

Twardzik DR, Todaro GJ, Reynolds FH Jr, Stephenson JR (1983) Similar transforming growth factors (TGFs) produced by cells transformed by different isolates of feline sarcoma virus. Virology 124:201–207

Twardzik DR, Brown JP, Ranchalis JE, Todaro GJ, Moss B (1985) Vaccinia virus-infected cells release a novel polypeptide functionally related to transforming and epidermal growth factors. Proc Natl Acad Sci USA 82:5300–5304

Ullrich A, Coussens L, Hayflick JS, Dull TJ, Gray A, Tam AW, Lee J, Yarden Y, Libermann TA, Schlessinger J, Downward J, Mayes ELV, Whittle N, Waterfield MD, Seeburg PH (1984) Human epidermal growth factor receptor cDNA sequence and aberrant expression of the amplified gene in A431 epidermoid carcinoma cells. Nature 309:418–425

Ullrich A, Bell JR, Chen EY, Herrera R, Petruzelli LM, Dull TJ, Gray A, Coussens L, Liao Y-C, Tsubokawa M, Mason A, Seeburg PH, Grunfeld C, Rosen OM, Ramachandran J (1985) Human insulin receptor and its relationship to the tyrosine kinase family of oncogenes. Nature 311:756–761

Ullrich A, Gray A, Tam AW, Yang-Feng T, Tsubokawa M, Collins C, Henzel W, Le Bon T, Kathuria S, Chen E, Jacobs S, Francke U, Ramachandran J, Fujita-Yamaguchi Y (1986) Insulin-like growth factor I receptor primary structure: comparison with insulin receptor suggests structural determinants that define functional specificity. EMBO J 5:2503–2512

Van de Vijver M, van de Bersselaar R, Devilee P, Cornelisse C, Peterse J, Nusse R (1987) Amplification of the *neu* (c-*erbB*-2) oncogene in human mammary tumors is relatively frequent and is often accompanied by amplification of the linked c-*erbA* oncogene. Mol Cell Biol 7:2019–2023

Van Ooyen A, Nusse R (1984) Structure and nucleotide sequence of the putative mammary oncogene *int*-1 proviral insertions leave the protein encoding domain intact. Cell 39:233–240

Van Snick J, Cayphas S, Vink A, Uyttenhove C, Coulie PG, Rubira MR, Simpson RJ (1986) Purification and NH_2-terminal amino acid sequence of a T-cell-derived lymphokine with growth factor activity for B-cell hybridomas. Proc Natl Acad Sci USA 83:9679–9683

Verbeek JS, de Ruyter O, Bloemers HPJ, Van de Ven WJM (1985) Molecular cloning and characterization of feline cellular genetic sequences homologous to the oncogene of the McDonough strain of feline sarcoma virus. Virology 141:322–327

Vlodavsky I, Folkman J, Sullivan R, Friedman R, Ishai-Michaeli R, Sasse J, Klagsbrun M (1987) Endothelial cell-derived basic fibroblast growth factor: synthesis and deposition into subendothelial extracellular matrix. Proc Natl Acad Sci USA 84:2292–2296

von Weizsäcker F, Beug H, Graf T (1986) Temperature-sensitive mutants of MH2 avian leukemia virus that map in the v-*mil* and v-*myc* genes of the virus. EMBO J 5:1521–1528

Watanabe S, Lazar E, Sporn MB (1987) Transformation of normal rat kidney (NRK) cells by an infectious retrovirus carrying a synthetic rat type α transforming growth factor gene. Proc Natl Acad Sci USA 84:1258–1262

Waterfield MD, Scrace GT, Whittle N, Stroobant P, Johnsson A, Wasteson Å, Westermark B, Heldin C-H, Huang JS, Deuel TF (1983) Platelet-derived growth factor is structurally related to the putative transforming protein $p28^{sis}$ of simian sarcoma virus. Nature 304:35–39

Westermark B, Heldin C-H (1988) Activation of protooncogenes coding for growth factors or growth factor receptors. In: Klein G (ed) Cellular oncogene activation. Dekker, New York, pp 149–180

Westermark B, Johnsson A, Paulsson Y, Betsholtz C, Heldin C-H, Herlyn M, Rodeck U, Koprowski H (1986) Human melanoma cell lines of primary and metastatic origin express the genes encoding the constituent chains of PDGF and produce a PDGF-like growth factor. Proc Natl Acad Sci USA 83:7197–7200

Westermark B, Betsholtz C, Johnsson A, Heldin C-H (1987) Acute transformation by simian sarcoma virus is mediated by an externalized PDGF-like growth factor. In: Kjelgaard NO, Forchhammer J (eds) Viral carcinogens. Munksgaard, Copenhagen, pp 445–457

Wheeler EF, Roussel MF, Hampe A, Walker MH, Fried VA, Look AT, Rettenmier CW, Sherr CJ (1986a) The amino-terminal domain of the v-fms oncogene product includes a functional signal peptide that directs synthesis of a transforming glycoprotein in the absence of FeLV *gag* sequences. J Virol 59:224–233

Wheeler EF, Rettenmier CW, Look AT, Sherr CJ (1986b) The v-*fms* oncogene induces factor independence and tumorigenicity in CSF-1 dependent macrophage cell line. Nature 324:377–380

Winkles JA, Friesel R, Burgess WH, Howk R, Mehlman T, Weinstein R, Maciag T (1987) Human vascular smooth muscle cells both express and respond to heparin binding growth factor I (endothelial cell growth factor). Proc Natl Acad Sci USA 84:7124–7128

Wong AJ, Bigner SH, Bigner DD, Kinzler KW, Hamilton SR, Vogelstein B (1987) Increased expression of the epidermal growth factor receptor gene in malignant gliomas is invariably associated with gene amplification. Proc Natl Acad Sci USA 84:6899–6903

Yamamoto T, Nishida T, Miyajima N, Kawai S, Ooi T, Toyoshima K (1983) The *erbB* gene of avian erythroblastisis virus is a member of the *src* gene family. Cell 35:71–78

Yamamoto T, Ikawa S, Akiyama T, Semba K, Nomura N, Miyajima N, Saito T, Toyoshima K (1986a) Similarity of protein encoded by the human c-*erb-B*-2 gene to epidermal growth factor receptor. Nature 319:230–234

Yamamoto T, Kamata N, Kawano H, Shimizu S, Kuroki T, Toyoshima K, Rikimaru K, Nomura N, Ishizaki R, Pastan I, Gamou S, Shimizu N (1986b) High incidence of amplification of the epidermal growth factor receptor gene in human squamous carcinoma cell lines. Cancer Res 46:414–416

Yarden Y, Escobedo JA, Kuang W-J, Yang-Feng TL, Daniel TO, Tremble PM, Chen EY, Ando ME, Harkins RN, Francke U, Fried VA, Ullrich A, Williams LT (1986) Structure of the receptor for platelet-derived growth factor helps define a family of closely related growth factor receptors. Nature 323:226–232

Yarden Y, Kuang W-J, Yang-Feng T, Coussens L, Munemitsu S, Dull TJ, Chen E, Schlessinger J, Francke U, Ullrich A (1987) Human protooncogene c-*kit:* a new cell surface receptor tyrosine kinase for an unidentified ligand. EMBO J 6:3341–3350

Ymer S, Tucker WQJ, Sanderson CJ, Hapel AJ, Campbell HD, Young IG (1985) Constitutive synthesis of interleukin-3 by leukaemia cell line WEHI-3B is due to retroviral insertion near the gene. Nature 317:255–258

Young D, Griffin JD (1986) Autocrine secretion of GM-CSF in acute myeloblastic leukemia. Blood 68:1178–1181

Zacchary I, Rozengurt E (1985) High affinity receptors for peptides of the bombesin family in Swiss 3T3 cells. Proc Natl Acad Sci USA 82:7616–7620

Zhou D, Battifora H, Yokota J, Yamamoto T, Cline MJ (1987) Association of multiple copies of the c-*erbB*-2 oncogene with spread of breast cancer. Cancer Res 47:6123–6125

Zilberstein A, Ruggieri R, Korn JH, Revel M (1986) Structure and expression of cDNA and genes for human interferon-beta-2, a distinct species inducible by growth-stimulatory cytokinase. EMBO J 5:2529–2537

CHAPTER 14

Signal Transduction in Proliferating Normal and Transformed Cells

M. J. O. WAKELAM

A. Introduction

The understanding of the early events induced in cells by growth factors has been facilitated by the use of cloned cell lines that can be made quiescent by either withdrawing serum growth factors from the culture medium or by allowing the cells to grow until the available growth factors are depleted and the cells are contact-inhibited. Removal of growth factors from an exponentially growing culture does not, however, stop proliferation immediately. Those cells in or beyond late G_1, and thus committed to division, complete the cell cycle (ZETTERBERG and LARSSON 1985). When the resulting progeny cells and the other cells in the culture enter early G_1, they progress no further. The production of mRNA in these arrested cells then falls, whilst the rate of degeneration of mRNA is unchanged (RUDLAND et al. 1975). The cultured cells are then said to be quiescent or G_0 cells; for further discussion of this transition see WHITFIELD et al. (1987). Following the addition of serum or defined growth factors, the quiescent cells are stimulated to enter the cell cycle. Using this experimental system, the early biochemical events occurring in the cell in response to growth stimulation can be investigated. Such studies have demonstrated several types of early change including alterations in intracellular ion concentrations, protein phosphorylation and production of second messengers.

B. Mitogen-Induced Ion Changes

I. pH Changes

When quiescent Swiss 3T3 cells are incubated with vasopressin and insulin or with platelet-derived growth factor (PDGF), intracellular pH is increased by about 0.2 units (SCHULDINER and ROZENGURT 1982). Mitogen-induced increases in pH have been detected in other cell types, including human fibroblasts (MOOLENAAR et al. 1983) and mouse thymocytes (HESKETH et al. 1985). HESKETH et al. (1985) also demonstrated that the Ca^{2+} ionophore A23187 and the tumour promoter TPA, which can both be used to stimulate mitogenesis in the absence of receptor stimulation, also induce an increase in cytoplasmic pH. The increase in pH observed in the mitogen-stimulated cells appears to be linked to other ionic changes. Thus, addition of growth factors, such as epidermal growth factor (EGF) or PDGF, causes activation of the plasma membrane Na^+/H^+ antiporter, which can exchange intracellular H^+ with extracellular Na^+. This antiport system is electroneutral and sensitive to amiloride. The internal Na^+ concentration is maintained by the exchange of intracellular Na^+ and extracellular K^+ through the oubain-sensitive Na^+/K^+ pump. Therefore, the changes in pH and in internal K^+ concentration in response to stimulation by mitogens are associated with a Na^+ cycle (see ROZENGURT 1985 for further discussion).

II. Changes in Ca^{2+}

The addition of serum or defined growth factors to quiescent cells has a rapid effect upon the intracellular concentration of free Ca^{2+}. These changes were originally detected as a rapid efflux of $^{45}Ca^{2+}$ from cells preloaded with the radiolabel in response to stimulation by growth factors or serum. LOPEZ-RIVAS and ROZENGURT (1983) demonstrated that stimulation of quiescent Swiss 3T3 cells with serum caused a detectable efflux of Ca^{2+} within seconds. This efflux is also stimulated by the fibroblast mitogens vasopressin and prostaglandin $F_{2\alpha}$ ($PGF_{2\alpha}$) but not by other mitogens such as EGF or by TPA (LOPEZ-RIVAS and ROZENGURT 1983; SMITH and SMITH 1984). Efflux is, at least in part, a consequence of stimulated release of Ca^{2+} from intracellular stores. Indeed, FRANTZ (1985) has demonstrated that stimulation with PDGF results in almost half of the cells' store of Ca^{2+} being lost within 5 min.

The development of fluorescent dyes for the measurement of cytoplasmic Ca^{2+} (see COBBOLD and RINK 1987 for review) has permitted the detection and measurement of rapid changes in Ca^{2+} concentration. Using this method, two types of increased internal free Ca^{2+} concentration were detected in thymocytes and fibroblasts (HESKETH et al. 1985). On exposure of Swiss 3T3 cells to $PGF_{2\alpha}$ and of thymocytes to Con A, there was a rapid increase in intracellular free Ca^{2+} concentration which peaked within 30 s and that was only partially sensitive to the removal of Ca^{2+} from the external culture medium. EGF, on the other hand, induced a slower rise in internal Ca^{2+} concentration which peaked after 2 min and was absolutely dependent on the presence of external Ca^{2+}. The Con A-induced rise in thymocytes declined after reaching a peak value of about 250 n*M* (from 100 n*M* in the control) to a new plateau of about 150 n*M*. When the experiment was repeated in the absence of external Ca^{2+}, the peak value obtained was 150 n*M*, and the Ca^{2+} concentration returned to the resting level within 8 min. Thus, two different mechanisms of increased Ca^{2+} appear to be operating, one involving release of internal stores of Ca^{2+} and the other involving entry of Ca^{2+} into the cell (see Sect. C.I.2.c for further discussion of these mechanisms).

III. Phosphorylation

Protein phosphorylation appears to be the common consequence of mitogenic stimulation by the different types of growth factors. However, the kinases responsible for this phosphorylation and the regulation of their activities are dependent upon the signal transduction system associated with each type of receptor. The kinases will be described in more detail in the following sections.

C. Stimulation of Second Messenger Metabolism

Whilst changes in mono- and divalent cation fluxes, an increase in cytoplasmic pH and changes in the phosphorylation patterns are early consequences of mitogenic stimulation and are probably early signals in the stimulation of DNA synthesis, they are in themselves the consequence of the generation of second messengers. There are two major second messenger-generating signal transduction pathways which appear to be involved in the regulation of proliferation, and their characteristics and roles will be discussed here.

I. The Inositol Phospholipid Pathways

Four inositol-containing phospholipids have been identified in mammalian cells. Three of these, phosphatidylinositol (PI), phosphatidylinositol-4-phosphate (PIP) and phosphatidylinositol-4,5-bisphosphate (PIP_2) are involved in the pathway under discussion (Fig. 1). The fourth, glycosylphosphatidylinositol (see Low et al. 1986), will not be considered here.

PIP and PIP_2 are minor cellular phospholipids in the eukaryotic cell, being found only in the inner face of the plasma membrane (MICHELL 1975). These three inositol phospholipids are involved in a cycle in which specific kinases are operating (see BERRIDGE and IRVINE 1984). Within cells inositol phospholipids

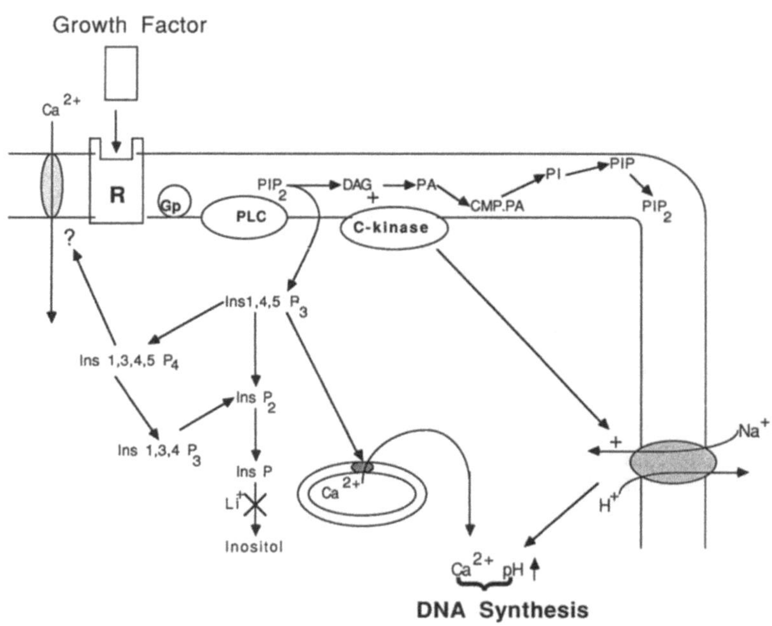

Fig. 1. Schematic representation of the inositol phospholipid pathway and its relation to the stimulation of DNA synthesis. Ins, inositol; PI, phosphatidylinositol; DAG, *sn*-1,2-diacylglycerol; Gp, G-protein; PLC, phospholipase C; CMPPA, cytidine monophosphate-phosphatidate

appear to exist in two separate pools: one that is hormone-sensitive and one that is hormone-insensitive (see MONACO 1987). However, the importance of these two pools and of any interaction that may occur between them is unclear.

Upon binding of certain agonists to their specific receptors, the hydrolysis of PIP_2 is stimulated. This involves cleavage of a phosphodiester bond by an inositide-specific phospholipase C to generate two products; *sn*-1,2-diacylglycerol (DAG) and inositol-1,4,5-trisphosphate (Ins 1,4,5 P_3). Whilst the enzyme directly responsible for catalysing this reaction remains to be identified, a number of PIP_2-specific phospholipase C isoenzymes have been identified (see BERRIDGE 1987 for review). The existence of multiple forms of inositide-specific phospholipase C is consistent with the observation that under certain conditions PI, rather than PIP_2, is preferentially hydrolysed (IMAI and GERSHENGORN 1986). Although the significance of this diversity has not been established, it is conceivable that the different forms of phospholipase C are involved in coupling to different receptors.

1. G-protein Coupling in the Inositol Lipid Pathway

Research into the potential role of G-proteins in the stimulation of PIP_2 hydrolysis is much less advanced than comparable studies on the adenylate cyclase transduction system. The first indication that G-proteins were involved in the stimulation of PIP_2 hydrolysis came from the studies of GOODHARDT et al. (1982) who demonstrated a shift in the binding curve for α_1-adrenergic receptors when non-hydrolysable GTP analogues were added to membrane preparations of hepatocytes. This work has been extended to demonstrate that such GTP analogues can stimulate the hydrolysis of PIP_2 in the absence of agonists in membrane preparations (see COCKCROFT 1987 for review).

Cholera and pertussis toxins can act as inhibitors of signal transduction by binding to cellular G-proteins. In the adenylate cyclase system the effects of these toxins have been clearly defined. However, their effects upon the inositol phospholipid system are less clear. In certain cell types there is an approximate 50% inhibition of agonist-stimulated PIP_2 hydrolysis by pertussis toxin (see MILLIGAN 1988 for review). In other cell types, however, such an inhibition is not observed, and some studies have demonstrated a small, but significant, increase in agonist-stimulated inositol phosphate generation in pertussis-toxin-treated cells. These studies have led to the recognition that at least two distinct Gp proteins are involved in coupling to the inositol phospholipid pathway.

Cholera toxin treatment of cells has also been shown to cause an inhibition of agonist-stimulated PIP_2 breakdown in some cell types (see e.g. LO and HUGHES 1987). One interpretation of these results is that there is a novel, cholera-toxin-sensitive G-protein involved in inositol lipid metabolism. Alternatively, the observed effects could be a consequence of interactions with the stimulatory G-protein (Gs) of the adenylate cyclase pathway (see Fig. 2). Whilst the latter explanation may explain the short-term effects of the toxin, recent work examining the effect of long-term stimulation of rat mesangial cells (GUILLON et al. 1988) and L6 myoblasts (GARDNER et al. 1989) by cholera toxin demonstrates that the toxin eventually causes a reduction in the binding of agonists to their receptors.

It is this indirect long-term effect rather than a direct involvement of a cholera-toxin-sensitive G-protein that is the probable cause of the observed inhibition of stimulated PIP_2 breakdown.

The role of G-proteins in the inositol phospholipid pathway will be considered further in Sect. D.III.2.

2. Functions of the Second Messengers

The importance of the inositol phospholipid signalling pathway in the activation of mitosis is demonstrated by experiments in which microinjection of PIP_2-specific antibodies into NIH 3T3 cells abolished the stimulation of DNA synthesis in response to PDGF and bombesin (MATUOKA et al. 1988). This antibody has also been found to inhibit yeast cell mitosis, emphasising the widespread importance of the inositol lipid pathway in the regulation of proliferation (UNO et al. 1988). The actions of the messenger products of lipid breakdown and their potential roles in the regulation of proliferation are considered here.

a) Inositol-1,4,5-trisphosphate

The first indication of a physiological role for Ins 1,4,5 P_3 came from the work of STREB et al. (1983). When these workers added purified Ins 1,4,5 P_3 to permeabilised pancreatic acinar cells, they detected a rapid release of Ca^{2+} from a vesicular store. These studies have been repeated many times on various cell types. There appears to be an Ins 1,4,5 P_3 receptor upon the smooth endoplasmic reticulum which regulates the release of stored Ca^{2+}. Chemical or biochemical modifications of Ins 1,4,5 P_3 generates molecules having no, or a reduced, effect in activating the receptor (IRVINE et al. 1984). From these studies it is clear that a pair of vicinal phosphates at the 4- and 5-positions on the inositol ring are essential for activity. Whilst the phosphate at the 1-position confers greater Ca^{2+}-releasing activity on the molecule, it is not essential for activity (IRVINE et al. 1984).

It is characteristic of second messenger molecules that they are formed and degraded quickly and exert their effects rapidly. Ca^{2+} release induced by Ins 1,4,5 P_3 occurs within seconds (see e.g. JACKSON et al. 1987), and the half-life of the molecule in liver cells has been estimated to be 1.2 s (KIRK et al. 1987). When Ins 1,4,5 P_3 was added to permeabilised Swiss 3T3 cells, approximately 50% of the stored Ca^{2+} was released within 1 min (BERRIDGE et al. 1984). This value is consistent with the time course of the depletion of the internal reserves of Ca^{2+} of fibroblasts in response to growth factors (FRANTZ 1985).

b) Inositol Tetrakisphosphate

Ins 1,4,5 P_3 is degraded by a specific 5-phosphatase to yield Ins 1,4 P_2. However, this is not the only metabolic pathway of Ins 1,4,5 P_3 removal. The messenger can be phosphorylated by an Ins 1,4,5 P_3 kinase to generate inositol-1,3,4,5-tetrakisphosphate (Ins 1,3,4,5 P_4). This compound is degraded by a 5-phosphatase, which may be the same enzyme that acts on Ins 1,4,5 P_3 (see MAJERUS et al.

1988), to yield Ins 1,3,4 P_3. This inositol phosphate is not a Ca^{2+}-releasing agent under physiological conditions (IRVINE et al. 1984).

Recent work has suggested that Ins 1,3,4,5 P_4 may also have a second messenger role. IRVINE and MOOR (1986) showed that the formation of the sea urchin egg fertilisation envelope, which is mediated by Ca^{2+} release, required the microinjection of both Ins 1,4,5 P_3 and Ins 1,3,4,5 P_4. This was demonstrated by making use of Ins 2,4,5 P_3 which can act at the IP_3 receptor to stimulate the release of Ca^{2+}, but, unlike Ins 1,4,5 P_3, it is not a substrate for IP_3 kinase and thus cannot be used to generate Ins 1,3,4,5 P_4 in the cell. These studies have been extended by MORRIS et al. (1987) using intracellular perfusion of lacrimal acinar cells. This work demonstrates that, in this cell type, both Ins 1,4,5 P_3 and Ins 1,3,4,5 P_4 are required to generate a Ca^{2+}-activated K^+ current.

Whilst the Ca^{2+} released from intracellular stores appears to be derived from an Ins 1,4,5 P_3-sensitive endoplasmic reticular store, the mechanism of Ca^{2+} entry into the cell that is stimulated by inositol-phosphate-generating agonists remains unclear. PUTNEY (1986) has proposed that Ca^{2+} enters the cell via the Ins 1,4,5 P_3-sensitive store, and IRVINE and MOOR (1987) suggest that Ins 1,3,4,5 P_4 modulates the entry of external Ca^{2+} into the store. It is possible that, under those instances in which Ins 1,3,4,5 P_4 generation is undetectable (see ALTMAN 1988), the regulation of Ca^{2+} entry into the Ins 1,4,5 P_3-sensitive store is uncoupled.

c) *sn*-1,2-Diacylglycerol

Whilst Ins 1,4,5 P_3 is released into the cytoplasm following activation of phospholipase C, the second product of inositol phospholipid hydrolysis, DAG, remains within the membrane. It is there that the lipid functions as the physiological activator of protein kinase C (NISHIZUKA 1984). It has been proposed that, following generation of DAG, protein kinase C is translocated from the cytosol to the membrane (see e.g. FARRAR et al. 1985; GUY et al. 1986). Whilst there is no doubt that upon generation of DAG in the plasma membrane more protein kinase C activity is found membrane-associated after fractionation, it is possible that this is not due to translocation. It is more likely that the enzyme is plasma membrane-associated and is simply bound more firmly by its interaction with DAG.

Protein kinase C can be activated by both Ca^{2+} and DAG. The lipid is believed to activate the enzyme by decreasing the K_m of the enzyme for Ca^{2+} (KISHIMOTO et al. 1980). In addition, prior elevation of intracellular Ca^{2+} can augment the activation by DAG (WOLF et al. 1985; DOUGHERTY and NIEDEL 1986).

The importance of protein kinase C in the stimulation of cell proliferation is demonstrated by the fact that the enzyme is the target of action of the tumour-promoting phorbol esters (CASTAGNA et al. 1982) and of teleocidin and aplysiatoxin (FUJIKI et al. 1984). These agents appear to act by mimicking the functions of DAG, and since they are not enzymatically removed, they cause sustained activation of protein kinase C.

Protein kinase C is not a single enzyme but exists in multiple forms that are encoded by a multigene family (NISHIZUKA 1988). It is possible that the different forms of the enzyme are regulated by different DAGs; for example the DAG generated by insulin action in BC_3H_I cells appears to have a dimyristoyl acyl backbone (SALTIEL and CUATRECASAS 1986), whereas that generated by PIP_2 breakdown is generally a 1-stearoyl, 2-arachidonyl derivative (NISHIZUKA 1984). Additionally, activation of protein kinase C can stimulate diametrially opposed processes such as proliferation and differentiation (BERRIDGE 1987), suggesting the involvement of different isozymes.

It is also important to note that the inositol phospholipid pathway is not the only source of DAG in cells. Some of the actions of insulin are mediated via the activation of a phospholipase C which hydrolyses an inositol glycan (SALTIEL and CUATRECASAS 1986). This generates a DAG which may be dimyristylated (see above). In several cell types growth-factor-stimulated phosphatidylcholine (PC) breakdown has also been detected (e.g. BESTERMAN et al. 1986). It is perhaps significant that TPA can also stimulate PC breakdown (MUIR and MURRAY 1987). This suggests that stimulation of protein kinase C by DAG could lead to further generation of DAG by activating PC breakdown. This is discussed further in Sect. D.III.2.

Whilst mitogenic stimulation has been demonstrated to involve activation of protein kinase only a few probable targets of this enzyme have been identified. A target of major importance is the Na^+/H^+ antiporter; the importance of this protein for proliferation was discussed in Sect. B.I. This protein is activated by a range of growth factors known to stimulate PIP_2 breakdown (HESKETH et al. 1985; L'ALLEMAIN et al. 1984; MOOLENAAR et al. 1983) and also by tumour-promoting phorbol esters (HESKETH et al. 1985; MOOLENAAR et al. 1983). Another target is vinculin (WERTH and PASTAN 1984); it is possible that phosphorylation of this molecule is involved in cytoskeletal reorganisation, an event also stimulated by tumour-promoting phorbol esters (SCHLIWA et al. 1984) and found to result from viral transformation (see DRIEDGER and BLUMBERG 1977).

A further proposed target of potential physiological importance is IP_3 phosphatase (CONNOLLY et al. 1986). This phosphorylation is thought to lead to a stimulation of activity of the enzyme which may lead to an attenuation of the inositol phospholipid-linked Ca^{2+} signal.

II. cAMP as a Signal for Mitosis?

Stimulation of receptors by another group of agonists (e.g. β-adrenergic agents) results in the generation of cAMP as a consequence of the action of a specific G-protein, Gs, which activates adenylate cyclase activity (see CASEY and GILMAN 1988 for review). Adenylate cyclase activity is also subject to negative regulation via receptor activation of inhibitory G-protein (Gi) by agonists such as α_2-adrenergic agents (see CASEY and GILMAN 1988 for review). cAMP is believed to have both positive and negative effects upon the proliferation of quiescent fibroblasts. It has been shown that raising cAMP levels in Swiss 3T3 cells reduces the rate of growth and inhibits serum-stimulated DNA synthesis (BURGER et al.

1972; KRAM et al. 1973). Considerable objections can be raised against these studies since the concentrations of cAMP used were very high, and the effects could be deemed non-specific.

In contrast ROZENGURT et al. (1981) found that if the intracellular levels of cAMP were raised by cholera toxin treatment of Swiss 3T3 cells, then serum-stimulated DNA synthesis was promoted rather than inhibited. The effects of the toxin upon both cAMP levels and DNA synthesis showed a similar dose-dependency. In support of a role for raised cAMP levels being the mediator of cholera-toxin-stimulated DNA synthesis, ROZENGURT et al. (1981) found that inhibitors of cyclic nucleotide phosphodiesterases such as RO-20-1724 were able to potentiate the toxin's effects both upon cAMP accumulation and DNA synthesis.

Contrary to this proposal, it has been shown that the cAMP level is higher in dense, non-proliferating cultures than in growing cells and that cAMP levels drop when growth-arrested cells are stimulated by mitogens (FROEHLICH and RACHMELER 1972; OTTEN et al. 1972). Furthermore, transformed cells which are not subject to density arrest tend to have lower cAMP levels (JOHNSON et al. 1971; SHEPPARD 1972).

It is probable that the changes in cAMP are linked to the stage in the cell cycle. When synchronised cells are mitogenically stimulated, a broad increase in cAMP is measurable in G_1 just preceding S phase, a small increase is detected in G_2, and lower levels are observed during mitosis (SHEPPARD and PRESCOTT 1972; COSTA et al. 1976). SMETS and VAN ROOY (1987) have found that the addition of cholera toxin to Swiss 3T3 cells for short periods during the G_0–G_1 transition increased the fraction of cells responding to serum stimulation, whereas its addition during late G1 inhibited the onset of DNA synthesis.

The explanation for the apparent paradox of a raised cAMP concentration having both growth promoting and inhibiting effects may lie in the existence of two distinct isozymes (I and II) of cAMP-dependent protein kinase. The C subunits of the two forms are identical, but each isozyme has a different R subunit (CORBIN et al. 1975; HOFFMAN et al. 1975). Since both isozymes have been found in most cells, selective modulation of these enzymes may control the precise effect that cAMP has in regulating mitosis. Each R subunit has two cAMP binding sites which can be distinguished by site-selective cAMP analogues (DOSKELAND 1978; RANNELS and CORBIN 1980).

Use of a range of cAMP analogues has suggested the involvement of the type II protein kinase in growth inhibition of human cancer cell lines (KATSAROS et al. 1987). The inhibition appears to be associated with an increase in the cellular levels of the RII subunit and a decrease in RI. Intriguingly, there was also an apparent fall in the level of $p21^{ras}$ proteins in the analogue-treated cells. TAGLIAFERRI et al. (1988) have found that activation of the RII subunit by site-selective analogues is also able to reverse the transformation of NIH 3T3 cells induced by the Harvey murine sarcoma virus.

Treatment of Kirsten murine sarcoma virus-transformed BALB/c 3T3 cells with dibutyryl cAMP has been found to inhibit the expression of transformation-related properties such as anchorage-independent growth (RIDGEWAY et al. 1988). These experiments strongly suggest that uncontrolled proliferation is asso-

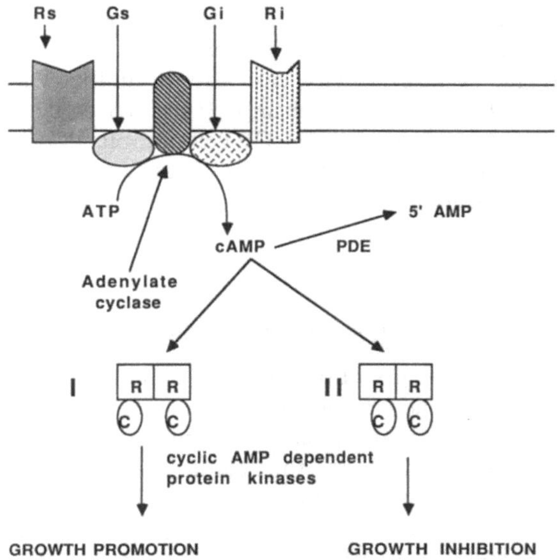

Fig. 2. Schematic representation of the regulation of cyclic AMP generation and its role in proliferation. Rs, stimulatory receptor; Ri, inhibitory receptor; Gs, stimulatory G-protein; Gi, inhibitory G-protein; R, C, regulatory and catalytic subunits of cyclic AMP-dependent protein kinase; PDE, cyclic AMP phosphodiesterase

ciated with a decrease in cellular cAMP concentrations. However, it is unclear whether the reduction in cAMP content is a consequence of reduced generation or of activated degradation. Insulin, a known co-mitogen, has been demonstrated to activate cAMP phosphodiesterase in a number of cells (see HOUSLAY 1985), and this may be one of the consequences of exposure to certain growth factors.

The target proteins of the protein kinase and the immediate metabolic consequences of the phosphorylation are unknown. However, the importance of cAMP in the regulation of cell proliferation is emphasised by the recent finding of an altered Gs and an increase in adenylate cyclase activity in growth-hormone-secreting pituitary adenomas (VALLAR et al. 1987). Figure 2 summarises the cAMP pathways and their potential involvement in the regulation of proliferation.

III. Interactions Between Signalling Pathways

Most studies on changes in signal transduction following mitogenic stimulation have examined one pathway in isolation; this is clearly not a physiologically accurate assessment of signalling pathway function. Whilst interactions between the inositol phospholipid and the cyclic nucleotide pathways have not been examined in any detail in mitogenic systems, results from other cell types emphasise the potential importance of such events. For example, in hepatocytes, the desensitisation of glucagon stimulated adenylate cyclase activity has been found to be mediated via activation of protein kinase C as a consequence of inositol phospholipid breakdown (WAKELAM et al. 1986a). Desensitisation of glucagon-stimulated adenylate cyclase activity can also be induced heterologously by stimulation with, for example, angiotensin II (MURPHY et al. 1987). Conversely, raising intracellular cAMP levels in NIH 3T3 cells inhibits $PGF_{2\alpha}$-stimulated PIP_2 breakdown (Wakelam, unpublished results). These experiments emphasise the need to consider the whole cell and the potential effects of different signals upon one another.

D. Involvement of Oncogenes with Signal Transduction

It is clear from the preceding sections that the actions of growth factors upon cells are mediated via the activation of one or more signal transduction pathways. Therefore, loss of normal regulation of the pathways will lead to a loss in regulation of mitogenesis, which can thus result in the development of a cancer cell. The first indication of this came from the identification of protein kinase C as the target of action of the tumour-promoting phorbol esters, teleocidins and aplysiatoxins (CASTAGNA et al. 1982; FUJIKI et al. 1984). This section examines the evidence for the effects of oncogene products upon signal transduction.

I. Autocrine Stimulation of Growth

The first demonstration of a direct involvement of an oncogene product with the inositol phospholipid pathway came from the determination of the sequence of the *sis* oncogene and the observation that it encoded a protein of very similar sequence to PDGF (DOOLITTLE et al. 1983; WATERFIELD et al. 1983). Expression of this oncogene will cause a marked increase in stimulation of PIP_2 breakdown via PDGF receptor activation; thus the oncogene can transform cells expressing the appropriate receptors (LEAL et al. 1985).

The transformation of cells by the *sis* oncogene is therefore an example of an autocrine stimulation of transformation. Many cancer cells are autonomous of growth factors, and this can, in some cases, be explained by the autocrine hypothesis (SPORN and ROBERTS 1985). The autocrines most commonly produced by tumour cells are the transforming growth factors, TGFα and TGFβ, PDGF-related peptides and bombesin-related peptides.

TGFα binds to, and acts through, the EGF receptor. The release and action of TGFα and its relationship to transformation are, however, unclear. In experiments with cells expressing temperature-sensitive oncogene products of mutant rodent sarcoma viruses (p37*mos* and p21 K-*ras*), TGFα was only secreted at permissive temperatures (OZANNE et al. 1980; DELARCO et al. 1981; KAPLAN et al. 1982). Thus, one of the actions of these two oncogenes may be to control TGFα synthesis at the transcriptional or translational level.

TGFβ (ANZANO et al. 1985) and PDGF-like molecules (BOWEN-POPE et al. 1984) are also released from rodent cells transformed by Moloney, Harvey or Kirsten sarcoma viruses. The PDGF-like molecules bind to the PDGF receptor(s) and activate inositol lipid hydrolysis and tyrosine kinase activity. The mode of action of TGFβ is, however, less clear. This polypeptide stimulates the growth of fibroblasts (TUCKER et al. 1983). However, it can also act as a potent inhibitor of growth in, for example, monkey kidney cells (TUCKER et al. 1984). In *myc*-transfected rat fibroblasts, TGFβ can stimulate anchorage-independent growth in the presence of PDGF, but it is anti-proliferative in the presence of EGF (ROBERTS et al. 1985). The precise effect that TGFβ has upon cellular signalling is unclear, although a recent report suggested that it may activate protein kinase C by an unidentified mechanism in immature rat brain (MARKOVAC and GOLDSTEIN 1988). This effect upon protein kinase C could explain the inhibition of EGF-stimulated growth through a mechanism involving phosphorylation and transmodulation of the EGF receptor.

The importance of autocrine loops in maintaining the malignant state is demonstrated by the use of the polyanionic drug suramin. This drug has been shown to block the binding of EGF, TGFα, FGF, and PDGF to their receptors and also to bring about the elution of these growth factors prebound to their receptors (BETSHOLTZ et al. 1986). The drug has been shown to revert the transformed morphology and to inhibit the growth of *sis*-transformed fibroblasts (BETSHOLTZ et al. 1986). On the other hand, the growth of the same cell type transformed by activated oncogenic *ras* is not affected by suramin (BETSHOLTZ et al. 1986). Thus, the growth and transformation of fibroblasts by the *sis* oncogene is dependent upon growth factor-receptor interaction, whereas transformation by viral *ras* is not.

The inhibition of growth of cancer cells by a receptor antagonist has also been accomplished with small cell lung carcinoma cells in culture. These "oat cell" cancers secrete gastrin-releasing peptide (GRP), which is thought to be the mammalian equivalent of the amphibian peptide bombesin (MOODY et al. 1981). This peptide is a potent mitogen and stimulates inositol phospholipid hydrolysis in both Swiss and NIH 3T3 cells (BROWN et al. 1984; WAKELAM et al. 1986b). A monoclonal antibody against bombesin inhibits growth of "oat cell" cultures, and the development of tumours in nude mice can be prevented by injection of the antibody (CUTTITTA et al. 1985).

II. Oncogene Products as Receptors

In addition to amplifying signal transduction pathways by increasing concentrations of agonists via autocrine generation, cells can be transformed by increasing receptor number or by modifying existing receptors. To this end some oncogene products have been found to be cell surface receptors or receptor-like molecules. This function of oncogene products is considered further in Chaps. 13–15.

III. Effects of Oncogene Products on Signal Transduction Generation

Since signal transduction pathways are tightly regulated, they are an ideal target for oncogene-induced loss of growth control.

1. Gene Products with Tyrosine Kinase Activity

The receptors for several hormones and growth factors (e.g. insulin and EGF) possess an intrinsic tyrosine kinase activity. A range of oncogene products (e.g. *src* and *ros*) also have an intrinsic tyrosine kinase activity, which suggests that they might act in the cell to generate the same type of message. An interaction between stimulated tyrosine kinase activity and inositol phospholipid metabolism has been suggested by a range of studies demonstrating that the activity of PI kinase is increased in cells stimulated either by PDGF or by the activation of $pp60^{c\text{-}src}$. WHITMAN et al. (1987) have identified two biochemically distinct PI kinases in fibroblasts, and one of these (type 1) is associated with activated tyrosine kinases (KAPLAN et al. 1987; COURTNEIDGE and HEBER 1987). These experiments suggested that activation of tyrosine kinases could lead to the amplification of PIP_2 breakdown by increasing the levels of PIP_2. However, WHITMAN et al. (1988) have now demonstrated that type 1 PI kinase catalyses the phosphorylation of PI on the 3 rather than on the 4 position of the inositol ring and thus generates a novel phosphoinositide. The relevance of this observation is unclear since the hydrolysis product of a PIP_2 having a phosphate at the 3 position is unlikely to be capable of mobilising Ca^{2+}. It may be that such a product has an as yet undetermined function or that the generated lipid itself has a regulatory role in proliferation.

JACKOWSKI et al. (1986) have demonstrated in CCL64 epithelial cells transformed by either the v-*fes* or the v-*fms* oncogene that there is an increase in the level of PIP_2 phospholipase C activity. These two oncogenes have an associated tyrosine kinase activity, suggesting that phospholipase C can be activated either directly or indirectly by such a phosphorylation. It may be that the novel phosphoinositide detected by WHITMAN et al. (1988) functions to regulate the activity of phospholipase C.

2. Involvement of *ras* Gene Products in Signal Transduction

The three *ras* genes in mammalian cells each encode 21000 molecular weight proteins which exhibit extensive sequence homology except in the C-terminal region. Within this region is a conserved cysteine residue that is post-translationally palmitoylated. This modification is essential both for the transforming ability of the proteins and for their association with the plasma membrane. The $p21^{ras}$ proteins are able to bind guanine nucleotides and have an intrinsic GTPase activity. These properties led to the proposal that they function as G-proteins in cells and, in particular, act in a Gp-like manner (see BARBACID 1987).

Several lines of evidence support the postulate that $p21^{ras}$ is coupled to PIP_2 hydrolysis. Transformation of BALB-3T3 fibroblasts with the EJ/T24-Ha-*ras* oncogene increases the responsiveness of phospholipase C to muscarinic stimulation (CHIARRUGI et al. 1986). An increase in the turnover of inositol phospholipids was detected by FLEISHMAN et al. (1986) in both NRK and NIH 3T3 cells transformed by three different *ras* genes. KAMATA et al. (1987) demonstrated that NIH 3T3 cells transformed by the Ki-MSV gene also display elevated inositol phospholipid turnover, whilst in a flat, untransformed *ras*-resistant variant, derived from the same transfection, no differences from the untransformed parental NIH 3T3 cells were observed. FLEISHMAN et al. (1986), PREISS et al. (1986), KAMATA et al. (1987) and LACAL et al. (1987a, b, c) have all shown that there are raised levels of DAG in cells transformed by *ras* genes. In the T15$^+$ cell line in which transformation is achieved by the overproduction of normal p21$^{N\text{-}ras}$ under the control of a steroid-inducible promotor, there is an increased generation of inositol phosphates in response to bombesin as compared with control cells (WAKELAM et al. 1986b). This increase is observed with no accompanying change in receptor number or in the receptors K_a and K_d. More recent experiments have demonstrated that this amplification in inositol phosphate generation is a consequence of amplified PIP_2 breakdown and that the bombesin-stimulated release of intracellular Ca^+ is also amplified (LLOYD et al. 1989). These results led us to suggest that p21$^{N\text{-}ras}$ could indeed function in a Gp-like manner, at least in this cell line. Support for this proposal is provided by the finding that the stimulation of PIP_2 hydrolysis by fluoraluminate ions, which directly activate G-proteins, is also amplified in the T15$^+$ cell line compared with control cells (WAKELAM, 1989).

Whilst transformation of NIH 3T3 cells with transforming mutants of N- or Ha-*ras* did not increase the apparent coupling of receptors to phospholipase C, it did result in an increase in the turnover of inositol phospholipids (WAKELAM et al. 1987; HANCOCK et al. 1988). This apparent increase in agonist-independent phosphoinositide turnover has been shown only to be induced by *ras* genes which are transforming. Deletion by site-directed mutagenesis of the palmitoylation site or increasing lengths of the C-terminal amino acids and modification of the putative effector-binding domain all inhibit the transforming capability of the protein products; they also abolish the effects of the *ras* gene products upon inositol phosphate generation (HANCOCK et al. 1988).

The transforming mutants of *ras* genes are characterised by having a greatly reduced GTPase activity. This major reduction in enzyme activity only became apparent once the accessory protein GAP (GTPase-activating protein) was identified (TRAHEY and MCCORMICK 1987). This cytosolic protein, which has been found in all cells examined, stimulates the GTPase activity of normal Ha-*ras* but is without effect upon its transforming mutants. GAP has an apparent molecular weight of 60 K and appears to interact with the effector region of *ras* (CALES et al. 1988). The biochemical functions of GAP are unclear. Whilst it is tempting to assign a role analogous to the Gs of adenylate cyclase or Gp of phospholipase C, its identity and function will have to await purification and cloning.

The data from a number of groups (FLEISCHMAN et al. 1986; KAMATA et al. 1987; HANCOCK et al. 1988) have demonstrated that transformation of NIH 3T3, NRK and COS-1 cells by mutant *ras* genes, whose protein products have a reduced GTPase activity, is accompanied by an increase in inositol phospholipid turnover. If p21^{ras} does function in a Gp-like manner, this reduced enzyme activity could result in the protein being in a permanent, GTP-bound state. This would result in persistent activation of the putative effector, phospholipase C, in a manner analogous to the action of cholera toxin. This toxin induces ADP-ribosylation of GSα with concomitant inhibition of its GTPase activity and thus results in persistent activation of adenylate cyclase.

The role of the *ras* proteins in amplifying inositol phospholipid turnover has, however, been questioned. PARRIES et al. (1987) found that in cells transformed by viral Ha- and Ki-*ras* the PDGF-stimulated inositol phosphate response was desensitised whilst the responses to other growth factors, in particular to bradykinin, were amplified. Receptor binding studies demonstrated that the number of high affinity bradykinin binding sites was increased in the Ki-MuSV-transformed cells whilst there was no change in PDGF binding. BLACK and WAKELAM (1989) have found that the inositol phospholipid response to $PGF_{2\alpha}$ is desensitised in NIH 3T3 cells transformed by N-, Ha- or Ki-*ras* genes. Further investigation by BLACK and WAKELAM (1989) has suggested that this desensitisation is mediated by increased protein kinase C activity and is a consequence of *ras*-induced increases in proliferation rather than a direct effect of *ras* itself, since a similar effect is observed in NIH 3T3 cells in logarithmic growth.

The difficulties in detecting *ras*-induced effects are further demonstrated by the finding that the bombesin stimulation of inositol phosphate generation in $T15^+$ cells is desensitised by high-density culturing (WAKELAM 1988). This desensitisation can be reversed by culturing in the presence of the receptor antagonist suramin, and it has thus been suggested (WAKELAM 1988) that the effect is induced by the secretion of autocrine growth factors by *ras*-transformed cells (OWEN and OSTROWSKI 1987). This effect of autocrine growth factors is probably mediated by protein kinase C since phorbol esters can mimic both the desensitisation induced by high-density culturing and inhibit the amplification of the basal rate of inositol phosphate generation induced by mutant *ras* transformation (Wakelam, manuscript in preparation). Whilst protein kinase C-catalysed phosphorylation of $p21^{ras}$ has not been demonstrated in vivo, JENG et al. (1987) have shown that v-Ha-*ras* can be phosphorylated in vitro by purified protein kinase C. These observations may go some way towards explaining the results of SEUWEN et al. (1988). These investigators found that transformation of CCL39 fibroblasts by mutant Ha-*ras* did not amplify an already negligible basal rate of inositol phosphate generation. In addition in a growth-factor-dependent clone of transformed cells, the inositol phosphate response to thrombin or serum was not amplified, whereas in a growth-factor-independent clone, these responses were observed to be desensitised.

Proliferation-related desensitisation mediated by protein kinase C activation may explain the variation in the observed effects upon inositol phospholipid metabolism in *ras*-transformed cells. It does not explain the consistent observation of an increased DAG content in all *ras*-transformed cells examined. WOLFMAN and MACARA (1987) found elevated DAG levels in mutant Ha- and Ki-*ras*-transformed cells without any observable increases in the steady state levels of inositol phosphates. This study also demonstrated that the basal level of phosphorylation of the protein kinase C substrate 80 K protein was significantly increased in all the *ras*-transformed cells examined. However, further phosphorylation of the protein in responses to phorbol esters was reduced in the *ras*-transformed cells, suggesting that DAG elevation induces partial down-regulation of protein kinase C activity in these cells. The fundamental importance of protein kinase C activity in *ras*-induced transformation has been demonstrated by LACAL et al. (1987a). This study showed that down-regulation of the enzyme in Swiss 3T3 cells markedly reduced the mitogenic activity of microinjected mutant Ha-*ras*. Co-injection of purified protein kinase C with the p21*ras* protein restored the mitogenic activity.

LACAL et al. (1987b) examined the effects of microinjection of Ha-*ras* p21 into *Xenopus* oocytes. Within 2 min of injection, effects upon inositol phospholipid metabolism were detectable. The increase in inositol phosphate generation reached a maximum after 6 min but then appeared to be desensitised. When DAG levels were examined in parallel experiments, a similar rapid increase was detected; however, the increased DAG generation was maintained for at least up to 20 min post-injection.

This work strongly suggests that the *ras* proteins do not simply affect the inositol phospholipid signal transduction pathway. LACAL et al. (1987c) have now detected changes in the metabolism of other lipids in NIH 3T3 cells transformed by Ha-*ras* genes. In these cells there was an apparent increase in the turnover of phosphatidylcholine (PC) and phosphatidylethanolamine (PE) but no change in phosphatidylserine metabolism. This suggests that the *ras* gene products may simultanously cause PIP_2 breakdown and PC breakdown. An alternative explanation is that $p21^{ras}$ initially activates the generation of DAG from PIP_2 and then switches to activate PC breakdown. Thus, in those studies in which no effect of *ras* transformation upon inositol phospholipid metabolism has been detected, there may be effects upon the breakdown of PC. Whilst it remains to be shown what the primary effector system coupled to $p21^{ras}$ is, it is possible to concludethat a major consequence of the expression of the protein is the activation of protein kinase C.

E. Conclusions

It is clear from the information presented in this review that effects upon signal transduction pathways can go some way towards explaining the effects of oncogenes upon cell proliferation. Figure 3 summarises some of the proposed roles for oncogene products in interactions with the inositol phospholipid pathway. However, our understanding of the roles of signal transduction pathways remains rudimentary. Consequently, until the pathways are fully defined, the identification of potential sites of therapeutic intervention will be difficult. A recent example of the increasing complexity of signal transduction is the finding of polyphosphoinositide synthesis that is dependent upon the state of differentiation in the nuclei of Friend cells (COCCO et al. 1987).

Acknowledgements. Work from my laboratory has been financed by grants from the Medical Research Council (UK) and the Cancer Research Campaign. Thanks are due to Simon Cook for the production of Fig. 1 and to Jane Fensome for typing this article.

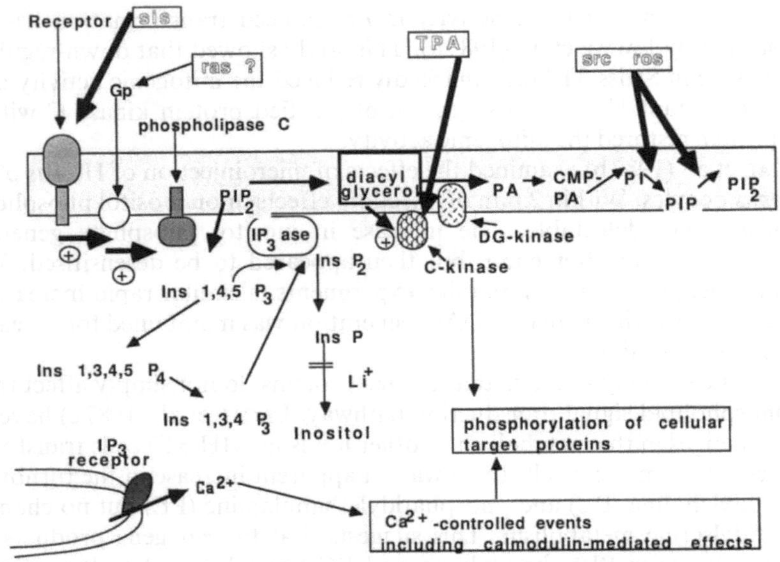

Fig. 3. Proposed sites of interaction of some examples of oncogenes with the inositol phospholipid pathway. Gp, G-protein; Ins, inositol; PI, phosphotidylinositol

References

Altman J (1988) Ins and outs of cell signalling. Nature 331:119–120

Anzano MA, Roberts AB, De Larco JE, Wakefield LM, Assoian RK, Roche NS, Smith JM, Lazarus JE, Sporn MB (1985) Increased secretion of type beta transforming growth factor accompanies viral transformation of cells. Mol Cell Biol 5:242–247

Barbacid M (1987) *ras* genes. Ann Rev Biochem 56:779–827

Berridge MJ (1987) Inositol lipids and cell proliferation. Biochim Biophys Acta 907:33–45

Berridge MJ, Irvine RF (1984) Inositol trisphosphate, a novel second messenger in cellular signal transduction. Nature 213:315–321

Berridge MJ, Heslop JP, Irvine RF, Brown KD (1984) Inositol trisphosphate formation and calcium mobilisation in Swiss 3T3 cells in response to platelet-derived growth factor. Biochem J 222:195–201

Besterman JM, Duronio V, Cuatrecasas P (1986) Rapid formation of diacylglycerol from phosphatidylcholine: a pathway for generation of a second messenger. Proc Natl Acad Sci USA 83:6785–6789

Betsholtz C, Johnsson A, Heldin C-H, Westermark B (1986) Efficient reversion of simian sarcoma virus-transformation and inhibition of growth factor-induced mitogenesis by suramin. Proc Natl Acad Sci USA 83:6440–6444

Blac FM, Wakelam MJO (1989) Dosensitization A prostaglandin $F_{2\alpha}$ stimulated inositol phosphate generation in NIH-3T3 fibroblasts transformed by overexpression of normal c-Ha-*ras*, c-Ki-*ras* and c-N-*ras* genes. Biochem J (in press)

Bowen-Pope DF, Vogel A, Ross R (1984) Production of platelet-derived growth factor-like molecules and reduced expression of platelet-derived growth factor receptors accompany transforming by a wide spectrum of agents. Proc Natl Acad Sci USA 81:2396–2400

Brown KD, Blay J, Irvine RF, Heslop JP, Berridge MJ (1984) Reduction of epidermal growth factor receptor affinity by heterologous ligands: evidence for a mechanism involving the breakdown of phosphoinositides and the activation of protein kinase C. Biochem Biophys Res Commun 123:377–384

Burger MM, Bombik BM, Breckenridge B McL, Shepphard JR (1972) Growth control and cyclic alterations of cyclic AMP in the cell cycle. Nature New Biol 239:161–162
Cales C, Hancock JF, Marshall CJ, Hall A (1988) The cytoplasmic protein GAP is implicated as the target for regulation by the *ras* gene product. Nature 332:548–551
Casey PJ, Gilman AG (1988) G protein involvement in receptor effector coupling. J Biol Chem 263:2577–2580
Castagna M, Takai Y, Kaibuchi K, Sano K, Kikkaw U, Nishizuka Y (1982) Direct activation of calcium-activated, phospholipid dependent protein kinase by tumour-promoting phorbol esters. J Biol Chem 257:7847–7851
Chiarrugi VP, Pasquali F, Vannucchi S, Ruggiero M (1986) Point mutated p21^{ras} couples a muscarinic receptor to calcium channels and polyphosphoinositide hydrolysis. Biochem Biophys Res Commun 141:591–599
Cobbold PH, Rink TJ (1987) Fluorescence and bioluminescence measurement of cytoplasmic free calcium. Biochem J 248:313–328
Cocco L, Gilmour RS, Ognibene A, Letcher A, Manzoli FA, Irvine RF (1987) Synthesis of polyphosphoinositides in nuclei of Friend cells. Biochem J 248:765–770
Cockroft S (1987) Polyphosphoinositide phosphodiesterase: regulation by a novel guanine nucleotide binding protein, Gp. Trends Biochem Sci 12:75–78
Connolly TM, Lawing WJ Jr, Majerus PW (1986) Protein kinase C phosphorylates human platelet inositol triphosphate 5′-monophosphatase, increasing phosphatase activity. Cell 46:951–958
Corbin JD, Keely SL, Park CR (1975) The distribution and dissociation of cyclic adenosine 3′:5′-monophosphate-dependent protein kinases in adipose, cardiac and other tissues. J Biol Chem 250:218–225
Costa M, Gerner EW, Russell DH (1976) G_1 specific increases in cyclic AMP levels and protein kinase activity in Chinese hamster ovary cells. Biochim Biophys Acta 425:246–255
Courtneidge S, Heber A (1987) An 81 kd protein complexed with middle T antigen and pp60$^{c\text{-}src}$: a possible phosphatidylinositol kinase. Cell 50:1031–1037
Cuttitta F, Carney DN, Mulshine J, Moody TW, Fedorki J, Fischler A, Minna JD (1985) Bombesin-like peptides can function as autocrine growth factors in human small-cell lung cancer. Nature 316:823–826
Delarco JE, Preston YA, Todaro GJ (1981) Properties of a sarcoma-growth-factor-like peptide from cells transformed by a temperature sensitive sarcoma virus. J Cell Physiol 109:143–152
Doolittle RF, Hunkapiller MW, Hood LH, Devare SG, Robbins KC, Aaronson SA, Antoniades HN (1983) Simian sarcoma virus oncogene, v-*sis*, is derived from the gene (or genes) encoding a platelet-derived growth factor. Science 221:275–280
Doskeland SO (1978) Evidence that rabbit muscle protein kinase has two kinetically distinct binding sites for adenosine 3′,5′-cyclic monophosphate. Biochem Biophys Res Commun 83:542–549
Dougherty RW, Niedel JE (1986) Cytosolic calcium regulates phorbol dibutyrate binding affinity in intact phagocytes. J Biol Chem 261:4097–4100
Driedger PE, Blumberg PM (1977) The effect of phorbol diesters on chicken embryo fibroblasts. Cancer Res 37:3257–3265
Farrar WL, Thomas TP, Anderson WB (1985) Altered cytosol/membrane enzyme redistribution on interleukin 3 activation of protein kinase C. Nature 315:235–237
Fleischman LF, Chahwala SB, Cantely L (1986) *ras*-transformed cells: altered levels of phosphatidylinositol 4,5-bisphosphate and catabolites. Science 231:407–410
Frantz CN (1985) Effect of platelet-derived growth factor on Ca^{2+} in 3T3 cells. Exp Cell Res 158:287–300
Froelich JE, Rachmeler M (1972) Effect of adenosine-3′-5′-cyclic monophosphate on cell proliferation. J Cell Biol 55:19–31
Fujiki H, Tanaka Y, Miyake R, Kikkawa U, Nishizuka Y (1984) Activation of calcium activated phospholipid dependent protein kinase (protein kinase C) by a new class of tomour promotors: teleocidin and debromoaplysiatoxin. Biochem Biophys Res Commun 120:339–343

Gardner SD, Milligan G, Rice JE, Wakelam MSO (1989) The effect of cholera toxin on the inhibition of vasopressin stimulated inositol phosphate generation is a cyclic AMP mediated effect at the level of receptor binding. Biochem J 259:679–684

Goodhardt M, Ferry N, Geynet P, Hanoune J (1982) Hepatic α_1-adrenergic receptors show agonist specific regulation by guanine nucleotides. J Biol Chem 257:11577–11583

Guillon G, Gallo-Payet N, Balestre MN (1988) Cholera toxin and ACTH modulation of inositol phosphate accumulation induced by vasopressin and angiotensin II in rat glomerulosa cells. Biochem J 253:765–775

Guy GR, Gordon J, Walker I, Michell RH, Brown G (1986) Redistribution of protein kinase C during mitogenesis of human B lymphocytes: Biochem Biophys Res Commun 135:146–153

Hancock JF, Marshall CJ, McKay IA, Gardner S, Houslay MD, Hall A, Wakelam MJO (1988) Mutant but not normal p21*ras* elevates inositol phospholipid breakdown in two different cell systems. Oncogene 3:187–193

Hesketh TR, Moore JP, Morris JDH, Taylor MV, Rogers J, Smith GA, Metcalfe JC (1985) A common sequence of calcium and pH signals in the mitogenic stimulation of eukaryotic cells. Nature 313:481–484

Hoffman F, Beavo JA, Bechtel PJ, Krebs EG (1975) Comparison of adenosine 3′,5′-monophosphate-dependent protein kinase from rabbit skeletal and bovine heart muscle. J Biol Chem 250:7795–7801

Houslay MD (1985) The insulin receptor and signal generation at the plasma membrane. Mol Asp Cell Reg 5:279–334

Imai A, Gershengorn MC (1986) Phosphatidylinositol 4,5-bisphosphate turnover is transient while phosphatidylinositol turnover is persistent in thyrotropin-releasing hormone-stimulated rat pituitary cells. Proc Natl Acad Sci USA 83:8540–8544

Irvine RF, Moor RM (1986) Microinjection of inositol (1,3,4,5) tetrakisphosphate activates sea urchin eggs by promoting Ca^{2+} entry. Biochem J 240:917–920

Irvine RF, Moor RM (1987) Inositol (1,3,4,5) tetrakisphosphate-induced activation of sea urchin eggs requires the presence of inositol trisphosphate. Biochem Biophys Res Commun 146:284–290

Irvine RF, Brown KD, Berridge MJ (1984) Specificity of inositol trisphosphate-induced calcium release from permeabilised Swiss-mouse 3T3 cells. Biochem J 222:269–272

Jackowski S, Rettenmier CW, Sherr CJ, Rock CO (1986) A guanine nucleotide-dependent phosphatidylinositol 4,5-diphosphate phospholipase C in cells transformed by the v-*fms* and v-*fes* oncogenes. J Biol Chem 261:4978–4985

Jackson TR, Hallam TJ, Downes CP, Hanley MR (1987) Receptor coupled events in bradykinin action: rapid production of inositol phosphates and regulation of cytosolic free Ca^{2+} in a neural cell line. EMBO J 6:49–54

Jeng AY, Srivastava SJK, Lacal JC, Blumberg PM (1987) Phosphorylation of *ras* oncogene product by protein kinase C. Biochem Biophys Res Commun 145:782–788

Johnson GS, Friedman RM, Pastan I (1971) Restoration of several morphological characteristics of normal fibroblasts in sarcoma cells treated with adenosine-3′:5′-cyclic monophosphate and its derivatives. Proc Natl Acad Sci USA 68:425–429

Kamata T, Sullivan NF, Wooten MW (1987) Reduced protein kinase C activity in a *ras* resistant cell line derived from Ki-MSV transformed cells. Oncogene 1:37–46

Kaplan DR, Whitman M, Schaffhausen B, Pallas DC, White M, Cantley L, Roberts RM (1987) Common elements in growth factor stimulation and oncogenic transformation: 85 kd phosphoprotein and phosphatidylinositol kinase activity. Cell 50:1021–1029

Kaplan PL, Anderson M, Ozanne B (1982) Transforming growth factor(s) production enables cells to grow in the absence of serum: an autocrine system. Proc Natl Acad Sci USA 79:485–489

Katsaros D, Tortora G, Tagliaferri P, Clair T, Ally S, Neckers L, Robins RK, Cho-Chung YS (1987) Site selective cyclic AMP analogues provide a new approach in the control of cancer cell growth. FEBS Lett 223:97–103

Kirk CJ, Michell RH, Parry J, Shears SB (1987) Inositol trisphosphate and tetrakisphosphate phosphomonoesterases of rat liver. Biochem Soc Trans 15:28–32

Kishimoto A, Takai Y, Mori T, Kikkawa U, Nishizuka Y (1980) Activation of calcium and phospholipid-dependent protein kinase by diacylglycerol, its possible relation to phosphatidylinositol turnover. J Biol Chem 255:2273–2276
Kram R, Mamont P, Tomkins GM (1973) Pleiotypic control by adenosine 3′:5′-cyclic monophosphate: a model for growth control in animals. Proc Natl Acad Sci USA 70:1432–1436
Lacal JC, Fleming TP, Warren BS, Blumberg PM, Aaronson S (1987a) Involvement of functional protein kinase C in the mitogenic response to the H-*ras* oncogene product. Mol Cell Biol 7:4146–4149
Lacal JC, de la Pena P, Moscat J, Garcia-Barreno P, Anderson PS, Aaronson SA (1987b) Rapid stimulation of diacylglycerol production in *Xenopus* oocytes by micro injection of H-*ras* p21. Science 238:533–536
Lacal JC, Moscat J, Aaronson SA (1987c) Novel source of 1,2-diacylglycerol elevated in cells transformed by Ha-*ras* oncogene. Nature 330:269–272
L'Allemain G, Paris S, Pouyssegur J (1984) Growth factor action and intracellular pH regulation in fibroblasts. J Biol Chem 259:5809–5815
Leal F, Williams LT, Robbins KC, Aaronson SA (1985) Evidence that the v-*sis* gene product transforms by interaction with the receptor for platelet-derived growth factor. Science 230:327–330
Lloyd AC, Davies SA, Crossley I, Whittaker M, Houslay MD, Hall A, Marshall CJ, Wakelam MJO (1989) Bombesin stimulation of inositol 1,4,5-trisphosphate generation and intracellular calcium release is amplified in a cell line overexpressing the N-*ras* proto-oncogene. Biochem J 260:813–819
Lo WWY, Hughes J (1987) A novel cholera toxin-sensitive G-protein (Gc) regulating receptor mediated phosphoinositide signalling in human pituitary clonal cells. FEBS Lett 220:327–331
Lopez-Rivas A, Rozengurt E (1983) Serm rapidly mobilises calcium from an intracellular pool in quiescent fibroblastic cells. Biochem Biophys Res Commun 114:240–247
Low MG, Ferguson MAJ, Fukerman AH, Silman I (1986) Covalently attached phosphatidylinositol as a hydrophobic anchor for membrane proteins. Trends Biochem Sci 11:212–215
Majerus PW, Connolly TM, Bansel VS, Inhorn RC, Ross TS, Lips DL (1988) Inositol phosphates: synthesis and degradation. J Biol Chem 263:3051–3054
Markovac J, Goldstein GW (1988) Transforming growth factor beta activates protein kinase C in microvessels isolated from immature rat brain. Biochem Biophys Res Commun 150:575–582
Matuoka K, Fukami K, Nakanishi P, Kawai S, Takenawa T (1988) Mitogenesis in response to PDGF and bombesin abolished by microinjection of antibody to PIP_2. Science 239:640–643
Michell RH (1975) Inositol phospholipids and cell surface receptor function. Biochim Biophys Acta 415:81–147
Milligan G (1988) Techniques used in the identification and analysis of function of pertussis-toxin-sensitive guanine nucleotide binding proteins. Biochem J 255:1–13
Monaco M (1987) Inositol metabolism in WRK-1 cells. Relationship of hormone-sensitive to -insensitive pools of phosphoinositides. J Biol Chem 262:13001–13006
Moolenaar WH, Tsien RY, van der Saag PT, De Laat SW (1983) Na^+/H^+ exchange and cycloplasmic pH in the action of growth factors in human fibroblasts. Nature 304:645–648
Morris AP, Gallacher DV, Irvine RF, Peterson OH (1987) Synergism of inositol trisphosphate and tetrakisphosphate in activating Ca^{2+}-dependent K^+ channels. Naure 330:653–655
Muir JG, Murray AW (1987) Bombesin and phorbol ester stimulate phosphatidylcholine hydrolysis by phospholipase C: evidence for a role of protein kinase C. J Cell Physiol 130:382–391
Murphy GJ, Hruby VJ, Trivedi D, Wakelam MJO, Houslay MD (1987) The rapid desensitization of glucagon-stimulated adenylate cyclase is a cyclic AMP-independent process that can be mimicked by hormones that stimulate inositol phospholipid metabolism. Biochem J 243:39–46

Nishizuka Y (1984) The role of protein kinase C in cell surface signal transduction and tumour promotion. Nature 308:693–697

Nishizuka Y (1988) The heterogeneity and differential expression of multiple species of the protein kinase C family. Biofactors 1:17–20

Otten J, Johnson GS, Pastan I (1972) Regulation of cell growth by cyclic adenosine 3′,5′-monophosphate. Effect of cell density and agents which alter cell growth on cyclic adenosine 3′,5′-monophosphate levels in fibroblasts. J Biol Chem 247:7082–7087

Owen RD, Ostrowski MC (1987) Rapid and selective alterations in the expression of cellular genes accompany conditional transcription of Ha-v-*ras* in NIH-3T3 cells. Molec Cell Biol 7:2512–2520

Ozanne B, Fulton RJ, Kaplan PL (1980) Kirsten murine sarcoma virus transformed cell lines and a spontaneously transformed rat cell line produce transforming factors. J Cell Physiol 105:163–180

Parries G, Hoebel R, Racker E (1987) Opposing effects of a *ras* oncogene on growth factor stimulated phosphoinositide hydrolysis: desensitization to platelet-derived growth factor and enhanced sensitivity to bradykinin. Proc Natl Acad Sci USA 84:2648–2652

Preiss J, Loomis CR, Bishop WR, Stein R, Niedel JE, Bell RM (1986) Quantitative measurement of *sn*-1,2-diacylglycerols in platelets, hepatocytes and *ras*- and *sis*-transformed normal rat kidney cells. J Biol Chem 261:8597–8600

Putney JW Jr (1986) A model for receptor regulated calcium entry. Cell Calcium 7:1–12

Rannels SR, Corbin JD (1980) Two different intrachain cAMP binding sites of cAMP-dependent protein kinases. J Biol Chem 255:7085–7088

Ridgeway AAG, DE Vouge MW, Mukerjee BB (1988) Dibutyryl cyclic AMP inhibits expression of transformation related properties in Kirsten murine sarcoma virus transformed Balb/c-3T3 cells despite continued presence of p21 v-Ki-*ras*. Biochem Cell Biol 66:54–65

Roberts AB, Anzona MA, Wakefield LM, Roche NS, Stern DF, Sporn MB (1985) Type beta transforming growth factor: a bifunctional regulator of cellular growth. Proc Natl Acad Sci USA 82:114–123

Rozengurt E (1985) The mitogenic response of cultured 3T3 cells: integration of early signal and synergistic effects in a unified framework. Mol Asp Cell Reg 5:429–452

Rozengurt E, Legg A, Strang G, Courtnay-Luck N (1981) Cyclic AMP, a mitogenic signal for Swiss 3T3 cells. Proc Natl Acad Sci USA 78:4392–4396

Rudland PS, Weil S, Hunter AR (1975) Changes in RNA metabolism and accumulation of presumptive messenger RNA during transition from the growing to the quiescent state of cultured mouse fibroblasts. J Mol Biol 96:745–766

Saltiel AR, Cuatrecasas P (1986) Insulin stimulates the generation from hepatic plasma membranes of modulators derived from an inositol glycolipid. Proc Natl Acad Sci USA 83:5793–5797

Schliwa M, Nakamura T, Porter KR, Euteneur U (1984) A tumour promotor induces rapid and coordinated reorganisation of actin and vinculin in cultured cells. J Cell Biol 99:1045–1059

Schuldiner S, Rozengurt E (1982) Na^+/H^+ antiport in Swiss 3T3 cells: mitogenic stimulation leads to cytoplasmic alkalinisation. Proc Natl Acad Sci USA 79:7778–7782

Seuwen K, Lagard A, Pouyssegur J (1988) Deregulation of hamster fibroblast proliferation by mutated *ras* oncogenes is not mediated by constitutive activation of phosphoinositide specific phospholipase C. EMBO J 7:161–168

Sheppard JR (1972) Difference in the cyclic adenosine 3′,5′-monophosphate levels in normal and transformed cells. Nature New Biol 236:14–16

Sheppard JR, Prescott DM (1972) Cyclic AMP levels in synchronised mammalian cells. Exp Cell Res 75:293–296

Smets LA, Van Rooy H (1987) Mitogenic and antimitogenic effects of cholera toxin-mediated cyclic AMP levels in 3T3 cells. J Cell Physiol 133:395–399

Smith JB, Smith L (1984) Rapid calcium mobilisation by vasopressin and prostaglandin $F_{2\alpha}$ independent of sodium influx in quiescent cells. Biochem Biophys Res Commun 123:803–809

Sporn MB, Roberts AB (1985) Autocrine growth factors and cancer. Nature 313:745–747

Streb H, Irvine RF, Berridge MJ, Schulz I (1983) Release of Ca^{2+} from a non-mitochondrial intracellular store in pancreatic accinar cells by inositol 1,4,5 trisphosphate. Nature 306:67–69

Tagliaferri P, Katsaros D, Clair T, Neckers L, Robins RK, Cho-Chung YS (1988) Reverse transformation of Harvey murine sarcoma virus transformed NIH-3T3 cells by site selective cyclic AMP analogues. J Biol Chem 263:409–416

Trahey M, McCormick F (1987) A cytoplasmic protein stimulates normal N-*ras* p21 GTPase but does not affect oncogenic mutants. Science 238:542–545

Tucker RF, Volkenant ME, Branum EL, Moses HL (1983) Comparison of intra- and extracellular transforming growth factors from nontransformed and chemically transformed mouse embryo cells. Cancer Res 43:1581–1586

Tucker RF, Shipley GD, Moses HL, Holley RW (1984) Growth inhibitor from BSC-1 cells closely related to platelet type beta transforming growth factor. Science 226:705–707

Uno I, Fukami K, Kato H, Takenawa T, Ishikawa T (1988) Essential role for phosphatidylinositol 4,5-bisphosphate in yeast cell proliferation. Nature 333:188–190

Vallar L, Spada A, Giannattasio G (1987) Altered Gs and adenylate cyclase activity in human pituitary adenomas. Nature 330:566–568

Wakelam MJO (1988) Inhibition of the amplified bombesin-stimulated inositol phosphate response in N-*ras* transformed cells by high density culturing. FEBS Lett 228:182–186

Wakelam MJO (1989) Amplification of fluorocaluminate-stimulated inositol phosphate generation in a cell line overexpressing the $p21^{Nras}$ gene. Biochem J 259:737–741

Wakelam MJO, Murphy GJ, Hruby VJ, Houslay MD (1986a) Activation of two signal-transduction systems in hepatocytes by glucagon. Nature 323:68–71

Wakelam MJO, Davies SA, Houslay MD, McKay I, Marshall CJ, Hall A (1986b) Normal $p21^{N\text{-}ras}$ couples bombesin and other growth factor receptors to inositol phosphate production. Nature 323:173–176

Wakelam MJO, Houslay MD, Davies SA, Marshall CJ, Hall A (1987) The role of $p21^{N\text{-}ras}$ in the coupling of growth factor receptors to inositol phospholipid turnover. Biochem Soc Trans 15:45–47

Waterfield MD, Scrace GT, Whittle N, Stroobant P, Johnsson A, Wateson A, Westermark B, Heldin CH, Huang JS, Deuel TF (1983) Platelet-derived growth factor is structurally related to the putative transforming protein p28 *sis* of simian sarcoma virus. Nature 304:35–39

Werth DK, Pastan I (1984) Vinculin phosphorylation in response to calcium and phorbol esters in intact cells. J Biol Chem 259:5264–5270

Whitfield JF, Durkin JP, Franks DJ, Kleine LP, Raptis L, Rixon RH, Sikorska M, Walker PR (1987) Calcium, cyclic AMP and protein kinase C-partners in mitogenesis. Cancer Metastasis Rev 5:205–250

Whitman M, Kaplan DR, Roberts TM, Cantley L (1987) Evidence for two distinct phosphatidylinositol kinases in fibroblasts. Biochem J 247:165–174

Whitman M, Downes CP, Keeler M, Keller T, Cantley L (1988) Type I phosphatidylinositol kinase makes a novel phospholipid phosphatidylinositol 3-phosphate. Nature 332:644–646

Wolf BA, Comens PG, Ackerman KE, Sherman WR, McDaniel ML (1985) The digitonin-permeabilized pancreatic islet model: effect of *myo*-inositol 1,4,5-trisphosphate on Ca^{2+} mobilisation. Biochem J 227:965–969

Wolfman A, Macara IG (1987) Elevated levels of diacylglycerol and decreased phorbol ester sensitivity in *ras*-transformed fibroblasts. Nature 325:359–369

Zetterberg A, Larsson O (1985) Kinetic analysis of regulatory events in G_1 leading to proliferation or quiescence of Swiss albino 3T3 cells. Proc Natl Acad Sci USA 82:5365–5369

CHAPTER 15

Effect of Oncogenes on Cell Differentiation

D. BOETTIGER and D. CHALMERS

A. Introduction

The processes of cell transformation in tissue culture and neoplastic transformation in vivo produce cells which are recognizable by their distinctive morphology and by their ability to proliferate under conditions which do not support, or severely limit, the proliferation of their normal counterparts. The altered morphology is recognized by the pathologist in tumor biopsies and is used not only for diagnosis of malignancy but also for tumor staging and prognostic prediction. In tissue culture the altered cell morphology is recognized as foci of cells which are morphologically distinct from the parental-type cells which surround the focus. This has been used for the development of assays for oncogenic viruses (TEMIN and RUBIN 1958) and for the identification of oncogenes derived from tumors and introduced into the indicator cells by DNA transfection (SHIH et al. 1979; COOPER 1982). In both cases the diagnosis or identification of the altered cells may be aided by the excess proliferation of these cells in relation to their normal counterparts. The altered morphology is an indication of changes in the synthesis of products which allow us to distinguish cell types. Typically, these are not products required for cell survival, although the synthesis of "house-keeping" products may be quantitatively altered. They reveal alterations in cell-type specific products or developmentally regulated products. Thus, the morphological changes are indicators of changes affecting the differentiation of the cell. This idea is strengthened by analysis of the synthesis of these differentiated cell products using biochemical and molecular biological techniques.

In an earlier chapter, the role of growth factors in neoplasia was considered. These factors also play a role in the determination of cell differentiation. In particular, hematopoietic growth factors such as CSF-1, IL-3, and others are used to induce cell proliferation, cell differentiation, and cell survival (SPOONCER et al. 1986; TUSHINSKI et al. 1982). In solid tissues, there is less information about differentiation-promoting effects of growth factors. However, TGF-β affects both proliferation and cell differentiation, depending on the particular system analyzed (MASSAGUE 1987; OLSON et al. 1986). As our systems of analysis of the other well-studied growth factors such as EGF and insulin expand, it is likely that cell differentiation will become a more important issue in these systems. In the meantime, most studies have focused on the proliferative effects of these factors since they are easier to measure in our current tissue culture systems.

In the in vivo situation, it is clear that there must be a tight linkage between cell differentiation and cell proliferation, as both are critical to the production of mature tissue. As cells approach the terminal phase of their differentiation lineage, they find ways to limit their further proliferation as part of their differentiation program. The extreme examples of this are particularly obvious: erythrocyte maturation involves extrusion of the cell nucleus, muscle development involves fusion of myocytes to produce multinucleate myotubes, and nerve cells produce long processes. For other cell lineages, there is a general reduction in division potential as cells approach their terminal phase. Some developmental biologists even equate terminal differentiation with inability to proliferate. The mechanisms which induce cells to stop proliferation as a correlate of differentiation are not well understood and may differ from cell type to cell type. There are a couple of suggestions concerning the mechanism which may be inserted here. Chondrocytes encase themselves in a thick collagenous matrix which one would expect to impose a physical restriction on their ability to divide. Macrophages require the continuous presence of one of the hematopoietic growth factors for their survival, so that by regulating the level of available factor, the total macrophage population is set or altered. In the production of neoplasia these controls which stop proliferation must be modified.

How do oncogenes fit into this pattern? The variety of oncogenes presented in an earlier chapter suggests a range of roles that they may assume in the alteration of cell differentiation. In some cases, it appears that it is the normal growth regulatory signals that may be constitutively stimulated (discussed in more detail below). For other cases the interactions involved are more difficult to define, probably because they affect regulatory pathways which have not yet been well elucidated. In addition, the exact effect of many oncogenes on cell differentiation has not been described in sufficient detail. For both of these reasons much of what is included in the sections below is speculative. A detailed analysis of parameters associated with cell differentiation, or so-called transformation parameters (HANAFUSA 1977), for any particular oncogene-target cell system reveals that the effects of the oncogenes are pleiotropic. The simplest model for pleiotropy is that the oncogene is acting on a regulatory pathway and that the diverse effects are downstream consequences of the alteration in the pathway. Furthermore, there is little evidence indicating that the affects of an encogene are restricted to a single pathway in the target cell. Hence, the patterns of effects of the oncogene may not be easy to interpret. One important area for present and future research will be to identify the regulatory systems on which individual oncogenes operate.

B. Classification of Oncogenes

Probably the ideal way to classify the oncogenes would be on the basis of the particular cell regulatory pathway(s) with which they interact. Little is known of the pathways and how they might be influenced by different oncogenes during transmission of a signal from the plasma membrane through the cytoplasm to the nucleus, resulting in a change of gene expression. Oncogenes are currently classified on the basis of their DNA sequence and potential function (Table 1). For example, within the class of tyrosine kinase oncogenes, there appear to be at least

Table 1. Oncogene classification scheme

Oncogene	Cell homologue	Function
sis	B subunit of PDGF	Peptide growth factor
int-1	*wingless* (*Drosophila*)	Secreted morphogen
int-2	FGF family	Peptide growth factor?
*erb*B	EGF receptor	Tyrosine kinase receptor
fms	CSF-1 receptor	Tyrosine kinase receptor
neu	EGF receptor family	Tyrosine kinase receptor
ros	Insulin receptor family	Tyrosine kinase receptor
src	*src* family	Membrane tyrosine kinase
fgr	*src* family (*src*2)	Membrane tyrosine kinase
fps/*fes*	Myeloid regulator?	Cytoplasm tyrosine kinase
abl	Lymphoid regulator?	Tyrosine kinase
mos	?	Cytoplasmic serine kinase
mil/*raf*	?	Cytoplasmic serine kinase
H-*ras*	G-protein family	Signal transduction
K-*ras*	G-protein family	Signal transduction
N-*ras*	G-protein family	Signal transduction
myc	?	Nucleic acid binding
fos	Transcription factor	Transcription regulator
myb	See c1 locus (maize)	Transcription regulator
ski	?	?
*erb*A	Thyroid hormone receptor	Steroid hormone binding
jun	AP-1	Transcription factor

PDGF, platelet-derived growth factor; FGF, fibroblast growth factor; EGF, epidermal growth factor; CSF, colony-stimulating factor.

three subgroups. The first class is derived from growth factor receptors that possess a ligand-binding domain on the cell exterior, a transmembrane domain and an internal tyrosine kinase domain. The second class is exemplified by the *src* protein, which appears to require association with the internal face of the plasma membrane to exert its transforming potential. Third, the *fps*/*fes* class may be able to function without the membrane association. While these differences portend alternative ways in which oncogenes might interact with a regulatory pathway, more information will be required to substantiate this. Even within each class the oncogenes may be "wired" to different signalling pathways that share the use of tyrosine phosphorylation as a molecular mechanism but which have very different consequences for the cell.

Finally, it is possible that several oncogenes may operate on the same signalling pathway but exert their effect at different steps in the pathway. There is some evidence that the viral *src* oncogene requires a cellular *ras* homologue to exert its transforming effect (Smith et al. 1986). In addition, the similar effects of *src* and *ras* on differentiating cell systems support the possibility that these oncogenes may act on the same pathway.

Conclusions on the possible effects of oncogenes on cell differentiation or on the ability of cells to express their differentiation program are best evaluated in experiments in which oncogenes have been introduced into pure populations of

Table 2. Developmental models for oncogene testing

System	Oncogene	Effect	Genes
Myogenesis (chick embryo)	*src*	s	Cell cycle withdrawal
	fps	s	m-myosin, m-actin,
	erbB	s	m-tropomyosin, AchR,
	myc	s	creatine-kinase
	ski	a	Fusion
Myogenesis (rat cell line)	*ras*	s	Fusion, m-myosin
Chondrogenesis (chick embryo)	*src*	s	Type II collagen, proteoglycan
	myc	o	No effect
	mil	s	Type II collagen, proteoglycan
Fibroblasts (chick embryo)	*scr*	s	Fibronectin, type I collagen
Fibroblasts (mouse 3T3)			
Lens (chick embryo)	*src*	s	Delta-crystallin, mp-28 laminin, type IV collagen
Keratinocytes	*ras*	s	Keratins, morphology
	src	s	Loss of epidermal growth factor
	fes	s	dependence
	abl	s	
	mos	s	
Neural retina	*src*	o	Extended proliferation
	myc	o	
Neural (PC12 cell line)	*src*	a	Neurite extension
	ras	a	Neurite extension
Cerebellar (rat newborn)	*src*	o	Extended proliferation morphology
Macrophage (chick embryo)	*src*	o	No effect
	myc	o?	Extended proliferation morphology (smaller)
	myb	s	Fc receptors, lipids, ATPase, acid-PO_4-ase
Erythrocyte	*ras*		Reduced erythropoietin required
	src		
	erbB		
	erbA		
B lymphocyte	*abl*	s	Apparent partial differentiation block at different stages

s, suppression, differentiation product synthesis turned-off.
a, activation or induction of differentiation.
o, no effect.

differentiated cells, usually using a retrovirus or retroviral vector. The use of cell lines, or cloning the cells following the introduction of an oncogene, can produce misleading results since secondary modifications of differentiation may take place during growth following cloning (more than 20 cell generations are required to obtain enough cells for most analyses) or have occurred in the establishment of cell lines. Table 2 gives some of the differentiation systems which have been used in this type of analysis. Since the number of systems and the number of oncogenes used in each system are limited, some interpolation has been given below to arrive at a best guess of mechanisms involved.

C. Growth Factors and Growth Factor Receptors

Some aspects of growth factors have been surveyed in an earlier chapter. However. the effects of growth factors on cells are not limited to effects on cell proliferation. Many, if not all, of these factors also alter cell differentiation. Since the effects on proliferation are easier to measure, this has been used as an end-point for assays of growth factors and for experiments aimed at determining the molecular mechanisms associated with their signal transduction. However, the effects of growth factors on differentiation have also been documented [see reviews on transforming growth factor (TGF) (MASSAGUE 1987) and fibroblast growth factor (FGF) (GOSPODAROWICZ et al. 1987)]. Recently, evidence has been presented for a role of growth factors not only in the late stages of development but also in the embryonic induction of mesoderm (KIMELMAN and KISCHNER 1987). In the hematopoietic system, differentiation and cell proliferation are governed by a series of "growth factors" (DEXTER 1984; METCALF 1986). While it is clear that they are essential for differentiation, the cells which respond to particular factors generally die in their absence. The actual linkages between proliferation, differentiation, and survival are not yet understood.

There are two classes of oncogenes which appear to have their effect via the peptide growth factor pathway: oncogenes derived from the growth factors themselves and those derived from the receptors for those factors. It is possible that some of the other oncogenes may be involved in the signal transduction associated with growth factor-receptor binding, but this remains to be demonstrated.

I. Growth Factors

The realization that growth factor genes could give rise to oncogenes was first demonstrated with the elucidation of the DNA sequence of the v-*sis* oncogene and its relationship to platelet-derived growth factor (PDGF) (WATERFIELD et al. 1983). Since this discovery there has been speculation that the genes encoding other growth factors could also become oncogenes and that "transformation" of cells could be accomplished by an autocrine mechanism in which the cell produces a growth factor which drives its own proliferation. Recently, this proposition has been tested using hematopoietic growth factors introduced into retroviral vectors (see below).

1. *sis*

The *sis* oncogene was identified in a retrovirus (simian sarcoma virus) isolated from a fibrosarcoma of a woolly monkey (THEILEN et al. 1980; ROBBINS et al. 1981). The protein coded by the *sis* oncogene is highly homologous to the β-chain of PDGF although the cellular gene had been truncated during retroviral transduction (DOOLITTLE et al. 1983; WATERFIELD et al. 1983). The mutations sustained in its transduction do not appear to be required for oncogenic potential since the cellular homologue of the β-chain of PDGF can transform cells in culture (GAZIT et al. 1984; CLARKE et al. 1984). Some evidence demonstrates that the *sis* product is not released from the transformed cells but remains membrane bound (ROBBINS et al. 1985). However, the action of *sis* in transformation appears to require interaction with the PDGF receptor, suggesting that it acts solely as a substitute for PDGF (JOHNSSON et al. 1986; LEAL et al. 1985). Thus, *sis*-mediated cell transformation involves an autocrine mechanism.

While the cellular homologue of *sis* is generally thought of as a growth factor, it may also affect cell differentiation. PDGF is involved in the development of arterial smooth muscle. In the development of atherosclerotic lesions, in which smooth muscle cells both migrate and proliferate, high levels of the cellular *sis* transcripts are expressed (BARRETT and BERDITT 1987). Cellular *sis* has also been implicated in the phenotypic changes which occur during the progression of granulocytic leukemia. Chronic granulocytic leukemia is characterized by a phase in which an excess of normal-looking granulocytes are produced; in later stages, an acute transformation occurs, resulting in increased proliferation of more immature blast cells which is accompanied by an increase in the level of cellular *sis* transcripts (ROMERO et al. 1986).

2. Fibroblast Growth Factor-like Oncogenes

Tumor induction by mammary tumor virus involves the activation of specific cellular genes, due to the integration of the viral provirus in an adjacent chromosomal site. *int*-2 is activated in this way in a number of mammary tumors. Sequence analysis of the cloned *int*-2 gene shows substantial homology to basic fibroblast growth factor (b-FGF), although *int*-2 encodes a protein about twice as long with an extended carboxyl terminus (DICKSON and PETERS 1987).

Using the transfection of human tumor DNA into an NIH 3T3 cell indicator system, two laboratories have reported isolation of oncogenes with significant homology to FGF (DELLI BOVI et al. 1987; TAIRA et al. 1987). It is suggested that these genes represent additional members of the FGF gene family.

3. *int*-1

int-1 is another gene activated by murine mammary tumor virus in some mammary tumors (NUSSE et al. 1984). It appears to be homologous to the *wingless* gene in *Drosophila* (RIJSEWIJK et al. 1987), which affects cell interactions involved in wing morphogenesis. While the actual mechanism of action has not been determined, several researchers have suggested that it functions as a growth factor (RIJSEWIJK et al. 1987; CABRERA et al. 1987). The *int*-1 gene exhibits very restricted expression also in vertebrates, confined largely to the neural tube and the testis (WILKINSON et al. 1987; SHACKLEFORD and VARMUS 1987).

4. Hematopoietic Growth Factors

Hematopoiesis is controlled by a series of peptide hormones which regulate the proliferation and differentiation of hematopoietic stem cells into the several hematopoietic lineages and which are required for cell viability (METCALF 1986; DEXTER 1984). The potential of these growth factors to act as oncogenes has been tested by construction of retroviral vectors capable of transferring the genes encoding these polypeptides into target hematopoietic cells which normally respond to a particular factor. A vector encoding interleukin 3 (IL-3) converted IL-3-dependent cell lines to factor independence. These factor-independent cell lines were tumorigenic, unlike the parental factor-dependent cell lines (HAPEL et al. 1987).

II. Growth Factor Receptors

The other defined component in the peptide hormone signalling pathway is the hormone receptor. In some cases the receptors are for known peptide hormones, and in others they bear close sequence homology to known receptors, but the specific ligands remain to be defined. Each of the oncogenic forms of the receptor genes encodes a tyrosine kinase, and each has sustained specific mutations which relax the normal controls on the level, and perhaps the specificity, of the kinase activity. Therefore, it is possible that these oncogenes function as a constitutive, or as an excessive, generator of the normal signal derived from the normal interaction of the receptor with its normal ligand, thus utilizing the normal signal transduction pathway for that signal; however, it is equally possible that the receptor kinase, freed of normal restraints, now phosphorylates novel targets.

1. *fms*

The relationship between the action of a viral oncogene and the normal growth factor pathway is best understood for the gene encoding the colony-stimulating factor (CSF-1) receptor, the cellular homologue of the *fms* oncogene. Transduction of the receptor gene into the McDonough strain of feline sarcoma virus produces a deletion in the C-terminus of the protein (COUSSENS et al. 1986), a region of the protein that may be involved in regulating the level of kinase activity. Notably, the protein encoded by the *fms* oncogene still binds to its ligand (CSF-1), but the binding no longer affects kinase activity (SACCA et al. 1986). Cells expressing *fms*, like cells transformed by other oncogenes in this class, thus have excess tyrosine kinase activity that may be capable of triggering signalling pathways. The question is which signalling pathway or pathways may be stimulated. One of the most revealing experiments on this point involves transfection of the gene for the human CSF-1 receptor into NIH 3T3 cells which allowed the cell proliferation to be driven by human CSF-1 (Sherr, personal communication). This demonstrates that the normal CSF-1 receptor can find suitable targets for its kinase even in non-hematopoietic cells and can still produce a signal for cell proliferation. The even more curious aspect of this system is that the transfected NIH 3T3 cells not only proliferate but also grow in suspension in agar (ROUSSEL et al., 1987). This is not a normal property of these cells, but it is characteristic of their transformed counterparts. Hence, both the growth and the phenotype of the cell are affected. Does this suggest that multiple pathways are stimulated, or just that the signalling pathway tapped by the CSF-1 receptor in NIH 3T3 cells is distinct from that tapped by other factors such as EGF which stimulates normal monolayer growth in NIH 3T3 cells?

2. *erbB*

The *erbB* oncogene is the major transforming gene of the avian erythroblastosis virus and is homologous to the epidermal growth factor (EGF) receptor (DOWNWARD et al. 1984). The viral oncogene is truncated at both the C-terminus and in the extracellular domains, producing product with constitutive tyrosine protein kinase activity which is incapable of binding its normal ligand, EGF (DOWNWARD et al. 1984; OLOFSSON et al. 1986; NG and PRIVALSKY 1986). The transforming ability of *erbB* is not dependant upon a membrane association, since deletions which no longer encode the transmembrane domain can still transform (BASSIRI and PRIVALSKY 1987). This raises the question as to whether the transforming ability of *erbB* depends on the normal EGF signalling pathway or on another effector mechanism.

Most of the growth-factor-receptor oncogenes have not been tested for their ability to manipulate the differentiation of cells. Experiments using *erbB*, however, have shown that it is able to prevent myogenic differentiation (FALCONE et al. 1985). The effect appears superficially similar to the effect of other tyrosine kinase oncogenes such as *src* and *fps/fes* which are discussed in more detail below. However, some caution must be exercised in this interpretation since the myogenic system is particularly sensitive to interruption of differentiation, and more than one mechanism has been described (FALCONE et al. 1985). The normal role of this receptor appears to include promotion of differentiation, since EGF itself has been shown to induce differentiation in a number of tissues during embryonic development (ADAMSON 1987).

3. Other

A number of additional oncogenes have been described, based both on their transduction into retroviruses and on their performance in the transfection of NIH 3T3 cells. The *ros* oncogene of the UR2 avian sarcomavirus is related to the insulin-receptor gene family (NECKAMEYER et al. 1986). The *neu* oncogene is a member of the EGF receptor/*erbB* family but appears to have a different distribution of expression (YAMAMOTO et al. 1986). The *kit* oncogene is more distantly related to the CSF-1 receptor family and probably represents the receptor for one of the hematopoietic growth factors (BESMER et al. 1986). The *met* oncogene was isolated from an osteosarcoma cell line and also bears the hallmarks of a growth factor receptor (DEAN et al. 1985). Relatively little is known about these last mentioned genes; in particular, it will be important to identify the normal ligand for the cellular homologues of these oncogenes.

D. Tyrosine Kinases

The tyrosine kinases form a large family of oncogenes which are often classed with the growth-factor-receptor oncogenes due to their related DNA sequences in the tyrosine kinase, ATP binding, and C-terminal regulatory domains. They differ from the growth factor receptors chiefly in the absence of an extracellular domain for ligand binding. The absence of this domain implies that neither the oncogenes nor their parental homologues are involved in the initial stages of signal transduction. However, the pleiotropic effects which these oncogenes exert on target cells suggest an involvement in signalling pathways comparable to those initiated by the growth factor receptors. They could be involved either as sequential elements in signal transduction or as regulators of the pathway.

The mutations which distinguish the tyrosine kinase oncogenes from their normal cellular homologues result in deregulated and elevated levels of tyrosine kinase activity. Even if the normal specificity of the enzyme is retained, the increased activity is likely to lead to the phosphorylation of substrates by the oncogenic form. Since mutations are sustained in the conversion of the cellular gene to the oncogene, it is reasonable to expect changes in substrate specificity due to changes in protein structure or subcellular location.

It is often assumed that all of the oncogenes with tyrosine kinase activity transform cells by similar mechanisms. This is based on the investigation of proteins which exhibit increased phosphorylation on tyrosine in the presence of these oncogenes (MAYTIN et al. 1984; SEFTON 1986; HUNTER and COOPER 1985). Unfortunately, at least in the case of the *src* oncogene, most of the described, putative targets for the src tyrosine kinase do not appear to be either necessary or sufficient to induce transformation (KAMPS et al. 1986). Hence, this evidence for commonality of mechanism is questionable. In fact, there are some indications that there are differences. It is possible to distinguish chicken embryo fibroblasts transformed by *src*, *fps*, *ros*, and *erbB* on the basis of their morphology in culture. This is also reflected in effects on synthesis of extracellular matrix components (NOTTER and BALDUZZI 1984).

I. Membrane-Associated Tyrosine Kinase Oncogenes

Based on the sub-cellular location of their normal cellular homologues, tyrosine kinase oncogenes may be separated into two families: the membrane associated *src* family and the non-(plasma)membrane associated kinases.

src, in both its cellular and oncogenic forms, is myristilated on the N-terminus, which serves to anchor the protein in the plasma membrane (BUSS et al. 1984; GARBER et al. 1983) where it localizes in the adhesion plaques (ROHRSCHNEIDER et al. 1983). Deletion of the N-terminal glycine produces a mutant gene product which is not myristilated, is not localized to the plasma membrane, and does not transform cells (CROSS et al. 1984; KAMPS et al. 1985, 1986; BUSS et al. 1986). Hence, the plasma membrane association is essential for transformation for the *src* group of oncogenes. Other oncogenes which are likely to be in this group include *fgr*, *lck*, *syn*, and *lyn* (NAHARRO et al. 1983; VORONOVA and SEFTON 1986; SEMBA et al. 1986; YAMANASHI et al. 1987).

1. Effects on Differentiating Systems

The *src* oncogene was originally defined as the transforming gene associated with Rous sarcoma virus (BRUGGE and ERIKSON 1977; STEHELIN et al. 1976b). The isolation of temperature-sensitive mutants of Rous sarcoma virus which were temperature-sensitive for cell transformation but not for virus replication provided the first demonstration that viral *src* was only a transformation gene (MARTIN 1970). Subsequently, a host sequence homologous with *src* was described (STEHELIN et al. 1976a). Unlike the viral replication genes which do have homologous genes in the chicken (normal host for Rous sarcoma virus), the *src* homologies were not limited to relatives of the chicken but have also been identified in a wide variety of both vertebrates and invertebrates (STEHELIN et al. 1976a; LEV et al. 1984). The evolutionary conservation suggests that the cellular *src* gene must play a central role, and it has been suggested that this role might be in development.

The availability of a large number of these temperature-sensitive mutants, which are still easier to manipulate than temperature-sensitive mutants isolated for other oncogenes, and the suggestion of a normal developmental role for *src* has led to the examination of the effects of the viral *src* oncogene on a variety of differentiated and differentiating systems.

a) Myogenesis

Tissue culture systems using chicken embryo breast muscle have been developed which allow efficient differentiation of myoblasts to produce mature myotubes which will contract spontaneously (BISCHOFF and HOLTZER 1967). The process typically takes about 5 days, and cultures consist of 80%–90% pure myogenic cells. These cells can be infected with a temperature-sensitive mutant of Rous sarcoma virus (ts-*src*). Infected cells grown at non-permissive temperature express the majority of the muscle-specific proteins including muscle-specific meromyosin (FISZMAN and FUCHS 1975), ceratin phosphokinase (HOLTZER et al. 1975), acetylcholine receptors (FALCONE et al. 1984), and muscle tropomyosin (ALEMA and TATO 1987) but the synthesis of these products is blocked in cells shifted to the permissive temperature. Since these products are not normally synthesized until the cells withdraw from the cell cycle prior to cell fusion, the absence of these proteins could be explained by a block do differentiation at the stage of cell cycle withdrawal. Once the ts-*src*-infected cells are shifted to the non-permissive temperature, they differentiate normally to produce contracting myotubes (FISZMAN and FUCHS 1975; TATO et al. 1983; HOLTZER et al. 1975). Hence, the effect of *src* is fully reversible. If these ts-*src*-infected myotubes are subjected to a second shift to the permissive temperature, the cells degenerate over a period of 3–5 days. Prior to the degeneration, there is a reduction in the rate of synthesis of the muscle-specific proteins (WEST and BOETTIGER 1982). *src* does not appear to interfere with the normal differentiation program since the myoblasts may be maintained for many cell generations at the permissive temperature as transformed cells and still differentiate normally following a shift to non-permissive temperatures (FALCONE et al. 1985; ALEMA and TATO 1987).

The transition from replicating myoblast to myocyte which has withdrawn from the cell cycle appears to be governed by *src* in the ts-*src*-infected myogenic cells. The process is independent of the cell cycle, since the muscle maturation following a shift to permissive temperature takes place equally in normal and mitomycin C-blocked cultures (FALCONE et al. 1984). This suggests that the cells are waiting for a triggering event, and *src* blocks the signal from the normal trigger. The effect can be mimicked by the addition of a monoclonal antibody to integrin. In the presence of the monoclonal antibody, the cells continue to proliferate just as they do when they express the *src* oncogene. The effect is also fully reversible on removal of the antibody (MENKO and BOETTIGER 1987). Integrin is located in adhesion plaques like *src* and can be phosphorylated by *src* (HIRST et al. 1986). Integrin provides the link between the cytoskeleton and the extracellular matrix (HORWITZ et al. 1985; TAMKUN et al. 1986; HYNES 1987). This suggests that it is the establishment of this link which is the triggering event in terminal myogenesis and that the *src* oncogene interferes with this process.

b) Chondrogenesis

Tissue culture model systems using chondroblasts derived from chicken embryo sternal cartilage have been developed which maintain the normal chondrocyte phenotype, allow proliferation, and provide essentially pure cultures of a single cell type (PACIFICI et al. 1977; ADAMS et al. 1982). Unlike the myogenic system, the chondrocytes do not withdraw from the cell cycle prior to the synthesis of their terminal differentiated products. Hence, the link between cell proliferation and cell differentiation is more relaxed, and effects on differentiation may be studied in cells with equal proliferation rates. Infection of these chondrocytes with ts-*src* virus results in a suppression of the chondroblast phenotype. Cells change from their typical spherical shape to attached, elongate cells, and there are decreases in the cartilage-specific products including the specific proteoglycan (PACIFICI et al. 1977) and type II collagen (ADAMS et al. 1982). In addition, there are increases in synthesis of products not normally associated with mature chondrocytes such as fibronectin and type I collagen (ALLEBACH et al. 1985). An examination of the molecular mechanisms responsible for these changes in gene regulation have revealed a complex pattern. The primary control for type II collagen and the cartilage-specific proteoglycan is at the level of mRNA and is presumably governed by changes in transcriptional initiation (ADAMS et al. 1985; ALLEBACH et al. 1985). Type I collagen mRNAs increase following transformation, and there are some processing differences (ALLEBACH et al. 1985). The translation of type I collagen is not dependent on the expression of the *src* oncogene but appears to be governed by cell shape. Synthesis occurs in attached chondrocytes (ADAMS et al. 1987b). In addition to changes in the synthesis of extracellular matrix components, there are prominent changes in the components of the cytoskeleton including actin, tubulin, and vimentin (ADAMS et al. 1982, 1987a). Again, the cell shape appears to play a role along with *src* in determining the rates of synthesis of these products, with regulation at both transcriptional and translational levels (ADAMS et al. 1987a, 1985).

c) Melanogenesis

Melanoblast cultures were derived by the culture of cells from the pigmented retina of the embryonic chicken eye. Infection of these cells with ts-*src* virus causes a loss of pigment at the permissive temperature and reaccumulation of pigment when shifted to the non-permissive temperature (BOETTIGER et al. 1977). Pigment loss at the permissive temperature involved the destruction of melanosomes in large autophagosomes and elimination from the cell by exocytosis. It appears that the stability of the mature phenotype was lost, resulting in the cells inability to retain the complex differentiated structures.

d) Lens Cells

A culture system which permits the differentiation of three-dimensional lentoid bodies from dispersed cells of the chicken embryonic lens has been developed (MENKO et al. 1982, 1984). This system has the advantage of providing both initiation of synthesis of specific cell products and concerted morphogenesis to produce a complex structure. Infection of these cultures with ts-*src* virus and incubation at the permissive temperature produces transformed cells which continue to proliferate, and, unlike normal lens cultures, these cells can be passaged several times in culture. Shifting of these cells to the non-permissive temperature results in the differentiation and formation of lentoid bodies as in control cultures (MENKO and BOETTIGER 1988). Hence, the process is fully reversible, and the effect is grossly similar to the other systems described above. An examination of the synthesis of the specific products of these cells revealed that the synthesis of the basement membrane components, laminin and type IV collagen, was reduced by *src*: in contrast, delta-crystallin (major crystallin in the chick) and MP28 (the major component of the lens gap junctions) continued to be synthesized at normal rates. However, there was no accumulation of these products in the cultures and no assembly of lens gap junctions (MENKO and BOETTIGER 1988). This pattern appears to draw together the patterns observed in the synthesis of extracellular matrix components by chondroblasts and the inability to maintain structure in melanoblasts.

e) Neural Cells

Two sources of cells have been used in these experiments: neural retina cells from the developing chicken embryo and rat cerebellar cells. In both types of experiments, cell proliferation was an important selective event. In the rat system, the inefficiency of infection with an avian virus necessitated the isolation of rare cell lines on the basis of growth (GIOTTA et al.1980). The chick neural retina has a very limited ability to proliferate under the culture conditions used (CALOTHY et al.1980; CRISANTI COMBES et al. 1982). In both cases the effect of temperature shifts of ts-*src*-infected cells resulted in a relatively modest change in the cells and only a limited effect on the expression of differentiated parameters. It is clear that the expression of the *src* oncogene is not sufficient to switch off the synthesis of many neuronal markers (CRISANTI et al. 1985; GIOTTA and COHN 1981). This may not be surprising because of the results obtained with melanoblasts and the lens cells described above, but it might be instructive to examine the synthesis of extracellular matrix components by these cells. The synthesis of extracellular matrix is altered by *src* in all systems which have been examined.

There is one context in which the effects of *src* on neural cells may be particularly interesting. The expression of the cellular *src* gene is high in neuronal cells (MANESS 1986; COTTON and BRUGGE 1983; GRADY et al. 1987), although this form of *src* is slightly different from that expressed in non-neuronal cells, due to an alternate splicing of the mRNA (BRUGGE et al. 1985; LYNCH et al. 1986). In the neuronal-derived PC-12 cells, the introduction of *src* has an effect similar to the addition of nerve growth factor, i.e. the induction of nerve process formation (ALEMA et al. 1985a). Hence, *src* may be involved in the regulation of neurite outgrowth and/or responses to nerve growth factor.

f) Macrophages

Macrophages represent another cell type in which the expression, of the cellular *src* gene or, more likely, the closely related cellular *fgr* (*src*2) gene is elevated (WILLMAN et al. 1987a, b; GEE et al. 1986). Experimentally, the *src* oncogene has been introduced into yolk sac-derived macrophages by infection with Rous sarcomavirus. While the *src* gene was expressed and exhibited a subcellular distribution similar to transformed fibroblasts, the infected macrophages failed to show altered morphology or altered expression of macrophage-specific markers (DURBAN and BOETTIGER 1981b; LIPSICH et al. 1984). This result implies that the *src* oncogene is limited in the cell types which it is able to transform, probably due to the absence of appropriate targets for the tyrosine kinase. This also illustrates what may become a common theme in the limitation of target cells for particular oncogenes. Individual oncogenes usually fail to transform cell types which normally express its cellular homologue at high levels.

II. Cytoplasmic Tyrosine Kinases

fps and *abl* are tyrosine kinase oncogenes, provisionally classed as cytoplasmic oncogenes. The basis of this classification is that the normal cellular homologues of these oncogenes may be expressed as cytoplasmic forms rather than being plasma membrane associated. Furthermore, there are results which suggest that, in some cases, the viral-derived oncogenes may transform in the absence of a membrane association. However, additional data will be required to establish that transformation is actually driven by a cytoplasmic form. In the virus from which these oncogenes were defined, both genes exist as fusions between the oncogene portion and viral *gag* gene sequences.

1. *fps*/*fes*

fps and *fes* represent the avian and feline forms of the same cellular gene which has been picked up by avian or feline retroviruses, respectively (SHIBUYA et al. 1982; SHIBUYA and HANAFUSA 1982; FEDELE et al. 1981). The normal cellular *fps* gene is expressed in the cytoplasm of myeloid cells (FELDMAN et al. 1983, 1985). In its viral oncogene form, *fps* is fused to the viral *gag* gene, which probably serves to translocate it to the plasma membrane (FELDMAN et al. 1985; MOSS et al. 1984). In any case, the fusion of viral *gag* sequences is sufficient to activate the oncogenic potential of cellular *fps* (FOSTER et al. 1985). In addition to being fused to the viral *gag* gene sequences, the viral *fps* oncogene in Fujinami sarcomavirus (the prototype *fps* viral oncogene) contains mutations in the *fps* sequences. These mutations also appear to be sufficient to activate the oncogenic potential of the cellular *fps* gene even in the absence of fusion to the viral *gag* gene (FOSTER and HANAFUSA 1983). Thus, the viral gene has been activated to an oncogenic form by two apparently independent mechanisms. While somewhat speculative, it appears possible that altering the subcellular location by translocation to the plasma membrane may convert it to a *src*-like transformation mechanism (see above). Alterations in the *fps* coding sequences may affect only the kinase activity or its regulation and leave the protein product with its normal cytoplasmic location. In the latter case, the membrane association may not be required for transformation, and it might be expected that other targets for the kinase activity may be involved in the transformation process. This speculation and the normal cytoplasmic localization of the cellular homologue provide the rationale for the separation of this class from the *src* class.

Studies of the effects of *fps* on differentiation have been limited to the viral *gag-fps* gene using the original viruses. In the myogenic differentiation system, the effect of *fps* and *src* in suppression of myogenic differentiation appears to be analogous (FALCONE et al. 1985). In fibroblasts, the targets for the tyrosine kinase activity of *fps* and *src* have not been distinguished (HUNTER and COOPER 1985). Hence, it may be that the viral fusion protein acts like the plasma membrane-associated kinases in these systems. It may be more instructive to test the *fps* gene separated from its viral *gag* tail to determine its effects on differentiation of the cytoplasmic form.

2. *abl*

The case with *abl* is similar. The viral oncogene is a fusion of *abl* with the viral *gag* gene (WITTE 1986). The murine gag protein is myristilated at the N-terminus, which serves to attach it to the plasma membrane (SCHULTZ and OROSZLAN 1984). Hence, the viral form of the *abl* oncogene is primarily membrane associated (WITTE 1986). The membrane association may affect the ability of *abl* to transform different types of cells since the gag portion of the fusion protein is essential for transformation of lymphocytes but does not appear to be required for the transformation of fibroblasts (PRYWES et al. 1983). Again this may be an indication of different pathways of transformation, one membrane associated and one not. The cellular form of the *abl* oncogene has been detected both in the cytoplasm and in association with the plasma membrane (WITTE 1986). This complexity may be explained by alternative splicing of the mRNA, which allows the production of four different polypeptides which differ at the N-terminus (BEN-NERIAH et al. 1986). The membrane-associated form is probably due to myristilation of the N-terminus of one of these forms.

The *abl* gene can transform a number of different tissues depending upon the virus it is transduced into (WANECK and ROSENBERG 1981; BOSS et al. 1979; WHITLOCK et al. 1983; SCOTT et al. 1986; PIERCE et al. 1985). Viruses carrying the *abl* gene when transfected into bone marrow cells can both stimulate cell proliferation and differentiation of erythroid cells. This effect can also be achieved by *ras* genes, but in the case of *abl* the phenomenon occurs independently of growth factor (WANECK et al. 1986). The *abl* gene has also been shown to transform hematopoietic lineages to IL-3 independence (MATHEY PREVOT et al. 1986; SACCA et al. 1986). These results suggest that the oncogene is involved in the transmission of signals in certain growth factor pathways. However, as in the case with *fps*, all the experiments have used the viral *gag-abl* fusion oncogene which is likely to exert its oncogenic potential by virtue of its association with the plasma membrane, as for the *src* class. Experiments with a cytoplasmic version of the oncogene will be required to determine whether or not the association with the plasma membrane is essential for transformation.

Expression of the *abl* cellular gene is developmentally regulated (MULLER et al. 1982). Expression occurs in nearly all adult tissues, but the level and form may vary between tissues; for example, only one type is formed in haploid cells of the mature testes (WOLGEMOUTH et al. 1986).

E. Serine-Threonine Kinases

In addition to the tyrosine kinases which are potentially cytoplasmic, there are two additional, related oncogenes which display serine-threonine kinase activity and also are localized to the cytoplasm. This family consists of the *mil/raf* and the *mos* oncogenes. These oncogenes are often classified as members of the *src* family on the basis of sequence homology; however, they are more distant relatives, and they appear to encode serine-threonine-specific protein kinases rather than thyrosine kinases (SETH et al. 1987; MOELLING et al. 1984). The sequence homology appears to be largely a reflection of the general relatedness of the catalytic domains of all protein kinases. In addition to their protein kinase activity, both these oncogenes will bind DNA and RNA in vitro (SETH et al. 1987; BUNTE et al. 1983). The cytoplasmic localization of these proteins is not based on strong direct evidence. Due to the very low levels of expression in normal tissues which have been studied, it has only been possible to localize the *mos* oncogene to the cytoplasm in cells transformed by Moloney sarcomavirus which encodes the prototypic *mos* oncogene (PAPKOFF et al. 1983). These oncogenes appear to be members of a larger gene family as additional, closely related genes have been identified by Southern blot analysis, e.g. *pks* (MARK et al. 1986) and *raf 2* (HUEBNER et al. 1986).

I. *mos*

The cellular homologue of the *mos* oncogene is unusual for its lack of introns and for its ability to transform cells in vitro, the latter implying that activating mutations which affect the protein are not required for oncogenic potential (BLAIR et al. 1981, 1986). The level of transcription of the cellular homologue is quite low and may be limited largely to testes and ovaries (PROPST and VANDE WOUDE 1985; GOLDMAN et al. 1987). Given its very low transcription rates, it may not be surprising that activation of the transforming potential of *mos* requires alteration of elements involved in the tight control of its transcription (MCGEADY et al. 1986; van der Hoorn et al. 1985). In addition to its expression in germ cells, there is some indication that it may function in early development. Transcripts have been detected in embryonal carcinoma cells (OGISO et al. 1986) and in limb bud chondrocytes (BOETTIGER, unpublished data).

Experimental introduction of the *mos* oncogene into differentiated cell types has employed either the Moloney sarcomavirus (MoMSV) or myeloproliferative sarcomavirus (MPSV). Both encode a *mos* oncogene as their transforming gene; MPSV *mos* is more closely related to the cellular *mos* gene. Infection of thyroid epithelial cells with MPSV resulted in loss of thyroglobulin secretion, loss of iodide uptake, and loss of dependence on six growth factors including thyrotropin (FUSCO et al. 1985). When a temperature-sensitive mutant of MPSV was used and shifted to non-permissive temperature, the transformed phenotype reverted, but there was no re-expression of the differentiation markers (FUSCO et al. 1985), as has been generally observed with the *src* systems described above. This raises the possibility that different methods of interfering with differentiation may be operating in this system. In fibroblasts, *mos* suppresses the synthesis of both fibronectin and type I collagen (SETOYAMA et al. 1985), which parallels the effect of *src* on fibroblasts. The mechanism of suppression may be different from *src*, however, since evidence for a direct effect on the collagen promoter has been presented (SCHMIDT et al. 1985).

In vivo MPSV induces primarily a myeloproliferative disease with eventual bone marrow failure in vivo (OSTERTAG et al. 1980). Part of the difference in tumor induction between this virus and the related MoMSV is due to differences in the LTR which affects the tissue-specific expression of the viral genes (STOCKING et al. 1986). The effect on the bone marrow stroma is paralleled in vitro. Long-term bone marrow cultures rapidly transform and hematopoiesis ceases following exposure to retroviruses carrying the *mos* oncogene (DEXTER et al. 1977).

II. *mil/raf*

The *raf* oncogen was found using a protocol designed to isolate retroviruses carrying new oncogenes which would produce carcinomas (RAPP and TODARO 1980; RAPP et al. 1983a, b). This virus induces adenocarcinomas, histiocytic leukemias, and fibrosarcomas (RAPP et al. 1983a, b). Analysis of the oncogene of this virus revealed its homology to the *mil* oncogene of the avian MH2 virus (SUTRAVE et al. 1984; JANSEN et al. 1984), which induces primarily histiocytic tumors. The gene products of these oncogenes are primarily cytoplasmic (BUNTE et al. 1983).

Both *mil* and *raf* have been reported to affect factor-dependent growth of myeloid cells, but there may be different mechanisms involved in different test systems. Avian macrophages are induced by *mil* to produce a macrophage growth factor which leads to autocrine stimulation of the infected cells. However, *myc* is also required in this system to drive cell proliferation (GRAF et al. 1986). IL-3-dependent cell lines of murine origin are rendered factor independent by expression of the *raf* oncogene (CLEVELAND et al. 1986). In this case, it is not an autocrine response and c-*myc* expression is elevated. The cooperation of *myc* with these oncogenes is underlined by the finding of *myc* in addition to *mil* in the MH2 virus from which *mil* was derived (SUTRAVE et al. 1984).

These oncogenes also affect the expression of differentiation markers in solid tissue cells. *raf* appears to function as a *trans*-activator for alphafetoprotein gene expression (VOGT et al. 1987). In the avian chondroblast system, *mil* suppressed the synthesis of the normal chondroblast extracellular matrix components even more strongly than *src* and also affected the reemergence of type 1 collagen expression (ALEMA et al. 1985b). Hence, these oncogenes can suppress differentiation in some of the same systems affected by *src* and *ras* genes, but to date exploration of their potential for altering the expression of differentiation has been very limited.

F. *ras* Oncogenes

The cellular homologue to the *ras* oncogene is a member of the G-protein group which binds and hydrolzyes GTP (MANNE et al. 1985; SWEET et al. 1984; GIBBS et al. 1984). Some members of this family function as regulatory subunits of the cAMP-dependent protein kinases. Thus, the *ras* oncogenes may be related in function to the protein kinase oncogenes as additional regulators of the process. As might be expected from this analysis, their effects on differentiation parallel the protein kinases, particularly the *src* oncogene. The family consists of three closely related oncogenes; K-*ras*, from Kirsten sarcomavirus; H-*ras*, from Harvey sarcomavirus; and N-*ras* isolated from a neuroblastoma line by DNA-mediated transformation (HARVEY 1964; LOWY and WILLUMSEN 1986). While there appears to be some tissue specificity for the expression of the cellular homologues of each of these oncogenes (LEON et al. 1987; MULLER et al. 1982; MULLER and VERMA 1984), they appear to substitute equally for each other in functional studies (BARBACID 1987; LOWY and WILLUMSEN 1986).

From the frequent association of increased levels of *ras* gene expression in tumor cells, it has been postulated that *ras* plays a direct role either in malignany or in cell proliferation. Microinjection of monoclonal antibodies to ras proteins prevent serum-stimulated NIH 3T3 cells from entering S phase (MULCAHY et al. 1985). Temperature-sensitive *ras* mutants have been used to reveal a possible second requirement for *ras* in the G2 phase of the cell cycle (DURKIN and WHITFIELD 1987). These experiments tend to support the hypothesis that *ras* is involved in cell proliferation. An examination of malignant cells and normal human embryos for levels of *ras* proteins shows a lack of correlation between the level of *ras* and the rate of cell proliferation. Immunohistochemical examination of the distribution of *ras* proteins in normal and malignant tissue has revealed *ras* in nearly all normal tissues. However, expression is restricted to cells at specific stages of differentiation and is not correlated with proliferation states (CHESA et al. 1987). The pattern of expression of *ras* in neoplastic tissue thus reflects the normal tissue from which it is derived (CHESA et al. 1987). It is also relevant to this argument that the highest levels of H-*ras* protooncogene product in mice occur in muscle and brain, tissues in which proliferation is at a minimum (LEON et al. 1987). In *Drosophila*, three *ras* homologues, which are developmentally regulated, have been identified. *Ras* is expressed uniformly in the embryo, in dividing cells (imaginal discs, gonads, and brain), in larvae, and in adult ovaries, brain, and ganglia (SEGAL and SHILO 1986).

Ras constructs and retroviruses encoding *ras* oncogenes have been used to transfer *ras* into a variety of mesenchymal and epithelial cell types. Many of the experiments focused only on the ability of *ras* to transform particular differentiated cell types or differentiated cell lines. Thus, *ras* has been used to transform bronchial epithelial cells (YOAKUM et al. 1985), intestinal epithelial cells (BUICK et al. 1987), thyroid epithelial cells (NAKAGAWA et al. 1987), and others. In a few cases differentiation parameters have been examined and are suppressed as in the case of human keratinocytes (WEISSMAN and AARONSON 1985) and myogenic cell lines (OLSON et al. 1987). Thus, it is clear that *ras* is able to transform a wide variety of different cell types. Superficially, the pattern of cell types appears to parallel that of *src*, through more extensive experiments will be required to confirm this. In the few cases in which differentiation parameters have been examined, both *ras* and *src* have the same effect (WEISSMAN and AARONSON 1985; OLSON et al. 1987), and since the other transformation systems are accompanied by morphological changes, it appears likely that differentiation parameters will also be suppressed in those systems. A striking example of the parallelism of *ras* and *src* is the induction of neurite extension of PC-12 cells which mimics the effects of nerve growth factor on these cells. This can be induced either by transformation by a retrovirus carrying the *src* oncogene (ALEMA et al. 1985a) or by microinjection of the p21 *ras* oncogene product (BAR SAGI and FERAMISCO 1985).

In hematopoietic systems *ras* does not appear to alter the phenotype of differentiated cells but affects the responses of cells either by extending their life span in culture, as for mast cells (REIN et al. 1985) and B lymphocytes (LICHTMAN et al. 1986), or by altering the response to growth factors as in the erythropoietin-independent differentiation of erythroid cells (PHARR et al. 1987). Unlike most kinase oncogenes, *ras* does not eliminate the requirement for growth-factor-sustained proliferation (REIN et al. 1985).

G. Nuclear Oncogenes

The area of nuclear oncogenes is one of the most active in oncogene research, particularly at the molecular level. With current interest in the isolation of transcription factors and the demonstration that expression of nuclear oncogenes is accompanied by altered expression of a variety of other host genes, there have been attempts to find a connection between these genes and direct, transcriptional regulation. Given the somewhat unsettled nature of the field and the rapid accumulation of data, it is somewhat risky to attempt a classification of groups of these oncogenes. The attempt presented here focuses more on the broad aspects of the biology of the cellular homologues of these oncogenes rather than on specific modes of oncogenic action or molecular mechanism.

I. Factor-Responsive Genes

This category is derived from the observations that following the treatment of a variety of cell types with growth factors or other hormones there is a transient expression of the cellular homologues of these oncogenes which follows within

Table 3. Some inducers of *myc* and *fos*

Inducer	Cells type	*myc/fos*		References
IL-2	T cell	M	F	KOVACS et al. (1986)
IL-2	T cell	M		REED et al. (1985)
IL-3	Hematopoietic cell lines	M	F	CONSCIENCE et al. (1986)
EGF	Fibroblasts	M	F	BRAVO et al. (1985); RAN et al. (1986)
PDGF	Fibroblasts	M	F	WOME et al. (1987); KRUIJER et al. (1984)
BCGF	B cell	M		LACY et al. (1986)
IGF-1	Muscle cell line		F	ONG et al. (1987)
NGF	PC-12, other neural cells		F	MILBRANDT (1986); KRUIJER et al. (1985); CURRAN and MORGAN (1985)
B-adrenergic	Submandibular gland		F	BARKA et al. (1986)
TSH	Thyroid	M		DERE et al. (1985)
Thyrotropic hormone	Thyroid cells		F	COLLETTA et al. (1986)
TPA	B cell	M		NILSSON et al. (1986)
LPS	Bladder epithelial cell		F	SKOUV et al. (1986)
	Macrophage	M	F	INTRONA et al. (1986)
Ca ionophore	HL-60, fibroblast	M		CHAPEKAR et al. (1987); RAN et al. (1986)
Retinoic acid	HL-60	M		CHAPEKAR et al. (1987)
Antigenic	B cell	M		LACY et al. (1986)
Antigenic	T cell		F	KAUFMANN et al. (1987)
Seizure	Central nervous system		F	MORGAN et al. (1987)

M, *myc* gene induced; F, *fos* gene induced; IL, interleukin; EGF, epidermal growth factor; PDGF, platelet-derived growth factor; TSH, Thyroid-stimulating hormone; BCGF, B-cell growth factor; IGF, insulin-like growth factor; NGF, nerve growth factor; TPA, 12-O-Tetradecanoyl phorbol 13-acetate; LPS, lipopoly saccharide.

minutes to hours after treatment (Table 3). The primary oncogenes in this category are *myc* and *fos*. *myb* might be considered as a possible member of the group, but its appearance is less regularly observed following growth factor treatment, and it usually follows after a longer time course than *myc* and *fos* (PAUZA 1987; STERN and SMITH 1986). Functional studies link *myb* to macrophage differentiation (see below), therefore, *myb* is considered in a separate category.

1. *myc*

Both the virus-derived *myc* oncogene and its normal cellular homologue appear to have the same ability to transform cells in vitro or induce macrophage tumors in vivo (COLE 1986; BAUMBACH et al. 1986). *myc* has been used with either a murine mammary tumor virus enhancer (LEDER et al. 1986) or an immunoglobin heavy chain enhancer (LANGDON et al. 1986) to construct transgenic mice. In both cases expression occurs largely in target tissues for the respective enhancer, but tumors which arise are monoclonal. This implies that *myc* expression is not sufficient to induce tumors in these cell types but that additional factors are required.

Due to its induction as an early response to stimulation by a variety of growth factors and its elevated expression in many tumors and tumor-derived cell lines, *myc* function has been related to control of the cell cycle. The introduction of *myc* into factor-independent hematopoietic cell lines relieves the factor requirement, and activation of the cellular *myc* homologue is a common element for all oncogene transductions which render these cells independent of growth factor (RAPP et al. 1985).

Cell differentiation in several lineages requires withdrawal from the cell cycle as an essential part of the process. Introduction of a *myc* gene into differentiating muscle prevents withdrawal from the cell cycle and expression of muscle-specific products (FALCONE et al. 1985). However, it is likely that only a transient drop in *myc* is required since *myc* expression increases in post-mitotic cells (ENDO and NADAL GINARD 1986). Erythroid cell differentiation exhibits a similar pattern. Expression of the cellular *myc* gene is transiently repressed at the stage of commitment to terminal differentiation and then reexpressed as differentiation proceeds (LACHMAN and SKOULTCHI 1984). Reduction of the reexpression with antisense RNA reduces the rate of differentiation (LACHMAN et al. 1986). At the critical commitment point, the cell also starts counting the terminal cell divisions, and the programming of that event may be analogous to the myogenic withdrawal from the cell cycle. Transgenic mice have been produced with *myc* coupled to an immunoglobulin heavy chain promoter. In these animals, the small, resting B lymphocytes are not seen (LANGDON et al. 1986). This may be explained by the difficulty of withdrawing from the cell cycle when high levels of *myc* are expressed, as in the previous examples.

The pattern of *myc* expression during development shows considerable specificity for cell type and differentiation stage, but this does not correlate with the rate of division in that cell population. In *Xenopus*, high levels of *myc* are found as maternal mRNA in the egg, and it is retained at moderale levels throughout cleavage (KING et al. 1986; TAYLOR et al. 1986). In human development, high levels are also found in the outer layers of the skin, not a location of high cell proliferation. Even during the course of the cell cycle in cell lines in culture, the level of *myc* synthesis appears to be constant throughout the cell cycle (THOMPSON et al. 1985; HANN et al. 1985).

While *myc* can affect the expression of cell differentiation in cases in which there is a tight coupling of cell division and differentiation, when this coupling is loose, it appears that *myc* does not alter the differentiated phenotype of the cell. In macrophages transformed by *myc*, the normal macrophage markers are still expressed (DURBAN and BOETTIGER 1981 b). Chondroblasts expressing *myc* also continue to synthesize type II collagen and cartilage-specific proteoglycan (ALEMA et al. 1985 b).

The actual mechanism by which *myc* affects cells is still unclear. The effects on cell differentiation are probably mediated via a link with the control of cell proliferation. This leaves open the question of the function of elevated *myc* in a number of terminally differentiated cell types. The effect on the cell cycle does not follow the general pattern exhibited by growth-factor-mediated stimulation of proliferation. More information will be required to understand the logic of *myc*.

2. *fos*

The cellular homologue of *fos* is able to transform cells in vitro (MILLER et al. 1984), indicating that mutations which affect the coding region are not critical to activate its oncogenic potential. However, transgenic mice which express *fos* in a number of different tissues do not develop an increased number of tumors (RUTHER et al. 1987). Thus, elevated *fos* does not represent a preneoplastic condition. The major effect in transgenic mice is on bone remodelling, in which the resorption process is reduced (RUTHER et al. 1987). This is consistent with the induction of bone tumors by the original *fos*-encoding retroviruses (FINKEL et al. 1966).

The most promising clue is the recent identification of a cellular gene which is regulated, apparently directly, by *fos*. In the differentiation of adipocytes the AP-2 gene is switched on. The *fos* gene product is part of a nucleoprotein complex found bound to the regulatory region of this gene (DISTEL et al. 1987).

II. Direct Gene Regulation

This class is distinct from the factor-responsive oncogenes in that they are not often induced in response to a variety of factors. The evidence for a direct role in the specific regulation of gene transcription is also more direct for these genes. The class consists of two genes, *erbA*, which appears to be the oncogenic homologue of the thyroid hormone receptor (SAP et al. 1986; WEINBERGER et al. 1987), and *jun*, which appears to be the homologue of the AP-1 transcription factor (BOHMANN et al. 1987).

1. *erbA*

The *erbA* oncogene received its name by virtue of its association with avian erythroblastosis virus, in which it enhanced the transforming potential of a second oncogene associated with the same virus, *erbB* (BEUG et al. 1982). *erbB* is able to produce erythroid transformation independently of *erbA*, while *erbA* is not able to induce transformation of erythroid cells by itself (FRYKBERG et al. 1983). Data from transformation experiments suggest that *erbA* assists in transformation by reducing the spontaneous rate of differentiation of transformed erythroblasts. Thus, the general effect is suppression of erythroid differentiation. *erbA* is not able to transform fibroblasts, but it does increase their proliferation and their life span in culture (GANDRILLON et al. 1987). Since little is known about the mechanism of increased life span in vitro, it is difficult to ascribe this to an action of *erbA* on the expression of differentiated cell functions or to a more direct effect on proliferation.

The demonstration of close sequence homology to known steroid hormone receptors and the ability of the oncogene to bind thyroid hormone have permitted the identification of *erbA* as a homologue to the thyroid hormone receptor (Sap et al. 1986; Weinberger et al. 1987). The steroid hormone receptors are involved with the transport of hormone to the nucleus and bind to specific regions of the DNA, causing the induction of hormone-responsive genes. On this basis the *erbA* gene has received reclassification as a nuclear oncogene with direct involvement in gene regulation. Thyroid hormone is known to have a variety of effects on development including the classic demonstration of its ability to induce metamorphosis in the frog (Hauser et al. 1986). Thyroid hormone also controls differentiation of rat adipocytes in culture (Chabot et al. 1987) and the development of skeletal muscle (Albis et al. 1987). Hence, it is likely that further investigations will reveal a variety of effects of this oncogene on differentiation, now that we have some idea of where to look.

2. *jun*

The *jun* oncogene was isolated from an avian sarcomavirus and is able to induce transformation in vitro and in vivo (Maki et al.1987). The *jun* oncogene product has a nuclear location and is related, by sequence homology, to a yeast transcription factor (GCN4) which regulates the expression of genes in amino acid biosynthesis (Bohmann et al. 1987; Arndt and Fink 1986). The *jun* gene is also related to human transcription factor AP-1, which controls gene expression in response to phorbol esters (Bohmann et al. 1987); both the AP-1 gene and the yeast GCN4 recognize and bind a specific palindromic sequence. Because of its direct effect on gene expression, it is likely that *jun* will also effect the expression of cell differentiation in some systems, but this remains to be determined.

III. Differentiation-Responsive Oncogenes

The oncogenes *myb* and *ski* are classified as differentiation responsive because they may play direct roles in control of differentiation expression. Very little information is available about *ski*, so inclusion with *myb* may be somewhat arbitrary.

1. *myb*

The *myb* oncogene has been identified by its association with avian myeloblastosis virus (Beug et al. 1979; Bister et al. 1982). It is localized in the nucleus in tumor cell lines expressing either the viral-encoded *myb* or its cellular homologue (Boyle et al. 1985; Klempnauer et al. 1984) and appears to have DNA binding capability (Moelling et al. 1985), although no sequence specificity has yet been demonstrated. While some researchers have tried to associate *myb* with cell cycle control (Thompson et al. 1986), the viral-encoded *myb* appears to affect only the expression of macrophage differentiation parameters (Beug et al. 1979; Durban and Boettiger 1981 b; Moscovici and Moscovici 1983; Gazzolo et al. 1979; Graf et al. 1981). The rate of proliferation either in vitro or in vivo of *myb*-transformed macrophages is about 3 days, which is similar to, or slower than, macrophages with a comparable level of cytodifferentiation (promonocyte) (Durban and Boettiger 1981 a). One of the remarkable properties of *myb* is its restriction of target cells to cells in the macrophage lineage (DURBAN 1980; SHEN ONG et al. 1987) (Olsen and Boettiger, unpublished data), although it may also function in the lymphocyte lineage. The ability of *myb* to switch-off specifically macrophage differentiation suggested that it might function as a suppressor of differentiation during the expansion of macrophage progenitors. Indeed, the expression of the cellular homologue reaches higher levels in these macrophage progenitors (DUPREY and BOETTIGER 1985) than in any other reported case (SHEINESS and GARDINIER 1984; EMILIA et al. 1986; COLL et al. 1983). Consistent with this hypothesis, suppression of c-*myb* expression with anti-sense oligonucleotides reduces the colony size for human bone marrow mononuclear cells in a standard hematoporetic colony assay (GEWIRTZ and CALABRETTA 1988). Also, in double infections with *myb* and *myc* retroviruses, it is the expression of *myb* that determines the cell (differentiation) phenotype (NESS et al. 1987).

2. *ski*

The *ski* oncogene was identified as the transforming gene of a retrovirus isolated by tissue culture passage of a non-transforming retrovirus (LI et al. 1986; STAVNEZER et al. 1981). The *ski* gene product has been localized to the nucleus (STAVNEEZER, personal communication). It is expressed at moderate levels throughout the early avian embryo (BOETTIGER, unpublished data). *ski* has the interesting property of inducing myogenic differentiation in cultures of quail embryo cells (ALEMA and TATO 1987).

H. Summary and Perspectives

I. Cell-Type Specificity of Transformation

Where it has been tested, all oncogenes are able to alter the differentiation of cells within their target tissue or range of susceptible cell types. Where a sufficiently wide range of cell types has been tested for susceptibility to transformation by a particular oncogene, there is both a class of susceptible and a class of resistant cell types. The range of susceptible cell types may be very restricted, as in the restriction of *myb* to transformation of cells in the monocytic lineage, or broad, as in the case of *src* or *ras*, in which few resistant cell types have been identified. The test systems which have been employed to test the ability of oncogenes to alter differentiation have generally been restricted to cells which are committed to a particular differentiation pathway and usually those in which some of the terminal differentiation markers are already expressed but prior to withdrawal from the cell cycle permanently. This reflects, on the one hand, the practical problems of obtaining sufficient quantities of stem cells, or multipotential cells, and the inability to distinguish particular differentiation markers for these cells. The cells are largely recognized only by the progeny they produce. On the other hand, gene transfer into cells which have withdrawn from the cell cycle is not practical using retroviral vectors or DNA transfection procedures. Thus, the interpretations are based on observation of this limited developmental window.

II. Function of Cellular Oncogenes

Each of the identified oncogenes has evolved from a normal cellular gene. This fact suggests that the function carried out by the oncogene and that carried out by its normal homologue share certain molecular mechanisms. With the expection of *ras*, the identified oncogenes do not appear to be present in yeast or other unicellular organisms (Levin and Bishop, personal communication), but they are widely distributed among multicellular organisms in the animal kingdom. This coincidence suggests that they evolved with the development of multicellular organisms. Hence, it is reasonable to speculate, as many have done, that they are utilized in the regulation of multicellular differentiation. The evolution of cellular specialization and the coordination of specialized cells must require new regulatory mechanisms not required by simpler organisms. How these mechanisms function is the fundamental question of developmental biology. Oncogenes are defined by their ability to cause tumors or neoplasia. Calculations based on epidemiological and genetic modelling studies of cancer have suggested that it usually arises as a multi-step process. For the oncogens described here, the process has been reduced to a single step. Elevated expression either in its normal cellular form or, more often, in a mutated, more oncogenic form is sufficient to induce transformation or tumors in appropriate target cells. The effect of these genes on the target cell is pleiotropic, suggesting an interference with basic control processes. From the point of producing a tumor or excess cellular proliferation, this interference with cellular controls or cellular signals can be of two general types: (a) production of a signal which induces growth or (b) removal of a signal which blocks growth.

III. Stimulation of Cell Growth

We assay growth factors in tissue culture systems that are relatively free of the constraints found in vivo. The real function of these factors is to achieve the proper balance of cell types and to respond to demands caused by cell loss, cell destruction, and proportional growth of the animal. As such, they tend often to couple both growth-promoting and differentiation-regulating functions. This has been explored in greatest detail for the hematopoietic growth factors which are involved in cell survival, cell proliferation, and cell differentiation. In vivo the stimulus provided by a growth factor is more complex than in the in vitro system. Cells may or may not express the receptors for particular growth factors as a function of differentiation, and cells may be faced with other developmental constraints. For example, in tissue culture, fibroblasts require more growth factor to induce proliferation in confluent cultures than in sparse or sub-confluent cultures. Thus, the nature of the physical contacts made by the cell influences its response to factors. Given the existence of these growth factor circuits and their potent ability to drive cell proliferation under some conditions, it is not surprising that growth factors and growth-factor receptors can act as oncogenes when released from normal constraints by mutation. One might also expect that the intracellular elements in the growth factor signalling pathway could also mutate to short-circuit the normal signal process. Perhaps some of the already known oncogenes have these functions. In both cases, the result of mutations of these functions is the constitutive signalling of cell proliferation.

IV. Stimulation of Cell Differentiation

Terminal differentiation is often equated with withdrawal from the cell cycle. While this is a rather extreme view, it is clear that many cell lineages drive the cells to terminal phenotypes which constrain the ability of the cell to divide. Red blood cells extrude their nuclei, muscle cells fuse to produce multinucleate myotubes, nerve cells grow long processes, chondroblasts are encased in a tight extracellular matrix. All these are the result of expression of differentiation and constrain cells from further proliferation for slightly different reasons. In some ways, these terminal types are analogous to the confluent plate of fibroblasts. Proliferation is constrained by factors which are independent of signals derived from growth factors. Perhaps these might be considered "structural" constraints. How are these established? How are they maintained? What signals are involved? These are important questions that need to be answered. Interference with these signals is a critical part of forcing these cells to divide. These are the developmental signals, signals to express differentiated phenotypes as opposed to growth signals. Much less is understood about what these signals are and how they function. However, the ability of many oncogenes to interfere with the expression of cell differentiation suggests that they interfere by disrupting these developmental signals.

In summary, the regulation of development to produce a complex, multicellular organism requires the coordination of many cells. Coordination of the developmental programs is accomplished, in part, by factors which stimulate growth and cell proliferation. Some oncogenes make these functions constitutive. Other aspects are coordinated by signals to differentiate and perhaps to maintain the differentiated state as well. Some oncogenes act on these developmental signals to disrupt the expression of cell differentiation. Finally, these two processes must be coupled, or interregulated, so that pure effects of either differentiation or cell proliferation will be difficult to observe.

References

Adams SL, Boettiger D, Focht RJ, Holtzer H, Pacifici M (1982) Regulation of the synthesis of extracellular matrix components in chondroblasts transformed by a temperature-sensitive mutant of Rous sarcoma virus. Cell 30:373–384

Adams SL, Pacifici M, Focht RJ, Allebach ES, Boettiger D (1985) Collagen synthesis in virus-transformed cells. Ann NY Acad Sci 460:202–213

Adams SL, Pacifici M, Boettiger D, Pallante KM (1987b) Modulation of fibronectin gene expression in chondrocytes by viral transformation and substrate attachment. J Cell Biol 105:483–488

Adamson E (1987) Oncogenes in development. Development 99:449–471

Albis A, Lenfant Guyot M, Janmot C, Chanoine C, Weinman J, Gallien CL (1987) Regulation by thyroid hormones of terminal differentiation in the skeletal dorsal muscle. I. Neonate mouse. Dev Biol 123:25–32

Alema S, Tato F (1987) Interactions of retroviral oncogenes with the differentiation program of myogenic cells. Adv Cancer Res 49:1–28

Alema S, Casalbore P, Agostini E, Tato F (1985a) Differentiation of PC12 phaeochromocytoma cells induced by v-*src* oncogene. Nature 316:557–559

Alema S, Tato F, Boettiger D (1985b) *myc* and *src* oncogenes have complementary effects on cell proliferation and expression of specific extracellular matrix components in definitive chondroblasts. Mol Cell Biol 5:538–544

Allebach ES, Boettiger D, Pacifici M, Adams SL (1985) Control of types I and II collagen and fibronectin gene expression in chondrocytes delineated by viral transformation. Mol Cell Biol 5:1002–1008

Arndt K, Fink G (1986) GCN4 protein, a positive transcription factor in yeast, binds general control promoters at all 5′ TGACTC 3′ sequences. Proc Natl Acad Sci USA 83:8516–8520

Barbacid M (1987) *ras* Genes. Annu Rev Biochem 56:779–827

Barka T, Gubits RM, van der Noen HM (1986) Beta-adrenergic stimulation of c-*fos* gene expression in the mouse submandibular gland. Mol Cell Biol 6:2984–2989

Barrett TB, Berditt EP (1987) *sis* (platelet derived growth factor B chain) gene transcript levels are elevated in human atherosclerotic lesions compared to normal artery. Proc Natl Acad Sci USA 84:1099–1103

Bar Sagi D, Feramisco JR (1985) Microinjection of the *ras* oncogene protein into PC12 cells induces morphological differentiation. Cell 42:841–848

Bassiri M, Privalsky M (1987) Transmembrane domain of AEV *erb* B oncogene protein is not required for partial manifestation of the transformed phenotype Virology 159:20–30

Baumbach WR, Keath EJ, Cole MD (1986) A mouse c-*myc* retrovirus transforms established fibroblast lines in vitro and induces monocyte-macrophage tumors in vivo. J Virol 59:276–283

Ben-Neriah Y, Bernards A, Paskind M, Daley GQ (1986) Alternative 5′ exons in c-abl mRNA. Cell 44:577–586

Besmer P, Murphy JE, George PC, Qiu FH, Bergold PJ, Lederman L, Snyder HW Jr, Brodeur D, Zuckerman EE, Hardy WD (1986) A new acute transforming feline retrovirus and relationship of its oncogene v-*kit* with the protein kinase gene family. Nature 320:415–421

Beug H, von Kirchbach A, Doderlein G, Conscience JF, Grag T (1979) Chicken cells transformed by seven strains of defective avian leukemia viruses display three distinct phenotypes of differentiation. Cell 18:375–390

Beug H, Pamieri S, Freudenstein C, Zentgraf H, Graf T (1982) Hormone-dependent terminal differentiation in vitro of chicken erythroleukemia cell transformed by ts mutants of avian erythroblastosis virus. Cell 28:907–919

Bischoff R, Holtzer H (1967) The effect of mitotic inhibitors on myogenesis in vitro. J Cell Biol 36:111–128

Bister K, Nunn M, Moscovici M, Perbal B, Baluda MA, Duesberg PH (1982) Acute leukemia virus E26 and avian myeloblastosis virus have related transformation-specific RNA sequences but different genetic structures, gene products, and oncogenic properties. Proc Natl Acad Sci USA 79:3677–3681

Blair DG, Oskarsson M, Wood TG, McClements WL, Vande Woude GG (1981) Activation of the transforming potential of a normal cell sequence: a molecular model for oncogenesis. Science 212:941–943

Blair DG, Oskarsson MK, Seth A, Dunn KJ, Dean M, Zweig M, Tainsky MA, Vande Woude GF (1986) Analysis of the transforming potential of the human homolog of *mos*. Cell 46:785–794

Boettiger D, Roby K, Brumbaugh J, Biehl J, Holtzer H (1977) Transformation of chicken embryo retinal melanoblasts by a temperature-sensitive mutant of Rous sarcoma virus. Cell 11:881–890

Bohmann D, Bos TJ, Admon A, Nishimira T, Vogt PK, Tijan R (1987) Human proto-oncogene c-*jun* encodes a DNA binding protein with structural and functional properties of transcription factor AP-1. Science 232:1386–1393

Boss M, Greaves M, Teich N (1979) Abelson virus transformed haematopoietic cell lines with pre-B cell characteristics. Nature 278:551–553

Boyle WJ, Lampert MA, Li AC, Baluda MA (1985) Nuclear compartmentalization of the v-*myb* oncogene product. Mol Cell Biol 5:3017–3023

Bravo R, Burckhardt J, Curran T, Muller R (1985) Stimulation and inhibition of growth by EGF in different A431 cell clones is accompanied by the rapid induction of c-*fos* and c-*myc* proto-oncogenes. EMBO J 4:1193–1197

Brugge JS, Erikson RL (1977) Identification of a transformation-specific antigen induced by avian sarcoma virus. Nature 269:346–348

Brugge JS, Cotton PC, Queral AE, Barrett JN, Nonner D, Keane RW (1985) Neurones express high levels of a structurally modified, activated form of pp60c-*src*. Nature 316:554–557

Buick RN, Filmus J, Quaroni A (1987) Activated H-*ras* transforms rat intestinal epithelial cells with expression of alpha-TGF. Exp Cell Res 170:300–309

Bunte T, Greiser Wilke I, Moelling K (1983) The transforming protein of the MC29-related virus CMII is a nuclear DNA-binding protein whereas MH2 codes for a cytoplasmic RNA-DNA binding polyprotein. EMBO J 2:1087–1092

Buss JE, Kamps MP, Sefton BM (1984) Myristic acid is attached to the transforming protein of Rous sarcoma virus during or immediately after synthesis and is present in both soluble and membrane-found of the protein. Mol Cell Biol 4:2697–2704

Buss JE, Kamps MP, Gould K, Sefton BM (1986) The absence of myristic acid decreases membrane binding of p60 src but does not affect tyrosine protein kinase activity. J Virol 58:468–474

Cabrera CV, Alonzo MC, Johnson P, Philips RG, Lawrence PA (1987) Phenocopies induced with antisense RNA identify the *wingless* gene. Cell 50:659–663

Calothy G, Poirier F, Dambrine G, Mignatti P, Combes P, Pessac B (1980) Expression of viral oncogenes in differentiating chick embryo neuroretinal cells infected with avian tumor viruses. Cold Spring Harbor Symp Quant Biol 44:(2-P)983–990

Chabot JG, Walker P, Pelletier G (1987) Thyroxine accelerates the differentiation of granular convoluted tubule cells and the appearance of epidermal growth factor in the submandibular gland of the neonatal mouse. A fine-structural immunocytochemical study. Cell Tissue Res 248:351–358

Chapekar MS, Hartman KD, Knode MC, Glazer RI (1987) Synergistic effect of retinoic acid and calcium ionophore differentiation, c-*myc* expression,a nd membrane tyrosine activity in human promyelocytic leukemia cell line HL-60. Mol Pharmacol 31:140–145

Chesa PG, Rettig WJ, Melamed MR, Old LJ, Niman HL (1987) Expression of p21*ras* in normal and malignant human tissues: lack of association with proliferation and malignancy. Proc Natl Acad Sci USA 84:3234–3238

Clarke MF, Westin E, Schmidt D, Josephs SF, Ratner L, Wong Staal F, Gallo RC, Reitz MS Jr (1984) Transformation of NIH 3T3 cells by a human c-*sis* cDNA clone. Nature 308:464–467

Cleveland JL, Jansen HW, Bister K, Fredrickson TN, Morse HC3d, Ihle JN, Rapp UR (1986) Interaction between *raf* and *myc* oncogenes in transformation in vivo and in vitro. J Cell Biochem 30:195–218

Cole MD (1986) The *myc* oncogene: its role in transformation and differentiation. Annu Rev Genet 20:361–384

Coll J, Saule S, Martin P, Raes MB, Lagrou C, Graf T, Simon IE, Stehelin D (1983) The cellular oncogenes c-*myc*, c-*myb*, and c-*erb* are transcribed in defined types of avian hematopoietic cells. Exp Cell Res 149:151–162

Colletta G, Cirafici AM, Vecchio G (1986) Induction of the c-*fos* oncogene by thyrotropic hormone in rat thyroid cells in culture. Science 233:458–460

Conscience JF, Verrier B, Martin G (1986) Interleukin-3-dependent expression of the c-*myc* and c-*fos* proto-oncogenes in hemopoietic cell lines. EMBO J 5:317–323

Cooper GM (1982) Cellular transforming genes. Science 217–801

Cotton PC, Brugge JS (1983) Neural tissues express high levels of the cellular *src* gene product pp60c-*src*. Mol Cell Biol 3:1157–1162

Coussens L, Van Beveren C, Smith D, Chen E, Mitchell RL, Isacke DM, Verma IM, Ullrich A (1986) Structural alteration of viral homologue of receptor proto-oncogene *fms* at carboxyl terminus. Nature 320:277

Crisanti Combes P, Lorinet AM, Girard A, Pessac B, Wasseff M, Calothy G (1982) Effects of Rous sarcoma virus on the differentiation of chick and quail neuroretina cells in culture. Adv Exp Med Biol 158:115–122

Crisanti P, Lorinet AM, Calothy G, Pessac B (1985) Glutamic acid decarboxylase activity is stimulated in quail retina neuronal cells transformed by Rous sarcoma virus and is regulated by pp60v-*src*. EMBO J 4:1467–1470

Cross FR, Garber EA, Pellman D, Hanafusa H (1984) A short sequence in the p60src N terminus is required for p60*ser* myristylation and membrane association and for cell transformation. Mol Cell Biol 4:1834–1842

Curran T, Morgan JI (1985) Superinduction of c-*fos* by nerve growth factor in the presence of peripherally active benzodiazepines. Science 229:1265–1268

Dean M, Park M, LeBeau M, Robins T, Diaz M, Rowley J, Blair D, Vande Woode GF (1985) The human *met* oncogene is related to the tyrosine kinase oncogenes. Nature 318:385–388

Delli Bovi P, Curatola AM, Kern FG, Greco A, Ittmann M, Basilico C (1987) An oncogene isolated by transfection of Kaposi's sarcoma DNA encodes a growth factor that is a member of the FGF family. Cell 50:729–737

Dere WH, Hirayu H, Rapoport B (1985) TSH and cAMP enhance expression of the *myc* proto-oncogene in cultured thyroid cells. Endocrinology 117:2249–2251

Dexter TM (1984) Blood cell development. The message in the medium (news). Nature 309:764–767

Dexter TM, Scott D, Teich NM (1977) Infection of bone marrow cells in vitro with FLV: effects on stem cell proliferation, differentiation and leukaemogenic capacity. Cell 12:355–364

Dickson C, Peters G (1987) Potential oncogene product related to growth factor. Nature 326:833

Distel RJ, Ro HS, Rosen BS, Groves DL, Spiegelman BM (1987) Nucleoprotein complexes that regulate gene expression in adipocyte differentiation: direct participation of c-*fos*. Cell 49:835–844

Doolittle RF, Hunkapiller MW, Hood LE, Devare SG, Robbins KC, Aaronson SA, Antoniades HN (1983) Simian sarcoma virus onc gene, v-*sis*, is derived from the gene (or genes) encoding a platelet-derived growth factor. Science 221:275–277

Downward J, Yarden Y, Mayes E, Scarce G, Totty N, Stockwell P, Ullrich A, Schlesinger J, Waterfield MD (1984) Close similarity of epidermal gowth factor receptor and v-*erbB* oncogene protein sequences. Nature 307:521–527

Duprey SP, Boettiger D (1985) Developmental regulation of c-*myb* in normal myeloid progenitor cells. Proc Natl Acad Sci USA 82:6937–6941

Durban EM (1980) Viral leukemogenesis: in vitro interactioned between avian myeloblastosis virus and differentiating hematopoietic cells. Ph D Thesis, Univ of Pennsylvania

Durban EM, Boettiger D (1981 a) Replicating differentiated macrophages can serve as target cells for transformation by avian myeloblastosis virus. J Virol 37:488–492

Durban EM, Boettiger D (1981 b) Differential effects of transforming avian RNA tumor viruses on avian macrophages. Proc Natl Acad Sci USA 78:3600–3604

Durkin JP, Whitfield JF (1987) The viral Ki-*ras* gene must be expressed in the G2 phase if the Kirsten sarcoma virus-infected NRK cells are to proliferate in serum-free medium. Mol Cell Biol 7:444–449

Emilia G, Donelli A, Ferrari S, Torelli U, Selleri L, Zucchini P, Moretti L, Venturelli D, Ceccherelli G (1986) Cellular levels of mRNA from c-*myc*, c-*myb* and c-*fes* onc-genes in normal myeloid and erythroid precursors of human bone marrow: an in situ hybridization study. Br J Haematol 62:287–292

Endo T, Nadal Ginard B (1986) Transcriptional and posttranscriptional control of c-*myc* myogenesis: its mRNA remains inducible in differentiated cells and does not suppress the differentiated phenotype. Mol Cell Biol 6:1412–1421

Falcone G, Boettiger D, Tato F, Alema S (1984) Role of cell division in differentiation of myoblasts infected with a temperature-sensitive mutant of Rous sarcoma virus. EMBO J 3:1327–1331

Falcone G, Tato F, Alema S (1985) Distinctive effects of the viral oncogenes *myc*, *erb*, *fps*, and *src* on the differentiation program of quail myogenic cells. Proc Natl Acad Sci USA 82:426–430
Fedele LA, Even J, Garon CF, Donner L, Sherr CJ (1981) Recombinant bacteriophage containing the integrated transforming provirus of Gardner-Arnstein feline sarcoma virus. Proc Natl Acad Sci USA 78:4036–4040
Feldman RA, Wang E, Hanafusa H (1983) Cytoplasmic localization of the transforming protein of Fujinami sarcoma virus: salt-sensitive association with sub-cellular components. J Virol 45:782–791
Feldman RA, Gabrilove JL, Tam JP, Moore MA, Hanafusa H (1985) Specific expression of the human cellular *fps/fes*-encoded protein NCP92 in normal and leukemic myeloid cells. Proc Natl Acad Sci USA 82:2379–2383
Finkel MP, Biskis BO, Jinkins PB (1966) Virus induction of osteosarcomas in mice. Science 151:698–701
Fiszman MY, Fuchs P (1975) Temperature-sensitive expression of differentiation in transformed myoblasts. Nature 254:429–431
Foster DA, Hanafusa H (1983) A *fps* gene without *gag* gene sequences transforms cells in culture and induces tumors in chickens. J Virol 48:744–751
Foster DA, Shibuya M, Hanafusa H (1985) Activation of the transformation potential of the cellular *fps* gene. Cell 42:105–115
Frykberg L, Palmeri S, Beug H, Graf T, Hayman MJ, Vennstrom B (1983) Transforming capacities of avain erythroblastosis virus mutants deleted in the *erbA* or *erbB* oncogenes. Cell 32:227–238
Fusco A, Portella G, Di Fiore PP, Berlingieri MT, Di Lauro R, Schneider AB, Vecchio G (1985) A *mos* oncogene-containing retrovirus, myeloproliferative sarcoma virus, transforms rat thyroid epithelial cells and irreversibly blocks their differentiation pattern. J Virol 56:284–292
Gandrillon O, Jurdic P, Benchaibi M, Xiao JH, Ghysdael J, Samarut J (1987) Expression of the v-*erbA* oncogene in chicken embryo fibroblasts stimulates their proliferation in vitro and enhances tumor growth in vivo. Cell 49:687–697
Garber EA, Krueger JG, Hanafusa H, Goldberg AR (1983) Only membrane-associated RSV *src* proteins have amino-terminally bound lipid. Nature 302:161–163
Gazit A, Igarashi H, Chiu IM, Srinivasan A, Yaniv A, Tronick SR, Robbins KC, Aaronson SA (1984) Expression of the normal human *sis*/PDGF-2 coding sequence induces cellular transformation. Cell 39:89–97
Gazzolo L, Moscovici C, Moscovici MG, Samarut J (1979) Response of hematopoietic cells to avian acute leukemia viruses: effects on the differentiation of the target cells. Cell 16:627–638
Gee CE, Griffin J, Sastre L, Miller LJ, Springer TA, Piwnica Worms H, Roberts TM (1986) Differentiation of myeloid cells is accompanied by increased levels of pp60c-*src* protein and kinase activity. Proc Natl Acad Sci USA 83:5131–5135
Gewirtz AM, Calabretta B (1988) A c-*myb* antisense oligo nucleotide inhibits normal human hematopoiesis in vitro. Science 242:1303–1306
Gibbs JB, Sigal IS, Poe M, Scolnick EM (1984) Intrinsic GTPase activity distinguishes normal and oncogenic *ras* p21 molecules. Proc Natl Acad Sci USA 81:5704–5708
Giotta GJ, Heitzmann J, Cohn M (1980) Properties of two temperature-sensitive Rous sarcoma virus transformed cerebellar cell lines. Brain Res 202:445–458
Giotta GJ, Cohn M (1981) The expression of glial fibrillary acidic protein in a rat cerebellar cell line. J Cell Physiol 107:219–230
Goldman DS, Kiessling AA, Millette CF, Cooper GM (1987) Expression of c-*mos* RNA in germ cells of male and female mice. Proc Natl Acad Sci USA 84:4509–4513
Gospodarowicz D, Ferrara N, Schweigerer L, Neufeld G (1987) Structural characterization and biological functions of fibroblast growth factor. Endocrine Rev 8:95–114
Grady EF, Schwab M, Rosenau W (1987) Expression of N-*myc* and c-*src* during the development of fetal human brain. Cancer Res 47:2931–2936
Graf T, von Kirchback A, Beug H (1981) Characterization of the hematopoietic target cells for AEV, MC29, and AMV avian leukemia viruses. Exp Cell Res 131:331–343

Graf T, von Weizsaecker F, Grieser S, Coll J, Stehelin D, Patschinsky T, Bister K, Bechade C, Calothy G, Leutz A (1986) v-*mil* induces autocrine growth and enhanced tumorigenicity in v-*myc*-transformed cells. Cell 45:357–364

Hanafusa H (1977) Cell transformation by RNA tumor viruses. In: Frankel-Conrat H, Wagner R (eds) Comprehensive virology, vol 10. Plenum Press, New York, pp 410–483

Hann SR, Thompson CB, Eisenman RN (1985) c-*myc* oncogene protein synthesis is independent of the cell cycle in human and avian cells. Nature 314:366–369

Hapel AJ, Vande Woode GF, Campbell HD, Yound IG, Robins T (1987) Generation of an autocrine leukemia using a retroviral expression vector carrying the interleukin-3 gene. Lymphokine Res 5:249–254

Harvey JJ (1964) An unidentified virus which causes the rapid production of tumors in mice. Nature 204:1104–1105

Hauser KF, Uray NJ, Gona AG (1986) Granule cell development in the fro cerebellum during spontaneous and thyroxine-induced metamorphosis. J Comp Neurol 253:185–196

Hauser KF, Uray NJ, Gona AG (1986) Granule cell development in the frog cerebellum during spontaneous and thyroxine-induced metamorphosis. J Comp Neurol 253:185–196

Hirst R, Horwitz A, Buck C, Rohrschneider L (1986) Phosphorylation of the fibronectin receptor complex in cells transformed by oncogenes that encode tyrosine kinases. Proc Natl Acad Sci USA 83:6470–6474

Holtzer H, Beihl J, Yeoh G, Meganathan R, Kaji A (1975) Effects of oncogenic virus on muscle differentiation. Proc Natl Acad Sci USA 72:4051–4055

Horwitz A, Duggan K, Greggs R, Decker C, Buck C (1985) The cell substrate attachment (CSAT) antigen has properties of a receptor for laminin and fibronectin. J Cell Biol 101:2134–2144

Huebner K, ar Rushdi A, Griffin CA, Isobe M, Kozak C, Emanuel BS, Nagarajan L, Cleveland JL, Bonner TI, Goldsborough MD et al. (1986) Actively transcribed genes in the *raf* oncogene group, located on the X chromosome in mouse and human. Proc Natl Acad Sci USA 83:3934–3938

Hunter T, Cooper JA (1985) Protein-tyrosine kinases. Annu Rev Biochem 54:897–930

Hynes RO (1987) Integrins: a family of cell surface receptors. Cell 48:549–554

Introna M, Hamilton TA, Kaufman RE, Adams DO (1986) Treatment of murine peritoneal macrophages with bacterial lipopolysaccharide alters expression of c-*fos* and c-*myc* oncogenes. J Immunol 137:2711–2715

Jansen HW, Lurz R, Bister K, Bonner TI, Mark GE, Rapp UR (1984) Homologous cell-deriveds oncogenes in avian carcinoma virus MH2 and murine sarcoma virus 3611. Nature 307:281–284

Johnsson A, Betsholtz C, Heldin CH, Westermark B (1986) The phenotypic characteristics of simian sarcoma virus-transformed human fibroblasts suggest that the v-*sis* gene product acts solely as a PDGF receptor agonist in cell transformation. EMBO J 5:1535–1541

Kamps MP, Buss JE, Sefton BM (1985) Mutation of NH_2-terminal glycine of p60*src* prevents both myristoylation and morphological transformation. Proc Natl Acad Sci USA 82:4625–4628

Kamps MP, Buss JE, Sefton BM (1986) Rous sarcoma virus transforming protein lacking myristic acid phosphorylates known polypeptide substrates without inducing transformation. Cell 45:105–112

Kaufmann Y, Silverman T, Levi BZ, Ozato K (1987) Induction of c-*ets* and c-*fos* gene expression upon antigenic stimulation of a T cell hybridoma with inducible cytolytic capacity. J Exp Med 166:810–815

Kimelman D, Kischner M (1987) Synergistic induction of mesoderm by FGF and TGF-B and the identification of an mRNA coding for FGF in early *Xenopus* embryo. Cell 51:869–877

King MW, Roberts JM, Eisenman RN (1986) Expression of the c-*myc* proto-oncogene during development of *Xenopus laevis*. Mol Cell Biol 6:4499–4508

Klempnauer KH, Symonds G, Evan GI, Bishop JM (1984) Subcellular localization of proteins encoded by oncogenes of avian myeloblastosis virus and avian leukemia virus E26 and by chicken c-*myb* gene. Cell 37:537–547

Kovacs EJ, Oppenheim JJ, Young HA (1986) Induction of c-*fos* and c-*myc* expression in T lymphocytes after treatment with recombinant interleukin 1-alpha. J Immunol 137:3649–3651

Kruijer W, Cooper JA, Hunter T, Verma IM (1984) Platelet-derived growth factor induces rapid but transient expression of the c-*fos* gene and protein. Nature 312:711–716

Kruijer W, Schubert D, Verma IM (1985) Induction of the proto-oncogene *fos* by nerve growth factor. Proc Natl Acad Sci USA 82:7330–7334

Lachman HM, Skoultchi AI (1984) Expression of c-*myc* changes during differentiation of mouse erythroleukaemia cells. Nature 310:592–594

Lachman HM, Cheng GH, Skoultchi AI (1986) Transfection of mouse erythroleukemia cells with *myc* sequences changes the rate of induced commitment to differentiate. Proc Natl Acad Sci USA 83:6480–6484

Lacy J, Sarkar SN, Summers WC (1986) Induction of c-*myc* expression in human B lymphocytes by B-cell growth factor and anti-immunoglobulin. Proc Natl Acad Sci USA 83:1458–1462

Langdon WY, Harris AW, Cory S, Adams JM (1986) The c-*myc* oncogene perturbs B lymphocyte development in E-mu-*myc* transgenic mice. Cell 47:11–18

Leal F, Williams LT, Robbins KC, Aaronson SA (1985) Evidence that the v-*sis* gene product transforms by interaction with the receptor for platelet-derived growth factor. Science 230:327–330

Leder A, Pattengale PK, Kuo A, Stewart TA, Leder P (1986) Consequences of widespread deregulation of the c-*myc* gene in transgenic mice: multiple neoplasms and normal development. Cell 45:485–495

Leon J, Guerrero I, Pellicer A (1987) Differential expression of the *ras* gene family in mice. Mol Cell Biol 7:1535–1540

Lev Z, Leibovitz N, Segev O, Shilo BZ (1984) Expression of the *src* and *abl* cellular oncogenes during development of *Drosophila melanogaster*. Mol Cell Biol 4:982–984

Li Y, Turck CM, Teumer JK, Stavnezer E (1986) Unique sequence, *ski*, in Sloan-Lettering avian retroviruses with properties of a new cell-derived oncogene. J Virol 57:1065–1072

Lichtman AH, Reynolds DS, Faller DV, Abbas AK (1986) Nature murine B lymphocytes immortalized by Kirsten sarcoma virus. Nature 324:489–491

Lipsich L, Brugge JS, Boettiger D (1984) Expression of the Rous sarcoma virus *src* gene in avian macrophages fails to elicit transformed cell phenotype. Mol Cell Biol 4:1420–1424

Lowy DR, Willumsen BM (1986) The *ras* gene family. Cancer Surv 5:275–289

Lynch SA, Brugge JS, Levine JM (1986) Induction of altered c-*src* product during neural differentiation of embryonal carcinoma cells. Science 234:873–876

Maki Y, Bos TJ, Davis C, Starbuck M, Vogt PK (1987) Avian sarcoma virus 17 carries the *jun* oncogene. Proc Natl Acad Sci USA 83:2848–2852

Maness PF (1986) pp60c-*src* encoded by the proto-oncogene c-*src* is a product of sensory neurons. J Neurosci Res 16:127–139

Manne V, Bekesi E, Kung HF (1985) Ha-*ras* proteins exhibit GTPase activity: point mutations that activate the Ha-*ras* gene products result in decreased GTPase activity. Proc Natl Acad Sci USA 82:376–380

Mark GE, Seeley TW, Shows TB, Mountz JD (1986) *Pks*, a *raf*-related sequence in humans. Proc Natl Acad Sci USA 83:6312–6316

Martin GS (1970) Rous sarcoma virus: a function required for maintenance of the transformed state. Nature 227:1021–1023

Marx JL (1987) The *fos* gene as "master switch" (news). Science 237:854–856

Massague J (1987) The TGF-beta family of growth and differentiation factors: Cell 49:437–438

Mathey Prevot B, Nabel G, Palacios R, Baltimore D (1986) Abelson virus abrogation of interleukin-3 dependence in a lymphoid cell line. Mol Cell Biol 6:4133–4135

Maytin EV, Balduzzi PC, Notter MF, Young DA (1984) Changes in the synthesis and phosphorylation of cellular proteins in chick fibroblasts transformed by two avian sarcoma viruses. J Biol Chem 259:12135–12143

McGeady ML, Wood TG, Maizel JV, Vande Woude GF (1986) Sequences upstream from the mouse c-*mos* oncogene may function as a transcription termination signal. DNA 5:289–298

Menko AS, Boettiger D (1987) Occupation of the extracellular matrix receptor integrin is a control point for myogenic differentiation. Cell 51:51–57

Menko AS, Boettiger D (1988) Inhibition of chick embryo lens differentiation and lens junction formation in culture by pp60v-*src*. Mol Cell Biol 8:1414–1420

Menko AS, Klulas KA, Quade B, Liu T-F, Johnson R (1982) Lens gap junctions in differentiating cultures of chick embryo lens cells. J Cell Biol 95:106a

Menko AS, Klukas K, Johnson R (1984) Chicken embryo lens cultures mimic differentiation in the lens. Dev Biol 103:129–141

Metcalf D (1986) Haemopoietic growth factors now cloned. Br J Haematol 62:409–412

Milbrandt J (1986) Nerve growth factor rapidly induces c-*fos* mRNA in PC12 rat pheochromocytoma cells. Proc Natl Acad Sci USA 83:4789–4793

Miller AD, Curran T, Verma IM (1984) c-*fos* protein can induce cellular transformation: a novel mechanism of activation of a cellular oncogene. Cell 36:51–60

Moelling K, Heimann B, Beimling P, Rapp UR, Sander T (1984) Serine- and threonine-specific protein kinase activities of purified *gag-mil* and *gag-raf* proteins. Nature 312:558–561

Moelling K, Pfaff E, Beug H, Beimling P, Bunte T, Schaller HE, Graf T (1985) DNA-binding activity is associated with purified *myb* proteins from AMV and E26 viruses and is temperature-sensitive for E26 ts mutants. Cell 40:983–990

Morgan JI, Cohen DR, Hempstead JL, Curran T (1987) Mapping patterns of c-*fos* expression in the central nervous system after seizure. Science 237:192–197

Moscovici MG, Moscovici C (1983) Isolation and characterization of a temperature-sensitive of avian myeloblastosis virus. Proc Natl Acad Sci 80:1421–1425

Moss P, Radke K, Carter VC, Young J, Gilmore T (1984) Cellular localization of the transforming protein of wild-type and temperature-sensitive Fujinami sarcoma virus. J Virol 52:557–565

Mulcahy LS, Smith MR, Stacey DW (1985) Requirement for *ras* proto-oncogene function growth of NIH 3T3 cells. Nature 313:241–243

Muller R, Verma IM (1984) Expression of cellular oncogenes. Curr Topics Micro Immunol 112:73–115

Muller R, Slamon DJ, Tremblay JM, Cline MJ, Verma IM (1982) Differential expression of cellular oncogenes during pre- and post-natal development of the mouse. Nature 299:640–644

Naharro G, Tronick SR, Rasheed S, Gardner MB, Aaronson SA, Robbins KC (1983) Molecular cloning of integrated Gardner-Rasheed feline sarcoma virus: genetic structure of its cell-derived sequence differs form that of other tyrosine kinase-coding *onc* genes. J Virol 47:611–619

Nakagawa T, Mabry M, de Bustros A, Ihle JN, Nelkin BD, Baylin SB (1987) Introduction of v-Ha-*ras* oncogene induces differentiation of cultured human medullary thyroid carcinoma cells. Proc Natl Acad Sci USA 84:5923–5927

Neckameyer WS, Shibuya M, Hsu MT, Wang LH (1986) Proto-oncogene c-*ros* codes for a molecule with structure common to those of growth factor receptors and displays tissue specific and developmentally regulated expression. Mol Cell Biol 6:1478–1486

Ness SA, Beug H, Graf T (1987) v-*myb* dominance over v-*myc* in doubly transformed chick myelomonocytic cells. Cell 51:41–50

Ng M, Privalsky ML (1986) Structural domains of the avian erythroblastosis virus *erbB* protein required for fibroblast transformation: dissection by in-frame insertional mutagenesis. J Virol 58:542–553

Nilsson K, Larsson LG, Carlsson M, Danersund A, Hellman L, Totterman T, Pettersson U (1986) Expression of c-*myc* and c-*fos* during phorbol ester induced differentiation of B-type chronic lymphocytic leukemia cells. Curr Topics Micro Immunol 132:280–289

Notter MF, Balduzzi PC (1984) Cytoskeletal changes induced by two avian sarcoma viruses: UR2 and Rous sarcoma virus. Virology 136:56–68
Nusse R, van Ooyen A, Cox D, Fung YK, Varmus H (1984) Mode of proviral activation of a putative mammary oncogene (*int*-1) on mouse chromosome 15. Nature 307:131–136
Ogiso Y, Matsumoto M, Morita T, Nishino H, Iwashima A, Matsushiro A (1986) Expression of c-*mos* proto-oncogene in undifferentiated teratocarcinoma cells. Biochem Biophys Res Commun 140:477–484
Olofsson B, Pizon V, Zahraoui A, Tavitian A, Therwath A (1986) Structure and expression of the chicken epidermal growth factor receptor gene locus. Eur J Biochem 160:261–266
Olson EN, Sternberg E, Hu JS, Spizz G, Wilcox C (1986) Regulation of myogenic differentiation by type beta transforming growth factor. J Cell Biol 103:1799–1805
Olson EN, Spizz G, Tainsky MA (1987) The oncogenic forms of N-*ras* or H-*ras* prevent skeletal myoblast differentiation. Mol Cell Biol 7:2104–2111
Ong J, Yamashita S, Melmed S (1987) Insulin-like growth factor I induces c-*fos* messenger ribonucleic acid in L6 rat skeletal muscle cells. Endocrinology 120:353–357
Ostertag W, Vehmeyer K, Fagg B et al. (1980) Myeloproliferative virus, a cloned murine sarcoma virus with spleen focus forming properties in adult mice. J Virol 33:573–582
Pacifici M, Boettiger D, Roby K, Holtzer H (1977) Transformation of chondroblasts by Rous sarcoma virus and the synthesis of the extracellular matrix. Cell 11:891–899
Papkoff J, Nigg EA, Hunter T (1983) The transforming protein of Moloney murine sarcoma virus is a soluble cytoplasmic protein. Cell 33:161–172
Pauza CD (1987) Regulation of human T-lymphocyte gene expression by interleukin 2: immediate-response genes include the proto-oncogene c-*myb*. Mol Cell Biol 7:342–348
Pharr PN, Ogawa M, Hankins WD (1987) In vitro retroviral transfer of *ras* genes to single hemopoietic progenitors. Exp Hematol 15:323–330
Pierce JH, DiFiore PP, Aaronson SA, Potter M, Pumphrey A, Scott A, Ihle JN (1985) Neoplastic transformation of mast cells by Abelson-MuLV: abrogation of IL-3 dependence by a non-autocrine mechanism. Cell 41:685–693
Propst F, Vande Woude GF (1985) Expression of c-*mos* proto-oncogene transcripts in mouse tissues. Nature 315:516–518
Prywes R, Foulkes JG, Rosenberg N, Baltimore D (1983) Sequences of the A-MuLV protein needed for fibroblast and lymphoid cell transformation. Cell 34:569–579
Ran W, Dean M, Levine RA, Henkle C, Campisi J (1986) Induction of c-*fos* and c-*myc* mRNA by epidermal growth factor or calcium ionophore is cAMP dependent. Proc Natl Acad Sci USA 83:8216–8220
Rapp UR, Todaro GJ (1980) Generation of oncogenic mouse type C viruses: in vitro selection of carcinoma-inducing variants. Proc Natl Acad Sci USA 77:624–628
Rapp UR, Reynolds FH, Stephenson JR (1983b) New mammalian transforming retrovirus: demonstration of a poly protein gene product. J Virol 45:914–924
Rapp UR, Goldsborough MD, Mark GE, Bonner TI, Groffen J, Reynolds FH Jr, Stephenson JR (1983a) Structure and biological activity of v-*raf*, a unique oncogene transduced by a retrovirus. Proc Natl Acad Sci USA 80:4218–4222
Rapp UR, Cleveland JL, Brigthman K, Scott A, Ihle JN (1985) Abrogation of IL-3 and IL-2 dependence by recombinant murine retroviruses expressing v-*myc* oncogenes. Nature 317:434–438
Reed JC, Sabath DE, Hoover Rg, Prystowsky MB (1985) Recombinant interleukin 2 regulates levels of c-*myc* mRNA in a murine T lymphocyte. Mol Cell Biol 5:3361–3368
Rein A, Keller J, Schultz AM, Holmes KL, Medicus R, Ihle JN (1985) Infection of immune mast cells by Harvey sarcoma virus: immortalization without loss of requirement for interleukin-3. Mol Cell Biol 5:2257–2264
Rijsewijk F, Scheermann M, Wagenaar E, Paren P, Weigel D, Nusse R (1987) The *Drosophila* homolog of the mouse mammary oncogene *int*-1 is identical to the segment polarity gene *wingless*. Cell 50:647–657

Robbins KC, Devare SG, Aaronson SA (1981) Molecular cloning of integrated simian sarcoma virus – genome organization of infectious clones. Proc Natl Acad Sci USA 78:2918–2922
Robbins KC, Leal F, Pierce JH, Aaronson SA (1985) The v-*sis*/PDGF-2 transforming gene product localizes to cell membranes but is not a secretory protein. EMBO J 4:1783–1792
Rohrschneider L, Rosok MJ, Gentry LE (1983) Molecular interaction of the *src* gene product with cellular adhesion plaques. Prog Nucleic Acid Res Mol Biol 29:233–244
Romero P, Blick M, Talpaz M, Murphy E, Hester J, Gutterman J (1986) C-*sis* and C-*abl* expression in chronic myelogenous leukemia and other hematologic malignancies. Blood 67:839–841
Roussel M, Dull TJ, Rettenmier CW, Ralph P, Ullrich A, Sherr CJ (1987) Transforming potential of the c-*fms* proto-oncogene (CSF-I receptor) Nature 325:549–552
Ruther U, Garber C, Komitowski D, Muller R, Wagner EF (1987) Deregulated c-*fos* expression interferes with normal bone development in transgenic mice. Nature 325:412–416
Sacca R, Stanley ER, Sherr CJ, Rettenmier CW (1986) Specific binding of the mononuclear phagocyte colony-stimulating factor CSF-1 to the product of the v-*fms* oncogene. Proc Natl Acad Sci USA 83:3331–3335
Sap J, Munoz A, Damm K, Goldberg Y, Ghysdael J, Leutz A, Beug H, Vennstrom B (1986) The c-*erbA* protein is a high-affinity receptor for thyroid hormone. Nature 324:636–640
Schmidt A, Setoyama C, de Crombrugghe B (1985) Regulation of a collagen gene promoter by the product of viral *mos* oncogene. Nature 314:286–289
Schultz AM, Oroszlan S (1984) Myristylation of *gag-onc* fusion proteins in mammalian transforming retroviruses. Virology 133:431–437
Scott ML, Davis MM, Feinberg MB (1986) Transformation of T-lymphoid cells by Abelson murine leukemia virus. J Virol 59:434–443
Sefton BM (1986) The viral tyrosine protein kinases. Curr Topics Micro Immunol 123:39–72
Segal D, Shilo BZ (1986) Tissue localization of *Drosophila melanogaster ras* transcripts during development. Mol Cell Biol 6:2241–2248
Semba K, Nishizawa M, Miyajima N, Yoshida MC, Sukegawa J, Yamanashi Y, Sasaki M, Yamamoto T, Toyoshima K (1986) *Yes*-related protooncogene, *syn*, belongs to the protein-tyrosine kinase family. Proc Natl Acad Sci USA 83:5459–5463
Seth A, Priel E, Vande Woude GF (1987) Nucleoside triphosphate-dependent DNA-binding properties of *mos* protein. Proc Natl Acad Sci USA 84:3560–3564
Setoyama C, Liau G, de Crombrugghe B (1985) Pleiotropic mutants of NIH 3T3 cells with altered regulation in the expression of both type I collagen and fibronectin. Cell 41:201–209
Shackleford GM, Varmus HE (1987) Expression of the proto-oncogene *int*-1 is restricted to postmeiotic male germ cells and the neural tube of mid-gestational embryos. Cell 50:89–95
Sheiness D, Gardinier M (1984) Expression of a proto-oncogene (proto-*myb*) in hemopoietic tissues of mice. Mol Cell Biol 4:1206–1212
Shen Ong GL, Holmes KL, Morse HC3rd (1987) Phorbol ester-induced growth arrest of murine myelomonocytic leukemic cells with virus-disrupted *myb* locus is not accompanied by decreased *myc* and *myb* expression. Proc Natl Acad Sci USA 84:199–203
Shibuya M, Hanafusa H (1982) Nucleotide sequence of Fujinami sarcoma virus: evolutionary relationship of its transforming gene with transforming genes of other sarcoma viruses. Cell 30:787–795
Shibuya M, Hanafusa H, Balduzzi PC (1982) Cellular sequences related to three new *onc* genes of avian sarcoma virus (*fps*, *yes*, *ros*) and their expression in normal and transformed cells. J Virol 42:143–152
Shih C, Shilo BZ, Goldfarb MP, Dannenberg A, Weinberg RA (1979) Passage of phenotypes of chemically transformed cells via DNA transfection of DNA and chromatin. Proc Natl Acad Sci USA 76:5714–5718

Skouv J, Christensen B, Skibshj I, Autrup H (1986) The skin tumor-promoter 12-*O*-tetradecanoylphorbol-13-acetate induces transcription of the c-*fos* proto-oncogene in human bladder epithelial cells. Carcinogenesis 7:331–333

Smith MR; DeGudicibus SJ, Stacey DW (1986) Requirement for c-*ras* proteins during viral oncogene transformation. Nature 320:540–543

Spooncer E, Heyworth CM, Dunn A, Dexter TM (1986) Self-renewal and differentiation of interleukin-3-dependent multipotent stem cells are modulated by stromal cells and serum factors. Differentiation 31:111–118

Stavnezer E, Gerhard DS, Binari RC, Balazas I (1981) Generation of transforming viruses in cultures of chicken fibroblasts infected with avian leukosis virus. J Virol 39:920–934

Stehelin D, Varmus HE, Bishop JM, Vogt PK (1976a) DNA related to normal transforming gene(s) of avian sarcoma viruses is present in normal avian DNA. Nature 260:170–173

Stehelin D, Guntaka RV, Varmus HE, Bishop JM (1976b) Purification of DNA complementary to nucleotide sequences required for transformation of fibroblasts by avian sarcoma viruses. J Mol Biol 101:349–365

Stern JB, Smith KA (1986) Interleukin-2 induction of T-cell G1 progression and c-*myb* expression. Science 233:203–206

Stocking C, Kollek R, Bergholz U, Ostertag W (1986) Point mutations in the U3 region of the long terminal repeat of Moloney murine leukemia virus determine disease specificity of the myeloproliferative sarcoma virus. Virology 153:145–149

Sutrave P, Bonner TI, Rapp UR, Jansen HW, Patschinsky T, Bister K (1984) Nucleotide sequence of avian retroviral oncogene v-*mil*: homologue of murine retroviral oncogene v-*raf*. Nature 309:85–88

Sweet RW, Yokoyama S, Kamata T, Feramisco JR, Rosenberg M, Gross M (1984) The product of *ras* is a GTPase and the T24 oncogenic mutant is deficient in this activity. Nature 311:273–275

Taira M, Yoshida T, Miyagawa K, Sakamoto H, Terada M, Sugimura T (1987) cDNA sequence of human transforming gene *hst* and identification of the coding sequence required for transforming activity. Proc Natl Acad Sci USA 84:2980–2984

Tamkun JW, DeSimone DW, Fonda D, Patel RS, Buck C, Horwitz AF, Hynes RO (1986) Structure of integrin, a glycoprotein involved in the transmembrane linkage between fibronectin and actin. Cell 46:271–282

Tato F, Alema S, Dlygosz A, Boettiger D, Holtzer H, Cossu G, Pacifici M (1983) Development of revertant myotubes in cultures of Rous sarcoma virus transformed avian myogenic cells. Differentiation 24:131–139

Taylor MV, Gasse M, Evan GI, Dathan N, Mechali M (1986) *Xenopus myc* proto-oncogene during development: expression as a stable maternal mRNA incoupled from cell division. EMBO J 5:3563–3570

Temin HM, Rubin H (1958) Characteristics of an assay for Rous sarcoma virus and Rous sarcoma cells in tissue culture. Virology 6:669–688

Theilen GH, Gould D, Fowler M, Dungworth DL (1970) C-type virus in tumor tissue of a woolly monkey (*Lagothrix* sp.) with fibrosarcoma. INCI 47:881–889

Thompson CB, Challoner PB, Neiman PE, Groudine M (1985) Levels of c-*myc* oncogene mRNA are invariant throughout the cell cycle. Nature 314:363–366

Thompson CB, Challoner PB, Neiman PE, Groudine M (1986) Expression of the c-*myb* proto-oncogene during cellular proliferation. Nature 319:374–380

Tushinski RJ, Oliver IT, Guilbert LJ, Stanley ER (1982) Survival of mononuclear phagocytes depends on a lineage-specific growth factor that the differentiated cells selectively destroy. Cell 28:81–91

van der Hoorn FA, Muller V, Pizer LI (1985) Sequences upstream of c-*mos* (*rat*) that block RNA accumulation in mouse cells do not inhibit in vitro transcription. Mol Cell Biol 5:406–409

Vogt TF, Solter D, Tilghman SM (1987) *Raf*, a *trans*-acting locus, regulates the alpha-fetoprotein gene in a cell-autonomous manner. Science 236:301–303

Voronova AF, Sefton BM (1986) Expression of a new tyrosine protein kinase is stimulated by retrovirus promoter insertion. Nature 319:682–685

Waneck GL, Rosenberg N (1981) Abelson leukemia virus induces lymphoid and erythroid colonies in infected fetal liver cultures. Cell 26:91–97

Waneck GL, Keyes L, Rosenberg N (1986) Abelson virus drives the differentiation of Harvey virus infected erythroid cells. Cell 44:337–344

Waterfield MD, Scarce GJ, Whittle N, Stroobant P, Johnson A, Wasteson A, Westermark B, Heldinm CH, Huang JS, Deuel TF (1983) Platelet derived growth factor is structurally related to the putative transforming protein p28 of simian sarcoma virus. Nature 304:35–38

Weinberger C, Giguere V, Hollenberg SM, Thompson C, Arriza J, Evans RM (1987) Human steroid receptors and *erb*-A gene products form a superfamily of enhancer-binding proteins. Clin Physiol Biochem 5:179–189

Weissman B, Aaronson SA (1985) Members of the *src* and *ras* oncogene families supplant the epidermal growth factor requirement of BALB/MK-2 keratinocytes and induce distinct alterations in their terminal differentiation program. Mol Cell Biol 5:3386–3396

West CM, Boettiger D (1982) Selective effect of Rous sarcoma virus *src* gene expression on contractille protein synthesis in chick embryo myotubes. Cancer Res 43:2042–2048

Whitlock CA, Ziegler SF, Triedman LJ, Stafford JI, Witte ON (1983) Differentiation of cloned populations of immature B cells after transformatin with Abelson murine leukemia virus. Cell 32:903–911

Wilkinson DG, Bailes JA, McMahon AP (1987) Expression of the proto-oncogene *int*-1 is restricted to specific neural cells in the developing embryo. Cell 50:79–88

Willman CL, Stewart CC, Griffith JK, Stewart SJ, Tomasi TB (1987a) Differential expression and regulation of the c-*src* and c-*fgr* protooncogenes in myelomonocytic cells. Proc Natl Acad Sci USA 84:4480–4484

Willman CL, Stewart CC, Lin H, Tomasi TB (1987b) Modulation of c-*fms* and c-*fgr* expression in normal monocytic cells stimulated to proliferate with CSF-1 monocytic colony-stimulating factor. J Cell Biochem [Suppl] 11A:201

Witte ON (1986) Functions of the *abl* oncogene. Cancer Surv 5:183–197

Wolgemouth DJ, Engelmejr E, Duggal RN, Gizang-Ginsber E, Mutter GL, Ponzetto C, Viviano C, Zakeri ZF (1986) Isolation of a mouse cDNA coding for a developmentally regulated, testis specific, transcript containing homeobox homology. EMBO J 5:1229–1235

Womer RB, Frick K, Mitchell CD, Ross AH, Bishayee S, Scher CD (1987) PDGF induces c-*myc* mRNA expression in MG 63 human osteosarcoma cells but does not stimulate cell replication. J Cell Phys 132:65–72

Yamamoto T, Ikawa S, Akiyama T, Semba K, Nomura N, Miyajima N, Saito T, Toyoshima K (1986) Similarity of protein encoded by the human c-*erbB*2 gene to the epidermal growth factor receptor. Nature 391:230–234

Yamanashi Y, Fukushige S, Semba K, Sukegawa J, Miyajima N, Matsubara K, Yamamoto T, Toyoshima K (1987) The *yes*-related cellular gene *lyn* encodes a possible tyrosine kinase similar to p56*lck*. Mol Cell Biol 7:237–243

Yoakum GH, Lechner JF, Gabrielson EW, Korba BE, Malan Shibley L, Willey JC, Valerio MG, Shamsuddin AM, Trump BF, Harris CC (1985) Transformation of human bronchial epithelial cells transfected by Harvey *ras* oncogene. Science 227:1174–1179

Subject Index

MIX
Papier aus verantwortungsvollen Quellen
Paper from responsible sources
FSC® C105338

If you have any concerns about our products,
you can contact us on
ProductSafety@springernature.com

In case Publisher is established outside the EU,
the EU authorized representative is:
Springer Nature Customer Service Center GmbH
Europaplatz 3, 69115 Heidelberg, Germany

Printed by Libri Plureos GmbH
in Hamburg, Germany